Two-dimensional Valleytronic Materials

From principles to device applications

Online at: https://doi.org/10.1088/978-0-7503-5562-9

Two-dimensional Valleytronic Materials

From principles to device applications

Sake Wang
College of Science, Jinling Institute of Technology, Nanjing, Jiangsu, China

Hongyu Tian
School of Physics and Electronic Engineering, Linyi University, Linyi, Shandong, China

IOP Publishing, Bristol, UK

ISBN 978-0-7503-5562-9 (ebook)
ISBN 978-0-7503-5560-5 (print)
ISBN 978-0-7503-5564-3 (myPrint)
ISBN 978-0-7503-5561-2 (mobi)

DOI 10.1088/978-0-7503-5562-9

Version: 20250501

IOP ebooks

British Library Cataloguing-in-Publication Data: A catalogue record for this book is available from the British Library.

Published by IOP Publishing, wholly owned by The Institute of Physics, London

IOP Publishing, No.2 The Distillery, Glassfields, Avon Street, Bristol, BS2 0GR, UK

US Office: IOP Publishing, Inc., 190 North Independence Mall West, Suite 601, Philadelphia, PA 19106, USA

To our beloved families.

Contents

Preface xi

Acknowledgments xiv

Author biographies xv

List of symbols xvi

List of abbreviations xviii

1 Introduction to two-dimensional (2D) materials and valleytronics **1-1**

1.1 Introduction 1-1
1.2 Electronic properties of graphene, definition of 'valley' 1-2
1.2.1 Tight-binding description of graphene 1-2
1.2.2 Effective Hamiltonian close to the Dirac point 1-8
1.3 Electronic properties of zigzag graphene nanoribbons 1-9
1.4 Spin–orbit coupling in 2D materials 1-13
1.4.1 SOC between atoms 1-13
1.4.2 Fourier transform 1-15
1.5 Molybdenum disulfide (MoS_2) and the transition-metal dichalcogenides 1-17
1.5.1 Thickness-dependent electronic band structure 1-18
1.5.2 Spin–orbit coupling and its effects 1-19
1.5.3 Effective Hamiltonian 1-21
1.6 Valley-polarized Landau levels in graphene 1-24
1.7 AB-stacked (Bernal phase) bilayer graphene 1-28
1.7.1 TB description 1-28
1.7.2 Close to the Dirac point 1-30
1.8 Electronic properties of AB-stacked bilayer zGNRs 1-31
1.9 Twisted bilayer graphene 1-33
1.10 Early theoretical proposal for a valleytronic application 1-37
References 1-42

2 Berry phase and valley Hall effect (VHE) **2-1**

2.1 Topology in solid state physics and basic concepts of the Berry phase 2-1
2.2 Pedagogical example: the two-level system 2-3
2.2.1 Two-level system and gauge freedom 2-3
2.2.2 Gauge-dependent Berry connections, gauge-independent Berry curvature and Chern number 2-5

2.3 Electron dynamics in an electric field 2-8
2.3.1 Anomalous velocity 2-8
2.3.2 Berry curvature: symmetry considerations 2-9
2.4 VHE in graphene 2-9
2.4.1 TB model 2-9
2.4.2 Low-energy model (massive 2D Dirac fermions) 2-12
2.4.3 Non-local measurement for detection of VHE 2-15
2.4.4 Experimental detection of VHE 2-17
2.5 Wavepacket and its orbital moment 2-19
2.6 Valley and spin Hall effect in monolayer TMDCs 2-20
2.7 Chern number and valley Chern number 2-22
2.8 Topological kink states 2-23
References 2-27

3 Strain effect on valley polarisation (VP) 3-1

3.1 Effect of uniaxial strain 3-1
3.1.1 Model 3-2
3.1.2 Deformation-induced gauge fields in terms of the strain tensor 3-2
3.1.3 Valley-polarized transmission across a sharp barrier 3-6
3.1.4 Valley-polarized transmission across a smooth strain barrier 3-12
3.2 Local Gaussian bump for valley filtering 3-16
3.3 Other proposals corresponding to strain 3-21
References 3-22

4 Other valley filter and valve effects 4-1

4.1 Trigonal warping effects 4-1
4.1.1 The zigzag case 4-3
4.1.2 The armchair case 4-4
4.2 Spin–orbit coupling effects 4-5
4.3 Ferrovalley materials with spontaneous VP 4-9
4.3.1 Two-band effective model including exchange interaction 4-9
4.3.2 Band structures from density functional theory (DFT) 4-11
4.3.3 Chirality-dependent optical bandgap and Berry curvatures 4-13
4.3.4 Discussion 4-14
4.4 Detection of VP with a superconducting contact 4-16
4.4.1 The physical origin of VP 4-16
4.4.2 The Dirac–Bogoliubov–de Gennes (DBdG) equation 4-17
4.4.3 Effective boundary conditions 4-18

4.4.4 The dispersion relation 4-20
4.4.5 The conductance 4-22
References 4-23

5 Valley-polarized current in polycrystalline systems 5-1

5.1 Line defect 5-1
5.1.1 Model 5-1
5.1.2 The physical mechanism 5-3
5.1.3 Other viewpoints on VP 5-7
5.1.4 Multiple line defects 5-9
5.1.5 Line defect superlattice 5-10
5.2 Other polycrystalline graphene 5-12
5.2.1 The transmission characteristics and VP 5-14
5.2.2 Optical-like behaviours 5-16
5.3 Line defects of MoS_2 5-17
5.3.1 The transmission characteristics 5-17
5.3.2 The VP properties 5-20
References 5-22

6 Valley current in multilayer and twisted systems 6-1

6.1 Effect of the perpendicular electric field on AB-stacked bilayer graphene 6-1
6.1.1 Biased bilayer (bulk) 6-1
6.1.2 Biased bilayer nanoribbons with zigzag edges 6-4
6.2 VHE in biased bilayer graphene 6-4
6.3 Domain wall induced with a voltage kink 6-6
6.4 Domain wall induced with a tilt-boundary edge states 6-9
6.4.1 Bilayer graphene with AB–BA tilt-boundary 6-9
6.4.2 Multilayer graphene 6-11
6.4.3 Moiré systems 6-12
References 6-13

7 Valley optoelectronics based on monolayer 2D materials 7-1

7.1 Interaction of an electron with an external electromagnetic field 7-1
7.2 Jones vector 7-6
7.3 Light absorption in hexagonal lattice system 7-9
7.3.1 Dipole vector 7-9
7.3.2 System with inversion symmetry (IS) 7-10

	7.3.3 System with broken IS	7-11
7.4	Optical selection rule in TMDCs	7-15
	7.4.1 Dipole approximation	7-15
	7.4.2 Conservation law of the pseudoangular momentum	7-16
	7.4.3 Excitons	7-18
	7.4.4 Valley excitons: valley-polarized photoluminescence (PL)	7-19
7.5	Exciton Hall effect	7-21
7.6	Tuning VP: valley Zeeman effect	7-23
	7.6.1 Defect states	7-28
	7.6.2 Electron doping	7-28
	7.6.3 Optical Stark effect	7-28
	7.6.4 Composition ratios in monolayer TMDC alloys	7-29
7.7	Harness valley degree of freedom in optoelectronic devices	7-29
	7.7.1 Valley light-emitting diodes (LEDs)	7-30
	7.7.2 Valleytronic transistor—experimental demonstration of VHE	7-38
	References	7-41

8 Valley optoelectronics based on vdW materials **8-1**

8.1	Detection of VHE in bilayer TMDCs	8-1
8.2	2D layered materials	8-4
	8.2.1 Fabrication of 2D layered materials	8-4
	8.2.2 Types and applications of 2D heterostructures	8-4
8.3	Interlayer excitions (IXs) in type-II van der Walls (vdW) heterostructures	8-6
8.4	Valley-polarized exciton	8-8
	8.4.1 Manipulation of VP	8-10
	8.4.2 Diffusion of valley-polarized IXs	8-11
8.5	Moiré IXs	8-13
8.6	Enhancement of valley Zeeman effect in vdW heterostructures	8-16
	8.6.1 Magnetic proximity effect from magnetic layered materials or substrates	8-16
	8.6.2 IXs	8-17
	References	8-18

Appendix A: The matrix elements of the graphene TB Hamiltonian **A-1**

Appendix B: Hermite polynomials **B-1**

Appendix C: The dipole vector of hexagonal systems **C-1**

Preface

The advent of two-dimensional (2D) materials has ushered in a new era of discovery and innovation in condensed matter physics and materials science. Among the many fascinating phenomena associated with these materials, valleytronics has emerged as a particularly compelling field, offering unique properties and potential applications in the realm of electronics and optoelectronics. This book, *Two-dimensional Valleytronic Materials: From principles to device applications*, aims to provide a comprehensive exploration of the principles and device applications in this dynamic area, bridging the gap between fundamental theories and practical implementations.

The journey of 2D materials began in 2004 with the isolation of graphene by Sir Andre Konstantin Geim and Sir Konstantin Sergeevich Novoselov, an achievement that earned them the Nobel Prize in Physics in 2010. Graphene's exceptional electronic, thermal, and mechanical properties spurred an explosion of research into other 2D materials, such as silicene, transition-metal dichalcogenides (TMDCs), hexagonal boron nitride, black phosphorus, etc. These materials exhibit a wide range of electronic properties, from metallic to semiconducting. They are also being integrated with other advanced systems, such as flexible electronics and hybrid devices, and have paved the way for new technological applications.

Valleytronics, a portmanteau of 'valley' and 'electronics,' leverages the unique electronic property of valleys—distinct extrema in the band structure of a material where charge carriers can reside. Unlike traditional electronics, which primarily rely on charge and spin, valleytronics taps into this additional degree of freedom (DOF), which can be manipulated and utilized for information processing. TMDCs, such as MoS_2, WS_2, WSe_2, and others, have been identified as more promising compared to other valleytronic materials such as graphene and silicene because of their direct bandgaps in monolayer form and the strong spin–orbit coupling of TMDCs, leading to distinct valley-specific electronic and optical properties. By manipulating and controlling these valleys, researchers are opening up new avenues for developing faster, more efficient, and more versatile electronic and optoelectronic devices. Their potential applications span from high-speed data processing to innovative light-emitting diodes (LEDs), making valleytronics a pivotal area of research in modern technology.

This book is structured to guide the reader through both the theoretical underpinnings and practical applications of valleytronic materials. We start with an in-depth introduction to the principles of valleytronics. This section covers the basics of 2D materials, including graphene, silicene, TMDCs, and other emerging materials. We discuss the electronic band structures of these materials, explaining how valleys are formed and their significance in the context of valleytronics. This foundational knowledge is crucial for understanding the subsequent discussions of device applications and technology integration.

One of the key strengths of this book is its interdisciplinary approach. Valleytronics is a field that draws from multiple scientific disciplines, including quantum physics, solid-state physics, electromagnetism, materials science, optics, and electrical engineering. To reflect this, the book progresses by exploring the main approaches taken by various disciplines to manipulate and utilize valley DOF in a logical sequence, as follows:

1. Geometric structures: valley-polarized current due to quantum point contact (section 1.10), deformation/strain (chapter 3, sections 5.2 and 7.7.1.2), line defects/polycrystalline systems (chapter 5), and multilayer systems (chapters 6 and 8), including twisted/moiré systems (sections 6.4.3 and 8.5);
2. Intrinsic electronic properties: valley-polarized current due to trigonal warping (sections 4.1 and 7.7.1.1), spin–orbit coupling (section 4.2), and ferrovalley materials (section 4.3);
3. Electrical techniques: valley-polarized current due to electrostatic potential (sections 1.10, 4.1, 4.2), the valley Hall effect (sections 2.4, 2.6, 4.3.3, 4.3.4, 6.1, 6.2, 7.7.2, and 8.1), the topological valley kink states (sections 2.8, 6.3, and 6.4), and carrier distribution shift in momentum space due to the electric field in the p–i–n junctions to induce valley polarization (section 7.7.1);
4. Magnetic techniques: valley-polarized Landau levels (sections 1.6 and 4.4), valley orbital magnetic moment (section 2.5), valley-polarized current due to magnetic exchange interaction (section 4.2), the valley Zeeman effect (sections 7.6 and 8.6), and magneto-optical Kerr effect to detect valley polarization (section 8.1);
5. Thermal/statistical techniques: valley-selective exciton Hall effect (section 7.5) and diffusion of valley-polarized interlayer excitons (section 8.4.2);
6. Optical techniques: valley-dependent optical selection rules (sections 4.3 and 8.4–8.6, and chapter 7) and magneto-optical Kerr effect to detect valley polarization (section 8.1);
7. Heterostructures: detection of valley Hall effect in monolayer graphene (section 2.4.4), valley-polarized current due to metal and superconductor (section 4.4), and long-lived valley interlayer excitons compared with monolayer excitons (sections 8.3, and 8.4);
8. Topology: valley Chern number (section 2.7) and topological valley kink states (sections 2.7, 2.8, 6.3, and 6.4).

We also inspect the interplay between these approaches. Each method is analyzed in detail, highlighting how it operates, its practical advantages, and the challenges associated with its implementation. These topics are designed to provide readers with a thorough understanding of how different techniques can be applied to achieve precise control over valleytronic properties.

In the latter part of the book, we shift our focus to practical applications. Here, we delve into the real-world implications of valleytronics, showcasing how theoretical concepts are being translated into functional devices. We cover a range of applications, including valleytronic LEDs and transistors. By presenting case studies and current research, we aim to illustrate the transformative potential of valleytronic materials and their impact on various technological fields.

As we navigate through the pages of this book, our goal is to offer a resource that is both informative and inspiring. We hope to provide readers with a clear and detailed understanding of 2D valleytronic materials and their applications, while also sparking curiosity and encouraging further exploration in this exciting field. Whether you are a researcher, engineer, graduate, or undergraduate student, we believe this book will serve as a valuable tool in your journey through the world of valleytronics.

In conclusion, *Two-dimensional Valleytronic Materials: From Principles to Device Applications* is a celebration of the remarkable advances in valleytronic research and a glimpse into the future of this transformative technology. As we explore the intricate principles and cutting-edge applications of valleytronics, we invite you to join us in this exploration and envision the possibilities that lie ahead. The field of valleytronics is rapidly evolving, and the insights presented in this book are intended to contribute to its continued growth and innovation.

SW wrote chapters 1, 2, and 6–8, sections 3.2, 3.3, and 4.1–4.3, subsections 3.1.1–3.1.3, and 5.1.1, and appendices A–C. HT wrote sections 4.4, 5.2, and 5.3 and subsections 3.1.4 and 5.1.2–5.1.5. Both authors cross-checked and proofread each other's contributions to maintain coherence and consistency.

Despite our efforts, the book inevitably contains typos, errors, and less comprehensible explanations. We would appreciate if the readers could inform us of any of those; please send your comments to IsaacWang@jit.edu.cn. We are excited to share this journey with you and look forward to the continued advancements that valleytronic materials will bring to the world of technology and beyond.

The authors
August, 2024

Acknowledgments

During the preparation of this book, we have benefited from many fruitful discussions. SW is indebted to (in alphabetical order) Dr Nguyen Tuan Hung, Dr Minglei Sun, Dr Muhammad Shoufie Ukhtary, Dr Juanjuan Wang, Prof Dr Jun Wang, Dr Yun Yang, Dr Haogang Zhu, and Dr Xinyang Zhu for many helpful discussions. HT is thankful to Prof Dr Chongdan Ren for many discussions and collaborations.

We are very grateful to IOP Publishing for making this book possible. We would especially like to thank Ms Phoebe Hooper, Ms Erika Radzvilaite, Ms Claire Roy, Ms Emily Tapp, and IOP Publishing eBooks Production Team for their help in setting up and completing the production of this book, as well as the processes in between.

The preparation of this book was funded by the National Natural Science Foundation of China (Nos. 12264059 and 11704165), the China Scholarship Council (No. 201908320001), the Natural Science Foundation of Jiangsu (No. BK20211002), Shandong Province (No. ZR2023MA027), as well as the Qinglan Project of Jiangsu Province of China.

Author biographies

S Wang

S Wang earned his PhD in physics at Southeast University, Nanjing, China, in 2016. During this period, he was awarded the national scholarship for doctoral students. He was a visiting scientist at Tohoku University, Sendai, Japan, from 2019 to 2021. Since 2021, he has been an associate professor at the Jinling Institute of Technology, Nanjing, China. His current interests focus on theoretical and computational studies of spin and valley transport, as well as valley optoelectronic devices in two-dimensional materials. He is a principal investigator for the National Science Foundation for Young Scientists of China. He has published 75 papers with more than 4000 citations; for 31 of these papers, he was the first author or corresponding author. Four of the latter papers are among the top 1% highly cited papers (Essential Science Indicators). He was ranked among the world's top 2% most-cited scientists of 2023 and 2024 by Stanford University. In addition, he has served as an Associate Editor of the *Journal of Superconductivity and Novel Magnetism* (Springer Publishing) since 2020 and as a guest editor of the *Journal of Physics D: Applied Physics* (IOP Publishing) since 2023.

H Tian

H Tian has been an associate professor at Linyi University, Linyi, China, since 2023. He completed his PhD in physics at Southeast University, Nanjing, China, in 2013. He has published over 40 papers; for 16 of these papers, he was the first author or the corresponding author. His research interests mainly focus on the electronic transport properties of two-dimensional systems, including spin and valley-related transport properties, as well as the transport properties of polycrystalline systems. In addition, he is also very interested in information science and artificial intelligence.

List of symbols

α	Angle with respect to the x-axis
γ_0	Nearest-neighbor (NN) hopping energy
γ_n	Berry phase of the nth band
Δ	Energy separation between subsequent modes of zigzag graphene nanoribbon at the $\boldsymbol{K}_{\pm}$ points
Δ_z	Energy difference between lattices
$\delta(x)$	Delta function
δ_{ij}	Kronecker delta
$\boldsymbol{\varepsilon}$	Strain tensor
θ	Polar angle
$\theta_{\boldsymbol{q}}$	Polar angle measured between $\boldsymbol{q}$ and the k_x-axis
κ	Valley index
λ	Wavelength
λ_{SO}	Intrinsic spin–orbit coupling (SOC) strength
μ	Chemical potential
μ_{B}	Bohr magneton
ρ	Resistivity
$\boldsymbol{\sigma}$	Pauli matrices
σ_{H}	Hall conductivity
σ_{v}	Valley Hall conductivity
σ_{xx}	Electrical conductivity
σ_+	Left-handed circularly polarized light
σ_-	Right-handed circularly polarized light
ϕ	Angle of incidence
φ	Azimuthal angle
Ψ	Wave function
Ω_C	Solid angle subtended by the curve C
$\boldsymbol{\Omega}_n$	Berry curvature of the nth band
ω	Angular frequency of an electromagnetic wave
a	Lattice constant
$\boldsymbol{a}_i$	Primitive lattice vectors of graphene
$\boldsymbol{a}_i^{\mathrm{d}}$	Primitive lattice vectors of graphene with a line defect
a_{C}	C–C bond length
$\boldsymbol{A}$	Vector field
$\mathscr{A}_n$	Berry connection of the nth band
$\boldsymbol{B}$	Magnetic field
$\boldsymbol{b}_i$	Reciprocal lattice vectors of graphene
C	Closed path
$\mathcal{C}$	Total Chern number
$\mathcal{C}_{\mathrm{v}}$	Valley Chern number
c	speed of light
$\boldsymbol{D}$	Dipole vector
e	Elementary charge
e_0	Magnitude of strain
$\boldsymbol{E}$	Electric field
$\mathcal{E}$	Energy
$\mathcal{E}_{\mathrm{C}}$	On-site potential of carbon atom

$\mathcal{E}_{\mathrm{F}}$ Fermi energy
f_{FD} Fermi–Dirac distribution function
G Conductance
g_{s} Spin degeneracy
H Hamiltonian
$\mathcal{H}$ Hermite polynomials
$\hbar$ Reduced Planck's constant
I Intensity of light
$\boldsymbol{J}$ Jones vector
$\boldsymbol{j}$ Electric current density
$\boldsymbol{j}_{\mathrm{v}}$ Valley current density
L Lateral distance
$\boldsymbol{M}$ Magnetic moment
M Transition matrix element
m Quantum number
m_{e} Mass of the electron
m^* Effective mass
N Number of unit cells in 2D graphene
N_{z} Number of zigzag chains across the zigzag graphene nanoribbon
$\boldsymbol{p}$ Momentum of an electron
P Valley polarization (VP)
$\boldsymbol{q}$ Wave vector measured from the $\boldsymbol{K}_{\pm}$ points
R_i NN B atoms relative to an A atom in graphene
R_{NL} Nonlocal resistance
t time
${}^{t}\mathsf{M}$ Transpose of M
U Potential
$\boldsymbol{v}$ Velocity
$\dot{\boldsymbol{v}}$ The first derivative of $\boldsymbol{v}$ with respect to time
v_{F} Fermi velocity
$\boldsymbol{v}_{\mathrm{a}}$ Anomalous velocity
V_{g} Gate voltage
V_{NL} Nonlocal voltage
W Width of a nanoribbon
∂_x $\frac{\partial}{\partial x}$ Partial derivative with respect to variable x
v^* Complex conjugate of complex variable v
$\mathsf{M}^{\dagger}$ Hermitian (conjugate) transpose of complex matrix M
$\dot{\boldsymbol{v}}$ First derivative of vector $\boldsymbol{v}$ with respect to time
$\otimes$ Kronecker product

List of abbreviations

1D	one-dimensional
2D	two-dimensional
2DEG	2D electron gas
3D	three-dimensional
aGNR	armchair graphene nanoribbon
AR	Andreev reflection
ARPES	Angle-resolved photoemission spectroscopy
AVHE	Anomalous valley Hall effect
BZ	Brillouin zone
CBM	Conduction band minimum
CVD	Chemical vapor deposition
DBdG	Dirac–Bogoliubov–de Gennes
DFT	Density functional theory
DOF	Degree of freedom
DOS	density of states
EDL	Electrical double-layer
EL	Electroluminescence
FET	Field-effect transistor
FMI	Ferromagnetic insulator
GB	Grain boundary
GNR	Graphene nanoribbon
hBN	Hexagonal boron nitride
IDB	Inversion domain boundaries
IS	Inversion symmetry
IX	Interlayer exciton
KR	Kerr rotation
LB	Lower band
LCP	Left-handed circularly polarized
LDGSL	Line-defect graphene superlattice
LED	Light-emitting diode
LL	Landau level
LLL	Lowest Landau level
MTGB	Mirror twin grain boundary
NEGF	Nonequilibrium Green's function
NN	Nearest neighbor
NNN	Next-nearest neighbor
PDMS	Polydimethylsiloxane
PL	Photoluminescence
PMF	Pseudomagnetic field
QHE	Quantum Hall effect
QPC	Quantum point contact
RCP	Right-handed circularly polarized
RSOC	Rashba spin–orbit coupling
SOC	spin–orbit coupling
STM	Scanning tunneling microscopy
STS	Scanning tunneling spectroscopy
SVL	Single vacancy line

TB	Tight binding
TBLG	Twisted bilayer graphene
TEM	Transmission electron microscopy
TMDC	Transition-metal dichalcogenide
UB	Upper band
VBM	Valence band maximum
vdW	van der Waals
VHE	Valley Hall effect
VP	Valley polarization
zGNR	Zigzag graphene nanoribbon

IOP Publishing

Two-dimensional Valleytronic Materials
From principles to device applications
Sake Wang and Hongyu Tian

Chapter 1

Introduction to two-dimensional (2D) materials and valleytronics

This chapter provides a comprehensive overview of the emerging field of 2D materials and their applications in valleytronics. The chapter begins with an introduction to the fundamental concepts and significance of 2D materials. It then explores the basic geometry and electronic properties of graphene, transition-metal dichalcogenides (TMDCs), and (twisted) bilayer graphene, focusing on their unique characteristics and potential applications. Following this, the chapter examines the role of spin–orbit coupling (SOC) in enhancing their electronic behavior. The discussion then extends to valley-polarized Landau levels (LLs) in graphene. Finally, the chapter reviews early theoretical proposals for valleytronic applications, emphasizing the potential of these 2D materials in developing advanced technologies.

1.1 Introduction

Valleytronics is an emerging field of study in condensed matter physics that explores the use of a novel degree of freedom (DOF) in electrons, known as the 'valley' DOF, for information processing. In certain materials, carriers can occupy different energy extrema (valleys) in momentum space, typically located at distinct points (such as the $\boldsymbol{K}_+$ and $\boldsymbol{K}_-$ points) in the material's Brillouin zone (BZ). The concept of valleytronics is analogous to spintronics, where the spin of electrons is used for information storage and manipulation, but here, the focus is on controlling the valley index of electrons.

Compared to bulk materials, 2D materials are more suited for valleytronics due to their distinct electronic properties that arise from their reduced dimensionality. In 2D materials, carriers can occupy well-separated valleys in momentum space, such as the $\boldsymbol{K}_+$ and $\boldsymbol{K}_-$ points, making it easier to manipulate and distinguish between these valleys. The strong SOC found in some 2D materials, such as TMDCs, links the spin and valley DOFs, enabling robust control over valley polarization (VP). Additionally, 2D

doi:10.1088/978-0-7503-5562-9ch1

materials often exhibit valley-selective optical properties, allowing precise control of valley states using circularly polarized light. These materials also have weaker dielectric screening, which enhances Coulomb interactions and stabilizes valley-dependent phenomena, making the valley DOF more accessible and easier to manipulate.

Understanding the basic geometry and electronic properties of 2D materials is essential before studying valleytronic applications because these properties determine how valley states can be controlled and utilized. Key electronic properties include the band structure, which defines the energy levels and the locations of valleys in momentum space, and the SOC, which can link spin and valley DOFs. These factors affect how effectively different valleys can be accessed and manipulated. Additionally, electronic properties influence how a material responds to external stimuli such as electric fields, magnetic fields, or optical excitation. This knowledge is critical for designing and optimizing devices that leverage valleytronic effects, such as VP and the valley Hall effect (VHE), to achieve the desired performance and functionality in applications such as information storage and processing.

1.2 Electronic properties of graphene, definition of 'valley'

Graphene is considered the first 2D material to have been experimentally isolated and studied in detail. While theoretical studies of 2D materials, such as Wallace's work on graphite's electronic band structure in 1947, laid the groundwork [1], it was the 2004 experimental breakthrough by Sir Andre Konstantin Geim and Sir Konstantin Sergeevich Novoselov that successfully isolated single layers of graphene flakes from graphite and characterized them [2]. This discovery highlighted graphene's exceptional properties and established it as the first 2D material to be extensively studied, sparking the development of other 2D materials. The early study of graphene included the observation of the integer quantum Hall effect in monolayers [3, 4] and bilayers [5]. The honeycomb lattice of graphene has also been used theoretically to study Dirac fermions in a condensed matter system [6, 7].

It has been noted that the results obtained from *ab initio* density functional theory (DFT) calculations and the tight-binding (TB) approximation are almost similar for graphene-like systems. However, the latter is simpler and physically transparent. In this book, we will explore the TB calculations, mentioning the DFT approach where possible. The TB model of electrons in multilayer graphene, which takes into account coupling between layers, became known as the Slonczewski–Weiss–McClure model [8, 9].

1.2.1 Tight-binding description of graphene

Graphene is a 2D structure; it serves as a building block for three-dimensional (3D) materials. Carbon atoms in graphene are periodically arranged in an infinite honeycomb (hexagonal) lattice. The valence electrons give rise to the $2s$, $2p_x$, $2p_y$, and $2p_z$ orbitals, which are important in forming covalent bonds. If we define the z-axis as the direction perpendicular to the 2D graphene sheet [10], three orbitals (s, p_x, and p_y) combine to form in-plane σ bonds. The remaining p_z orbital, pointing along the z-axis, forms π bonds with neighboring p_z orbitals through the $pp\pi$ interaction. Due to their different symmetries, the orbitals forming the σ bonds are

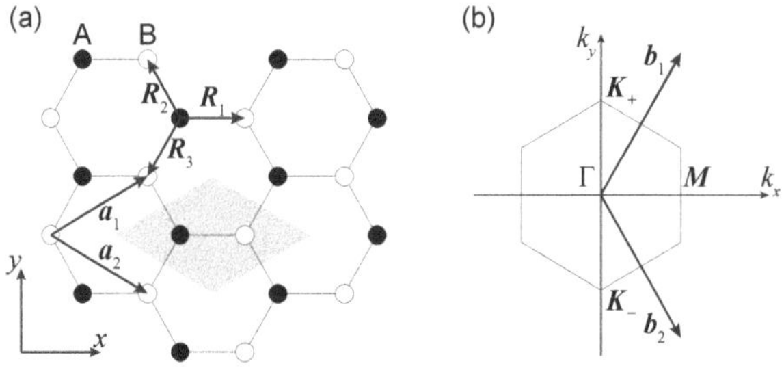

Figure 1.1. (a) Real space of a honeycomb graphene lattice. The primitive unit cell is shaded in gray, with two sublattices A (black dots) and B (white dots). $\boldsymbol{a}_1$ and $\boldsymbol{a}_2$ are the primitive lattice vectors. $\boldsymbol{R}_1$, $\boldsymbol{R}_2$, and $\boldsymbol{R}_3$ are vectors pointing from an A atom to its nearest B atoms. (b) First BZ with reciprocal lattice vectors $\boldsymbol{b}_1$ and $\boldsymbol{b}_2$. The high-symmetry points, $\Gamma = \boldsymbol{0}$, $\boldsymbol{K}_\pm = \pm\frac{4\pi}{3a}\hat{\boldsymbol{y}}$, and $\boldsymbol{M} = \frac{2\pi}{\sqrt{3}a}\hat{\boldsymbol{x}}$, are indicated.

decoupled from the π bonds: the three orbitals (s, p_x, and p_y) are even with respect to the graphene sheet, while the p_z orbital is odd [1].

The bonding (occupied) and antibonding (unoccupied) σ bands are largely separated in energy, by more than 12 eV compared to the π bands. Therefore, the π bands can completely describe the low-energy electronic structures and provide important insights into understanding the transport and optical properties.

In figure 1.1(a), we show a real-space 2D honeycomb graphene lattice based on two atoms, labeled A and B. In the x, y coordinates, the primitive lattice vectors are

$$\boldsymbol{a}_1 = \frac{1}{2}a\left(\sqrt{3}\,\hat{\boldsymbol{x}} + \hat{\boldsymbol{y}}\right),\ \boldsymbol{a}_2 = \frac{1}{2}a\left(\sqrt{3}\,\hat{\boldsymbol{x}} - \hat{\boldsymbol{y}}\right), \tag{1.1}$$

where $a = |\boldsymbol{a}_1| = |\boldsymbol{a}_2| = \sqrt{3}\,a_{\mathrm{C}} \approx 2.46$ Å is the lattice constant, while the distance between two neighboring carbon atoms $a_{\mathrm{C}} = 1.42$ Å. Note that this honeycomb structure does not form a Bravais lattice [11], as it is impossible to connect the A and B atoms using a lattice vector $n_1\boldsymbol{a}_1 + n_2\boldsymbol{a}_2$, where n_1 and n_2 are integers. The A and B atoms are nonequivalent. In the following, we will often refer to them as the 'A (B) sublattice'.

The three nearest neighbor (NN) B atoms relative to an A atom are:

$$\boldsymbol{R}_1 = \frac{a}{\sqrt{3}}\hat{\boldsymbol{x}},\ \ \boldsymbol{R}_2 = \frac{a}{2}\left(-\frac{1}{\sqrt{3}}\hat{\boldsymbol{x}} + \hat{\boldsymbol{y}}\right),\ \ \boldsymbol{R}_3 = -\frac{a}{2}\left(\frac{1}{\sqrt{3}}\hat{\boldsymbol{x}} + \hat{\boldsymbol{y}}\right). \tag{1.2}$$

The Wigner–Seitz primitive cell of the reciprocal lattice is known as the first BZ, which is shown in figure 1.1(b). By defining a unit vector $\boldsymbol{a}_3 = \hat{\boldsymbol{z}}$ that points in the direction perpendicular to the plane of the 2D graphene lattice, the reciprocal lattice vectors can be derived as follows [11]:

$$\boldsymbol{b}_1 = 2\pi\frac{\boldsymbol{a}_2 \times \boldsymbol{a}_3}{\boldsymbol{a}_1 \cdot (\boldsymbol{a}_2 \times \boldsymbol{a}_3)} = \frac{2\pi}{a}\left(\frac{1}{\sqrt{3}}\hat{\boldsymbol{x}} + \hat{\boldsymbol{y}}\right), \tag{1.3}$$

$$\boldsymbol{b}_2 = 2\pi\frac{\boldsymbol{a}_3 \times \boldsymbol{a}_1}{\boldsymbol{a}_1 \cdot (\boldsymbol{a}_2 \times \boldsymbol{a}_3)} = \frac{2\pi}{a}\left(\frac{1}{\sqrt{3}}\hat{\boldsymbol{x}} - \hat{\boldsymbol{y}}\right). \tag{1.4}$$

We can also find the reciprocal lattice vector $\boldsymbol{b}_3 = 2\pi\frac{\boldsymbol{a}_1 \times \boldsymbol{a}_2}{\boldsymbol{a}_1\cdot(\boldsymbol{a}_2 \times \boldsymbol{a}_3)} = 2\pi\hat{z}$, perpendicular to the plane of the 2D reciprocal lattice. Therefore, $\boldsymbol{b}_i$ satisfy

$$\boldsymbol{b}_i \cdot \boldsymbol{a}_j = 2\pi\delta_{ij}, \tag{1.5}$$

where δ_{ij} is the Kronecker delta symbol: $\delta_{ij} = 0$ for $i \neq j$ and $\delta_{ij} = 1$ for $i = j$. The two primitive lattice vectors $\boldsymbol{a}_1$ and $\boldsymbol{a}_2$ make an angle of 30° with each other, while the reciprocal lattice vectors $\boldsymbol{b}_1$ and $\boldsymbol{b}_2$ make an angle of 120° with each other.

We can now discover the electronic band structure using the TB model [8]. Since graphene exhibits translational symmetry, its wave function satisfies Bloch's theorem. Two Bloch functions, constructed from p_z orbitals for the two inequivalent A and B atoms, give the basis functions for 2D graphene. The TB wave function $\Psi(\boldsymbol{k}, \boldsymbol{r})$ of one electron in the hexagonal lattice system is represented by the linear combination of the Bloch orbital $p_z^{\mathrm{A}}(\boldsymbol{k}, \boldsymbol{r})$ of the A atom and the Bloch orbital $p_z^{\mathrm{B}}(\boldsymbol{k}, \boldsymbol{r})$ of the B atom.

$$\Psi(\boldsymbol{k}, \boldsymbol{r}) = c_{\mathrm{A}}(\boldsymbol{k})p_z^{\mathrm{A}}(\boldsymbol{k}, \boldsymbol{r}) + c_{\mathrm{B}}(\boldsymbol{k})p_z^{\mathrm{B}}(\boldsymbol{k}, \boldsymbol{r}), \tag{1.6}$$

where $\boldsymbol{k}$ is the electron wave vector, c_{A} (c_{B}) is a coefficient to be determined,

$$p_z^{\mathrm{A}}(\boldsymbol{k}, \boldsymbol{r}) = \frac{1}{\sqrt{N}}\sum_{\boldsymbol{R}_{\mathrm{A}}} \mathrm{e}^{\mathrm{i}\boldsymbol{k}\cdot\boldsymbol{R}_{\mathrm{A}}} p_z(\boldsymbol{r} - \boldsymbol{R}_{\mathrm{A}}), \tag{1.7}$$

$$p_z^{\mathrm{B}}(\boldsymbol{k}, \boldsymbol{r}) = \frac{1}{\sqrt{N}}\sum_{\boldsymbol{R}_{\mathrm{B}}} \mathrm{e}^{\mathrm{i}\boldsymbol{k}\cdot\boldsymbol{R}_{\mathrm{B}}} p_z(\boldsymbol{r} - \boldsymbol{R}_{\mathrm{B}}), \tag{1.8}$$

N is the number of unit cells in the 2D graphene, and $\boldsymbol{R}_{\mathrm{A}}$ ($\boldsymbol{R}_{\mathrm{B}}$) is the position of the A (B) atom. The wave functions p_z^{A} and p_z^{B} are normalized, and the overlap between p_z^{A} and p_z^{B} is negligible [12]. The Bloch sums now form an orthonormal set:

$$\left\langle p_z^{\alpha}(\boldsymbol{k}) | p_z^{\beta}(\boldsymbol{k}') \right\rangle = \delta_{\boldsymbol{k},\boldsymbol{k}'}\delta_{\alpha,\beta}, \tag{1.9}$$

where α, $\beta = \mathrm{A}, \mathrm{B}$. Using these orthogonality relations in the Schrödinger equation, $H(\boldsymbol{k})\Psi(\boldsymbol{k}, \boldsymbol{r}) = \mathcal{E}(\boldsymbol{k})\Psi(\boldsymbol{k}, \boldsymbol{r})$, one obtains a 2×2 eigenvalue problem,

$$\begin{pmatrix} H_{\mathrm{AA}}(\boldsymbol{k}) & H_{\mathrm{AB}}(\boldsymbol{k}) \\ H_{\mathrm{BA}}(\boldsymbol{k}) & H_{\mathrm{BB}}(\boldsymbol{k}) \end{pmatrix} \begin{pmatrix} c_\xi^{\mathrm{A}}(\boldsymbol{k}) \\ c_\xi^{\mathrm{B}}(\boldsymbol{k}) \end{pmatrix} = \mathcal{E}_\xi(\boldsymbol{k}) \begin{pmatrix} c_\xi^{\mathrm{A}}(\boldsymbol{k}) \\ c_\xi^{\mathrm{B}}(\boldsymbol{k}) \end{pmatrix}. \tag{1.10}$$

Here, $\xi = +$ or $-$ can be the conduction band or the valence band. The matrix elements of the Hamiltonian are given by

$$H_{\mathrm{AA}}(\boldsymbol{k}) = \frac{1}{N}\sum_{\boldsymbol{R}_{\mathrm{A}}}\sum_{\boldsymbol{R}'_{\mathrm{A}}} \left\langle \mathrm{e}^{\mathrm{i}\boldsymbol{k}\cdot\boldsymbol{R}'_{\mathrm{A}}} p_z\left(\boldsymbol{r} - \boldsymbol{R}'_{\mathrm{A}}\right) | H | \mathrm{e}^{\mathrm{i}\boldsymbol{k}\cdot\boldsymbol{R}_{\mathrm{A}}} p_z(\boldsymbol{r} - \boldsymbol{R}_{\mathrm{A}}) \right\rangle, \tag{1.11}$$

$$H_{\mathrm{AB}}(\boldsymbol{k}) = \frac{1}{N}\sum_{\boldsymbol{R}_{\mathrm{A}}}\sum_{\boldsymbol{R}_{\mathrm{B}}} \left\langle \mathrm{e}^{\mathrm{i}\boldsymbol{k}\cdot\boldsymbol{R}_{\mathrm{A}}} p_z(\boldsymbol{r} - \boldsymbol{R}_{\mathrm{A}}) \middle| H \middle| \mathrm{e}^{\mathrm{i}\boldsymbol{k}\cdot\boldsymbol{R}_{\mathrm{B}}} p_z(\boldsymbol{r} - \boldsymbol{R}_{\mathrm{B}}) \right\rangle. \tag{1.12}$$

After some manipulations (see appendix A), in which we restrict ourselves to the interactions between the first NNs only,[1] we obtain

$$\begin{aligned} H_{AA}(\boldsymbol{k}) &= H_{BB}(\boldsymbol{k}) \simeq \mathcal{E}_C, \\ H_{AB}(\boldsymbol{k}) &= H_{BA}(\boldsymbol{k})^* = -\gamma_0 \sum_{i=1}^{3} e^{i\boldsymbol{k}\cdot\boldsymbol{R}_i} = -\gamma_0 f(\boldsymbol{k}), \end{aligned} \tag{1.13}$$

where $\mathcal{E}_C$ is the on-site potential of the carbon atom, $f(\boldsymbol{k})$ is the sum of the phase factors for the three NNs with the relative vector $\boldsymbol{R}_i$, and

$$\gamma_0 = -\left\langle p_z(\boldsymbol{r} + \boldsymbol{R}_i) \middle| H \middle| p_z(\boldsymbol{r}) \right\rangle \quad (i = 1, 2, 3) \tag{1.14}$$

is the NN hopping energy. The TB model in this section cannot be used to determine the values of the parameter γ_0. They must be determined either by other theoretical methods, such as DFT, or by comparing the TB model with experimental results, which reveals that γ_0 ranges from 2.9 to 3.1 eV [15–17]. It should be noted that the main qualitative results presented here remain unchanged regardless of the specific parameter values [18].

If we take the on-site potential of a carbon atom $\mathcal{E}_C$ as the reference level by setting it to zero, the Hamiltonian matrix in equation (1.10) can be written as

$$H(\boldsymbol{k}) = \begin{pmatrix} 0 & -\gamma_0 f(\boldsymbol{k}) \\ -\gamma_0 f(\boldsymbol{k})^* & 0 \end{pmatrix}. \tag{1.15}$$

The eigenvalues of the above Hamiltonian yield two eigenenergies:

$$\begin{aligned} \mathcal{E}_{\pm}(\boldsymbol{k}) &= \pm \gamma_0 |f(\boldsymbol{k})| \\ &= \pm \gamma_0 |e^{i\boldsymbol{k}\cdot\boldsymbol{R}_1} + e^{i\boldsymbol{k}\cdot\boldsymbol{R}_2} + e^{i\boldsymbol{k}\cdot\boldsymbol{R}_3}| \\ &= \pm \gamma_0 \sqrt{3 + 2\cos k_y a + 4\cos\frac{\sqrt{3}k_x a}{2}\cos\frac{k_y a}{2}}. \end{aligned} \tag{1.16}$$

The corresponding symmetric wave functions (eigenvectors) are

$$\begin{aligned} |\Psi_+(\boldsymbol{k})\rangle &= \begin{pmatrix} c_+^A(\boldsymbol{k}) \\ c_+^B(\boldsymbol{k}) \end{pmatrix} = \frac{1}{\sqrt{2}} \begin{pmatrix} -\sqrt{\dfrac{f(\boldsymbol{k})}{w(\boldsymbol{k})}} \\ \sqrt{\dfrac{f(\boldsymbol{k})^*}{w(\boldsymbol{k})}} \end{pmatrix}, \\ |\Psi_-(\boldsymbol{k})\rangle &= \begin{pmatrix} c_-^A(\boldsymbol{k}) \\ c_-^B(\boldsymbol{k}) \end{pmatrix} = \frac{1}{\sqrt{2}} \begin{pmatrix} \sqrt{\dfrac{f(\boldsymbol{k})}{w(\boldsymbol{k})}} \\ \sqrt{\dfrac{f(\boldsymbol{k})^*}{w(\boldsymbol{k})}} \end{pmatrix}, \end{aligned} \tag{1.17}$$

[1] Other terms, such as the next-NN contributions, can be considered in the calculation [13, 14]. Generally, they have a small effect on the electronic band structure and cannot change the physics discussed in this book.

where

$$w(\boldsymbol{k}) = |f(\boldsymbol{k})| = \sqrt{f(\boldsymbol{k})^* f(\boldsymbol{k})} = \sqrt{3 + 2\cos k_y a + 4\cos\frac{\sqrt{3}k_x a}{2}\cos\frac{k_y a}{2}}. \quad (1.18)$$

It is evident that the use of inversion symmetry (IS), i.e., switching the A and B atoms, changes the valence (conduction) band wave function to a (negative) complex conjugate. This supports the fact that for general molecular orbitals, the valence (conduction) band wave function of a molecule is made up from a symmetric (antisymmetric) combination of atomic orbitals.

The energy band structure described by equation (1.16) is plotted in figure 1.2. We can see that the bandgap closes, i.e. $\mathcal{E}(\boldsymbol{k}) = 0$, at the six corners of the BZ. The six corners are known as K points, which are also called the Dirac points. The vicinities of these Dirac points are referred to as 'valleys', which stem from the similarity of the shape of the dispersion relation to a valley. Similar to the two unequal sublattices A and B in real space, there are two unequal Dirac points $\boldsymbol{K}_+$ and $\boldsymbol{K}_-$ in $\boldsymbol{k}$-space. As such, the $\boldsymbol{K}_\pm$ valley index $\kappa = \pm 1$ may be treated like the electron spin index $s = \pm 1$, where $\uparrow = 1$ and $\downarrow = -1$ for the electron [19]. These unequal $\boldsymbol{K}_+$ and $\boldsymbol{K}_-$ points form the valley pseudospin DOF in graphene. Their two valleys form a time-reversed symmetric pair, meaning that under time-reversal symmetry (TRS), the valley of an electron is inverted [20]. In figure 1.1(b), the coordinates of the $\boldsymbol{K}_+$ and $\boldsymbol{K}_-$ points are [21]:

$$\boldsymbol{K}_+ = \frac{4\pi}{3a}\hat{\boldsymbol{y}}, \quad \boldsymbol{K}_- = -\frac{4\pi}{3a}\hat{\boldsymbol{y}}. \quad (1.19)$$

It is possible to connect two of the other corners of the BZ to $\boldsymbol{K}_+$ using a reciprocal lattice vector; this indicates the other two are equivalent to $\boldsymbol{K}_+$. In addition, it is

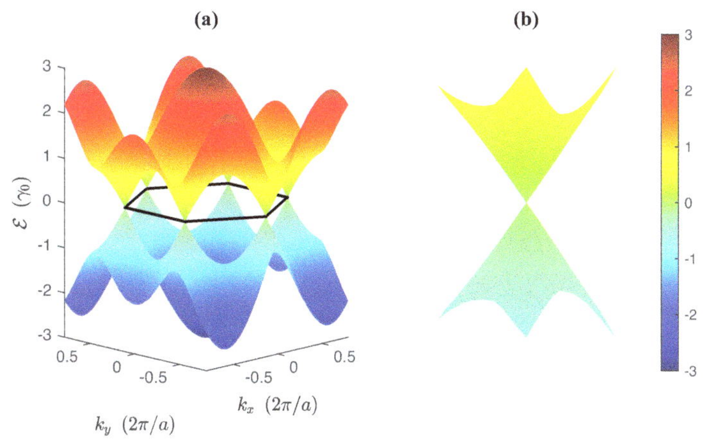

Figure 1.2. (a) Plot of the band structure of graphene based on equation (1.16). The first BZ is denoted by the black hexagon. The conduction and valence bands touch at the six corners of the first BZ. (b) An enlarged view of the vicinity of $\mathcal{E}(\boldsymbol{k}) = 0$, showing the linear dispersion relation, which is called a 'Dirac cone.'

possible to connect the remaining two corners to $\boldsymbol{K}_-$ using a reciprocal lattice vector, thus the remaining two corners are equivalent to $\boldsymbol{K}_-$. However, it is not possible to connect $\boldsymbol{K}_+$ and $\boldsymbol{K}_-$ with a reciprocal lattice vector. Another popular choice [22] for the coordinates of $\boldsymbol{K}_+$ and $\boldsymbol{K}_-$ in the literature is to set

$$\boldsymbol{K}_+ = \frac{2\pi}{\sqrt{3}a}\left(\hat{\boldsymbol{x}} - \frac{1}{\sqrt{3}}\hat{\boldsymbol{y}}\right), \quad \boldsymbol{K}_- = \frac{2\pi}{\sqrt{3}a}\left(\hat{\boldsymbol{x}} + \frac{1}{\sqrt{3}}\hat{\boldsymbol{y}}\right). \tag{1.20}$$

The physics and results obtained in the rest of this book are unchanged by this alteration. Other high-symmetry points are the center of the first BZ, the Γ point, and the midpoint of the side of the hexagon, the M point. The coordinate of the $\boldsymbol{M}$ point labeled in figure 1.1(b) is

$$\boldsymbol{M} = \frac{2\pi}{\sqrt{3}a}\hat{\boldsymbol{x}}. \tag{1.21}$$

For electrically neutral graphene, the chemical potential, i.e., the Fermi energy, is located at the Dirac points ($\mathcal{E}_\mathrm{F} = 0$). As a result, the valence band is fully occupied, while the conduction band is empty. Graphene displays a metallic character with a zero bandgap. However, since the Fermi surface shrinks to one point at $\mathcal{E}_\mathrm{F} = 0$, the term semimetallic or zero-gap/gapless semiconductor [23] is usually employed [22].

We emphasize that, for most experiments and potential applications where the quantum properties of carriers may be exploited, only the part of the dispersion near the Fermi energy ($\lessapprox$ 1 eV) is important and interesting [24]. Therefore, a proper description of this part of the band structure is essential to understand and theoretically model experiments. To this end, we set the origin of the $\boldsymbol{k}$-space to the $\boldsymbol{K}_\pm$ point by defining the wave vector $\boldsymbol{q}$ to measure the deviation from the $\boldsymbol{K}_\pm$ point, i.e. $\boldsymbol{k} = \boldsymbol{K}_\pm + \boldsymbol{q}$. We then expand equation (1.16) in the vicinity of $\boldsymbol{K}_\pm$ and obtain the linear low-energy dispersion relation

$$\mathcal{E}_\pm(\boldsymbol{q}) = \pm\frac{\sqrt{3}}{2}\gamma_0 a|\boldsymbol{q}| = \pm\hbar v_\mathrm{F} q, \tag{1.22}$$

where $\hbar$ is the reduced Planck's constant and

$$v_\mathrm{F} = \frac{\sqrt{3}}{2\hbar}\gamma_0 a \approx 1.0 \times 10^6 \ \mathrm{m\ s^{-1}} \tag{1.23}$$

is the Fermi velocity of graphene. Equation (1.22) tells us that in the vicinity of the corners of the first BZ, the carrier energy is linearly proportional to the wave vector measured from the Dirac point. This region is called the 'Dirac cone', as plotted in figure 1.2(b). This feature of the electronic spectrum has numerous experimental confirmations [25].

1.2.2 Effective Hamiltonian close to the Dirac point

1.2.2.1 Linear approximation

In this book, we usually treat valleys as independent, so we do not consider any valley-connecting processes. The effective Hamiltonian close to the Dirac points $\boldsymbol{K}_+$ and $\boldsymbol{K}_-$ is obtained in the same way as equation (1.22). We replace $\boldsymbol{k}$ with $\boldsymbol{K}_\pm + \boldsymbol{q}$ in the Hamiltonian (1.15) to set the origin of the $\boldsymbol{k}$-space at the $\boldsymbol{K}_\pm$ point; a linear expansion then gives

$$H_{\boldsymbol{K}_\pm}(\boldsymbol{q}) = \frac{\sqrt{3}}{2}\gamma_0 a \begin{pmatrix} 0 & \chi(q_x \pm \mathrm{i}q_y) \\ \chi^*(q_x \mp \mathrm{i}q_y) & 0 \end{pmatrix}, \tag{1.24}$$

where $\chi = \mathrm{e}^{\mathrm{i}\pi/6}$. The phase $\pi/6$ can be excluded by a unitary transformation of the basis functions. Finally, the effective Hamiltonian close to the $\boldsymbol{K}_\pm$ point reads:

$$H_{\boldsymbol{K}_\pm}(\boldsymbol{q}) = \hbar v_\mathrm{F} \begin{pmatrix} 0 & q_x \pm \mathrm{i}q_y \\ q_x \mp \mathrm{i}q_y & 0 \end{pmatrix} = v_\mathrm{F}(p_x\sigma_x \mp p_y\sigma_y), \tag{1.25}$$

where $\boldsymbol{p} = \hbar\boldsymbol{q}$ and the Pauli matrices $\boldsymbol{\sigma}$ are defined as usual [26, 27]:

$$\sigma_0 = \begin{pmatrix} 1 & 0 \\ 0 & 1 \end{pmatrix}, \quad \sigma_x = \begin{pmatrix} 0 & 1 \\ 1 & 0 \end{pmatrix}, \quad \sigma_y = \begin{pmatrix} 0 & -\mathrm{i} \\ \mathrm{i} & 0 \end{pmatrix}, \quad \sigma_z = \begin{pmatrix} 1 & 0 \\ 0 & -1 \end{pmatrix}. \tag{1.26}$$

Note that only the x- and y-components of $\boldsymbol{q}$ appear in equation (1.25). The corresponding eigenfunctions are

$$\Psi_{\boldsymbol{K}_\pm,\xi}(\boldsymbol{q}) = \frac{1}{\sqrt{2}} \begin{pmatrix} \xi \dfrac{q_x \pm \mathrm{i}q_y}{q} \\ 1 \end{pmatrix} = \frac{1}{\sqrt{2}} \begin{pmatrix} \xi \mathrm{e}^{\pm\mathrm{i}\theta_q} \\ 1 \end{pmatrix}, \tag{1.27}$$

where $\xi = \pm 1$ corresponds to the sign of the eigenenergies from equation (1.22), and $\theta_{\boldsymbol{q}} = \arctan(q_y/q_x)$ is the polar angle measured between $\boldsymbol{q}$ and the k_x-axis in $\boldsymbol{k}$-space.

1.2.2.2 Beyond the linear approximation

In the previous section, results were obtained by taking the lowest-order expansion of the TB Hamiltonian (1.15) near the Dirac points. The band structure of monolayer graphene, shown in figure 1.2, is approximately linear in the vicinity of $\boldsymbol{K}_\pm$. However, this simple picture is perturbed by additional contributions to the Hamiltonians. If one considers the next order—quadratic term, after similar manipulations, one finds

$$H_{\boldsymbol{K}_\pm}(\boldsymbol{q}) = \hbar v_\mathrm{F} \begin{pmatrix} 0 & q_x \pm \mathrm{i}q_y + \dfrac{\sqrt{3}}{12}\mathrm{i}(q_x \mp \mathrm{i}q_y)^2 a \\ q_x \mp \mathrm{i}q_y - \dfrac{\sqrt{3}}{12}\mathrm{i}(q_x \pm \mathrm{i}q_y)^2 a & 0 \end{pmatrix}. \tag{1.28}$$

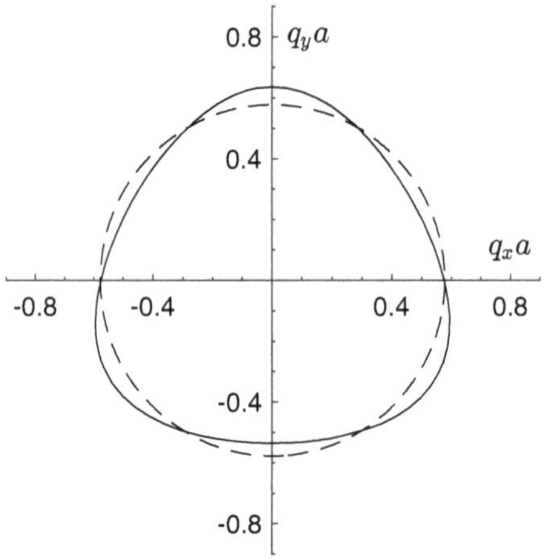

Figure 1.3. TW in monolayer graphene. The solid line shows the isoenergetic line at $\gamma_0/2$ in the vicinity of $\boldsymbol{K}_+$ using equation (1.29), while the dashed line shows the circular isoenergetic line (Fermi circle) obtained from the linear dispersion relation in equation (1.22), which neglects TW.

Compared with Hamiltonian (1.25), the additional terms in Hamiltonian (1.28) correspond to trigonal warping (TW) [29, 30, 47]. Diagonalization of the above Hamiltonian yields the eigenenergies

$$\begin{aligned} \mathcal{E}_{\boldsymbol{K}_\pm,\xi}(\boldsymbol{q}) &= \xi\hbar v_{\mathrm{F}}\sqrt{q^2 \pm \frac{a}{2\sqrt{3}}\left(3q_x^2 q_y - q_y^3\right) + \frac{a^2}{48}q^4} \\ &= \xi\hbar v_{\mathrm{F}} q\sqrt{1 \pm \frac{qa}{2\sqrt{3}}\sin(3\theta_q) + \frac{(qa)^2}{48}}. \end{aligned} \tag{1.29}$$

It is worth noting that the $\propto\sin(3\theta_q)$ term, resulting from the threefold rotational (trigonal) symmetry (C_3) of the crystal lattice, introduces the TW effect [30], causing the loss of isotropy in the dispersion relation. Mathematically, it is initially introduced into Hamiltonian (1.15) in the representation (1.13). The triangular deformation of the Fermi circle becomes stronger as the prefactor qa of $\sin(3\theta_q)$ becomes larger. Experimentally, according to the findings of angle-resolved photoemission spectroscopy (ARPES) data [31], the TW effect becomes relevant when the energy is larger than 0.6 eV. Figure 1.3 shows the TW of a Fermi circle of energy $\gamma_0/2$ near $\boldsymbol{K}_+$. The presence of the $\pm$ sign in front of the $\propto\sin(3\theta_q)$ term means that the orientation of the TW in the two nonequivalent valleys is reversed, as required by TRS.

1.3 Electronic properties of zigzag graphene nanoribbons

Graphene nanoribbons (GNRs) are strips of graphene with a finite width. They can be obtained by cutting a graphene sheet along a certain direction. Two typical directions are possible, as shown in figure 1.4: armchair (blue) and zigzag (red) [32–34]; these directions have a difference of 30° between them. In addition to these two cases, more complex shapes are possible, including chevron, chiral, and cove shapes, etc [35, 36].

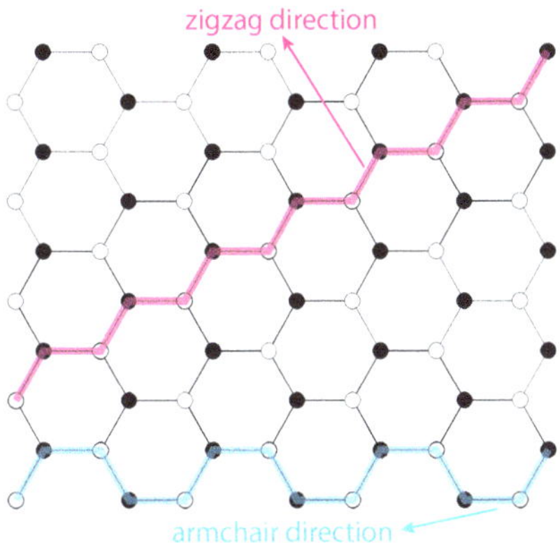

Figure 1.4. Illustration of armchair (blue) and zigzag (red) directions in a honeycomb lattice.

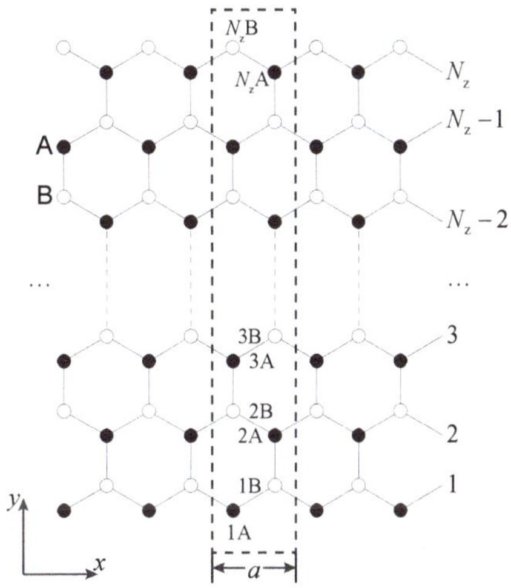

Figure 1.5. The structure of zGNR consists of N_Z zigzag chains. Black and white dots identify carbon atoms belonging to two different graphene sublattices, A and B. The dashed rectangle represents the one-dimensional (1D) unit cell of zGNR. The labeling of the atoms follows the numbering of the zigzag lines. The lattice constant for the zGNR is a.

Nowadays, among various synthetic approaches, bottom-up synthetic methods, such as on-surface polymerizations and solution-phase polymerizations, represent promising ways to control the width, length, and edge structure of GNRs [35, 36].

In GNRs with zigzag edges, known as zigzag GNRs (zGNRs), two valleys related to each Dirac cone are well separated in momentum space. On the other hand, in GNRs with armchair edges, namely armchair GNRs (aGNRs), the low-energy spectrum is described as the superposition of two nonequivalent Dirac points of graphene; this means an absence of two well-separated valley structures [37]. Since this book emphasizes the utilization of the valley DOF, we only discuss zGNRs.

The atomic structure of a zGNR is presented in figure 1.5. It is assumed that all dangling bonds at the edges are terminated by hydrogen atoms [38]. We follow the naming conventions in references [34] and [39] for the labeling of zigzag lines and atoms in the unit cell. zGNRs are classified by the number of zigzag chains, N_z, across the ribbon width, and we designate this type of zGNR as N_z-zGNR.

Similar to the previous treatment for 2D infinite graphene, based on $\{1A, 1B, 2A, 2B, \ldots, N_zA, N_zB\}$, the naming system used for the sites in the unit cell labeled in figure 1.5, the TB Hamiltonian of the nanoribbon is constructed as a k-dependent tridiagonal $2N_z \times 2N_z$ matrix with alternating off-diagonal elements [22, 40],

$$H_{\mathrm{zGNR}} = \gamma_0 \begin{pmatrix} 0 & 2\cos\frac{ka}{2} & 0 & 0 & \ldots & 0 & 0 \\ 2\cos\frac{ka}{2} & 0 & 1 & 0 & \ldots & 0 & 0 \\ 0 & 1 & 0 & 2\cos\frac{ka}{2} & \ldots & 0 & 0 \\ 0 & 0 & 2\cos\frac{ka}{2} & 0 & \ldots & 0 & 0 \\ \ldots & \ldots & \ldots & \ldots & \ldots & 1 & 0 \\ 0 & 0 & 0 & 0 & 1 & 0 & 2\cos\frac{ka}{2} \\ 0 & 0 & 0 & 0 & 0 & 2\cos\frac{ka}{2} & 0 \end{pmatrix}, \tag{1.30}$$

where the nonzero elements are

$$\sum_{i=1}^{2} e^{ik\cdot \boldsymbol{R}_{\mathrm{zGNR},i}} = \begin{cases} e^{ik\frac{a}{2}} + e^{-ik\frac{a}{2}} = 2\cos\frac{ka}{2} & \text{if NN sites are on the same zigzag chain,} \\ e^{ik\cdot 0} = 1 & \text{if NN sites are on the different zigzag chain,} \end{cases} \tag{1.31}$$

where $\boldsymbol{R}_{\mathrm{zGNR},i}$ corresponds to the vector connecting NNs.

The eigenvalues of the Hamiltonian (1.30) give the energy bands of a zGNR. In figure 1.6, we present the energy band structure of 14-zGNR for $k \in \left[-\frac{\pi}{a}, \frac{\pi}{a}\right]$. The energy bands show some typical features:

1. The Dirac points of the 2D graphene are folded to $k = \pm\frac{2\pi}{3a}$, as will be explained later.
2. The formation of flat and degenerate bands, originating from edge states [34, 39, 41–43] at $\mathcal{E} = 0$, between the Dirac points and the border of the BZ, i.e. $\frac{2\pi}{3a} \leqslant |k| \leqslant \frac{\pi}{a}$, corresponds to 1/3 of the possible k's in the zGNR. Such edge states can be observed in ultrahigh vacuum scanning tunneling microscopy (STM) [44, 45].
3. At $k = \pm\frac{2\pi}{3a}$, the eigenenergies are given (to a very good approximation for large N_z and $|n_{\mathrm{D}}| \ll N_z$ with $n_{\mathrm{D}} \in \mathbf{N}$) by the expression [40]

$$\mathcal{E}_n \approx \left(n_{\mathrm{D}} + \frac{1}{2}\right)\frac{\pi}{N_z + 1/2}\gamma_0 = \frac{\sqrt{3}}{2}\gamma_0 a\left(n_{\mathrm{D}} + \frac{1}{2}\right)\frac{\pi}{W}, \tag{1.32}$$

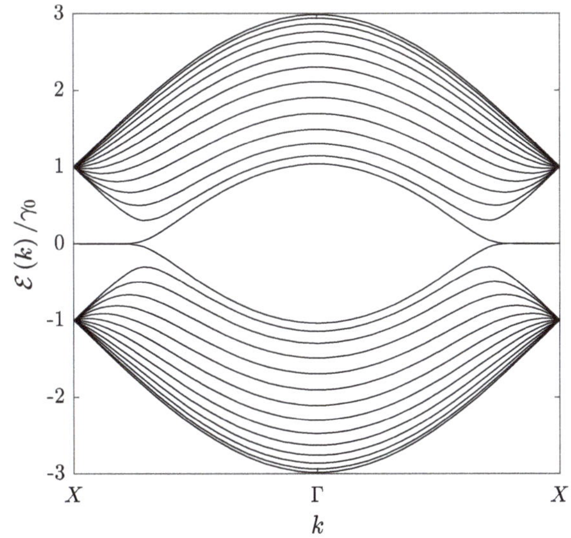

Figure 1.6. Band structure of 14-zGNR for $k \in [-\frac{\pi}{a}, \frac{\pi}{a}]$.

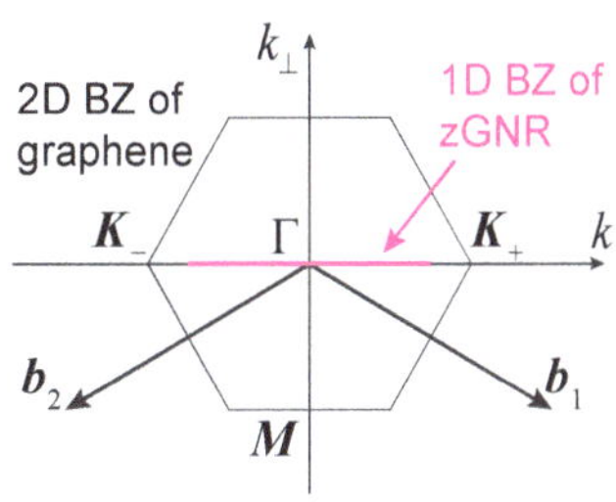

Figure 1.7. Two-dimensional BZ of graphene and 1D BZ of zGNR (red line segment) with length $2\pi/a$.

with W as the width of the nanoribbon. The energy separation between subsequent modes at $k = \pm\frac{2\pi}{3a}$ is [46, 47]

$$\Delta = \frac{\pi}{N_z + 1/2}\gamma_0 = \frac{\sqrt{3}}{2}\gamma_0 a \frac{\pi}{W}. \tag{1.33}$$

We can see that the eigenenergies nearest the reference energy $\mathcal{E} = 0$ are $\Delta/2$ and $-\Delta/2$, which correspond to $n_\mathrm{d} = 0$ and -1 in equation (1.32), respectively. Also, the energies are proportional to γ_0 and inversely proportional to W.

We now explain the folding (projection) of the Dirac points of 2D graphene to 1D GNR. Similar to 1D carbon nanotubes [48], for a 1D zGNR, as shown in figure 1.7(a), the wave vector along the direction perpendicular to the x-axis, $k_\perp$, is quantized, while the wave vector along the x-axis, k, remains continuous. Thus, defining a set of cutting lines in the BZ of 2D graphene, the energy bands of 1D zGNR consist of a set of 1D energy dispersions, which are the energy dispersion of 2D graphene on the cutting lines. Further, as depicted in figure 1.5, the structure is rotated 90° clockwise relative to the structure shown in figure 1.1(a). Therefore, the 2D BZ of graphene shown in figure 1.7 is also rotated 90° clockwise with respect to figure 1.1(b). The band structure of a zGNR may therefore be roughly understood as the 2D graphene band structure projected onto the 1D BZ of zGNR. For example, according to equation (1.20), the Dirac points $\boldsymbol{K}_{\pm}$ in the 2D BZ of graphene are folded to $k = \mp\frac{2\pi}{3a}$ in the 1D BZ of zGNR [34, 49], resulting in a gapless feature at $k = \pm\frac{2\pi}{3a}$, as shown in figure 1.6. This projection is similar to the Cheshire cat in *Alice in Wonderland* (1951) [50] or the idea of the IBM 8-bar logo [51].

1.4 Spin–orbit coupling in 2D materials

In the above sections, we discussed the electronic properties of graphene and zGNR. However, in 2D systems such as graphene or hexagonal boron nitride (hBN), it is not easy to change the energy bandgap [52]. Therefore, researchers began to consider materials such as silicene, germanene, and stanene [53], whose energy gaps can be modulated by an external electric field. These systems have a nonplanar structure, i.e. an out-of-plane deformed hexagonal lattice. Thus, applying an electric field perpendicular to the 2D layer allows researchers to control the on-site energy difference between the two atoms in the unit cell. A hexagonal lattice system having such a structure is called a buckled structure. Because of the heavier atoms present in these substances compared to carbon atoms, we cannot ignore the spin–orbit interaction [54–57]. In this section, we discuss how to introduce the SOC term into the Hamiltonian [54, 55, 58–65].

1.4.1 SOC between atoms

In the process of electron hopping between atoms, the electron acquires a force $\boldsymbol{F}$ from the nuclear Coulomb potential $\Phi(r)$. This force is called the spin–orbit interaction. If the momentum of the hopping electron is $\boldsymbol{p}$, its value can be determined by [66]

$$\begin{aligned} H_{\mathrm{SO}} &\propto (\boldsymbol{\nabla}\Phi(r) \times \boldsymbol{p}) \cdot \boldsymbol{s} \\ &= (-\boldsymbol{F} \times \boldsymbol{p}) \cdot \boldsymbol{s} \\ &= (-\boldsymbol{F}_{\parallel} \times \boldsymbol{p}) \cdot \boldsymbol{s} + (-\boldsymbol{F}_{\perp} \times \boldsymbol{p}) \cdot \boldsymbol{s} \\ &= H_{\parallel} + H_{\perp}, \end{aligned} \tag{1.34}$$

where $\boldsymbol{s}$ is the vector of Pauli matrices of spin. Here, the force $\boldsymbol{F}$ is divided into a component $\boldsymbol{F}_{\parallel}$ and a component $\boldsymbol{F}_{\perp}$, parallel and perpendicular to the atomic layer, respectively. The two terms in equation (1.34) are both SOCs, but the first item $H_{\parallel}$ is called the intrinsic SOC, and the second item $H_{\perp}$ is called the Rashba SOC (RSOC). For graphene without a buckled structure, $\boldsymbol{F}_{\perp} = 0$, and the RSOC term vanishes.

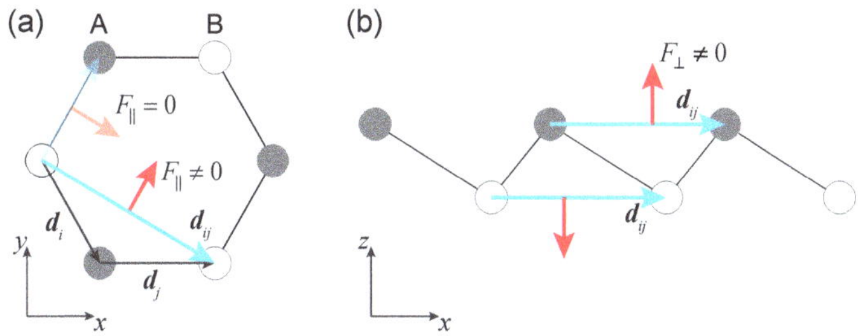

Figure 1.8. Schematic diagram of the force acting on an electron in the process of hopping to the NNN. The cyan arrow indicates the electron hopping direction, and the red arrow indicates the experienced by the hopping electron. (a) The intrinsic SOC arises from the force $\boldsymbol{F}_{\parallel}$ parallel with the 2D plane. Specifically, the NN force vanishes, while the NNN force is nonzero. (b) Side view of the buckled lattice structure. The RSOC arises from the force $\boldsymbol{F}_{\perp}$ normal to the 2D plane (z direction). Because of the buckled structure, the direction of $\boldsymbol{F}_{\perp}$ is reversed depending on whether the electron is hopping between sublattices A or B.

First, we consider the SOC term $H_{\parallel}$. Because of the mirror symmetry of the structure with respect to any NN bond (figure 1.1(a)), the intrinsic SOC of the NN is zero, as shown by the light red arrow in figure 1.8(a). When we consider the next-NN (NNN), $\boldsymbol{d}_{ij}$, neither $\boldsymbol{p}$ nor nonzero $\boldsymbol{F}_{\parallel}$ (darker red arrow) has a z-component. Therefore, $\boldsymbol{F}_{\parallel} \times \boldsymbol{p}$ has only the z-component. Considering the Pauli matrices $\boldsymbol{s} = (s_x, s_y, s_z)$ and $\boldsymbol{p} = -\mathrm{i}\hbar\boldsymbol{\nabla} \equiv -\mathrm{i}\gamma_2 \boldsymbol{d}_{ij}$ with an undetermined parameter γ_2, the SOC term becomes

$$H_{\parallel} = \mathrm{i}\gamma_2(\boldsymbol{F}_{\parallel} \times \boldsymbol{d}_{ij}) \cdot \boldsymbol{s} = \mathrm{i}\frac{\lambda_{\mathrm{SO}}}{3\sqrt{3}}\nu_{ij}s_z, \tag{1.35}$$

where λ_{SO} is the intrinsic SOC strength, $\nu_{ij} = \frac{\boldsymbol{d}_i \times \boldsymbol{d}_j}{|\boldsymbol{d}_i \times \boldsymbol{d}_j|} \cdot \hat{\boldsymbol{z}}$ with $\boldsymbol{d}_i$ and $\boldsymbol{d}_j$ are the two nearest bonds connecting the NNN $\boldsymbol{d}_{ij}$. If the direction of hopping is reversed, then the momentum is reversed; thus, the sign should be negative, i.e. $\nu_{ij} = -\nu_{ji}$. The Pauli matrix s_z indicates that the intrinsic SOC term has only matrix elements between the same spins, and the sign of the matrix element is reversed when the spin is reversed.

Next, in figure 1.8(b), we consider the Rashba term $H_{\perp}$. We define the unit vector pointing to the NNN as $\hat{\boldsymbol{d}}_{ij} = -\hat{\boldsymbol{d}}_{ji} = \frac{\boldsymbol{d}_{ij}}{|\boldsymbol{d}_{ij}|} = d_x\hat{\boldsymbol{x}} + d_y\hat{\boldsymbol{y}}$. Using the vector algebra relations and the fact that $\boldsymbol{F}_{\perp}$ has only the z-component allows us to obtain the following from equation (1.34):

$$H_{\perp} = \mathrm{i}\gamma_1(\boldsymbol{s} \times \hat{\boldsymbol{d}}_{ij}) \cdot F_{\perp}\hat{\boldsymbol{z}} = -\mathrm{i}\frac{2}{3}\lambda_{\mathrm{R}}\mu_{ij}(\boldsymbol{s} \times \hat{\boldsymbol{d}}_{ij})_z = -\mathrm{i}\frac{2}{3}\lambda_{\mathrm{R}}\mu_{ij}(s_x d_y - s_y d_x). \tag{1.36}$$

Here, γ_1 is an undetermined parameter, λ_{R} is the RSOC strength, and $\mu_{ij} = 1\ (-1)$ for the hopping between A (B) sites because of the buckled structure. The term

$$s_x d_y - s_y d_x = \begin{pmatrix} 0 & d_y + \mathrm{i}d_x \\ d_y - \mathrm{i}d_x & 0 \end{pmatrix} = \begin{pmatrix} 0 & \mathrm{e}^{\mathrm{i}\left(\frac{\pi}{2}-\theta_d\right)} \\ \mathrm{e}^{-\mathrm{i}\left(\frac{\pi}{2}-\theta_d\right)} & 0 \end{pmatrix}, \tag{1.37}$$

where θ_d is the angle that $\boldsymbol{d}_{ij}$ makes with the positive x-axis. We can conclude from equation (1.37) that the Rashba term has only off-diagonal matrix elements between different spins.

From the above, the NNN TB Hamiltonian of the silicene system can be given as [58, 67]

$$H_{\mathrm{Si}} = -\gamma_0 \sum_{\langle i,j\rangle\alpha} c_{i\alpha}^{\dagger} c_{j\alpha} + \mathrm{i}\frac{\lambda_{\mathrm{SO}}}{3\sqrt{3}} \sum_{\langle\langle i,j\rangle\rangle\alpha\beta} \nu_{ij} c_{i\alpha}^{\dagger} s_{\alpha\beta}^{z} c_{j\beta} - \mathrm{i}\frac{2}{3}\lambda_{\mathrm{R}} \sum_{\langle\langle i,j\rangle\rangle\alpha\beta} \mu_{ij} c_{i\alpha}^{\dagger} (\boldsymbol{s} \times \hat{\boldsymbol{d}}_{ij})_{\alpha\beta}^{z} c_{j\beta}, \tag{1.38}$$

where $c_{i\alpha}^{\dagger}$ creates an electron with spin α localized at atom i. The first term represents the usual NN hopping term in equation (1.15), where we sum over all pairs $\langle i, j\rangle$ of NN sites. The second (third) term represents the intrinsic (Rashba) SOC, where we sum over all pairs $\langle\langle i, j\rangle\rangle$ of NNN sites.

1.4.2 Fourier transform

We now perform a Fourier transformation to obtain the Hamiltonian in momentum space [58, 67]. First, we consider the intrinsic SOC term $H_{\parallel}$. Since this term becomes zero when $\alpha \neq \beta$, we only need to consider the case of $\alpha = \beta$. In honeycomb lattices, an atom has six NNNs [12]:

$$\begin{aligned} H_{\mathrm{SO}}^{\mathrm{A}\uparrow\mathrm{A}\uparrow}(\boldsymbol{k}) &= \mathrm{i}\frac{\lambda_{\mathrm{SO}}}{3\sqrt{3}} \sum_{\langle\langle i,j\rangle\rangle} \nu_{ij} \mathrm{e}^{\mathrm{i}\boldsymbol{k}\cdot\boldsymbol{d}_{ij}} \\ &= \mathrm{i}\frac{\lambda_{\mathrm{SO}}}{3\sqrt{3}} [\mathrm{e}^{\mathrm{i}\boldsymbol{k}\cdot\boldsymbol{a}_1} - \mathrm{e}^{-\mathrm{i}\boldsymbol{k}\cdot\boldsymbol{a}_1} - \mathrm{e}^{\mathrm{i}\boldsymbol{k}\cdot\boldsymbol{a}_2} + \mathrm{e}^{-\mathrm{i}\boldsymbol{k}\cdot\boldsymbol{a}_2} - \mathrm{e}^{\mathrm{i}\boldsymbol{k}\cdot(\boldsymbol{a}_1 - \boldsymbol{a}_2)} + \mathrm{e}^{-\mathrm{i}\boldsymbol{k}\cdot(\boldsymbol{a}_1 - \boldsymbol{a}_2)}] \\ &= \frac{2\lambda_{\mathrm{SO}}}{3\sqrt{3}} \{-\sin(\boldsymbol{k}\cdot\boldsymbol{a}_1) + \sin(\boldsymbol{k}\cdot\boldsymbol{a}_2) + \sin[\boldsymbol{k}\cdot(\boldsymbol{a}_1 - \boldsymbol{a}_2)]\} \\ &= \frac{2\lambda_{\mathrm{SO}}}{3\sqrt{3}} \left[\sin(k_y a) - 2\cos\frac{\sqrt{3}k_x a}{2} \sin\frac{k_y a}{2}\right] \\ &= \gamma_{\mathrm{SO}}(\boldsymbol{k}). \end{aligned} \tag{1.39}$$

For hopping between B atoms, the direction of the force $\boldsymbol{F}_{\parallel}$ is reversed; thus, the result changes sign. When the spin is reversed, the result also changes sign because of the Pauli matrix s_z. Therefore, on the basis of $\{|\mathrm{A}\rangle, |\mathrm{B}\rangle\} \otimes \{\uparrow, \downarrow\} = \{|\mathrm{A}\uparrow\rangle, |\mathrm{A}\downarrow\rangle, |\mathrm{B}\uparrow\rangle, |\mathrm{B}\downarrow\rangle\}$, the intrinsic SOC term reads

$$H_{\mathrm{SO}}(\boldsymbol{k}) = \mathrm{diag}[\gamma_{\mathrm{SO}}(\boldsymbol{k}), -\gamma_{\mathrm{SO}}(\boldsymbol{k}), -\gamma_{\mathrm{SO}}(\boldsymbol{k}), \gamma_{\mathrm{SO}}(\boldsymbol{k})]. \tag{1.40}$$

Next, we consider the RSOC term $H_{\perp}$. Here, we only need to focus on the case of different spin combinations, i.e. $\alpha \neq \beta$.

$$
\begin{aligned}
&H_{\mathrm{R}}^{\mathrm{A}\uparrow\mathrm{A}\downarrow}(\boldsymbol{k}) \\
&= -\mathrm{i}\frac{2}{3}\lambda_{\mathrm{R}}\sum_{\langle\langle i,j\rangle\rangle}(\boldsymbol{s}\times\hat{\boldsymbol{d}}_{ij})^{z}_{\uparrow\downarrow}\mathrm{e}^{\mathrm{i}\boldsymbol{k}\cdot\boldsymbol{d}_{ij}} \\
&= -\mathrm{i}\frac{2}{3}\lambda_{\mathrm{R}}\Big[\mathrm{e}^{\mathrm{i}\left(\frac{\pi}{2}-\frac{\pi}{6}\right)}\mathrm{e}^{\mathrm{i}\boldsymbol{k}\cdot\boldsymbol{a}_1} - \mathrm{e}^{\mathrm{i}\left(\frac{\pi}{2}-\frac{\pi}{6}\right)}\mathrm{e}^{-\mathrm{i}\boldsymbol{k}\cdot\boldsymbol{a}_1} + \mathrm{e}^{\mathrm{i}\left(\frac{\pi}{2}+\frac{\pi}{6}\right)}\mathrm{e}^{\mathrm{i}\boldsymbol{k}\cdot\boldsymbol{a}_2} \\
&\quad - \mathrm{e}^{\mathrm{i}\left(\frac{\pi}{2}+\frac{\pi}{6}\right)}\mathrm{e}^{-\mathrm{i}\boldsymbol{k}\cdot\boldsymbol{a}_2} + \mathrm{e}^{\mathrm{i}\left(\frac{\pi}{2}-\frac{\pi}{2}\right)}\mathrm{e}^{\mathrm{i}\boldsymbol{k}\cdot(\boldsymbol{a}_1-\boldsymbol{a}_2)} - \mathrm{e}^{\mathrm{i}\left(\frac{\pi}{2}-\frac{\pi}{2}\right)}\mathrm{e}^{-\mathrm{i}\boldsymbol{k}\cdot(\boldsymbol{a}_1-\boldsymbol{a}_2)}\Big] \qquad (1.41)\\
&= \frac{4}{3}\lambda_{\mathrm{R}}\left\{\mathrm{e}^{\mathrm{i}\frac{\pi}{3}}\sin(\boldsymbol{k}\cdot\boldsymbol{a}_1) + \mathrm{e}^{\mathrm{i}\frac{2\pi}{3}}\sin(\boldsymbol{k}\cdot\boldsymbol{a}_2) + \sin[\boldsymbol{k}\cdot(\boldsymbol{a}_1-\boldsymbol{a}_2)]\right\} \\
&= \frac{4}{3}\lambda_{\mathrm{R}}\left[\sin(k_y a) + \cos\frac{\sqrt{3}k_x a}{2}\sin\frac{k_y a}{2} + \mathrm{i}\sqrt{3}\sin\frac{\sqrt{3}k_x a}{2}\cos\frac{k_y a}{2}\right] \\
&= \gamma_{\mathrm{R}}(\boldsymbol{k}).
\end{aligned}
$$

For hopping between B atoms, the force $\boldsymbol{F}_{\perp}$ acts in the opposite direction, and the sign of μ_{ij} is reversed. In addition, as can be seen from equation (1.37), exchanging the spins results in a complex conjugate. Therefore, we have

$$
H_{\mathrm{R}}(\boldsymbol{k}) = \begin{pmatrix} 0 & \gamma_{\mathrm{R}}(\boldsymbol{k}) & 0 & 0 \\ \gamma_{\mathrm{R}}(\boldsymbol{k})^* & 0 & 0 & 0 \\ 0 & 0 & 0 & -\gamma_{\mathrm{R}}(\boldsymbol{k}) \\ 0 & 0 & -\gamma_{\mathrm{R}}(\boldsymbol{k})^* & 0 \end{pmatrix}. \qquad (1.42)
$$

Finally, similar to section 1.2.2, we expand the intrinsic SOC $\gamma_{\mathrm{SO}}(\boldsymbol{k})$ and RSOC $\gamma_{\mathrm{R}}(\boldsymbol{k})$ close to the Dirac points $\boldsymbol{K}_\kappa$ to the second order by replacing $\boldsymbol{k}$ with $\boldsymbol{K}_\kappa + \boldsymbol{q}$; this yields

$$
\gamma_{\mathrm{SO},\boldsymbol{K}_\kappa}(\boldsymbol{q}) = -\kappa\lambda_{\mathrm{SO}}\left[1 - \left(\frac{qa}{2}\right)^2\right], \qquad (1.43)
$$

$$
\gamma_{\mathrm{R},\boldsymbol{K}_\kappa}(\boldsymbol{q}) = -\lambda_{\mathrm{R}}\left[(\mathrm{i}q_x + q_y)a + \kappa\frac{\sqrt{3}}{4}(q_x + \mathrm{i}q_y)^2a^2\right]. \qquad (1.44)
$$

At the $\boldsymbol{K}_\pm$ points, $\gamma_{\mathrm{SO}}(\boldsymbol{K}_\kappa) = -\kappa\lambda_{\mathrm{SO}}$, $\gamma_{\mathrm{R}}(\boldsymbol{K}_\kappa) = 0$. Therefore, to the first order, the intrinsic SOC and RSOC terms read

$$
H_{\mathrm{SO},\boldsymbol{K}_\kappa} = -\kappa\lambda_{\mathrm{SO}}\sigma_z \otimes s_z, \qquad (1.45)
$$

$$
H_{\mathrm{R},\boldsymbol{K}_\kappa}(\boldsymbol{q}) = -a\lambda_{\mathrm{R}}\sigma_z \otimes (q_y s_x - q_x s_y). \qquad (1.46)
$$

We can see that, up to the first order of qa, the intrinsic SOC term is momentum independent. Also, SOC couples the spin DOFs to the valley DOFs, which allows us to discover interesting spin–valley-coupled physics in silicene, germanene, and stanene.

1.5 Molybdenum disulfide (MoS_2) and the transition-metal dichalcogenides

The TMDCs are a broad class of inorganic layered materials [68] of the form MX_2, where M stands for a transition-metal (TM) element from group IV (Ti, Zr, Hf, etc.), group V (V, Nb, Ta, etc.), or group VI (Mo, W, etc.), and X stands for a chalcogen (S, Se, Te) [69–76]. The naturally stable TMDCs usually exhibit 2H crystal structure phases [76]. Industrial and commercial TMDC bulk crystals and powders with 2H or 3R phases are often grown through chemical vapor transport methods [19, 69, 77–80]. The letters H and R represent hexagonal and rhombohedral crystal symmetries, respectively, while the Arabic numeral denotes the number of layers within each unit cell [81]. Thus, 2H signifies two layers per unit cell in a trigonal prismatic structure. The metal has an oxidation state of +4, and the chalcogen's oxidation state is −2 [71]. Among the ~60 materials categorized as TMDCs, two thirds possess layered structures that can potentially undergo exfoliation to the monolayer limit [69, 82, 83]. This is because of the presence of strong covalent bonds within a single layer [63, 77] and weak van der Waals (vdW) forces between layers [84].

Monolayer TMDCs have received considerable attention in recent years [69, 85–90]. These are ~0.7 nm thick [75, 76, 91] graphene-like hexagonal crystal lattices consisting of three atomic layers; the bottom and top layers are chalcogens (X) and the middle layer is the TM (M), as shown in figure 1.9 [88, 92]. The TM atom is coordinated by six chalcogen atoms in a trigonal prismatic geometry [77, 93]. Considering the M atom as the inversion center, an X atom points towards an empty location. Therefore, the IS is explicitly broken for a monolayer of MX_2. The bottom and top layers are inert [71], which leads to vdW bonding between layers in multilayer form. Akin to graphene, the valleys of energy–momentum dispersion are located at the six corners of the hexagonal BZ, which includes two sets of three equivalent points. The two nonequivalent valleys, $\boldsymbol{K}_+$ and $\boldsymbol{K}_-$, form a time-reversal symmetric pair with each other. These valleys are involved in direct transitions in monolayer TMDCs [75, 76, 94]. Therefore, the TMDCs are also candidate materials for use in valleytronics [76, 77, 85, 95]. The observation of many interesting electrical and optical properties in TMDCs [96] emphasizes the importance of studying their band structures [97].

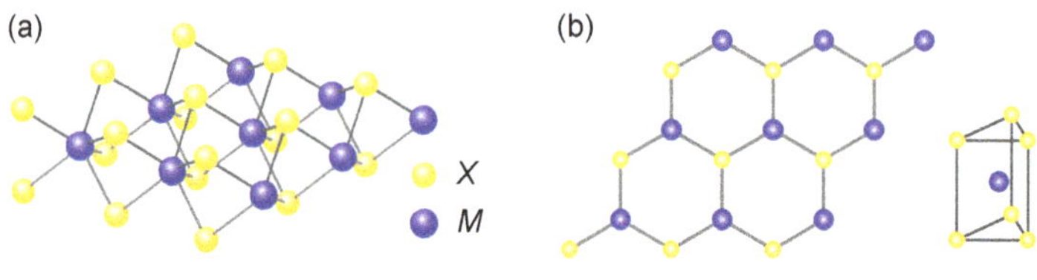

Figure 1.9. Lattice structure of monolayer TMDCs (MX_2): (a) Perspective view. A layer of TM atoms (M, violet spheres) is sandwiched between two layers of chalcogen atoms (X, yellow spheres). (b) Top view (left panel) and unit cell (right panel), which demonstrates the hexagonal crystal structure with trigonal prismatic metal coordination and broken IS.

1.5.1 Thickness-dependent electronic band structure

One of the most impressive characteristics of TMDCs is the modification of the band structure as the number of layers is varied [76, 99–102]. In figure 1.10, we present the electronic band structures of a typical TMDC, in the forms of bulk and monolayer MoS_2 crystal, calculated by DFT. For bulk MoS_2, the calculations reveal an indirect bandgap [103] of 1.2 eV located between the valence-band maximum (VBM), situated at the center of the BZ (Γ point), and a conduction band minimum (CBM) near the midpoint between the Γ and $\boldsymbol{K}_\pm$ points [77, 101, 104–108]. The arrow illustrates the indirect optical transition between these points. As the number of layers decreases from the bulk to the monolayer limit, the lowest sub-band of the conduction band moves upward monotonically with a total increment of 0.6 eV, increasing the bandgap [97].

This evolution of the electronic band structure can be explained by the origins of the electronic states at the Γ and $\boldsymbol{K}_\pm$ points [75, 77, 109–112]. Since the band states in the $\boldsymbol{K}_\pm$ valleys mainly originate from the $d_{x^2-y^2} \pm id_{xy}$ (valence band) and d_{z^2} orbitals (conduction band) of the Mo atoms in the middle of the sandwich, they are relatively unaffected by interlayer interactions in the out-of-plane direction. The direct bandgap in the $\boldsymbol{K}_\pm$ valleys only increases by about 0.05–0.1 eV [98, 113, 114]. On the other hand, compared to the band states in the $\boldsymbol{K}_\pm$ valleys produced by the d orbitals of Mo atoms, the conduction band states at the Γ point, which are contributed by the hybridization between the p_z orbitals of S atoms and the d_{z^2}

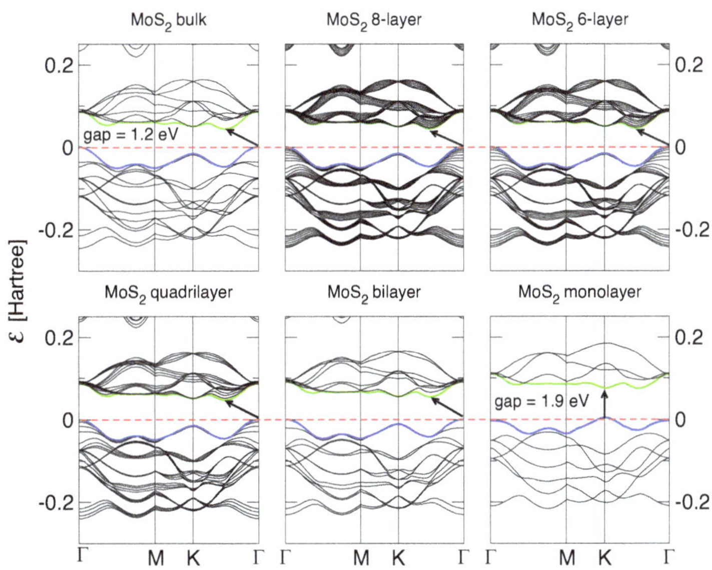

Figure 1.10. Evolution of the band structure of MoS_2 as the number of layers decreases, as analyzed using DFT calculations. Red dashed lines indicate the positions of the Fermi level. The lowest conduction (highest valence) sub-bands are marked by green (blue) and the arrows indicate the lowest energy transitions. MoS_2 becomes a direct bandgap semiconductor when thinned down to a monolayer. Reprinted figure with permission from [98]. Copyright (2011) by the American Physical Society.

orbitals of Mo atoms [73, 100], should be more influenced by interlayer interactions [75, 115]. Thus, the bands at Γ are more affected by a decrease in the number of layers [75, 113]. To summarize, as the thickness reduces, the indirect bandgap gradually opens up, while the direct bandgap remains almost unchanged because of the increasing quantum confinement effect along the *z*-axis [98, 116], long-range coulomb effects [75, 76, 99, 114, 117, 118], and the resultant variation of hybridizations between the *d* orbitals on the Mo atom sites and the Mo–S *s*–*p* orbitals [115, 119]. Finally, in the monolayer, the indirect bandgap at the Γ point is wider than that in the $\boldsymbol{K}_{\pm}$ valleys [120], and the smallest bandgap is thus the direct bandgap at the $\boldsymbol{K}_{\pm}$ points [69, 70, 75, 76, 94, 98, 100, 102, 109, 110, 113, 121–128] of about 1.9 eV. In most of the other TMDCs that have been studied, similar band structure behaviors have also been observed [129], i.e., in $MoSe_2$ [130], WS_2 [98], and WSe_2 [131].

This thickness-dependent electronic band structure has been directly confirmed by ARPES [72, 132] and investigated through optical measurements. Experiments have demonstrated significantly higher photoluminescence (PL) in monolayers compared to their bulk counterparts, namely an enhancement of up to four orders of magnitude [75, 76, 109, 110, 133–135] and an absorption of 5%–10% of the incident light for monolayer MoS_2, $MoSe_2$, and WS_2 (~1 nm thick) [96, 136]. In bulk TMDCs, band-edge PL is a phonon-assisted emission process, thus resulting in very weak indirect band emission and negligible PL quantum yield. Similar evidence for a direct bandgap transition has also been reported for all the group VIB monolayer TMDCs [137], such as monolayer $MoSe_2$, WS_2, and WSe_2 [98, 122, 126, 134, 138–144], showing that TMDCs are more practical than graphene, which has a zero bandgap. This makes monolayer TMDCs ideal candidates for optical emitters, photovoltaic cells, and optoelectronic devices [119].

1.5.2 Spin–orbit coupling and its effects

When TMDCs are thinned down to monolayer form, i.e., when the effects of interlayer vdW interaction are absent, SOC has very prominent effects on the electronic properties compared to those of graphene [111, 145, 146]. This is attributed to the much heavier metal elements [77, 124, 147, 148] in the TMDC systems, as the SOC strength is directly proportional to the fourth power of the atomic number [149]. The strong SOC and breaking of IS modify the electronic band structure of monolayer TMDCs; in this respect, they are unlike graphene and silicene.

Calculated band structures of monolayer MX_2, including the effects of SOC, are presented in figure 1.11. We have assigned the names v_1 and v_2 to the split bands of the VBM at $\boldsymbol{K}_{\pm}$. The strong SOC in monolayer TMDCs induces band splitting in both the valence and conduction bands. For instance, in the valence band, spin splits exist at $\Delta\mathcal{E}_{SO} \sim 148$, 183, 426, and 456 meV, corresponding to the $\boldsymbol{K}_{\pm}$ points of monolayer MoS_2, $MoSe_2$, WS_2, and WSe_2 [72, 77, 111, 121, 124, 137, 148, 150, 151], respectively, while in the conduction band, there is a much smaller spin split (on the order of 3 to 50 meV) of the opposite sign for MoX_2 and WX_2 at the $\boldsymbol{K}_{\pm}$ points [73, 93, 145, 146, 152–157].

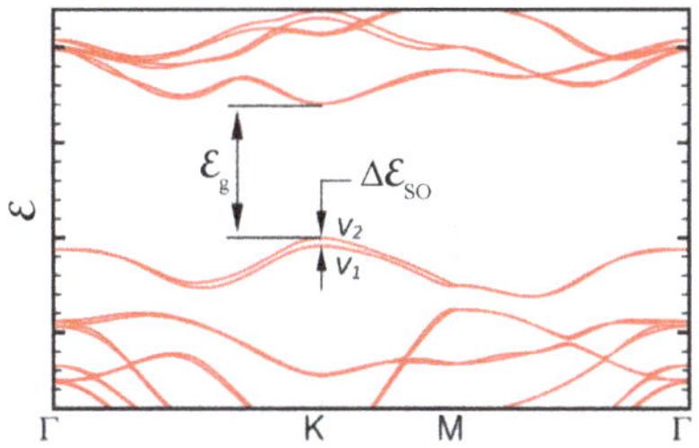

Figure 1.11. Typical band structure of a monolayer MX_2 calculated using DFT, where $\mathcal{E}_g$ represents the quasiparticle bandgap. The VBM at the $\boldsymbol{K}_{\pm}$ points is split due to SOC. The lower and upper spin-polarized sub-bands are denoted by v_1 and v_2, respectively. Reprinted figure with permission from [124]. Copyright (2012) by the American Physical Society.

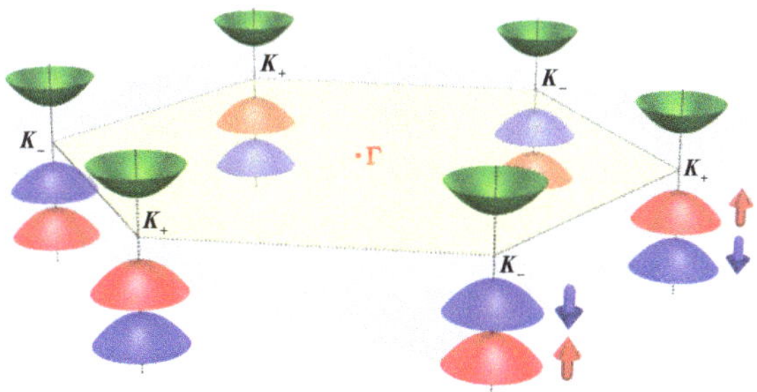

Figure 1.12. A schematic drawing of the conduction band, the spin split valence bands and the spin polarization at the band edges located at the $\boldsymbol{K}_{\pm}$ points in the BZ of monolayer TMDCs. The red and blue bands correspond to the spin-up and spin-down bands, respectively. Reprinted figure with permission from [77]. Copyright (2012) by the American Physical Society.

It should be noted that the spin splitting vanishes at the other high-symmetry $\boldsymbol{k}$-points, i.e. the Γ and M points [97]. At the Γ point, spin degeneracy arises from TRS, while at the M point, it is prevented by both TRS and translational symmetry [72–74, 94, 111, 148, 158–160]. Compared with the $\boldsymbol{K}_{\pm}$ points, these high-symmetry points have high energy and can be disregarded when working within a low-energy regime [19].

The stronger SOC compared with that of graphene, in combination with the previously mentioned IS breaking, gives rise to a thought-provoking effect—strong spin–valley coupling (also known as spin–momentum coupling). That is, for the $\boldsymbol{K}_+$ and $\boldsymbol{K}_-$ points, the magnitude of spin splitting, $\Delta\mathcal{E}_{SO}$, is the same, but the directions of the spin for the upper and lower split states have opposite signs [147, 161, 162] because of Kramers's degeneracy theorem describing TRS [77, 163, 164]. Figure 1.12 illustrates this statement, showing that the upper and lower split states for $\boldsymbol{K}_+$ ($\boldsymbol{K}_-$) are spin up (down) and down (up), respectively [145, 146, 148, 155, 165–168]. This splitting is much larger than the thermal energy at room temperature (~25 meV). Thus, the $\boldsymbol{K}_+$ and $\boldsymbol{K}_-$ valleys in the valence band are nonequivalent, even at room temperature [157]. This allows us to control the VP. One of the most important results of spin–valley coupling is that the flipping of the spin/valley index suppresses both spin and valley relaxations, and that spin–valley locking forbids the isolation of either the spin index or the valley index [161]. This leads to long-lived valley and spin polarization [63]. Furthermore, electron spin is present in many valley-dependent phenomena, such as VP and the VHE, due to spin–valley coupling [77]. Conversely, it is also possible to apply a magnetic field to lift valley degeneracy [169, 170]. This is an important distinguishing feature of TMDCs compared to other systems, such as GaAs or CdTe quantum wells, where the band edges occur at the Γ point and both bands remain spin degenerate [171]. Their spin–valley coupling at the valence band edges [77] makes TMDCs an appealing choice for beyond-electronics applications and thus will form the basis for developing integrated valleytronic and spintronic applications in the future.

1.5.3 Effective Hamiltonian

As with graphene, the TB approach provides insight into many of the key aspects of the physics of TMDC materials. Because the *d* orbitals arise from metal atoms, the construction of the TB Hamiltonian for monolayer TMDCs is much more complicated than for graphene. There are several papers describing the TB approach for monolayer TMDCs in detail [96, 152, 153, 172–178], but here, we only present the two-band $\boldsymbol{k} \cdot \boldsymbol{p}$ Hamiltonian [77, 145, 146, 152, 153, 168, 179], which offers a useful midpoint between the computationally expensive studies of TB models and the simplistic analytics offered by a Dirac model [147]. This allows us to compact the Hamiltonian to just the dynamics of single carriers around the $\boldsymbol{K}_{\pm}$ valleys, thus including terms that account for electron–hole number asymmetry, particle–hole symmetry, and TW [19].

1.5.3.1 $\boldsymbol{k} \cdot \boldsymbol{p}$ theory

As we mentioned earlier, the lowest conduction (c) band-edge state is produced by d_{z^2} orbitals and the highest valance (v) band is due to the hybridization of $d_{x^2-y^2}$ and d_{xy} orbitals. The basis functions at the band edges have the form [77, 96, 97, 111, 174, 179]

$$|\hat{\phi}_{\mathrm{c}}\rangle = |d_{z^2}\rangle, \left|\hat{\phi}_{\mathrm{v}}^{\kappa}\right\rangle = \frac{1}{\sqrt{2}}\left(|d_{x^2-y^2}\rangle + \mathrm{i}\kappa|d_{xy}\rangle\right). \tag{1.47}$$

The valence band wave functions for the two valleys, $|\hat{\phi}_{\mathrm{v}}^{+}\rangle$ and $|\hat{\phi}_{\mathrm{v}}^{-}\rangle$, are related by TRS. As a starting point, without considering SOC, we can express the two-band $\boldsymbol{k} \cdot \boldsymbol{p}$ Hamiltonian with a bandgap of Δ_z to the first order of $\boldsymbol{q}$ as follows [63, 77]:

$$H_0^{\kappa} = a\gamma(\kappa q_x \hat{\sigma}_x + q_y \hat{\sigma}_y) + \frac{\Delta_z}{2}\hat{\sigma}_z, \tag{1.48}$$

where a, γ, and $\hat{\boldsymbol{\sigma}}$ are the lattice constant, the effective hopping integral, and Pauli matrices for the two basis functions in equation (1.47), respectively [77].

In TMDCs, the nature of the heavy TM *d* orbitals provides strong SOC that lifts the spin degeneracy within electronic bands. When we consider SOC, which splits the highest valence band, the effective $\boldsymbol{k} \cdot \boldsymbol{p}$ Hamiltonian can be further written as [77]

$$\begin{aligned} H_{\mathrm{TMDC}}^{\kappa} &= a\gamma(\kappa q_x \hat{\sigma}_x + q_y \hat{\sigma}_y) + \frac{\Delta_z}{2}\hat{\sigma}_z - \kappa s \lambda_{\mathrm{SO}} \frac{\hat{\sigma}_z - 1}{2} \\ &= \begin{pmatrix} \dfrac{\Delta_z}{2} & a\gamma(\kappa q_x - \mathrm{i} q_y) \\ a\gamma(\kappa q_x + \mathrm{i} q_y) & -\dfrac{\Delta_z}{2} + \kappa s \lambda_{\mathrm{SO}} \end{pmatrix}, \end{aligned} \tag{1.49}$$

where $2\lambda_{\mathrm{SO}}$ is the spin splitting at the VBM induced by SOC and broken IS. s denotes the spin index, which remains a good quantum number. The effective parameters of the $\boldsymbol{k} \cdot \boldsymbol{p}$ model can be extracted from the first-principles band structures in the neighborhood of the $\boldsymbol{K}_{\pm}$ points. It is straightforward to derive the eigenenergies, as follows:

$$\mathcal{E}_{\xi}^{\kappa}(\boldsymbol{q}) = \frac{\xi}{2}\sqrt{4a^2\gamma^2 q^2 + (\Delta_z - \kappa s \lambda_{\mathrm{SO}})^2} + \kappa s \frac{\lambda_{\mathrm{SO}}}{2}. \tag{1.50}$$

The terms with κs evince the spin–valley locking effect and indicate that the spin splits are opposite at the $\boldsymbol{K}_{+}$ ($\kappa = 1$) and $\boldsymbol{K}_{-}$ ($\kappa = -1$) points. At exactly $\boldsymbol{K}_{\kappa}$, i.e. $\boldsymbol{q} = 0$, the band edges of the conduction and valence bands are given by

$$\begin{aligned} \mathcal{E}_{\mathrm{c}}^{\kappa}(0) &= \frac{\Delta_z}{2}, \\ \mathcal{E}_{\mathrm{v}}^{\kappa}(0) &= -\frac{\Delta_z}{2} + \kappa s \lambda_{\mathrm{SO}}, \end{aligned} \tag{1.51}$$

respectively, indicating a larger split in the valence band.

1.5.3.2 TW effect

Similar to graphene, as discussed in section 1.2.2.2, the TW effect also appears in TMDCs due to the symmetry of the crystal lattice. Here, we introduce a model given by Fang *et al* [174] that captures the details of the conduction and valence band dispersions near the $\boldsymbol{K}_{\pm}$ points. We can define a 2×2 spinless effective Hamiltonian as follows for monolayer TMDC on the basis of equation (1.47). According to equation (20) in reference [174], the effective Hamiltonian reads

$$\begin{aligned} &H_{\mathrm{TMDC,\,TW}}^{\kappa}(\boldsymbol{q}) \\ &= \frac{f_0}{2}\hat{\sigma}_z + f_1 a(\kappa q_x \hat{\sigma}_x + q_y \hat{\sigma}_y) + (qa)^2(f_2 + f_3 \hat{\sigma}_z) + f_4 a^2\left(q_x^2 \hat{\sigma}_x - q_y^2 \hat{\sigma}_y - 2\kappa q_x q_y \hat{\sigma}_y\right), \end{aligned} \tag{1.52}$$

where f_i ($i = 0, \ldots, 4$) is a real parameter in eV. The first term gives the conduction and valence band energies, $\frac{f_0}{2}$ and $-\frac{f_0}{2}$, respectively, at the Dirac point. The energy corresponding to a bandgap equals f_0. The second term with f_1 is the massive Dirac fermion model for TMDCs [73, 74] presented in equation (1.48). In the third term, f_2 breaks the electron–hole symmetry and leads to different energy dispersions between the conduction and valence bands. The last term, together with the second term as off-diagonal terms, $\hat{\sigma}_x$ and $\hat{\sigma}_y$, account for the TW effect, as we will show in detail below. For WSe_2, in which the valley optoelectronic effect was first observed and investigated [180], Fang *et al* give the lattice constant as $a = 3.32$ Å and the band parameters f_i fitted by a DFT study as [174]: $f_0 = 1.5455$, $f_1 = 1.1894$, $f_2 = 0.1184$, $f_3 = -0.0064$, and $f_4 = -0.0627$ in eV. Employing these terms up to the desired order, the effective band structure of a monolayer TMDC may be tailored to the problem at hand.

The eigenvalues of the effective Hamiltonian yield the energy dispersion near the $\boldsymbol{K}_\pm$ point:

$$\mathcal{E}^{\kappa}_{\xi}(\boldsymbol{q}) = f_2(qa)^2 + \xi\sqrt{\left(f_3^2 + f_4^2\right)(qa)^4 + 2\kappa f_1 f_4(qa)^3\cos(3\theta_{\boldsymbol{q}}) + \left(f_0 f_3 + f_1^2\right)(qa)^2 + \left(\frac{f_0}{2}\right)^2}. \quad (1.53)$$

In order to understand the effect of TW, we expand equation (1.53) near $\boldsymbol{q} = 0$ and take the terms of qa up to the fourth order; the result is given by

$$\mathcal{E}^{\kappa}_{\mathrm{c}}(\boldsymbol{q}) = \frac{f_0}{2} + (A_2 + f_2)(qa)^2 + A_3(qa)^3 + A_4(qa)^4, \quad (1.54)$$

$$\mathcal{E}^{\kappa}_{\mathrm{v}}(\boldsymbol{q}) = -\left[\frac{f_0}{2} + (A_2 - f_2)(qa)^2 + A_3(qa)^3 + A_4(qa)^4\right], \quad (1.55)$$

where $A_2 = \frac{f_1^2}{f_0} + f_3 = 0.9089$ eV, $A_3 = \frac{2\kappa f_1 f_4}{f_0}\cos(3\theta_{\boldsymbol{q}}) = -0.0965\kappa\cos(3\theta_{\boldsymbol{q}})$ eV, and $A_4 = \frac{f_4^2}{f_0} - \frac{2f_1^2 f_3}{f_0^2} - \frac{f_1^4}{f_0^3}$. As we can see, $A_3 \propto \cos(3\theta_{\boldsymbol{q}})$ in the third-order term produces the TW effect. When the third term including $(qa)^3$ is larger than the second term including $(qa)^2$, TW becomes important. As can be seen in equations (1.54) and (1.55), the TW effect is highlighted in the valence band rather than the conduction band [85, 168] with the help of f_2. This situation can be understood by considering that the second-order term for the conduction and valence bands, $|A_2 + f_2| > |A_2 - f_2|$. It can also be understood by inspecting equation (1.53) that the first term $f_2(qa)^2$ and the first term $(f_3^2 + f_4^2)(qa)^4$ inside the square root cancel each other when equation (1.53) takes a '−' sign ($\xi = -1$), leaving the contribution of the TW term $\propto\cos(3\theta_{\boldsymbol{q}})$ relatively strengthened. However, the coefficient of the TW term A_3 is much smaller than the second-order term $A_2 \pm f_2$. The difference in the TW effect between the valence and conduction bands is crucial to our understanding of circularly polarized electroluminescence in electrically switchable multilayer and monolayer WSe_2 p–n junctions [180], as will be discussed in section 7.7.1.1.

In figure 1.13, we show the equi-energy contour plots of both (a) valence and (b) conduction bands. The TW distortion (seen as separation from the dashed black circle) is more obvious in the valence band than in the conduction band, which indicates a pronounced TW effect. The TW effect increases with increasing distance from the $\boldsymbol{K}_\pm$ points. The narrower adjacent equi-energy lines in figure 1.13(b) correspond to the lighter effective mass of electrons compared with that of holes in figure 1.13(a) [181].

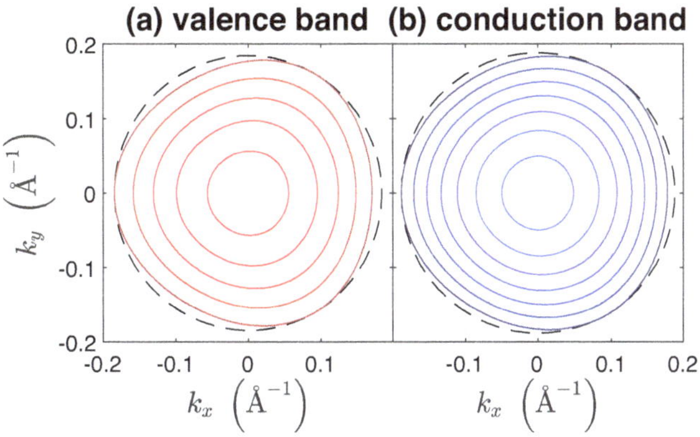

Figure 1.13. Isoenergy contour plots of the (a) valence and (b) conduction bands of monolayer WSe_2 around the $\boldsymbol{K}_-$ point. The innermost contour represents an energy of ± 0.80 eV, and the energy difference between adjacent contour lines is 0.05 eV. The $\boldsymbol{K}_+$ point can be obtained by applying a time-reversal operation. The dashed black circle tangent to the outermost contour is plotted to make it easier to identify the TW effect, which is more pronounced in the valence band.

1.6 Valley-polarized Landau levels in graphene

As in a 2D electron gas (2DEG), the motion of relativistic charges in graphene immersed in a strong magnetic field is also quantized, which results in discrete energy levels known as Landau levels (LLs) [182]. Historically, scientists have examined such LLs by using the TB [183, 184] and effective models [185–189]. The TB approach is more general and valid for a wider energy range. However, in this section, we will show that a magnetic field produces VP of the zeroth LL for each sublattice in monolayer graphene. Therefore, only low-energy levels are of interest and the system can be efficiently described by the effective model discussed in section 1.2.2.

Before considering quantum mechanics in detail, it is intuitive to consider the classical motion of a graphene electron with charge $-e$ under a magnetic field $\boldsymbol{B}$. The classical equation of motion for an electron in band ξ is

$$\hbar \frac{\mathrm{d}\boldsymbol{k}}{\mathrm{d}t} = -\frac{e}{c} v_\xi(\boldsymbol{q}) \times \boldsymbol{B}, \tag{1.56}$$

where c is the speed of light and

$$v_\xi(\boldsymbol{q}) = \frac{1}{\hbar} \frac{\partial \mathcal{E}_\xi(\boldsymbol{q})}{\partial \boldsymbol{k}} = \frac{\xi v_\mathrm{F} \boldsymbol{q}}{|\boldsymbol{q}|}. \tag{1.57}$$

This gives the cyclotron frequency [188, 190]

$$\omega_\mathrm{c}(\mathcal{E}) = \frac{eBv_\mathrm{F}^2}{c\mathcal{E}}, \tag{1.58}$$

which is inversely proportional to the energy $\mathcal{E}$ and the speed of light c.

In quantum mechanics, when the magnetic field, $\boldsymbol{B} = (0, 0, B)$, is applied perpendicular to the monolayer graphene sheet, we can choose $\boldsymbol{A} = (0, Bx, 0)$ as the induced vector potential in the Landau gauge due to $\nabla \times \boldsymbol{A} = \boldsymbol{B}$. By inserting $\boldsymbol{A}$ into the effective Hamiltonian close to the $\boldsymbol{K}_{\pm}$ points in equation (1.25), we obtain [185–188, 191, 192]

$$H_{\boldsymbol{K}_{\pm}} = v_{\mathrm{F}} \begin{pmatrix} 0 & p_x \pm \mathrm{i}\left(p_y + \dfrac{e}{c}A_y\right) \\ p_x \mp \mathrm{i}\left(p_y + \dfrac{e}{c}A_y\right) & 0 \end{pmatrix}. \tag{1.59}$$

This expression has additional twofold degeneracy due to spin, since we disregard the spin of the electron. For a single electron, this degeneracy can be lifted by the Zeeman interaction [193].

Let us first focus on the $\boldsymbol{K}_{+}$ valley by solving the Dirac equation of motion

$$H_{\boldsymbol{K}_{+}}\Psi(\boldsymbol{r}) = \mathcal{E}\Psi(\boldsymbol{r}). \tag{1.60}$$

This yields

$$v_{\mathrm{F}} \begin{pmatrix} 0 & p_x + \mathrm{i}\left(p_y + \dfrac{e}{c}A_y\right) \\ p_x - \mathrm{i}\left(p_y + \dfrac{e}{c}A_y\right) & 0 \end{pmatrix} \begin{pmatrix} \psi_{\mathrm{A}}(x, y) \\ \psi_{\mathrm{B}}(x, y) \end{pmatrix} = \mathcal{E} \begin{pmatrix} \psi_{\mathrm{A}}(x, y) \\ \psi_{\mathrm{B}}(x, y) \end{pmatrix}, \tag{1.61}$$

where $\psi_{\mathrm{A}}(x, y)$ and $\psi_{\mathrm{B}}(x, y)$ are the wave functions related to sublattices A and B, respectively, in graphene. The eigenstates and eigenvalues of equation (1.61) are then determined by the following relations:

$$v_{\mathrm{F}}\left[p_x + \mathrm{i}\left(p_y + \frac{e}{c}A_y\right)\right]\psi_{\mathrm{B}}(x, y) = \mathcal{E}\psi_{\mathrm{A}}(x, y), \tag{1.62}$$

$$v_{\mathrm{F}}\left[p_x - \mathrm{i}\left(p_y + \frac{e}{c}A_y\right)\right]\psi_{\mathrm{A}}(x, y) = \mathcal{E}\psi_{\mathrm{B}}(x, y). \tag{1.63}$$

Using the Landau gauge $\boldsymbol{A} = (0, Bx, 0)$, which preserves translational invariance in the y direction, we can write the eigenstates in terms of states that are plane waves in the y direction:

$$\psi_{\mathrm{A}}(x, y) = \mathrm{e}^{-iky}X_{\mathrm{A}}(x), \tag{1.64}$$

$$\psi_{\mathrm{B}}(x, y) = \mathrm{e}^{-iky}X_{\mathrm{B}}(x). \tag{1.65}$$

We obtain

$$\mathrm{i}^{|n|}X_{\mathrm{A}}(x) = \phi_{|n|} \tag{1.66}$$

$$-\mathrm{i}^{|n|-1}X_{\mathrm{B}}(x) = \phi_{|n|-1}, \tag{1.67}$$

where $n \in \mathbf{N}$ is the LL index, and

$$\phi_{|n|} = (2^{|n|}|n|!\sqrt{\pi}\, l_B)^{-\frac{1}{2}} \exp\left[-\frac{1}{2}\left(\frac{x - l_B^2 k}{l_B}\right)^2\right]\mathcal{H}_{|n|}\left(\frac{x - l_B^2 k}{l_B}\right) \tag{1.68}$$

is the standard Landau wave function for particles with nonrelativistic parabolic dispersion relation in the nth LL, with magnetic length

$$l_B = \sqrt{\frac{\hbar}{eB}} \tag{1.69}$$

and Hermite polynomials $\mathcal{H}_{|n|}(x)$. The first few Hermite polynomials are given in appendix B [26, 27, 194, 195].

Therefore, the wave function for the $\boldsymbol{K}_+$ point can be expressed as

$$\Psi_{\boldsymbol{K}_+} = \frac{C_n}{\sqrt{L}}\begin{pmatrix} \mathrm{i}^{|n|}\phi_{|n|} \\ \mathrm{sgn}\,(n)\mathrm{i}^{|n|-1}\phi_{|n|-1} \end{pmatrix}\mathrm{e}^{-\mathrm{i}ky}, \tag{1.70}$$

where L^2 is the area of the system,

$$C_n = \begin{cases} 1 & \text{if } n = 0 \\ 1/\sqrt{2} & \text{if } n \neq 0, \end{cases} \tag{1.71}$$

and the sign function is defined by

$$\mathrm{sgn}\,(n) = \begin{cases} 0 & \text{if } n = 0 \\ n/|n| & \text{if } n \neq 0. \end{cases} \tag{1.72}$$

Correspondingly, for the $\boldsymbol{K}_-$ point, we have

$$\Psi_{\boldsymbol{K}_-} = \frac{C_n}{\sqrt{L}}\begin{pmatrix} \mathrm{sgn}\,(n)\mathrm{i}^{|n|-1}\phi_{|n|-1} \\ \mathrm{i}^{|n|}\phi_{|n|} \end{pmatrix}\mathrm{e}^{-\mathrm{i}ky}. \tag{1.73}$$

The eigenenergies are denoted by

$$\mathcal{E}_n = \mathrm{sgn}\,(n)\frac{\hbar v_\mathrm{F}}{l_B}\sqrt{2|n|}, \tag{1.74}$$

and are only specified by n. Here, each LL is fourfold degenerate due to spin and valley degeneracies. In figure 1.14, we illustrate the quantized LL. In the presence of a magnetic field, the linear dispersion relation in pristine graphene evolves to a sequence of discrete LLs, each corresponding to a peak in the density of states (DOS). The LL index $n > 0$ (<0) represents electrons (holes), while $n = 0$ represents the Dirac point. Notice that in contrast to the case of conventional 2DEGs, which have equidistant LLs [26], the LLs in graphene have a square root dependency on

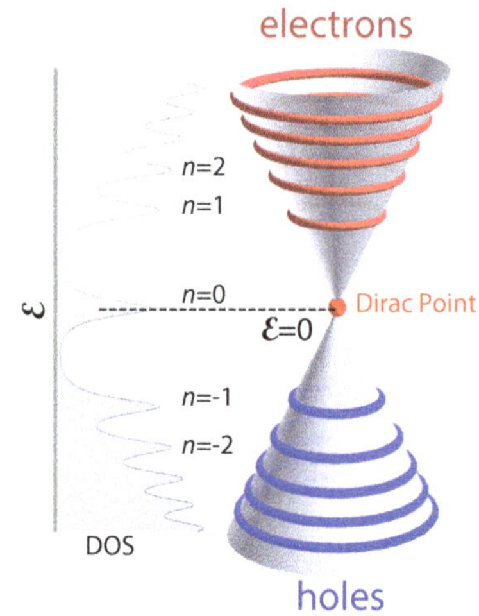

Figure 1.14. Illustration of quantized energy levels in graphene and their signature in the density of states (DOS). Right side: In the presence of a magnetic field, the Dirac cone of pristine graphene no longer has a continuum of energy but discrete levels given by equation (1.74). The LL index $n = 0$ represents the Dirac point, while $n > 0$ (<0) represents electrons (holes), which are denoted by red (blue) rings. Left side: The corresponding DOS. For each LL, there is a peak in the DOS which is broadened by electron–electron interactions in ideal systems. In the presence of disorder, the LLs are further broadened. For neutral graphene, the Fermi level (dashed line) is exactly at the Dirac point ($\mathcal{E} = 0$), and the shaded area represents filled states. Reproduced from [190], with permission from Springer Nature.

the LL index, and the largest energy separation is between the zeroth and the first LLs. For large n, the energy separation can be written in terms of the classical cyclotron frequency as

$$\mathcal{E}_{n+1} - \mathcal{E}_n \sim \frac{\hbar v_\mathrm{F}}{l_B}\frac{1}{\sqrt{2n}} = \hbar\omega_\mathrm{c}(\mathcal{E}_n). \tag{1.75}$$

Specifically, according to equations (1.70) and (1.73), the wave functions at the LL with index $n \neq 0$ should always have nonzero amplitudes in both sublattices A and B. However, for LLs with index $n = 0$, the eigenfunctions read

$$\Psi_{\boldsymbol{K}_+} = \frac{\mathrm{e}^{-\mathrm{i}k_y}}{\sqrt{L}}\begin{pmatrix}\phi_0\\0\end{pmatrix}, \tag{1.76}$$

$$\Psi_{\boldsymbol{K}_-} = \frac{\mathrm{e}^{-\mathrm{i}k_y}}{\sqrt{L}}\begin{pmatrix}0\\\phi_0\end{pmatrix}. \tag{1.77}$$

Hence, in this special case, electrons at the $\boldsymbol{K}_+$ point are distributed on sublattice A, and electrons at $\boldsymbol{K}_-$ are distributed on sublattice B [18, 183, 196]. This property of the wave functions for the LLs in graphene makes $n = 0$ LL very special for various magnetic applications of graphene.

In experiments, a direct way to study the quantized LLs is through scanning tunneling spectroscopy (STS), as was demonstrated on highly oriented pyrolytic graphite [197, 198] and adsorbate-induced 2DEGs formed by depositing Cs atoms on an InSb(110) surface [199].

1.7 AB-stacked (Bernal phase) bilayer graphene

It is important and interesting to expand the family of 2D materials beyond monolayer graphene. One method of doing this is to manufacture few-layer carbon systems, of which the plainest and most straightforward is bilayer graphene. Bilayer graphene consists of two coupled monolayer graphenes. It provides an attractive platform with which to explore topological valley physics.

Bilayer graphene can exist in three forms: AA-, AB-stacked (or the Bernal phase) and twisted bilayer. The simplest form is the AA bilayer [200–202], where each carbon atom in the second layer is precisely positioned above the corresponding atom in the first layer. However, this structure has the worst stability [203] and the least amount of available research among all forms, with only a few authors reporting the manufacture of AA-stacked forms [200, 204–206]. In the AB-stacked bilayer [8, 9, 16, 207, 208] or Bernal phase [209], as drawn in figure 1.15 [18, 190, 210–214], the layers align in such a way that two atoms, B1 (B atoms from the bottom layer) and A2 (A atoms from the top layer), are vertically aligned, and we refer to this pair of sites as the 'dimer sites' [215]. In contrast, the other two atoms, A1 (A atoms from the bottom layer) and B2 (B atoms from the top layer), do not have a counterpart in the other layer, and they align vertically with the centers of the hexagons in the other layer; these sites are called 'non-dimer sites' [215]. Both AA- and AB-stacked bilayers share the same unit cell as monolayer graphene.

1.7.1 TB description

Let us consider the most common AB-stacked bilayer graphene. We will explore its electronic properties using the procedure shown in section 1.2. As can be seen from figure 1.15, there are four atoms in the unit cell, a pair A1, B1 from the bottom layer and a pair A2, B2 from the top layer. The primitive lattice vectors $\boldsymbol{a}_1$ and $\boldsymbol{a}_2$, as well as the lattice constant a, are the same as those in monolayer graphene, as

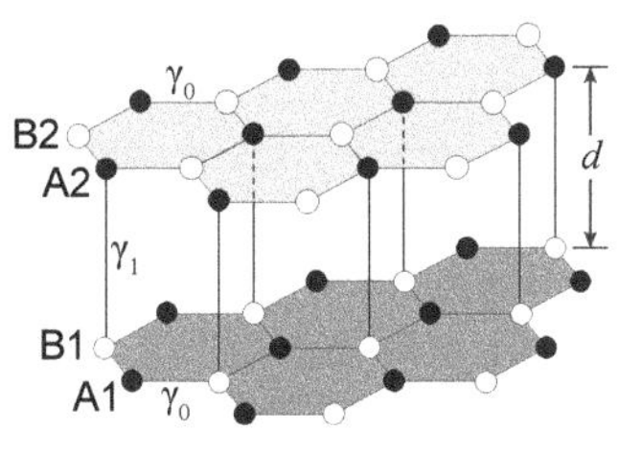

Figure 1.15. Schematic representation of a perspective view of AB-stacked bilayer graphene with an interlayer distance of d. The B1 atom from the bottom layer and the A2 atom from the top layer are aligned vertically, with a hopping parameter of γ_1. Within each layer, NN hopping is represented by γ_0.

discussed in section 1.2.1. The area of the unit cell in the xOy-plane does not change compared with monolayer graphene. The interlayer distance d is much larger than the carbon bond length. X-ray reflectivity experiments and first-principles calculations performed for bilayer graphene epitaxially grown on SiC [216] or in vacuum [217] suggest that $d \approx 3.35$Å [212, 215, 218], as in graphite [24]. Since there are four carbon atoms in each unit cell, and if the TB model only includes the p_z orbital per each carbon site, a 4×4 Hamiltonian can describe the system [219]. This results in four bands, instead of the two bands in monolayer graphene shown in figure 1.2.

To model the AB bilayer graphene, we adopt the Slonczewski–Weiss–McClure model [8, 9], and we first recall the approximations used in section 1.2.1 for each individual layer. In addition, we take into account interlayer hopping γ_1 between the orbitals of the dimer sites. In the description of the band structure of bulk graphite, several interlayer coupling parameters other than γ_1 were introduced [16, 220]. However, they do not change the qualitative nature of the low-energy spectrum [220] and will be neglected in the rest of this book [221].

In terms of the basis set {A1, B1, A2, B2}, the simplest possible Hamiltonian matrix for AB bilayer graphene is

$$H_{\mathrm{B:AB}}(\boldsymbol{k}) = \begin{pmatrix} 0 & -\gamma_0 f(\boldsymbol{k}) & 0 & 0 \\ -\gamma_0 f(\boldsymbol{k})^* & 0 & \gamma_1 & 0 \\ 0 & \gamma_1 & 0 & -\gamma_0 f(\boldsymbol{k}) \\ 0 & 0 & -\gamma_0 f(\boldsymbol{k})^* & 0 \end{pmatrix}, \tag{1.78}$$

in which the upper left (lower right) 2×2 submatrix describes the bottom (top) layer and the coupling between the respective sublattices A and B. The remaining two off-diagonal elements represent interlayer hopping. An estimation of the TB parameters for AB bilayer graphene yields values of [25, 215, 222–226] 2.9 eV $\leqslant \gamma_0 \leqslant$ 3.16eV and 0.3 eV $\leqslant \gamma_1 \leqslant$ 0.4 eV. The values of these parameters and their dependence on the interatomic distance have been extensively studied [9, 225, 227–230]. Here, the elements containing γ_1 do not have a factor of $f(\boldsymbol{k})$ because they describe the coupling between dimer sites, which are vertically aligned; thus, the hopping only has a vertical component [231].

Diagonalising Hamiltonian (1.78) easily yields four bands, which can be categorized into two pairs:

$$\begin{aligned} \mathcal{E}_{1,2}(\boldsymbol{k}) &= \pm \left\{ \sqrt{[\gamma_0 |f(\boldsymbol{k})|]^2 + \frac{\gamma_1^2}{4}} - \frac{\gamma_1}{2} \right\}, \\ \mathcal{E}_{3,4}(\boldsymbol{k}) &= \pm \left\{ \sqrt{[\gamma_0 |f(\boldsymbol{k})|]^2 + \frac{\gamma_1^2}{4}} + \frac{\gamma_1}{2} \right\}, \end{aligned} \tag{1.79}$$

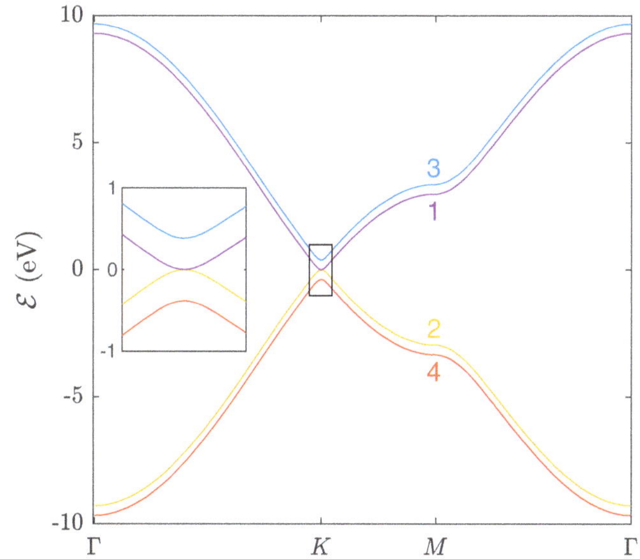

Figure 1.16. The band structure of AB bilayer graphene is plotted along the $\boldsymbol{k}$-space trajectory $\Gamma \to K \to M \to \Gamma$. The band indices are labeled by the adjacent numbers. The inset, which magnifies the area bounded by the black box, shows that two parabolic bands 1 and 2 touch each other at the $\boldsymbol{K}_\pm$ points. Plots were made using parameters $\gamma_0 = 3.16$ eV and $\gamma_1 = 0.381$ eV [218, 232].

with their corresponding wave functions

$$\psi_{1,2}(\boldsymbol{k}) = \frac{1}{2\sqrt{[\gamma_0|f(\boldsymbol{k})|]^2 - \frac{\gamma_1}{2}|\mathcal{E}_{1,2}(\boldsymbol{k})|}}\begin{pmatrix} \mp\gamma_0 f(\boldsymbol{k}) \\ |\mathcal{E}_{1,2}(\boldsymbol{k})| \\ \mp|\mathcal{E}_{1,2}(\boldsymbol{k})| \\ \gamma_0 f(\boldsymbol{k})^* \end{pmatrix},$$
$$\psi_{3,4}(\boldsymbol{k}) = \frac{1}{2\sqrt{[\gamma_0|f(\boldsymbol{k})|]^2 + \frac{\gamma_1}{2}|\mathcal{E}_{3,4}(\boldsymbol{k})|}}\begin{pmatrix} \mp\gamma_0 f(\boldsymbol{k}) \\ |\mathcal{E}_{3,4}(\boldsymbol{k})| \\ \pm|\mathcal{E}_{3,4}(\boldsymbol{k})| \\ -\gamma_0 f(\boldsymbol{k})^* \end{pmatrix}. \tag{1.80}$$

The four bands are plotted in figure 1.16. Over most of the BZ, the two conduction bands (1 and 3) are separated from each other by an energy on the order of γ_1, while the same is true for the two valence bands (2 and 4).

1.7.2 Close to the Dirac point

In the vicinity of the Dirac points $\boldsymbol{K}_\pm$, equation (1.79) can be expanded using the same approach as that used for monolayer graphene in section 1.2.2, as follows:

$$\mathcal{E}^{\kappa}_{1,2}(\boldsymbol{q}) = \pm\frac{3(\gamma_0 a)^2}{4\gamma_1}|\boldsymbol{q}|^2 = \pm\frac{(\hbar v_{\mathrm{F}})^2}{\gamma_1}|\boldsymbol{q}|^2, \tag{1.81}$$

$$\mathcal{E}^{\kappa}_{3,4}(\boldsymbol{q}) = \pm\left[\gamma_1 + \frac{3(\gamma_0 a)^2}{4\gamma_1}|\boldsymbol{q}|^2\right] = \pm\left[\gamma_1 + \frac{(\hbar v_{\mathrm{F}})^2}{\gamma_1}|\boldsymbol{q}|^2\right], \tag{1.82}$$

where we used equation (1.23) for v_F, which is valid near the $\boldsymbol{K}_\pm$ points. If the AB bilayer is undoped, at the $\boldsymbol{K}_\pm$ points, bands 1 and 2 touch exactly at the Fermi energy with parabolic dispersion [233–235], unlike the linear dispersion in the monolayer graphene [236, 237]. The other two bands, 3 and 4, are shifted away from the neutral point by interlayer hopping $\gamma_1 = 0.381\text{eV}$ in each direction [238].

It is also helpful to discuss the effective Hamiltonian of the AB-stacked bilayer [218], which was first analysed in [239] and proved useful [5, 240–244]. We may neglect bands 3 and 4 because $|\mathcal{E}_{3,4}(\boldsymbol{k})| \geqslant \gamma_1$. Furthermore, one can obtain from equation (1.80) that, at the $\boldsymbol{K}_\pm$ points, the wave functions for bands 1 and 2 (3 and 4) are nonzero only at non-dimer (dimer) sites. Therefore, in general, only the electrons at the non-dimer sites participate in the low-energy process. Consequently, it is possible to write an effective Hamiltonian using the basis set {A1, B2} following the procedure given in section 2.3.1 of reference [218], as follows:

$$H_{\text{B:AB}}(\boldsymbol{q}) = -\frac{\hbar^2}{2m}\begin{pmatrix} 0 & (q_x + \mathrm{i}\kappa q_y)^2 \\ (q_x - \mathrm{i}\kappa q_y)^2 & 0 \end{pmatrix} \tag{1.83}$$

with $m = \frac{\gamma_1}{2v_F^2} = 0.032m_e$, where m_e is free-electron mass. We note that in contrast to monolayer graphene, the pseudospin in AB bilayer graphene characterizes the layer DOF.

Similarly, we can derive the effective Hamiltonian for the BA-stacked bilayer [218, 245]. In terms of the basis set {B1, A2}, it reads [246]

$$H_{\text{B:BA}}(\boldsymbol{q}) = -\frac{\hbar^2}{2m}\begin{pmatrix} 0 & (q_x - \mathrm{i}\kappa q_y)^2 \\ (q_x + \mathrm{i}\kappa q_y)^2 & 0 \end{pmatrix}. \tag{1.84}$$

As we can see,

$$H_{\text{B:AB}}(\boldsymbol{q}) = H_{\text{B:BA}}(\boldsymbol{q})^*. \tag{1.85}$$

We can explain this geometrically. The AB-stacked bilayer can be converted to the BA-stacked bilayer by switching sublattices A and B. During the sublattice switch, the only complex element in equation (1.78), namely $f(\boldsymbol{k})$, becomes $f(\boldsymbol{k})^*$.

1.8 Electronic properties of AB-stacked bilayer zGNRs

In figure 1.17, we present the geometric structure of AB-stacked zGNR. It extends infinitely in the longitudinal zigzag (x) direction with a finite width in the transverse armchair (y) direction. Each layer consists of N zigzag lines. White (red) colors represent carbon atoms in the bottom (top) layer. Since there are $4N$ carbon atoms in a unit cell, the single-orbital TB Hamiltonian is a $4N \times 4N$ Hermitian matrix. In terms of the basis set $\{1A_1, 1B_1, 2A_1, 2B_1, \ldots, NA_1, NB_1, 1A_2, 1B_2, 2A_2, 2B_2, \ldots, NA_2, NB_2\}$, we find that we can treat each group of $2N$ atoms within the same layer as one entity. Therefore, the Hamiltonian can then be divided into 2×2 blocks:

$$H_{\text{Bz}}(k) = \begin{pmatrix} h_{11}(k) & h_{12} \\ h_{21} & h_{22}(k) \end{pmatrix}, \tag{1.86}$$

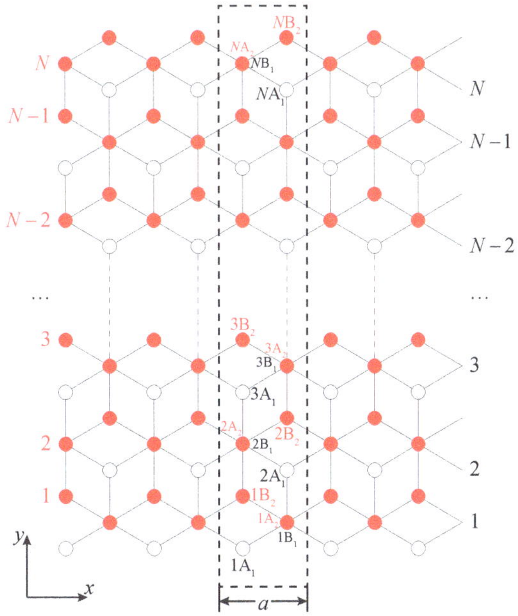

Figure 1.17. Structure of AB-stacked bilayer zGNR. Each layer consists of N zigzag lines. The 1D unit cell with the lattice constant a is represented by the dashed rectangle. White (red) colors represent carbon atoms and their corresponding labels in the bottom (top) layer. The dimer sites containing B atoms from the bottom layer and A atoms from the top layer are aligned vertically. The labeling of atoms is determined by the numbering of zigzag lines and layers (1 and 2 for the bottom and the top layers, respectively). In this case, the A (B) atom in the ith zigzag line in layer j is labeled as iA_j (iB_j). In addition, the font used to label the dimer sites is smaller than for the non-dimer sites.

such that each block is a $2N \times 2N$ matrix with $h_{21} = h_{12}^{\dagger}$. Since each layer is formed from one zGNR, as shown in figure 1.5, $h_{11}(k)$ and $h_{22}(k)$ have the same form as the Hamiltonian H_{zGNR} shown in equation (1.30).

We can now build the block h_{12}, which contains interlayer interaction γ_1. From figure 1.17, the dimer sites, or the sites between which the interlayer hopping γ_1 takes place, are the B atoms from the bottom layer and A atoms from the top layer, both of which are in the nth zigzag line. Therefore, the only nonzero elements in h_{12} are

$$(h_{i+1,i})_{12} = \gamma_1, \tag{1.87}$$

with i as an odd number.

After diagonalising the Hamiltonian (1.86), we plotted the band structure of a bilayer zGNR for $k \in [-\frac{\pi}{a}, \frac{\pi}{a}]$ in figure 1.18. It reveals that bilayer zGNRs and monolayer zGNRs share some similar electronic properties, such as edge states localized at the zigzag edges [247, 248]. The four edge states, two per edge, correspond to the four partly flat bands at $\mathcal{E}(k) = 0$ for $\frac{2\pi}{3a} \leqslant |k| \leqslant \frac{\pi}{a}$. In the ribbon, the overlap of four edge states leads to slight dispersion and nondegeneracy.

We note that for BA-stacked zGNRs, the sites between which the interlayer hopping γ_1 takes place change to the A atoms from the bottom layer and B atoms

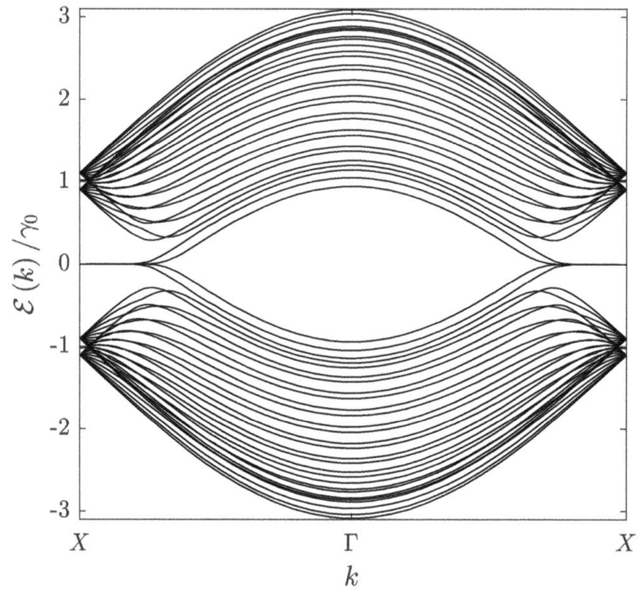

Figure 1.18. Band structure of an AB-stacked zGNR for $k \in \left[-\frac{\pi}{a}, \frac{\pi}{a}\right]$. The parameters are $N = 14$ and $\gamma_1 = 0.2\gamma_0$ [196, 247].

from the top layer. Therefore, we have two options for correcting Hamiltonian (1.86):

1. The nonzero elements in h_{12} and h_{21} change to

$$(h_{i,i+1})_{12} = (h_{i+1,i})_{21} = \gamma_1, \tag{1.88}$$

with i as an odd number.

2. Taking the Hermitian transpose of both h_{12} and h_{21}.

1.9 Twisted bilayer graphene

In the previous section, we discussed the AB- (Bernal-) stacked bilayer graphene, which is known to be the most stable structure. On the other hand, the weak vdW interaction between layers allows for variation in the stacking arrangement, depending on the fabrication process, resulting in a moiré pattern (figure 1.19(a)). The field of moiré systems has recently become one of the most fashionable topics in condensed matter physics [171, 250]. If either a twist angle or a lattice mismatch is present between two layers, a periodic moiré superlattice appears because of lattice mismatch [251–253]. This moiré period can be much larger than the intrinsic lattice constant of the two layers, leading to a small moiré miniband.

One typical example is twisted bilayer graphene (TBLG), which consists of two rotationally stacked monolayer graphenes observed in epitaxially grown multilayer graphenes [254–261]. After the discovery of graphene, researchers started investigating the electronic properties of TBLG [262] both theoretically [263–272] and experimentally [273–289]. Theoretical research predicts that, unlike Bernal stacked bilayer

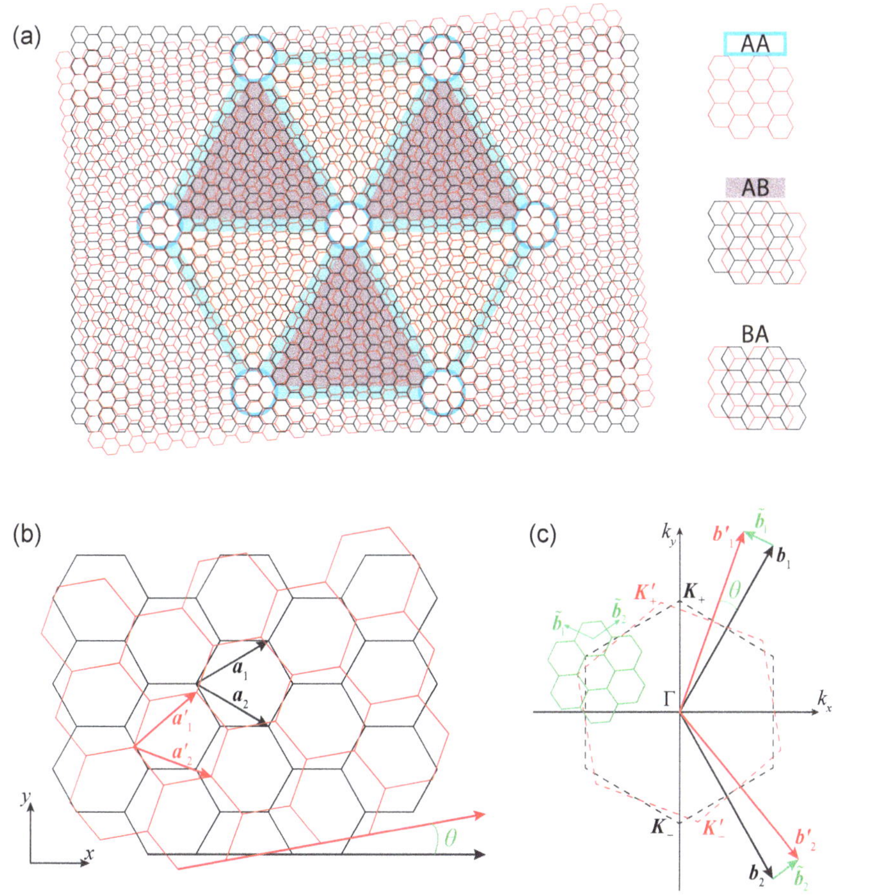

Figure 1.19. Moiré superlattice in twisted bilayer graphene (TBLG). (a) Two hexagonal lattices of graphene are twisted relative to each other in real space, leading to three primary stacking configurations: AA (blue outline), AB (violet triangular domain), and BA (yellow triangular domain) [249]. The green lines indicate their boundaries. (b) TBLG with a twist angle of θ. The primitive lattice vectors of the bottom (top) layer are $\boldsymbol{a}_1$ and $\boldsymbol{a}_2$ ($\boldsymbol{a}_1'$ and $\boldsymbol{a}_2'$). (c) Two BZs of the bottom and top layers overlap to form a superlattice BZ in reciprocal space. The reciprocal lattice vectors of the bottom layer, top layer, and moiré structure are $\boldsymbol{b}_1$, $\boldsymbol{b}_i'$, and $\tilde{\boldsymbol{b}}_i$ respectively, and their relationship is $\tilde{\boldsymbol{b}}_i = \boldsymbol{b}_i' - \boldsymbol{b}_i$. The elements in bottom layer, the top layer, and the moiré structure are shown in black, red, and green, respectively.

graphene, multilayer twisted graphene retains a linear dispersion relation similar to that of monolayer graphene due to weak interlayer coupling [290]. This has also been confirmed experimentally using ARPES, STM, and STS [260, 291, 292]. Further investigations have shown that bilayer graphene configurations possess superconducting and insulating states [274, 293–297]. Such investigations are important, since the preparation of graphene by chemical vapor deposition (CVD) seems to favor the formation of twisted layers [298, 299].

The distinct periodicity of the two layers causes an interference effect that leads to the formation of a moiré pattern, similar to the beat effect [300], as shown in figure 1.19(a). The moiré interference pattern includes three primary local stacking configurations: AA (blue outline), AB (violet background) and BA (yellow background). Such a pattern can also emerge at the interface of two materials with similar lattice structures. The best-known example is graphene on hBN, which also has similar hexagonal lattices with slightly larger lattice constants (1.8% [301–304]) [305–309], resulting in a moiré period of 14 nm.

Here, we present a geometric theory of 2D moiré structure without considering atomic relaxation [310–312]. Figure 1.19(b) schematically depicts a close-up of a TBLG lattice model, which is the building block for multilayer graphene. The elements in the bottom and top layers are shown in black and red, respectively. The top layer is rotated anticlockwise with respect to the bottom layer by a twist angle of θ. Generally, except for some special values of θ, the resulting structure is incommensurate and called a quasicrystal [312]. For theory and simulation, the special cases, i.e. strict periodic repetition of some large supercell [215], are important because they allow us to utilize well-developed solid-state theory for crystals [263, 264, 266, 267, 269, 313–315]. We denote the primitive lattice vectors of the bottom layer using $\boldsymbol{a}_1$ and $\boldsymbol{a}_2$, as given in equation (1.1), and those of the top layer using $\boldsymbol{a}_1'$ and $\boldsymbol{a}_2'$. The vectors of the top layer and those of the bottom layer are, in general, connected by the transformation

$$\boldsymbol{a}_i' = \mathsf{R}\boldsymbol{a}_i \tag{1.89}$$

with a 2×2 transformation matrix R. The corresponding reciprocal lattice vectors $\boldsymbol{b}_1$, $\boldsymbol{b}_2$ and $\boldsymbol{b}_1'$, $\boldsymbol{b}_2'$ for the bottom and top layers, respectively, are derived from the orthogonality relations (1.5); they read

$$\boldsymbol{b}_i \cdot \boldsymbol{a}_j = \boldsymbol{b}_i' \cdot \boldsymbol{a}_j' = 2\pi\delta_{ij}, \tag{1.90}$$

as discussed in section 1.2.1. Therefore,

$$\begin{aligned} \boldsymbol{b}_1 &= \frac{2\pi}{a_{1x}a_{2y} - a_{1y}a_{2x}}(a_{2y}\hat{\boldsymbol{x}} - a_{2x}\hat{\boldsymbol{y}}), \\ \boldsymbol{b}_2 &= \frac{2\pi}{a_{1x}a_{2y} - a_{1y}a_{2x}}(-a_{1y}\hat{\boldsymbol{x}} + a_{1x}\hat{\boldsymbol{y}}). \end{aligned} \tag{1.91}$$

We further define the transformation of the reciprocal space lattice vectors of the bottom and the top layers as

$$\boldsymbol{b}_i' = \mathsf{S}\boldsymbol{b}_i. \tag{1.92}$$

The relationship between transformation in real space R and transformation in reciprocal space S can be derived by considering equation (1.89), substituting equation (1.91) into (1.92), and solving the four orthogonality relations $\boldsymbol{b}_i' \cdot \boldsymbol{a}_j' = 2\pi\delta_{ij}$ in equation (1.90) for four unknowns (four elements in the 2×2 matrix S). This yields:

$$\mathsf{R}^{\mathrm{t}}\mathsf{S} = 1 \tag{1.93}$$

or

$$\mathsf{S} = {}^{t}(\mathsf{R}^{-1}) = ({}^{t}\mathsf{R})^{-1} \tag{1.94}$$

where ${}^{t}\mathsf{S}$ denotes the transpose of matrix S.

The reciprocal lattice vector of the moiré structure $\tilde{\boldsymbol{b}}_i$ is given by the slight difference between $\boldsymbol{b}_i$ and $\boldsymbol{b}_i'$ [49],

$$\tilde{\boldsymbol{b}}_i = \boldsymbol{b}_i' - \boldsymbol{b}_i = (\mathsf{S} - 1)\boldsymbol{b}_i. \tag{1.95}$$

This corresponds to real-space periodicity vectors $\tilde{\boldsymbol{a}}_i$, called moiré lattice vectors. Similar to the application of orthogonality in equation (1.93), we have

$$\tilde{\boldsymbol{a}}_i = \mathsf{T}\boldsymbol{a}_i \tag{1.96}$$

with

$$\mathsf{T}\,{}^{t}(\mathsf{S} - 1) = 1. \tag{1.97}$$

Thus,

$$\mathsf{T} = [{}^{t}(\mathsf{S} - 1)]^{-1} = (\mathsf{R}^{-1} - 1)^{-1} = (1 - \mathsf{R})^{-1}\mathsf{R}. \tag{1.98}$$

In our special case, R is given by the rotation matrix; it reads

$$\mathsf{R} = \begin{pmatrix} \cos\theta & -\sin\theta \\ \sin\theta & \cos\theta \end{pmatrix}. \tag{1.99}$$

Here, bilayer graphene with a twist angle of θ is identical to that for $\theta + 120°$ due to the C_3 symmetry, while graphene with a twist angle of $-\theta$ is the mirror image of that with θ. Hence, we should consider a range of $0° < \theta < 60°$ for the twist angle in our structural analysis [316].

From equation (1.94), we can find the transformation of the reciprocal space lattice vectors

$$\mathsf{S} = \mathsf{R}. \tag{1.100}$$

This indicates that the rotation of the reciprocal space lattice vectors of the top layer relative to the bottom layer is the same as in real space [212, 317], which is intuitively shown in figures 1.19(b) and (c). Further, the transformation T reads

$$\mathsf{T} = \begin{pmatrix} -\dfrac{1}{2} & -\dfrac{1}{2}\cot\dfrac{\theta}{2} \\ \dfrac{1}{2}\cot\dfrac{\theta}{2} & -\dfrac{1}{2} \end{pmatrix}. \tag{1.101}$$

Thus, the primitive lattice vectors of TBLG in real space are

$$\begin{aligned} \tilde{\boldsymbol{a}}_1 &= \frac{a}{4}\left[\left(-\cot\frac{\theta}{2} - \sqrt{3}\right)\hat{\boldsymbol{x}} + \left(\sqrt{3}\cot\frac{\theta}{2} - 1\right)\hat{\boldsymbol{y}}\right], \\ \tilde{\boldsymbol{a}}_2 &= \frac{a}{4}\left[\left(\cot\frac{\theta}{2} - \sqrt{3}\right)\hat{\boldsymbol{x}} + \left(\sqrt{3}\cot\frac{\theta}{2} + 1\right)\hat{\boldsymbol{y}}\right]. \end{aligned} \tag{1.102}$$

This leads to the period of the moiré pattern or the moiré wavelength [254, 289]:

$$\lambda = |\tilde{\boldsymbol{a}}_1| = |\tilde{\boldsymbol{a}}_2| = \frac{a}{2\sin\frac{\theta}{2}}. \tag{1.103}$$

Moiré patterns are often observed experimentally in multilayer materials, especially in atomic-scale images of bilayer graphene obtained using STM, transmission electron microscopy (TEM), and scanning TEM [259, 318, 319]. The value of the twist angle θ is determined from the measured moiré period λ [215].

If the twist angle θ is small, the primitive lattice vectors of the top layer $\boldsymbol{a}_i'$ are very close to those of the bottom layer $\boldsymbol{a}_i$, and the corresponding reciprocal lattice vectors $\boldsymbol{b}_i'$ and $\boldsymbol{b}_i$ are also similar. Therefore, the reciprocal lattice vectors of TBLG $\tilde{\boldsymbol{b}}_i$ become quite small compared with those of pristine graphene. Consequently, the primitive lattice vectors $\tilde{\boldsymbol{a}}_i$ of TBLG in real space are considerably larger than the primitive lattice vectors of pristine graphene [317].

Figure 1.19(c) illustrates the BZ reduction in TBLG. The valleys of the bottom and top layers are located at the BZ corners $\boldsymbol{K}_\pm = \pm(\boldsymbol{b}_1 - \boldsymbol{b}_2)/3$ and $\boldsymbol{K}_\pm' = \pm(\boldsymbol{b}_1' - \boldsymbol{b}_2')/3$, respectively. When the moiré period is much greater than the atomic period a, the interlayer interaction only occurs between close momenta $\boldsymbol{k}$ and $\boldsymbol{k}'$ with $|\boldsymbol{k} - \boldsymbol{k}'| \ll 2\pi/a$ [49]. In this case, $\boldsymbol{K}_\pm$ and $\boldsymbol{K}_\pm'$ do not hybridize, allowing us to treat these distant valleys as independent subsystems [275, 276]. The two valleys are independently folded into [11, 320] the reduced BZ (small green hexagon) spanned by the moiré reciprocal lattice vectors $\tilde{\boldsymbol{b}}_i$.

1.10 Early theoretical proposal for a valleytronic application

Valley filters [47], serving as generators of valley-polarized current and permitting carriers of only a specific valley to pass through, are an important type of valleytronic device. Early in 2007, Rycerz *et al* [47] proposed the utilization of the valley DOF to control an electronic device, taking advantage of the suppression of intervalley scattering in low-temperature pure samples [321–323]. The valley is used in much the same way as the electron spin in spintronics or quantum computing. A key ingredient for valleytronics would be a controllable way of occupying a single valley in graphene, thereby producing a VP [108, 324–326]. They proposed a valley filter based on a ballistic point contact junction with zigzag edges.

For zGNRs, only a single valley contributes to the lowest propagating mode [46], as denoted by the filled circle in the middle dispersion relation of the narrow region in figure 1.20(a). Following the discussion in figure 1.7, we can label $k = \frac{2\pi}{3a}$ as $\boldsymbol{K}_-$ and $k = \frac{4\pi}{3a}$ as $\boldsymbol{K}_+$. A detailed dispersion relation can be seen in figure 1.21, where it is important to note that the lowest propagating mode ranges from $-\frac{3\Delta}{2}$ to $\frac{3\Delta}{2}$. When $\mathcal{E}_\mathrm{F}$ lies in $(0, \frac{3\Delta}{2}]$, an electron from the vicinity of $\boldsymbol{K}_+$ ($\boldsymbol{K}_-$) only has a rightward (leftward) group velocity, as can be seen from the positive (negative) slope of the dispersion relation [320, 327]. The direction of the propagating mode is swapped when $\mathcal{E}_\mathrm{F}$ changes sign because the slope of the dispersion relation changes sign for the valence band.

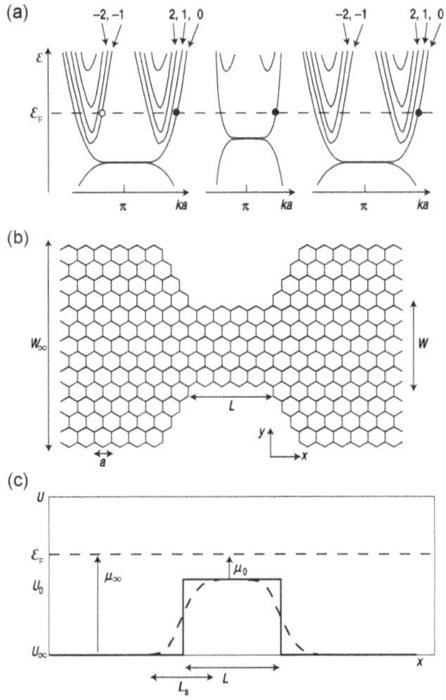

Figure 1.20. Schematic diagram of a valley filter based on zGNR. (a) Dispersion relation in the wide and narrow regions in (b) according to the electrostatic potential in (c). An electron in the first valley $\boldsymbol{K}_+$ (modes $n = 0, 1, 2, \ldots$) transmits (filled circle), whereas an electron in the second valley $\boldsymbol{K}_-$ (modes $n = -1, -2, \ldots$) reflects (cannot transmit, open circle). (b) Honeycomb lattice of carbon atoms in a strip containing a constriction with length L and width W. (c) Variation of the electrostatic potential along the strip for abrupt (solid line) and smooth potential barriers (dashed line). The polarity of the valley filter is reversed when the potential height, U_0, in the constriction crosses the Fermi energy, $\mathcal{E}_F$. Reproduced from [47], with permission from Springer Nature.

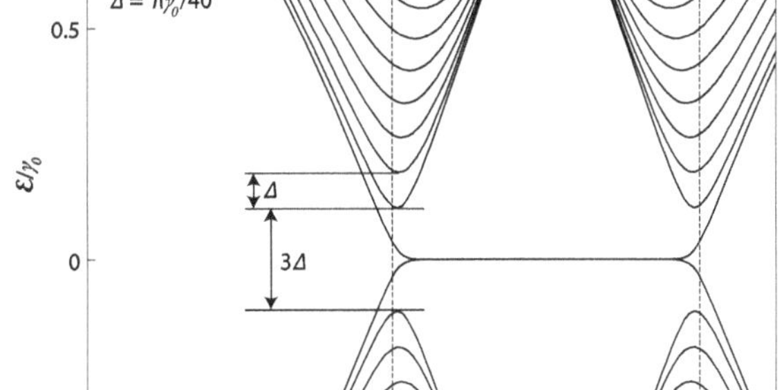

Figure 1.21. Dispersion relation of a zGNR. The spacing of the low-lying modes approaches $\Delta \equiv \frac{\sqrt{3}}{2}\gamma_0 a\pi / W$ for large $W \gg a$. The vertical lines mark the valley centers at $k = \frac{2\pi}{3a}$ and $\frac{4\pi}{3a}$. Reproduced from [47], with permission from Springer Nature.

When a 2D geometry consisting of a quantum point contact (QPC) [328] is constructed in a graphene sheet, as shown in figure 1.20(b), a current is passed through the QPC when a voltage difference is applied between the wide regions on the opposite sides of the constriction. To calculate the VP of the transmitted current, in figure 1.20(a), for a particular Fermi energy $\mathcal{E}_F$, one can define modes $n = 1, 2, \ldots, N$ lying in the $\boldsymbol{K}_+$ valley (with wave vector $k \in (\frac{\pi}{a}, \frac{2\pi}{a})$), whereas modes $n = -1, -2, \ldots, -N$ lie in the $\boldsymbol{K}_-$ valley (with wave vector $k \in (0, \frac{\pi}{a})$). The zeroth mode, $n = 0$, lies in a single valley restricted by the direction of propagation. Therefore, the wide regions support $2N + 1$ propagation modes at the Fermi energy and form a basis for the transmission matrix, t. The conductance of the constriction is determined by the Landauer formula:

$$G = \frac{2e^2}{h} \sum_{n=-N}^{N} T_n, \tag{1.104}$$

with

$$T_n = \sum_{m=-N}^{N} |t_{nm}|^2, \tag{1.105}$$

where $2e^2/h$ is the fundamental quantum of conductance, the factor of two accounts for the contributions from both the spin-up and spin-down electrons, and t_{nm} is the required amplitude for the electron to be transmitted from mode m of the left lead into mode n of the right lead. Thus, the incoming electron at the left lead is transmitted with a probability T_n to mode n at the right lead. The VP of the transmitted current is quantified by

$$P = \frac{T_0 + \sum_{n=1}^{N}(T_n - T_{-n})}{\sum_{n=-N}^{N} T_n}, \tag{1.106}$$

where one considers that for rightward propagation, the modes $n \in \mathbb{N}^{0+}$ lie in the $\boldsymbol{K}_+$ valley, while the modes $n \in \mathbb{N}^-$ lie in the $\boldsymbol{K}_-$ valley, as illustrated in figure 1.20(a). The polarization $P \in [-1, 1]$, with $P = 1$ (-1) if the transmitted current is fully contributed by the $\boldsymbol{K}_+$ ($\boldsymbol{K}_-$) valley.

In the numerical calculation, the widths of the wide and narrow regions are $W_\infty = 70\sqrt{3}\,a$ and $W = 20\sqrt{3}\,a$, respectively. The variation of the electrostatic potential along the graphene sheet is illustrated schematically in figure 1.20(c). The electrochemical potential in the wide regions is set to $\mathcal{E}_F - U_\infty \equiv \mu_\infty = \gamma_0/3$. From equation (1.32), one can work out that this electrochemical potential lies between $n_D = 14$ and 15, corresponding to $2n_D + 1 = 29$ propagating modes. The electrochemical potential $\mathcal{E}_F - U_0 \equiv \mu_0$ in the narrow region is measured in units of the mode spacing Δ.

Figure 1.22 shows the numerical results; parts (a) and (b) show the conductance and the VP, respectively, as a function of the electrochemical potential, μ_0, in the narrow region. For positive μ_0, as illustrated in figures 1.20(a) and (b), the current flows entirely within the conduction band, and conductance plateaus of quantized conductance occur at odd multiples of $2e^2/h$, as predicted by Peres *et al* [329, 330].

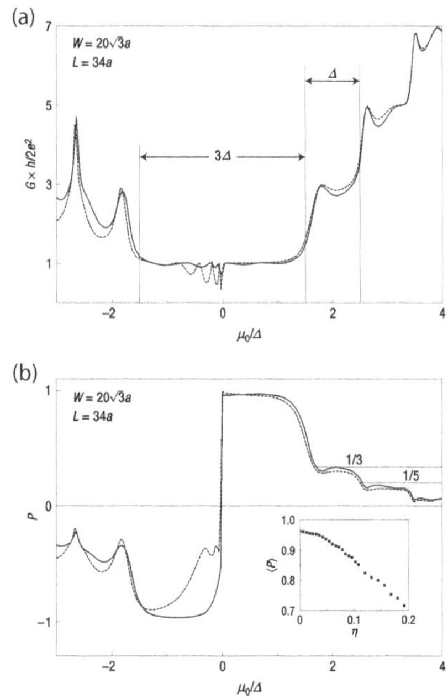

Figure 1.22. (a) Conductance and (b) VP for the valley filter in figure 1.20 as a function of the electrochemical potential in the narrow region. Abrupt ($L_s = 0$) and smooth ($L_s = 8a$) potential steps are represented by solid and dashed lines, respectively. The inset in the bottom panel shows the degradation of the VP as a function of a randomly introduced fraction, η, of vacancies at the edges of the constriction at $\mu_0 = 0.25\Delta$. The result, $\langle P \rangle$, is averaged over a few hundred random configurations of vacancies. Reproduced from [47], with permission from Springer Nature.

Upon comparing the solid and dashed lines, it is clear that smoothing the potential step improves the flatness of the plateaus. The plateaus in the conductance at $G = \frac{2e^2}{h}(2n + 1)$ correspond to plateaus in the VP at $P = 1/(2n + 1)$. Specifically:

1. On the lowest $n = 0$ plateau, and for $0 < \mu_0 \lesssim \Delta$, as illustrated in figure 1.20 (a), an electron in the $\boldsymbol{K}_+$ valley is transmitted (filled circle), whereas an electron in the $\boldsymbol{K}_-$ valley is reflected (open circle). This means that one can address the two valleys individually as independent DOFs for the conduction electrons. The calculated polarization is more than 95%.
2. On the second $n = 1$ plateau, and for $3\Delta/2 \lesssim \mu_0 \lesssim 5\Delta/2$, the narrow region supports two propagating modes in the $\boldsymbol{K}_+$ valley and one propagating mode in the $\boldsymbol{K}_-$ valley. This leads to conductance $G = 3 \times \frac{2e^2}{h}$ with polarization $P = \frac{1}{3}$.
3. On the third $n = 2$ plateau, and for $5\Delta/2 \lesssim \mu_0 \lesssim 7\Delta/2$, one more propagating mode is added to both $\boldsymbol{K}_+$ and $\boldsymbol{K}_-$ valleys in the narrow region, compared with $n = 1$. This results in a conductance $G = 5 \times \frac{2e^2}{h}$ and a polarization $P = \frac{1}{5}$.
4. Similarly, on the fourth $n = 3$ plateau, and for $7\Delta/2 \lesssim \mu_0 \lesssim 9\Delta/2$, there is a conductance $G = 7 \times \frac{2e^2}{h}$ with polarization $P = \frac{1}{7}$.

For negative μ_0, the current makes a transition from the conduction band in the wide regions to the valence band in the narrow region. This interband transition destroys the conductance quantization in the studied QPC—except on the first plateau, which remains rather flat over the entire interval $\mu_0 \in (-3\Delta/2, 3\Delta/2)$. The resonances at negative μ_0 are due to quasi-bound states in the valence band [331, 332]. The polarity of the valley filter flips with some degradation in quality (especially for the smooth potential) as the slope of the dispersion relation in the valence band changes sign. In this situation, for $-\frac{3\Delta}{2} \lesssim \mu_0 < 0$, an electron in the $\boldsymbol{K}_-$ valley is transmitted, whereas an electron in the $\boldsymbol{K}_+$ valley is reflected, resulting in a VP close to -1. The polarization can also be switched by changing the direction of propagation in accordance with TRS.

Rycerz *et al* also investigated the robustness of the quality of the valley filter against edge imperfections. They randomly introduced a fraction, η, of vacancies among the sites at the edges of the constriction. The inset of figure 1.22(b) shows the resulting degradation of the polarization (averaged over a few hundred random configurations of vacancies) at $\mu_0 = 0.25\Delta$. The VP stays above 95% if a few percent of vacancies are introduced, and the VP still stays above 85% if 10% of the atoms at the edge are removed.

The results of Rycerz *et al* were recently confirmed experimentally in reference [333], the authors of which used a GNR with a nonuniform width in conjunction with a p–n junction.

One caveat of proposal [47] is that the valley filtering properties rely on well-defined graphene zigzag edges, which are difficult to control in experiments. Later, Rycerz [334] complemented their previous study [47] by constructing a constriction with generic edges, evolving from zigzag to armchair upon ‘rotation’ of the crystallographic orientation of the system axis. Numerical calculation showed that the valley filter effect is surprisingly robust against changes in the crystallographic orientation of the system axis unless a point contact has perfect armchair edges. Theoretical calculations have supported other, more complex valleytronic devices, including double-bended zGNR/aGNR/zGNR [335], two semi-infinite zGNRs with different widths [336], T junctions [337, 338], L junctions [339], cross junctions [340], silicene [341], a double electrostatic potential [342], a MoS_2 zigzag nanoribbon with a nonuniform potential profile [343], and a graphene–silicene–graphene heterojunction [344]. Regrettably, the schemes proposed in the original proposal [47] and subsequent works [334–336] rely on a QPC in a nanoribbon to achieve the generation and detection of the valley-polarized current. The major difficulty in the experimental realization of this approach comes from the necessity of tailoring the graphene samples into the nanoribbons and QPCs with zigzag edges needed to support edge states, limiting their practical applications.

Based on the lattice model presented in figure 1.20, Luo *et al* [345] proposed a simple mechanism to detect the valley filtering effect by depositing a superconductor electrode on the right-hand side of the wide region. Andreev reflection (AR) occurs at the graphene–superconductor interface, and the incident electron and the reflected hole should come from different valleys [346] because of the conservation of TRS. In this device, the Fermi energy in the narrow region is fixed near the flat band, so that the incident electron and the reflected hole are in different bands. For instance, when the incident $\boldsymbol{K}_+$ electron is from the conduction band, the corresponding reflected $\boldsymbol{K}_-$ hole should be from the valence band in the narrow region. However, the AR process is hindered by the lack of a leftward propagating mode in the valence band of the $\boldsymbol{K}_-$ valley in the narrow region. This suppression of AR leads to the valley filter.

References

[1] Wallace P R 1947 The band theory of graphite *Phys. Rev.* **71** 622–34

[2] Novoselov K S, Geim A K, Morozov S V, Jiang D, Zhang Y, Dubonos S V, Grigorieva I V and Firsov A A 2004 Electric field effect in atomically thin carbon films *Science* **306** 666

[3] Novoselov K S, Geim A K, Morozov S V, Jiang D, Katsnelson M I, Grigorieva I V, Dubonos S V and Firsov A A 2005 Two-dimensional gas of massless Dirac fermions in graphene *Nature* **438** 197–200

[4] Zhang Y, Tan Y-W, Stormer H L and Kim P 2005 Experimental observation of the quantum Hall effect and Berry's phase in graphene *Nature* **43** 201–4

[5] Novoselov K S, McCann E, Morozov S V, Fal'ko V I, Katsnelson M I, Zeitler U, Jiang D, Schedin F and Geim A K 2006 Unconventional quantum Hall effect and Berry's phase of 2π in bilayer graphene *Nat. Phys.* **2** 177–80

[6] Semenoff G W 1984 Condensed-matter simulation of a three-dimensional anomaly *Phys. Rev. Lett.* **53** 2449–52

[7] Haldane F D M 1988 Model for a quantum Hall effect without Landau levels: condensed-matter realization of the "parity anomaly" *Phys. Rev. Lett.* **61** 2015–8

[8] Slonczewski J C and Weiss P R 1958 Band structure of graphite *Phys. Rev.* **109** 272–9

[9] McClure J W 1957 Band structure of graphite and de Haas-van Alphen effect *Phys. Rev.* **108** 612–8

[10] Wang S, Tian H, Luo Y, Yu J, Ren C, Sun C, Xu Y and Sun M 2019 First-principles calculations of aluminium nitride monolayer with chemical functionalization *Appl. Surf. Sci.* **481** 1549–53

[11] Ashcroft N W and Mermin N D 1976 *Solid State Physics* (Boston, MA: Cengage Learning)

[12] Reich S, Maultzsch J, Thomsen C and Ordejón P 2002 Tight-binding description of graphene *Phys. Rev.* B **66** 035412

[13] Sasaki K, Murakami S and Saito R 2006 Stabilization mechanism of edge states in graphene *Appl. Phys. Lett.* **88** 113110

[14] Peres N M R, Guinea F and Castro Neto A H 2006 Electronic properties of disordered two-dimensional carbon *Phys. Rev.* B **73** 125411

[15] Peres N M R 2010 Colloquium: the transport properties of graphene: an introduction *Rev. Mod. Phys.* **82** 2673–700

[16] Dresselhaus M S and Dresselhaus G 2002 Intercalation compounds of graphite *Adv. Phys.* **51** 1–186

[17] Toy W W, Dresselhaus M S and Dresselhaus G 1977 Minority carriers in graphite and the H-point magnetoreflection spectra *Phys. Rev.* B **15** 4077–90

[18] Raza H (ed) 2012 *Graphene Nanoelectronics: Metrology, Synthesis, Properties and Applications* NanoScience and Technology (Heidelberg: Springer)

[19] Brooks M 2019 Spin qubits in two-dimensional semiconductors *PhD Thesis* University of Konstanz, Konstanz
[20] Schaibley J R, Yu H, Clark G, Rivera P, Ross J S, Seyler K L, Yao W and Xu X 2016 Valleytronics in 2D materials *Nat. Rev. Mater.* **1** 16055
[21] Wolfram T and Ellialtıoğlu Ş 2014 *Applications of Group Theory to Atoms, Molecules, and Solids* (Cambridge: Cambridge University Press)
[22] Foa Torres L E F, Roche S and Charlier J-C 2020 *Introduction to Graphene-Based Nanomaterials: From Electronic Structure to Quantum Transport* 2nd edn (Cambridge: Cambridge University Press)
[23] Tsidilkovski I M 2012 *Electron Spectrum of Gapless Semiconductors* Springer Series in Solid-State Sciences **vol 116** (Berlin: Springer Science)
[24] Mucha-Kruczyński M 2013 *Theory of Bilayer Graphene Spectroscopy* Springer Theses (Berlin: Springer Science & Business Media)
[25] Castro Neto A H, Guinea F, Peres N M R, Novoselov K S and Geim A K 2009 The electronic properties of graphene *Rev. Mod. Phys.* **81** 109–62
[26] Gasiorowicz S 2003 *Quantum Physics* 3rd edn (New York: Wiley)
[27] Griffiths D J and Schroeter D F 2018 *Introduction to Quantum Mechanics* 3rd edn (Cambridge: Cambridge University Press)
[28] Ando T, Nakanishi T and Saito R 1998 Berry's phase and absence of back scattering in carbon nanotubes *J. Phys. Soc. Jpn.* **67** 2857–62
[29] McCann E, Kechedzhi K, Vladimir I, Fal'ko H S, Ando T and Altshuler B L 2006 Weak-localization magnetoresistance and valley symmetry in graphene *Phys. Rev. Lett.* **97** 146805
[30] Saito R, Dresselhaus G and Dresselhaus M S 2000 Trigonal warping effect of carbon nanotubes *Phys. Rev.* B **61** 2981–90
[31] Zhou S Y, Gweon G H, Graf J, Fedorov A V, Spataru C D, Diehl R D, Kopelevich Y, Lee D H, Louie S G and Lanzara A 2006 First direct observation of Dirac fermions in graphite *Nat. Phys.* **2** 595–9
[32] Brey L and Fertig H A 2006 Edge states and the quantized Hall effect in graphene *Phys. Rev.* B **73** 195408
[33] Brey L and Fertig H A 2006 Electronic states of graphene nanoribbons studied with the Dirac equation *Phys. Rev.* B **73** 235411
[34] Nakada K, Fujita M, Dresselhaus G and Dresselhaus M S 1996 Edge state in graphene ribbons: nanometer size effect and edge shape dependence *Phys. Rev.* B **54** 17954–61
[35] Yano Y, Mitoma N, Ito H and Itami K 2020 A quest for structurally uniform graphene nanoribbons: synthesis, properties, and applications *J. Org. Chem.* **85** 4–33
[36] Yano Y, Mitoma N, Ito H and Itami K 2021 Correction to a quest for structurally uniform graphene nanoribbons: synthesis, properties, and applications *J. Org. Chem.* **86** 4372–3
[37] Wakabayashi K, Takane Y, Yamamoto M and Sigrist M 2009 Electronic transport properties of graphene nanoribbons *New J. Phys.* **11** 095016
[38] Wang S, Hung N T, Tian H, Islam M S and Saito R 2021 Switching behavior of a heterostructure based on periodically doped graphene nanoribbon *Phys. Rev. Appl.* **16** 024030
[39] Wakabayashi K, Fujita M, Ajiki H and Sigrist M 1999 Electronic and magnetic properties of nanographite ribbons *Phys. Rev.* B **59** 8271–82
[40] Cresti A, Grosso G and Parravicini G P 2008 Electronic states and magnetotransport in unipolar and bipolar graphene ribbons *Phys. Rev.* B **77** 115408

[41] Cresti A, Nemec N, Biel B, Niebler G, Triozon F, Cuniberti G and Roche S 2008 Charge transport in disordered graphene-based low dimensional materials *Nano Res.* **1** 361–94

[42] Fujita M, Wakabayashi K, Nakada K and Kusakabe K 1996 Peculiar localized state at zigzag graphite edge *J. Phys. Soc. Jpn.* **65** 1920–3

[43] Wakabayashi K and Sigrist M 2000 Zero-conductance resonances due to flux states in nanographite ribbon junctions *Phys. Rev. Lett.* **84** 3390–3

[44] Kobayashi Y, Fukui K-i, Enoki T, Kusakabe K and Kaburagi Y 2005 Observation of zigzag and armchair edges of graphite using scanning tunneling microscopy and spectroscopy *Phys. Rev.* B **71** 193406

[45] Kobayashi Y, Fukui K-i, Enoki T and Kusakabe K 2006 Edge state on hydrogen-terminated graphite edges investigated by scanning tunneling microscopy *Phys. Rev.* B **73** 125415

[46] Cresti A, Grosso G and Parravicini G P 2008 Valley-valve effect and even-odd chain parity in p-n graphene junctions *Phys. Rev.* B **77** 233402

[47] Rycerz A, Tworzydło J and Beenakker C W J 2007 Valley filter and valley valve in graphene *Nat. Phys.* **3** 172–5

[48] Saito R, Dresselhaus G and Dresselhaus M S 1998 *Physical Properties of Carbon Nanotubes* (London: World Scientific)

[49] Enoki T and Ando T (ed) 2020 *Physics and Chemistry of Graphene: Graphene to Nanographene* 2nd edn (Singapore: Jenny Stanford Publishing)

[50] Geronimi C, Jackson W and Luske H 1951 *Alice in Wonderland* (animated film), Walt Disney Company

[51] IBM IBM Design Language—8-Bar: https://www.ibm.com/design/language/ibm-logos/8-bar/

[52] Wang S, Pratama F R, Ukhtary M S and Saito R 2020 Independent degrees of freedom in two-dimensional materials *Phys. Rev.* B **101** 081414(R)

[53] Rachel S and Ezawa M 2014 Giant magnetoresistance and perfect spin filter in silicene, germanene, and stanene *Phys. Rev.* B **89** 195303

[54] Boettger J C and Trickey S B 2007 First-principles calculation of the spin-orbit splitting in graphene *Phys. Rev.* B **75** 121402(R)

[55] Boettger J C and Trickey S B 2007 Erratum: first-principles calculation of the spin-orbit splitting in graphene [Phys. Rev. B 75, 121402(R) (2007)] *Phys. Rev.* B **75** 199903

[56] Gmitra M, Konschuh S, Ertler C, Ambrosch-Draxl C and Fabian J 2009 Band-structure topologies of graphene: spin-orbit coupling effects from first principles *Phys. Rev.* B **80** 235431

[57] Konschuh S, Gmitra M and Fabian J 2010 Tight-binding theory of the spin-orbit coupling in graphene *Phys. Rev.* B **82** 245412

[58] Liu C-C, Jiang H and Yao Y 2011 Low-energy effective Hamiltonian involving spin-orbit coupling in silicene and two-dimensional germanium and tin *Phys. Rev.* B **84** 195430

[59] Van Tuan D 2015 *Charge and Spin Transport in Disordered Graphene-based Materials* Springer Theses 1st edn (Cham: Springer)

[60] Kochan D, Irmer S and Fabian J 2017 Model spin-orbit coupling Hamiltonians for graphene systems *Phys. Rev.* B **95** 165415

[61] Ghalamkari K 2017 Genshi-sō busshitsu no hikari kyūshū ni okeru barē hen kyoku [Valley polarization in optical absorption of atomic layer materials] *PhD Thesis* Tohoku University, Sendai

[62] Yao Y, Ye F, Qi X-L, Zhang S-C and Fang Z 2007 Spin-orbit gap of graphene: first-principles calculations *Phys. Rev.* B **75** 041401
[63] Xu X, Yao W, Xiao D and Heinz T F 2014 Spin and pseudospins in layered transition metal dichalcogenides *Nat. Phys.* **10** 343–50
[64] Abdelouahed S, Ernst A, Henk J, Maznichenko I V and Mertig I 2010 Spin-split electronic states in graphene: effects due to lattice deformation, Rashba effect, and adatoms by first principles *Phys. Rev.* B **82** 125424
[65] Xu X, Liu C, Sun Z, Cao T, Zhang Z, Wang E, Liu Z and Liu K 2018 Interfacial engineering in graphene bandgap *Chem. Soc. Rev.* **47** 3059–99
[66] Min H, Hill J E, Sinitsyn N A, Sahu B R, Kleinman L and MacDonald A H 2006 Intrinsic and Rashba spin-orbit interactions in graphene sheets *Phys. Rev.* B **74** 165310
[67] Ezawa M 2012 A topological insulator and helical zero mode in silicene under an inhomogeneous electric field *New J. Phys.* **14** 033003
[68] Wang S, Sun M and Hung N T 2024 Advanced inorganic semiconductor materials *Inorganics* **12**
[69] Wang Q H, Kalantar-Zadeh K, Kis A, Coleman J N and Strano M S 2012 Electronics and optoelectronics of two-dimensional transition metal dichalcogenides *Nat. Nanotechnol.* **7** 699–712
[70] Kumar A and Ahluwalia P K 2012 Electronic structure of transition metal dichalcogenides monolayers 1H-MX_2 (M = Mo, W; X = S, Se, Te) from ab-initio theory: new direct band gap semiconductors *Eur. Phys. J.* B **85** 186
[71] Chhowalla M, Shin H S, Eda G, Li L-J, Loh K P and Zhang H 2013 The chemistry of two-dimensional layered transition metal dichalcogenide nanosheets *Nat. Chem.* **5** 263–75
[72] Zhang Y *et al* 2014 Direct observation of the transition from indirect to direct bandgap in atomically thin epitaxial $MoSe_2$ *Nat. Nanotechnol.* **9** 111–5
[73] Kormányos A, Burkard G, Gmitra M, Fabian J, Zólyomi V, Drummond N D and Fal'ko V 2015 $\mathbf{k}\cdot\mathbf{p}$ theory for two-dimensional transition metal dichalcogenide semiconductors *2D Mater.* **2** 022001
[74] Kormányos A, Burkard G, Gmitra M, Fabian J, Zólyomi V, Drummond N D and Fal'ko V 2015 Corrigendum: $\mathbf{k}\cdot\mathbf{p}$ theory for two-dimensional transition metal dichalcogenide semiconductors (2015 2D Mater. 2 022001) *2D Mater.* **2** 049501
[75] Splendiani A, Sun L, Zhang Y, Li T, Kim J, Chim C-Y, Galli G and Wang F 2010 Emerging photoluminescence in monolayer MoS_2 *Nano Lett.* **10** 1271–5
[76] Mak K F, Lee C, Hone J, Shan J and Heinz T F 2010 Atomically thin MoS_2: a new direct-gap semiconductor *Phys. Rev. Lett.* **105** 136805
[77] Xiao D, Liu G-B, Feng W, Xu X and Yao W 2012 Coupled spin and valley physics in monolayers of MoS_2 and other group-VI dichalcogenides *Phys. Rev. Lett.* **108** 196802
[78] Saito R, Tatsumi Y, Huang S, Ling X and Dresselhaus M S 2016 Raman spectroscopy of transition metal dichalcogenides *J. Phys.: Condens. Matter* **28** 353002
[79] Yan J, Xia J, Wang X, Liu L, Kuo J-L, Tay B K, Chen S, Zhou W, Liu Z and Shen Z X 2015 Stacking-dependent interlayer coupling in trilayer MoS_2 with broken inversion symmetry *Nano Lett.* **15** 8155–61
[80] Hue T *et al* 2014 Observing grain boundaries in CVD-grown monolayer transition metal dichalcogenides *ACS Nano* **8** 11401–8
[81] Wu L 2021 Investigating lattice vibrational and excitonic properties of two-dimensional heterostructures of transition metal dichalcogenides *PhD Thesis* Nanyang Technological University, Singapore

[82] Wilson J A and Yoffe A D 1969 The transition metal dichalcogenides discussion and interpretation of the observed optical, electrical and structural properties *Adv. Phys.* **18** 193–335
[83] Amsterdam S 2021 Electronic coupling and morphology in organic—two dimensional heterostructures *PhD Thesis* Northwestern University, Evanston
[84] Novoselov K S, Jiang D, Schedin F, Booth T J, Khotkevich V V, Morozov S V and Geim A K 2005 Two-dimensional atomic crystals *Proc. Natl. Acad. Sci. USA* **102** 10451–3
[85] Wang S, Ukhtary M S and Saito R 2020 Strain effect on circularly polarized electroluminescence in transition metal dichalcogenides *Phys. Rev. Res.* **2** 033340
[86] Zhang Z, Chen P, Duan X, Zang K, Luo J and Duan X 2017 Robust epitaxial growth of two-dimensional heterostructures, multiheterostructures, and superlattices *Science* **357** 788–92
[87] Chhowalla M, Liu Z and Zhang H 2015 Two-dimensional transition metal dichalcogenide (TMD) nanosheets *Chem. Soc. Rev.* **44** 2584–6
[88] Wang S, Ren C, Tian H, Yu J and Sun M 2018 MoS_2/ZnO van der Waals heterostructure as a high-efficiency water splitting photocatalyst: a first-principles study *Phys. Chem. Chem. Phys.* **20** 13394–9
[89] Wang S, Tian H, Ren C, Yu J and Sun M 2018 Electronic and optical properties of heterostructures based on transition metal dichalcogenides and graphene-like zinc oxide *Sci. Rep.* **8** 12009
[90] Liu Y, Zhang S, He J, Wang Z M and Liu Z 2019 Recent progress in the fabrication, properties, and devices of heterostructures based on 2D materials *Nano-Micro Lett.* **11** 13
[91] Lu H-Z, Yao W, Xiao D and Shen S-Q 2013 Intervalley scattering and localization behaviors of spin-valley coupled Dirac fermions *Phys. Rev. Lett.* **110** 016806
[92] Li M, Shi J, Liu L, Yu P, Xi N and Wang Y 2016 Experimental study and modeling of atomic-scale friction in zigzag and armchair lattice orientations of MoS_2 *Sci. Technol. Adv. Mater.* **7** 189–99
[93] Wang G, Chernikov A, Glazov M M, Heinz T F, Marie X, Amand T and Urbaszek B 2018 Colloquium: excitons in atomically thin transition metal dichalcogenides *Rev. Mod. Phys.* **90** 021001
[94] Jin W *et al* 2013 Direct measurement of the thickness-dependent electronic band structure of MoS_2 using angle-resolved photoemission spectroscopy *Phys. Rev. Lett.* **111** 106801
[95] Liu Y, Gao Y, Zhang S, He J, Yu J and Liu Z 2019 Valleytronics in transition metal dichalcogenides materials *Nano Res.* **12** 2695–711
[96] Diebold A and Hofmann T 2021 *Optical and Electrical Properties of Nanoscale Materials* Springer Series in Materials Science **vol 318** (Cham: Springer)
[97] Wang Z M (ed) 2014 *MoS_2: Materials, Physics, and Devices* Lecture Notes in Nanoscale Science and Technology **vol 21** (Cham: Springer)
[98] Kuc A, Zibouche N and Heine T 2011 Influence of quantum confinement on the electronic structure of the transition metal sulfide TS_2 *Phys. Rev.* B **83** 245213
[99] Ellis J K, Lucero M J and Scuseria G E 2011 The indirect to direct band gap transition in multilayered MoS_2 as predicted by screened hybrid density functional theory *Appl. Phys. Lett.* **99** 261908
[100] Li T and Galli G 2007 Electronic properties of MoS_2 nanoparticles *J. Phys. Chem.* C **111** 16192–6

[101] Yun W S, Han S W, Hong S C, Kim I G and Lee J D 2012 Thickness and strain effects on electronic structures of transition metal dichalcogenides: 2H-MX_2 semiconductors (M = Mo, W; X = S, Se, Te) *Phys. Rev.* B **85** 033305

[102] Hong J, Lee C, Park J-S and Shim J H 2016 Control of valley degeneracy in MoS_2 by layer thickness and electric field and its effect on thermoelectric properties *Phys. Rev.* B **93** 035445

[103] Bromley R A, Murray R B and Yoffe A D 1972 The band structures of some transition metal dichalcogenides. III. Group VIA: trigonal prism materials *J. Phys. C: Solid State Phys.* **5** 759

[104] Lu J, Liu H, Tok E S and Sow C-H 2016 Interactions between lasers and two-dimensional transition metal dichalcogenides *Chem. Soc. Rev.* **45** 2494–515

[105] Mattheiss L F 1973 Energy bands for 2H-$NbSe_2$ and 2H-MoS_2 *Phys. Rev. Lett.* **30** 784–7

[106] Zhao W, Ribeiro R M, Toh M, Carvalho A, Kloc C, Castro Neto A H and Eda G 2013 Origin of indirect optical transitions in few-layer MoS_2, WS_2, and WSe_2 *Nano Lett.* **13** 5627–34

[107] Wu Z *et al* 2016 Even–odd layer-dependent magnetotransport of high-mobility Q-valley electrons in transition metal disulfides *Nat. Commun.* **7** 12955

[108] Yao W, Xiao D and Niu Q 2008 Valley-dependent optoelectronics from inversion symmetry breaking *Phys. Rev.* B **77** 235406

[109] Eda G, Yamaguchi H, Voiry D, Fujita T, Chen M and Chhowalla M 2011 Photoluminescence from chemically exfoliated MoS_2 *Nano Lett.* **11** 5111–6

[110] Eda G, Yamaguchi H, Voiry D, Fujita T, Chen M and Chhowalla M 2012 Correction to photoluminescence from chemically exfoliated MoS_2 *Nano Lett.* **12** 526

[111] Tang C S, Yin X and Wee A T S (ed) 2023 *Two-Dimensional Transition-Metal Dichalcogenides: Phase Engineering and Applications in Electronics and Optoelectronics* (Weinheim: Wiley-VCH)

[112] Cappelluti E, Roldán R, Silva-Guillén J A, Ordejón P and Guinea F 2013 Tight-binding model and direct-gap/indirect-gap transition in single-layer and multilayer MoS_2 *Phys. Rev.* B **88** 075409

[113] Kadantsev E S and Hawrylak P 2012 Electronic structure of a single MoS_2 monolayer *Solid State Commun.* **152** 909–13

[114] Ganatra R and Zhang Q 2014 Few-layer MoS_2: a promising layered semiconductor *ACS Nano* **8** 4074–99

[115] Xie H 2019 Probing excitonic mechanics in suspended and strained transition metal dichalcogenides monolayers *PhD Thesis* The Pennsylvania State University, University Park, PA

[116] Palacios-Berraquero C 2018 *Quantum Confined Excitons in 2-Dimensional Materials* Springer Theses (Cham: Springer)

[117] Th Böker R, Severin A, Müller C, Janowitz R, Manzke D, Voß D, Krüger P, Mazur A and Pollmann J 2001 Band structure of MoS_2, $MoSe_2$ and α-$MoTe_2$: angle-resolved photoelectron spectroscopy and *ab initio* calculations *Phys. Rev.* B **64** 235305

[118] Ataca C, Şahin H and Ciraci S 2012 Stable, single-layer MX_2 transition-metal oxides and dichalcogenides in a honeycomb-like structure *J. Phys. Chem.* C **116** 8983–99

[119] Zeng H and Cui X 2015 An optical spectroscopic study on two-dimensional group-VI transition metal dichalcogenides *Chem. Soc. Rev.* **44** 2629–42

[120] Zhao L, Shang Q, Li M, Liang Y, Li C and Zhang Q 2021 Strong exciton-photon interaction and lasing of two-dimensional transition metal dichalcogenide semiconductors *Nano Res.* **14** 2021

[121] Cheiwchanchamnangij T and Lambrecht W R L 2012 Quasiparticle band structure calculation of monolayer, bilayer, and bulk MoS_2 *Phys. Rev.* B **85** 205302

[122] Zeng H *et al* 2013 Optical signature of symmetry variations and spin-valley coupling in atomically thin tungsten dichalcogenides *Sci. Rep.* **3** 1608

[123] Schmidt H, Giustiniano F and Eda G 2015 Electronic transport properties of transition metal dichalcogenide field-effect devices: surface and interface effects *Chem. Soc. Rev.* **44** 7715–36

[124] Ramasubramaniam A 2012 Large excitonic effects in monolayers of molybdenum and tungsten dichalcogenides *Phys. Rev.* B **86** 115409

[125] Lebègue S and Eriksson O 2009 Electronic structure of two-dimensional crystals from ab initio theory *Phys. Rev.* B **79** 115409

[126] Roldán R, Silva-Guillén J A, López-Sancho M P, Guinea F, Cappelluti E and Ordejón P 2014 Electronic properties of single-layer and multilayer transition metal dichalcogenides MX_2 (M = Mo, W and X = S, Se) *Ann. Phys.* **526** 347–57

[127] Ramakrishna Matte H S S, Gomathi A, Manna A K, Late D J, Datta R, Pati S K and Rao C N R 2010 MoS_2 and WS_2 analogues of graphene *Angew. Chem.* **49** 4059–62

[128] Castro Neto A H 2001 Charge density wave, superconductivity, and anomalous metallic behavior in 2D transition metal dichalcogenides *Phys. Rev. Lett.* **86** 4382–5

[129] Kam K K and Parkinson B A 1982 Detailed photocurrent spectroscopy of the semiconducting group VIB transition metal dichalcogenides *J. Phys. Chem.* **86** 463–7

[130] Liu L, Kumar S B, Ouyang Y and Guo J 2011 Performance limits of monolayer transition metal dichalcogenide transistors *IEEE Trans. Electron Devices* **58** 3042–7

[131] Ding Y, Wang Y, Ni J, Shi L, Shi S and Tang W 2011 First principles study of structural, vibrational and electronic properties of graphene-like MX_2 (M=Mo, Nb, W, Ta; X=S, Se, Te) monolayers *Physica B: Condens. Matter* **406** 2254–60

[132] Zhang Y, Suzuki R and Iwasa Y 2017 Potential profile of stabilized field-induced lateral p–n junction in transition-metal dichalcogenides *ACS Nano* **11** 12583–90

[133] Jariwala D, Marks T J and Hersam M C 2017 Mixed-dimensional van der Waals heterostructures *Nat. Mater.* **16** 170–81

[134] Zhao W, Ghorannevis Z, Chu L, Toh M, Kloc C, Tan P-H and Eda G 2013 Evolution of electronic structure in atomically thin sheets of WS_2 and WSe_2 *ACS Nano* **7** 791–7

[135] Radisavljevic B, Radenovic A, Brivio J, Giacometti V and Kis A 2011 Single-layer MoS_2 transistors *Nat. Nanotechnol.* **6** 147–50

[136] Bernardi M, Ataca C, Palummo M and Grossman J C 2017 Optical and electronic properties of two-dimensional layered materials *Nanophotonics* **6** 479–93

[137] Bao D 2020 Optical study of excitonic complexes in transition metal dichalcogenides monolayers *PhD Thesis* Nanyang Technological University, Singapore

[138] Tongay S, Zhou J, Ataca C, Lo K, Matthews T S, Li J, Grossman J C and Wu J 2012 Thermally driven crossover from indirect toward direct bandgap in 2D semiconductors: $MoSe_2$ versus MoS_2 *Nano Lett.* **12** 5576–80

[139] Sánchez-Royo J F *et al* 2014 Electronic structure, optical properties, and lattice dynamics in atomically thin indium selenide flakes *Nano Res.* **7** 1556–68

[140] Ruppert C, Aslan O B and Heinz T F 2014 Optical properties and band gap of single- and few-layer $MoTe_2$ crystals *Nano Lett.* **14** 6231–6

[141] Yi Y, Chen Z, Yu X-F, Zhou Z-K and Li J 2019 Recent advances in quantum effects of 2D materials *Adv. Quantum Technol.* **2** 1800111

[142] Voß D, Krüger P, Mazur A and Pollmann J 1999 Atomic and electronic structure of WSe_2 from *ab initio* theory: bulk crystal and thin film systems *Phys. Rev.* B **60** 14311–7
[143] Wickramaratne D, Zahid F and Lake R K 2014 Electronic and thermoelectric properties of few-layer transition metal dichalcogenides *J. Chem. Phys.* **140** 124710
[144] Kolobov A V and Tominaga J 2016 *Two-Dimensional Transition-Metal Dichalcogenides* Springer Series in Materials Science **vol 239** (Cham: Springer)
[145] Kormányos A, Zólyomi V, Drummond N D and Burkard G 2014 Spin-orbit coupling, quantum dots, and qubits in monolayer transition metal dichalcogenides *Phys. Rev.* X **4** 011034
[146] Kormányos A, Zólyomi V, Drummond N D and Burkard G 2014 Erratum: spin-orbit coupling, quantum dots, and qubits in monolayer transition metal dichalcogenides [Phys. Rev. X 4, 011034 (2014)] *Phys. Rev.* X **4** 039901
[147] Liu G-B, Xiao D, Yao Y, Xu X and Yao W 2015 Electronic structures and theoretical modelling of two-dimensional group-VIB transition metal dichalcogenides *Chem. Soc. Rev.* **44** 2643–63
[148] Zhu Z Y, Cheng Y C and Schwingenschlögl U 2011 Giant spin-orbit-induced spin splitting in two-dimensional transition-metal dichalcogenide semiconductors *Phys. Rev.* B **84** 153402
[149] Herman F, Kuglin C D, Cuff K F and Kortum R L 1963 Relativistic corrections to the band structure of tetrahedrally bonded semiconductors *Phys. Rev. Lett.* **11** 541–5
[150] Tang C S *et al* 2019 Three-dimensional resonant exciton in monolayer tungsten diselenide actuated by spin–orbit coupling *ACS Nano* **13** 14529–39
[151] Miwa J A, Ulstrup S, Sørensen S G, Dendzik M, Čabo A G, Bianchi M, Lauritsen J V and Hofmann P 2015 Electronic structure of epitaxial single-layer MoS_2 *Phys. Rev. Lett.* **114** 046802
[152] Liu G-B, Shan W-Y, Yao Y, Yao W and Xiao D 2013 Three-band tight-binding model for monolayers of group-VIB transition metal dichalcogenides *Phys. Rev.* B **88** 085433
[153] Liu G-B, Shan W-Y, Yao Y, Yao W and Xiao D 2014 Erratum: Three-band tight-binding model for monolayers of group-VIB transition metal dichalcogenides [Phys. Rev. B 88, 085433 (2013)] *Phys. Rev.* B **89** 039901
[154] Wang Z, Zhao L, Mak K F and Shan J 2017 Probing the spin-polarized electronic band structure in monolayer transition metal dichalcogenides by optical spectroscopy *Nano Lett.* **17** 740–6
[155] Kośmider K, González J W and Fernández-Rossier J 2013 Large spin splitting in the conduction band of transition metal dichalcogenide monolayers *Phys. Rev.* B **88** 245436
[156] Feng W, Yao Y, Zhu W, Zhou J, Yao W and Xiao D 2012 Intrinsic spin Hall effect in monolayers of group-VI dichalcogenides: a first-principles study *Phys. Rev.* B **86** 165108
[157] Rasmita A and Gao W-B 2021 Opto-valleytronics in the 2D van der Waals heterostructure *Nano Res.* **14** 2021
[158] Koroteev Yu M, Bihlmayer G, Gayone J E, Chulkov E V, Blügel S, Echenique P M and Hofmann Ph 2004 Strong spin-orbit splitting on Bi surfaces *Phys. Rev. Lett.* **93** 046403
[159] Alidoust N *et al* 2014 Observation of monolayer valence band spin-orbit effect and induced quantum well states in MoX_2 *Nat. Commun.* **5** 4673
[160] Alidoust N *et al* 2014 Erratum: Observation of monolayer valence band spin-orbit effect and induced quantum well states in MoX_2 *Nat. Commun.* **5** 5136
[161] Mitin V V, Kochelap V A, Dutta M and Stroscio M A 2019 *Introduction to Optical and Optoelectronic Properties of Nanostructures* (Cambridge: Cambridge University Press)

[162] Gong Z, Liu G-B, Yu H, Xiao D, Cui X, Xu X and Yao W 2013 Magnetoelectric effects and valley-controlled spin quantum gates in transition metal dichalcogenide bilayers *Nat. Commun.* **4** 2053

[163] Molina-Sánchez A, Hummer K and Wirtz L 2015 Vibrational and optical properties of MoS_2: from monolayer to bulk *Surf. Sci. Rep.* **70** 554–86

[164] Wang G *et al* 2015 Spin-orbit engineering in transition metal dichalcogenide alloy monolayers *Nat. Commun.* **6** 10110

[165] Cazalilla M A, Ochoa H and Guinea F 2014 Quantum spin Hall effect in two-dimensional crystals of transition-metal dichalcogenides *Phys. Rev. Lett.* **113** 077201

[166] Klinovaja J and Loss D 2013 Spintronics in MoS_2 monolayer quantum wires *Phys. Rev.* B **88** 075404

[167] Song Y and Dery H 2013 Transport theory of monolayer transition-metal dichalcogenides through symmetry *Phys. Rev. Lett.* **111** 026601

[168] Kormányos A, Zólyomi V, Drummond N D, Rakyta P, Burkard G and Fal'ko V I 2013 Monolayer MoS_2: trigonal warping, the Γ valley, and spin-orbit coupling effects *Phys. Rev.* B **88** 045416

[169] Li Y *et al* 2014 Valley splitting and polarization by the Zeeman effect in monolayer $MoSe_2$ *Phys. Rev. Lett.* **113** 266804

[170] MacNeill D, Heikes C, Mak K F, Anderson Z, Kormányos A, Zólyomi V, Park J and Ralph D C 2015 Breaking of valley degeneracy by magnetic field in monolayer $MoSe_2$ *Phys. Rev. Lett.* **114** 037401

[171] Ciarrocchi A, Tagarelli F, Avsar A and Kis A 2022 Excitonic devices with van der Waals heterostructures: valleytronics meets twistronics *Nat. Rev. Mater.* **7** 449–64

[172] Szulakowska L 2020 Electron-electron interactions and optical properties of two-dimensional nanocrystals *PhD Thesis* University of Ottawa, Ottawa

[173] Roldán R, López-Sancho M P, Guinea F, Cappelluti E, Silva-Guillén J A and Ordejón P 2014 Momentum dependence of spin–orbit interaction effects in single-layer and multi-layer transition metal dichalcogenides *2D Mater.* **1** 034003

[174] Fang S, Kuate Defo R, Shirodkar S N, Lieu S, Tritsaris G A and Kaxiras E 2015 Ab initio tight-binding Hamiltonian for transition metal dichalcogenides *Phys. Rev.* B **92** 205108

[175] Silva-Guillén J Á, San-Jose P and Roldán R 2016 Electronic band structure of transition metal dichalcogenides from ab initio and Slater–Koster tight-binding model *Appl. Sci.* **6** 284

[176] Vajpey D S 2016 Energy dispersion model using tight binding theory *PhD Thesis* Rochester Institute of Technology, Rochester

[177] Doran N J, Titterington D J, Ricco B and Wexler G 1978 A tight binding fit to the bandstructure of 2H-$NbSe_2$ and NbS_2 *J. Phys. C: Solid State Phys.* **11** 685

[178] Pearce A J, Mariani E and Burkard G 2016 Tight-binding approach to strain and curvature in monolayer transition-metal dichalcogenides *Phys. Rev.* B **94** 155416

[179] Rostami H, Moghaddam A G and Asgari R 2013 Effective lattice Hamiltonian for monolayer MoS_2: tailoring electronic structure with perpendicular electric and magnetic fields *Phys. Rev.* B **88** 085440

[180] Zhang Y J, Oka T, Suzuki R, Ye J T and Iwasa Y 2014 Electrically switchable chiral light-emitting transistor *Science* **344** 725–8

[181] Zhang C *et al* 2016 Systematic study of electronic structure and band alignment of monolayer transition metal dichalcogenides in van der Waals heterostructures *2D Mater.* **4** 015026

[182] McClure J W 1956 Diamagnetism of graphite *Phys. Rev.* **104** 666–71
[183] Andrei Bernevig B, Hughes T L, Zhang S-C, Chen H-D and Wu C 2006 Band collapse and the quantum Hall effect in graphene *Int. J. Mod. Phys.* B **20** 3257–78
[184] Ho J H, Lai Y H, Chiu Y H and Lin M F 2008 Landau levels in graphene *Physica E: Low Dimens. Syst. Nanostruct.* **40** 1722–5
[185] Shon N H and Ando T 1998 Quantum transport in two-dimensional graphite system *J. Phys. Soc. Jpn.* **67** 2421–9
[186] Ando T 2005 Theory of electronic states and transport in carbon nanotubes *J. Phys. Soc. Jpn.* **74** 777–817
[187] Chakraborty T, Peeters F and Sivan U (ed) 2002 *Nano-Physics & Bio-Electronics: A New Odyssey* 1st edn (Amsterdam: Elsevier)
[188] Zheng Y and Ando T 2002 Hall conductivity of a two-dimensional graphite system *Phys. Rev.* B **65** 245420
[189] Komeilizadeh N 2007 Investigations of the quantum Hall effect in graphene *PhD Thesis* University of Windsor, Windsor
[190] Aoki H and Dresselhaus M S (ed) 2014 *Physics of Graphene* NanoScience and Technology (Cham: Springer International Publishing)
[191] Apalkov V M and Chakraborty T 2006 Fractional quantum Hall states of Dirac electrons in graphene *Phys. Rev. Lett.* **97** 126801
[192] Koshino M and Ando T 2007 Splitting of the quantum Hall transition in disordered graphenes *Phys. Rev.* B **75** 033412
[193] Abergel D S L, Apalkov V, Berashevich J, Ziegler K and Chakraborty T 2010 Properties of graphene: a theoretical perspective *Adv. Phys.* **59** 261–482
[194] Hassani S 2009 *Mathematical Methods: For Students of Physics and Related Fields* 2nd edn (New York: Springer)
[195] Arfken G B, Weber H J and Harris F E (ed) 2013 *Mathematical Methods for Physicists: A Comprehensive Guide* 7th edn (Boston, MA: Academic Press)
[196] Carmelo J M P, dos Santos J M B L, Vieira V R and Sacramento P D (ed) 2007 *Strongly Correlated Systems, Coherence and Entanglement* (Singapore: World Scientific)
[197] Matsui T, Kambara H, Niimi Y, Tagami K, Tsukada M and Fukuyama H 2005 STS observations of Landau levels at graphite surfaces *Phys. Rev. Lett.* **94** 226403
[198] Li G and Andrei E Y 2007 Observation of Landau levels of Dirac fermions in graphite *Nat. Phys.* **3** 623–7
[199] Hashimoto K, Sohrmann C, Wiebe J, Inaoka T, Meier F, Hirayama Y, Römer R A, Wiesendanger R and Morgenstern M 2008 Quantum Hall transition in real space: from localized to extended states *Phys. Rev. Lett.* **101** 256802
[200] Liu Z, Suenaga K, Harris P J F and Iijima S 2009 Open and closed edges of graphene layers *Phys. Rev. Lett.* **102** 015501
[201] Ho J H, Lu C L, Hwang C C, Chang C P and Lin M F 2006 Coulomb excitations in AA- and AB-stacked bilayer graphites *Phys. Rev.* B **74** 085406
[202] Charlier J-C, Michenaud J-P, Gonze X and Vigneron J-P 1991 Tight-binding model for the electronic properties of simple hexagonal graphite *Phys. Rev.* B **44** 13237–49
[203] Rakhmanov A L, Rozhkov A V, Sboychakov A O and Nori F 2012 Instabilities of the AA-stacked graphene bilayer *Phys. Rev. Lett.* **109** 206801
[204] Lee J-K, Lee S-C, Ahn J-P, Kim S-C, Wilson J I B and John P 2008 The growth of AA graphite on (111) diamond *J. Chem. Phys.* **129** 234709

[205] Lauffer P, Emtsev K V, Graupner R, Seyller Th, Ley L, Reshanov S A and Weber H B 2008 Atomic and electronic structure of few-layer graphene on SiC(0001) studied with scanning tunneling microscopy and spectroscopy *Phys. Rev.* B **77** 155426
[206] Borysiuk J, Sołtys J and Piechota J 2011 Stacking sequence dependence of graphene layers on SiC $(000\bar{1})$—experimental and theoretical investigation *J. Appl. Phys.* **109** 093523
[207] McCann E, Abergel D S L and Fal'ko V I 2007 Electrons in bilayer graphene *Solid State Commun.* **143** 110–5
[208] Ando T and Koshino M 2009 Field effects on optical phonons in bilayer graphene *J. Phys. Soc. Jpn.* **78** 034709
[209] Bernal J D and Bragg W L 1924 The structure of graphite *Proc. R. Soc. Lond. Ser.* A **106** 749–73
[210] Liu J-M and Lin I-T 2018 *Graphene Photonics* (Cambridge: Cambridge University Press)
[211] Gong J R (ed) 2011 *Graphene: Synthesis, Characterization, Properties and Applications* (Rijeka: IntechOpen)
[212] Zhang M (ed) 2019 *Graphene-Like 2D Materials* Handbook of Graphene **vol 3** (Hoboken, NJ: Wiley)
[213] Lu C L, Chang C P, Huang Y C, Lu J M, Hwang C C and Lin M F 2006 Low-energy electronic properties of the AB-stacked few-layer graphites *J. Phys.: Condens. Matter* **18** 5849
[214] Lu C L, Chang C P, Huang Y C, Chen R B and Lin M L 2006 Influence of an electric field on the optical properties of few-layer graphene with AB stacking *Phys. Rev.* B **73** 144427
[215] Rozhkov A V, Sboychakov A O, Rakhmanov A L and Nori F 2016 Electronic properties of graphene-based bilayer systems *Phys. Rep.* **648** 1–104
[216] Varchon F *et al* 2007 Electronic structure of epitaxial graphene layers on SiC: effect of the substrate *Phys. Rev. Lett.* **99** 126805
[217] Trickey S B, Müller-Plathe F, Diercksen G H F and Boettger J C 1992 Interplanar binding and lattice relaxation in a graphite dilayer *Phys. Rev.* B **45** 4460–8
[218] McCann E and Koshino M 2013 The electronic properties of bilayer graphene *Rep. Prog. Phys.* **76** 056503
[219] Servet O 2021 *Electronic Properties of Rhombohedral Graphite* Springer Theses (Cham: Springer Nature)
[220] Koshino M and Ando T 2009 Electronic structures and optical absorption of multilayer graphenes *Solid State Commun.* **149** 1123–7
[221] Mikhailov S (ed) 2011 *Physics and Applications of Graphene: Theory* (Rijeka: IntechOpen)
[222] Malard L M, Nilsson J, Elias D C, Brant J C, Plentz F, Alves E S, Castro Neto A H and Pimenta M A 2007 Probing the electronic structure of bilayer graphene by Raman scattering *Phys. Rev.* B **76** 201401
[223] Zhang L M, Li Z Q, Basov D N, Fogler M M, Hao Z and Martin M C 2008 Determination of the electronic structure of bilayer graphene from infrared spectroscopy *Phys. Rev.* B **78** 235408
[224] Li Z Q, Henriksen E A, Jiang Z, Hao Z, Martin M C, Kim P, Stormer H L and Basov D N 2009 Band structure asymmetry of bilayer graphene revealed by infrared spectroscopy *Phys. Rev. Lett.* **102** 037403
[225] Brandt N B, Chudinov S M and Ponomarev Y G (ed) 1988 *Semimetals: 1. Graphite and its Compounds* Modern Problems in Condensed Matter Sciences **vol 20.1** (Amsterdam: Elsevier)

[226] Misu A, Mendez E E and Dresselhaus M S 1979 Near infrared reflectivity of graphite under hydrostatic pressure. I. Experiment *J. Phys. Soc. Jpn.* **47** 199–207
[227] Nozières P 1958 Cyclotron resonance in graphite *Phys. Rev.* **109** 1510–21
[228] Dresselhaus M S and Mavroides J G 1964 The Fermi surface of graphite *IBM J. Res. Dev.* **8** 262–7
[229] Soule D E, McClure J W and Smith L B 1964 Study of the Shubnikov-de Haas effect. Determination of the Fermi surfaces in graphite *Phys. Rev.* **134** A453–70
[230] Dillon R O, Spain I L, Woollam J A and Lowrey W H 1978 Galvanomagnetic effects in graphite—I: low field data and the densities of free carriers *J. Phys. Chem. Solids* **39** 907–22
[231] Prikoszovich K 2022 Valley-filters using kink states in bilayer graphene *PhD Thesis* Technische Universität Wien, Wien
[232] Kuzmenko A B, Crassee I, van der Marel D, Blake P and Novoselov K S 2009 Determination of the gate-tunable band gap and tight-binding parameters in bilayer graphene using infrared spectroscopy *Phys. Rev.* B **80** 165406
[233] Guinea F, Castro Neto A H and Peres N M R 2006 Electronic states and Landau levels in graphene stacks *Phys. Rev.* B **73** 245426
[234] Koshino M and Ando T 2006 Transport in bilayer graphene: calculations within a self-consistent Born approximation *Phys. Rev.* B **73** 245403
[235] Nilsson J, Castro Neto A H, Peres N M R and Guinea F 2006 Electron-electron interactions and the phase diagram of a graphene bilayer *Phys. Rev.* B **73** 214418
[236] Partoens B and Peeters F M 2006 From graphene to graphite: electronic structure around the K point *Phys. Rev.* B **74** 075404
[237] Partoens B and Peeters F M 2007 Normal and Dirac fermions in graphene multilayers: tight-binding description of the electronic structure *Phys. Rev.* B **75** 193402
[238] Das A, Chakraborty B, Piscanec S, Pisana S, Sood A K and Ferrari A C 2009 Phonon renormalization in doped bilayer graphene *Phys. Rev.* B **79** 155417
[239] McCann E and Fal'ko V I 2006 Landau-level degeneracy and quantum Hall effect in a graphite bilayer *Phys. Rev. Lett.* **96** 086805
[240] Abergel D S L and Fal'ko V I 2007 Optical and magneto-optical far-infrared properties of bilayer graphene *Phys. Rev.* B **75** 155430
[241] Katsnelson M I, Novoselov K S and Geim A K 2006 Chiral tunnelling and the Klein paradox in graphene *Nat. Phys.* **2** 620–5
[242] Kechedzhi K, Fal'ko V I, McCann E and Altshuler B L 2007 Influence of trigonal warping on interference effects in bilayer graphene *Phys. Rev. Lett.* **98** 176806
[243] Zhang F, MacDonald A H and Mele E J 2013 Valley Chern numbers and boundary modes in gapped bilayer graphene *Proc. Natl. Acad. Sci. USA* **110** 10546–51
[244] Katsnelson M I 2006 Minimal conductivity in bilayer graphene *Eur. Phys. J.* B **52** 151–3
[245] Mucha-Kruczyński M, McCann E and Fal'ko V I 2010 Electron–hole asymmetry and energy gaps in bilayer graphene *Semicond. Sci. Technol.* **25** 033001
[246] Wang Z, Cheng S, Liu X and Jiang H 2021 Topological kink states in graphene *Nanotechnology* **32** 402001
[247] Castro E V, Peres N M R, Lopes dos Santos J M B, Castro Neto A H and Guinea F 2008 Localized states at zigzag edges of bilayer graphene *Phys. Rev. Lett.* **100** 026802
[248] Rhim J-W and Moon K 2008 Edge states of zigzag bilayer graphite nanoribbons *J. Phys.: Condens. Matter* **20** 365202

[249] Bal G, Cazeaux P, Massatt D and Quinn S 2023 Mathematical models of topologically protected transport in twisted bilayer graphene *Multiscale Model. Simul.* **21** 1081–21

[250] Pacchioni G 2020 Valleytronics with a twist *Nat. Rev. Mater.* **5** 480

[251] Andrei E Y and MacDonald A H 2020 Graphene bilayers with a twist *Nat. Mater.* **19** 1265–75

[252] Aggarwal D, Narula R and Ghosh S 2023 A primer on twistronics: a massless Dirac fermion's journey to moiré patterns and flat bands in twisted bilayer graphene *J. Phys.: Condens. Matter* **35** 143001

[253] Cai L and Yu G 2021 Fabrication strategies of twisted bilayer graphenes and their unique properties *Adv. Mater.* **33** 2004974

[254] Beyer H, Müller M and Schimmel T 1999 Monolayers of graphite rotated by a defined angle: hexagonal superstructures by STM *Appl. Phys.* A **68** 163–6

[255] Berger C *et al* 2006 Electronic confinement and coherence in patterned epitaxial graphene *Science* **312** 1191–6

[256] Hass J, Feng R, Millán-Otoya J E, Li X, Sprinkle M, First P N, de Heer W A, Conrad E H and Berger C 2007 Structural properties of the multilayer graphene/4H-SiC(000$\bar{1}$) system as determined by surface x-ray diffraction *Phys. Rev.* B **75** 214109

[257] Hass J, Varchon F, Millán-Otoya J E, Sprinkle M, Sharma N, de Heer W A, Berger C, First P N, Magaud L and Conrad E H 2008 Why multilayer graphene on 4H-SiC(000$\bar{1}$) behaves like a single sheet of graphene *Phys. Rev. Lett.* **100** 125504

[258] Li G, Luican A, Lopes dos Santos J M B, Castro Neto A H, Reina A, Kong J and Andrei E Y 2010 Observation of Van Hove singularities in twisted graphene layers *Nat. Phys.* **6** 109–13

[259] Miller D L, Kubista K D, Rutter G M, Ruan M, de Heer W A, First P N and Stroscio J A 2010 Structural analysis of multilayer graphene via atomic moiré interferometry *Phys. Rev.* B **81** 125427

[260] Luican A, Li G, Reina A, Kong J, Nair R R, Novoselov K S, Geim A K and Andrei E Y 2011 Single-layer behavior and its breakdown in twisted graphene layers *Phys. Rev. Lett.* **106** 126802

[261] Xhie J, Sattler K, Ge M and Venkateswaran N 1993 Giant and supergiant lattices on graphite *Phys. Rev.* B **47** 15835–41

[262] Hicks J *et al* 2011 Symmetry breaking in commensurate graphene rotational stacking: comparison of theory and experiment *Phys. Rev.* B **83** 205403

[263] Lopes dos Santos J M B, Peres N M R and Castro Neto A H 2012 Continuum model of the twisted graphene bilayer *Phys. Rev.* B **86** 155449

[264] Lopes dos Santos J M B, Peres N M R and Castro Neto A H 2007 Graphene bilayer with a twist: electronic structure *Phys. Rev. Lett.* **99** 256802

[265] Bistritzer R and MacDonald A H 2011 Moiré bands in twisted double-layer graphene *Proc. Natl. Acad. Sci. USA* **108** 12233–7

[266] Shallcross S, Sharma S, Kandelaki E and Pankratov O A 2010 Electronic structure of turbostratic graphene *Phys. Rev.* B **81** 165105

[267] Shallcross S, Sharma S, Kandelaki E and Pankratov O A 2010 Publisher's note: electronic structure of turbostratic graphene [Phys. Rev. B 81, 165105 (2010)] *Phys. Rev.* B **81** 239904

[268] Cisternas E and Correa J D 2012 Theoretical reproduction of superstructures revealed by STM on bilayer graphene *Chem. Phys.* **409** 74–8

[269] Mele E J 2012 Interlayer coupling in rotationally faulted multilayer graphenes *J. Phys. D: Appl. Phys.* **45** 154004

[270] Shallcross S, Sharma S and Pankratov O A 2008 Quantum interference at the twist boundary in graphene *Phys. Rev. Lett.* **101** 056803
[271] Mele E J 2011 Band symmetries and singularities in twisted multilayer graphene *Phys. Rev.* B **84** 235439
[272] Mele E J 2010 Commensuration and interlayer coherence in twisted bilayer graphene *Phys. Rev.* B **81** 161405
[273] Cao Y *et al* 2018 Correlated insulator behaviour at half-filling in magic-angle graphene superlattices *Nature* **556** 80–4
[274] Cao Y, Fatemi V, Fang S, Watanabe K, Taniguchi T, Kaxiras E and Jarillo-Herrero P 2018 Unconventional superconductivity in magic-angle graphene superlattices *Nature* **556** 43–50
[275] Brihuega I, Mallet P, González-Herrero H, Trambly de Laissardière G, Ugeda M M, Magaud L, Gómez-Rodríguez J M, Ynduráin F and Veuillen J-Y 2012 Unraveling the intrinsic and robust nature of van Hove singularities in twisted bilayer graphene by scanning tunneling microscopy and theoretical analysis *Phys. Rev. Lett.* **109** 196802
[276] Brihuega I, Mallet P, González-Herrero H, Trambly de Laissardière G, Ugeda M M, Magaud L, Gómez-Rodríguez J M, Ynduráin F and Veuillen J-Y 2012 Publisher's note: unraveling the intrinsic and robust nature of van Hove singularities in twisted bilayer graphene by scanning tunneling microscopy and theoretical analysis [Phys. Rev. Lett. 109, 196802 (2012)] *Phys. Rev. Lett.* **109** 209905
[277] Meng L, Zhang Y, Yan W, Feng L, He L, Dou R-F and Nie J-C 2012 Single-layer behavior and slow carrier density dynamic of twisted graphene bilayer *Appl. Phys. Lett.* **100** 091601
[278] Yan W, Liu M, Dou R-F, Meng L, Feng L, Chu Z-D, Zhang Y, Liu Z, Nie J-C and He L 2012 Angle-dependent van Hove singularities in a slightly twisted graphene bilayer *Phys. Rev. Lett.* **109** 126801
[279] Brown L, Hovden R, Huang P, Wojcik M, Muller D A and Park J 2012 Twinning and twisting of tri- and bilayer graphene *Nano Lett.* **12** 1609–15
[280] Havener R W, Zhuang H, Brown L, Hennig R G and Park J 2012 Angle-resolved Raman imaging of interlayer rotations and interactions in twisted bilayer graphene *Nano Lett.* **12** 3162–7
[281] Righi A, Costa S D, Chacham H, Fantini C, Venezuela P, Magnuson C, Colombo L, Bacsa W S, Ruoff R S and Pimenta M A 2011 Graphene Moiré patterns observed by umklapp double-resonance Raman scattering *Phys. Rev.* B **84** 241409
[282] Righi A, Venezuela P, Chacham H, Costa S D, Fantini C, Ruoff R S, Colombo L, Bacsa W S and Pimenta M A 2013 Resonance Raman spectroscopy in twisted bilayer graphene *Solid State Commun.* **175-6** 13–7
[283] Robinson J T, Schmucker S W, Diaconescu C B, Long J P, Culbertson J C, Ohta T, Friedman A L and Beechem T E 2013 Electronic hybridization of large-area stacked graphene films *ACS Nano* **7** 637–44
[284] Liu J-B *et al* 2015 Observation of tunable electrical bandgap in large-area twisted bilayer graphene synthesized by chemical vapor deposition *Sci. Rep.* **5** 15285
[285] Poncharal P, Ayari A, Michel T and Sauvajol J-L 2008 Raman spectra of misoriented bilayer graphene *Phys. Rev.* B **78** 113407
[286] Ni Z, Wang Y, Yu T, You Y and Shen Z 2008 Reduction of Fermi velocity in folded graphene observed by resonance Raman spectroscopy *Phys. Rev.* B **77** 235403
[287] Carozo V *et al* 2013 Resonance effects on the Raman spectra of graphene superlattices *Phys. Rev.* B **88** 085401

[288] Othmen R, Arezki H, Ajlani H, Cavanna A, Boutchich M, Oueslati M and Madouri A 2015 Direct transfer and Raman characterization of twisted graphene bilayer *Appl. Phys. Lett.* **106** 103107
[289] Rong Z Y and Kuiper P 1993 Electronic effects in scanning tunneling microscopy: Moiré pattern on a graphite surface *Phys. Rev.* B **48** 17427–31
[290] Latil S, Meunier V and Henrard L 2007 Massless fermions in multilayer graphitic systems with misoriented layers: ab initio calculations and experimental fingerprints *Phys. Rev.* B **76** 201402
[291] Sprinkle M *et al* 2009 First direct observation of a nearly ideal graphene band structure *Phys. Rev. Lett.* **103** 226803
[292] Kim K S, Walter A L, Moreschini L, Seyller T, Horn K, Rotenberg E and Bostwick A 2013 Coexisting massive and massless Dirac fermions in symmetry-broken bilayer graphene *Nat. Mater.* **12** 887–92
[293] MacDonald A H 2019 Bilayer graphene's wicked, twisted road *Physics* **12** 12
[294] Balents L, Dean C R, Efetov D K and Young A F 2020 Superconductivity and strong correlations in moiré flat bands *Nat. Phys.* **16** 725–33
[295] Chen G *et al* 2019 Signatures of tunable superconductivity in a trilayer graphene moiré superlattice *Nature* **572** 215–9
[296] Yankowitz M, Chen S, Polshyn H, Zhang Y, Watanabe K, Taniguchi T, Graf D, Young A F and Dean C R 2019 Tuning superconductivity in twisted bilayer graphene *Science* **363** 1059–64
[297] Po H C, Zou L, Vishwanath A and Senthil T 2018 Origin of Mott insulating behavior and superconductivity in twisted bilayer graphene *Phys. Rev.* X **8** 031089
[298] Reina A, Jia X, Ho J, Nezich D, Son H, Bulovic V, Dresselhaus M S and Kong J 2009 Large area, few-layer graphene films on arbitrary substrates by chemical vapor deposition *Nano Lett.* **9** 30–5
[299] Reina A, Jia X, Ho J, Nezich D, Son H, Bulovic V, Dresselhaus M S and Kong J 2009 Layer area, few-layer graphene films on arbitrary substrates by chemical vapor deposition *Nano Lett.* **9** 3087
[300] San-Jose P, Gutiérrez-Rubio A, Sturla M and Guinea F 2014 Spontaneous strains and gap in graphene on boron nitride *Phys. Rev.* B **90** 075428
[301] Xue J, Sanchez-Yamagishi J, Bulmash D, Jacquod P, Deshpande A, Watanabe K, Taniguchi T, Jarillo-Herrero P and LeRoy B J 2011 Scanning tunnelling microscopy and spectroscopy of ultra-flat graphene on hexagonal boron nitride *Nat. Mater.* **10** 282–5
[302] Yankowitz M, Xue J, Cormode D, Sanchez-Yamagishi J D, Watanabe K, Taniguchi T, Jarillo-Herrero P, Jacquod P and LeRoy B J 2012 Emergence of superlattice Dirac points in graphene on hexagonal boron nitride *Nat. Phys.* **8** 382–6
[303] Ponomarenko L A *et al* 2013 Cloning of Dirac fermions in graphene superlattices *Nature* **497** 594–7
[304] Hunt B *et al* 2013 Massive Dirac fermions and Hofstadter butterfly in a van der Waals heterostructure *Science* **340** 1427–30
[305] Dean C R *et al* 2010 Boron nitride substrates for high-quality graphene electronics *Nat. Nanotechnol.* **5** 722–6
[306] Dean C R *et al* 2013 Hofstadter's butterfly and the fractal quantum Hall effect in moiré superlattices *Nature* **497** 598–602

[307] Yu G L *et al* 2014 Hierarchy of Hofstadter states and replica quantum Hall ferromagnetism in graphene superlattices *Nat. Phys.* **10** 525–9
[308] Yankowitz M, Xue J and LeRoy B J 2014 Graphene on hexagonal boron nitride *J. Phys.: Condens. Matter* **26** 303201
[309] Chen G *et al* 2019 Evidence of a gate-tunable Mott insulator in a trilayer graphene moiré superlattice *Nat. Phys.* **15** 237–41
[310] Hermann K 2012 Periodic overlayers and moiré patterns: theoretical studies of geometric properties *J. Phys.: Condens. Matter* **24** 314210
[311] Hermann K 2017 *Crystallography and Surface Structure: An Introduction for Surface Scientists and Nanoscientists* 2nd edn (Weinheim: Wiley)
[312] Katsnelson M I 2020 *The Physics of Graphene* 2nd edn (Cambridge: Cambridge University Press)
[313] Carlsson J M, Ghiringhelli L M and Fasolino A 2011 Theory and hierarchical calculations of the structure and energetics of [0001] tilt grain boundaries in graphene *Phys. Rev.* B **84** 165423
[314] Sato K, Saito R, Cong C, Yu T and Dresselhaus M S 2012 Zone folding effect in Raman G-band intensity of twisted bilayer graphene *Phys. Rev.* B **86** 125414
[315] Uchida K, Furuya S, Iwata J-I and Oshiyama A 2014 Atomic corrugation and electron localization due to Moiré patterns in twisted bilayer graphenes *Phys. Rev.* B **90** 155451
[316] Hirayama Y, Hirakawa K and Yamaguchi H (ed) 2022 *Quantum Hybrid Electronics and Materials* Quantum Science and Technology (Singapore: Springer)
[317] Weston A 2022 *Atomic and Electronic Properties of 2D Moiré Interfaces* Springer Theses (Cham: Springer)
[318] Pong W-T and Durkan C 2005 A review and outlook for an anomaly of scanning tunnelling microscopy (STM): superlattices on graphite *J. Phys. D: Appl. Phys.* **38** R329
[319] Latychevskaia T, Escher C and Fink H-W 2019 Moiré structures in twisted bilayer graphene studied by transmission electron microscopy *Ultramicroscopy* **197** 46–52
[320] Tian H, Ren C and Wang S 2022 Valleytronics in two-dimensional materials with line defect *Nanotechnology* **33** 212001
[321] Morozov S V, Novoselov K S, Katsnelson M I, Schedin F, Ponomarenko L A, Jiang D and Geim A K 2006 Strong suppression of weak localization in graphene *Phys. Rev. Lett.* **97** 016801
[322] Morpurgo A F and Guinea F 2006 Intervalley Scattering, long-range disorder, and effective time-reversal symmetry breaking in graphene *Phys. Rev. Lett.* **97** 196804
[323] Gorbachev R V, Tikhonenko F V, Mayorov A S, Horsell D W and Savchenko A K 2007 Weak localization in bilayer graphene *Phys. Rev. Lett.* **98** 176805
[324] Akhmerov A R and Beenakker C W J 2007 Detection of valley polarization in graphene by a superconducting contact *Phys. Rev. Lett.* **98** 157003
[325] Xiao D, Yao W and Niu Q 2007 Valley-contrasting physics in graphene: magnetic moment and topological transport *Phys. Rev. Lett.* **99** 236809
[326] Gunawan O, Shkolnikov Y P, Vakili K, Gokmen T, De Poortere E P and Shayegan M 2006 Valley susceptibility of an interacting two-dimensional electron system *Phys. Rev. Lett.* **97** 186404
[327] Ferry D K, Goodnick S M and Bird J 2009 *Transport in Nanostructures* 2nd edn (Cambridge: Cambridge University Press)
[328] van Houten H and Beenakker C 1996 Quantum point contacts *Phys. Today* **49** 22–7

[329] Peres N M R, Castro Neto A H and Guinea F 2006 Conductance quantization in mesoscopic graphene *Phys. Rev.* B **73** 195411

[330] Peres N M R, Castro Neto A H and Guinea F 2006 Erratum: Conductance quantization in mesoscopic graphene [Phys. Rev. B 73, 195411 (2006)] *Phys. Rev.* B **73** 239902

[331] Pereira J M, Mlinar V, Peeters F M and Vasilopoulos P 2006 Confined states and direction-dependent transmission in graphene quantum wells *Phys. Rev.* B **74** 045424

[332] Silvestrov P G and Efetov K B 2007 Quantum dots in graphene *Phys. Rev. Lett.* **98** 016802

[333] Handschin C, Makk P, Rickhaus P, Maurand R, Watanabe K, Taniguchi T, Richter K, Liu M-H and Schönenberger C 2017 Giant valley-isospin conductance oscillations in ballistic graphene *Nano Lett.* **17** 5389–93

[334] Rycerz A 2008 Nonequilibrium valley polarization in graphene nanoconstrictions *Phys. Stat. Sol. (a)* **205** 1281–9

[335] Zhang Z Z, Chang K and Chan K S 2008 Resonant tunneling through double-bended graphene nanoribbons *Appl. Phys. Lett.* **93** 062106

[336] Li M, Cai Z-L, Feng Z-B and Zhao Z-Y 2021 Valley-resolved transport in zigzag graphene nanoribbon junctions *Commun. Theor. Phys.* **73** 115701

[337] Chan K S 2018 Detection of valley currents in graphene nanoribbons *Phys. Lett.* A **382** 534–9

[338] Zhang Q and Chan K S 2021 Pure valley current generation in graphene nanostructure *Phys. Lett.* A **386** 126990

[339] Chan K S 2018 Valley dependent transport in graphene L junction *Physica E: Low Dimens. Syst. Nanostruct.* **99** 160–8

[340] Li R, Lin Z and Chan K S 2020 Valley polarized transport in graphene cross-junctions *Superlattices Microstruct.* **146** 106647

[341] Tsai W-F, Huang C-Y, Chang T-R, Lin H, Jeng H-T and Bansil A 2013 Gated silicene as a tunable source of nearly 100% spin-polarized electrons *Nat. Commun.* **4** 1500

[342] Wu X, Meng H, Zhang H, Bai Y and Xu X 2018 Valley-spin filtering through a nonmagnetic resonant tunneling structure in silicene *J. Phys. D: Appl. Phys.* **51** 265302

[343] Gut D, Prokop M, Sticlet D and Nowak M P 2020 Valley polarized current and resonant electronic transport in a nonuniform MoS_2 zigzag nanoribbon *Phys. Rev.* B **101** 085425

[344] Shen M, Zhang Y-Y, An X-T, Liu J-J and Li S-S 2014 Valley polarization in graphene-silicene-graphene heterojunction in zigzag nanoribbon *J. Appl. Phys.* **115** 233702

[345] Luo K, Zhou T and Chen W 2017 Probing the valley filtering effect by Andreev reflection in a zigzag graphene nanoribbon with a ballistic point contact *Phys. Rev.* B **96** 245414

[346] Beenakker C W J 2006 Specular Andreev reflection in graphene *Phys. Rev. Lett.* **97** 067007

IOP Publishing

Two-dimensional Valleytronic Materials
From principles to device applications
Sake Wang and Hongyu Tian

Chapter 2

Berry phase and valley Hall effect (VHE)

There are some properties of some forms of solid-state materials which are known to be robust to crystal defects and field noise. Scientists refer to these as topological properties, which are characterised by certain topological invariants. These invariants can be calculated and are inherent to the material, typically associated with the crystal symmetries and interactions [1–5].

This chapter delves into the intricate concepts of the Berry phase and the VHE, two phenomena that have become fundamental in the study of modern condensed matter physics. The Berry phase, a geometric phase acquired by a quantum system, plays a pivotal role in understanding various electronic properties, including polarization, orbital magnetism, and the anomalous Hall effect. A pedagogical example of the two-level system is used to illustrate the basic principles of the Berry phase and its geometric significance in quantum mechanics. Next, the chapter introduces electron dynamics under an electric field, leading to a discussion of the VHE in monolayer graphene and TMDCs, where electrons in different valleys deflect in opposite directions, driven by the Berry curvature. The concept of a wavepacket and its associated orbital magnetic moment is then explored. Finally, the chapter covers advanced topological concepts such as the Chern number and valley Chern number, which quantify the topological order in valley systems, and introduces topological kink states, which emerge at the interface between regions with different topological phases. These concepts underscore the deep connections between topology, Berry phase, and valley physics, and their implications for novel electronic phenomena in 2D materials.

2.1 Topology in solid state physics and basic concepts of the Berry phase

In 1980, it was discovered that a 2DEG at a semiconducting interface, exposed to a strong magnetic field ($\sim$15 T) at helium temperature, exhibits the quantum Hall effect (QHE) [6]. Due to the strong magnetic field, electrons in the bulk of the 2DEG

doi:10.1088/978-0-7503-5562-9ch2

were confined into set circular motions, making the 2DEG insulating in the bulk. Meanwhile, at the edges of the 2DEG, a quantised number of states were delocalised, and the 2DEG is conducting along the edges [7]. Further analysis of the band structure revealed that the number of edge state channels was correlated to the Chern number, a topological invariant of the occupied bands. Such behaviour is known as a topological insulator, and the associated edge states are robust against local perturbations [3, 8]. Since the discovery of the QHE, plentiful other topological properties of materials have been proposed and demonstrated. These include, for example, fractional QHE [9, 10], ultracold fermions [11], photonic crystals [12, 13], the quantum spin Hall effect/VHE [14, 15], a form of the QHE where the direction of the edge conductance is dependent on the spin/valley of the carrier, etc [16–18].

When investigating the topological properties of a material, it is useful to employ a property of the material known as the Berry phase with its associated Berry connection and Berry curvature [1, 19]. In quantum mechanics, when a Hamiltonian goes around a closed path in parameter space $\boldsymbol{R} = (R_1, R_2, R_3, \ldots)$ adiabatically, i.e.,

$$\begin{aligned} H &= H(\boldsymbol{R}), \\ \boldsymbol{R} &= \boldsymbol{R}(t), \end{aligned} \tag{2.1}$$

the wavefunction acquires a geometric phase called the Berry phase, which depends only on the path's geometry [20]. The Berry phase enclosed by a loop C can be expressed as a path integral in the parameter space [21]:

$$\gamma_n(C) = \int_C \mathrm{d}\boldsymbol{R} \cdot \mathscr{A}_n(\boldsymbol{R}). \tag{2.2}$$

Here, $n = 1, 2, 3, \ldots$is the band index, and $\mathscr{A}_n(\boldsymbol{R})$ is a vector-valued function named Berry connection,

$$\mathscr{A}_n(\boldsymbol{R}) = \mathrm{i}\langle \psi_n(\boldsymbol{R}) | \nabla | \psi_n(\boldsymbol{R}) \rangle, \tag{2.3}$$

with $|\psi_n(\boldsymbol{R})\rangle$ as an orthonormal basis function obtained from the eigenstates of $H(\boldsymbol{R})$. By applying Stokes' theorem[1], a new parameter defining the curvature of the Berry connection known as the Berry curvature, is formed:

$$\Omega_n(\boldsymbol{R}) = \boldsymbol{\nabla}_{\boldsymbol{R}} \times \mathscr{A}_n(\boldsymbol{R}). \tag{2.5}$$

The Berry connection and curvature behave like a pseudovector potential and pseudomagnetic field of the system, respectively, with the identity $\boldsymbol{\nabla}_{\boldsymbol{R}} \cdot \Omega_n(\boldsymbol{R}) = 0$ holding.

[1] Stokes' theorem or curl theorem: The line integral of a vector field $\boldsymbol{A}$ around a closed path C is equal to the surface integral of the curl of $\boldsymbol{A}$ on any surface whose only edge is C. Mathematically,

$$\oint_C \boldsymbol{A} \cdot \mathrm{d}\boldsymbol{l} = \int_S \nabla \times \boldsymbol{A} \cdot \mathrm{d}\boldsymbol{S}. \tag{2.4}$$

The direction of the normal to the infinitesimal area $\mathrm{d}\boldsymbol{S}$ of the surface S relates to the direction of integration around C by the right-hand rule.

2.2 Pedagogical example: the two-level system

Up to now, our discussions on the Berry phase, Berry curvature and their related concepts remain general. In this section, we consider a concrete yet simple example, a two-level system. The purpose of this study is threefold.

1. It visualizes the basic concepts and several important properties of the Berry phase.
2. The Hamiltonian describes several physical systems in condensed matter physics that have been discussed for their Berry phase effect. Examples include spin–orbit coupled systems [22, 23], spin 1/2 particle in a magnetic field [20, 24], and 2D systems [25, 26] examined in this book, such as equations (1.15), (1.25), etc [27–31].
3. We will repeatedly exploit the conclusions later in this book.

Therefore, it would be necessary to walk through this model.

2.2.1 Two-level system and gauge freedom

Consider the general Hamiltonian of a two-level system [1, 21, 32–35],

$$H = h_x\sigma_x + h_y\sigma_y + h_z\sigma_z = \boldsymbol{h} \cdot \boldsymbol{\sigma}, \tag{2.6}$$

where $\boldsymbol{\sigma}$ are Pauli matrices. Note the absence of a term proportional to identity matrix σ_0, which fails to affect adiabatic phases. By solving the Hamiltonian in equation (2.6), we get the eigenenergies $\pm|\boldsymbol{h}| = \pm h$ and the corresponding eigenstates [32],

$$|\psi_\pm\rangle = \frac{1}{\sqrt{2h(h \mp h_z)}}\begin{pmatrix} h_x - \mathrm{i}h_y \\ \pm h - h_z \end{pmatrix}. \tag{2.7}$$

Please note that we show the above eigenstates here, but we will use them only in later sections. In this section, we consider a practical representation of H as the Bloch sphere, as shown in figure 2.1. Under spherical polar coordinates, $\boldsymbol{h}$ reads

$$\boldsymbol{h} = |\boldsymbol{h}|(\sin\theta\cos\varphi, \sin\theta\sin\varphi, \cos\theta) \tag{2.8}$$

with its polar angle $\theta \in [0, \pi]$ and azimuthal angle $\varphi \in [0, 2\pi)$. The two eigenstates, with eigenenergies $|\boldsymbol{h}|$ and $-|\boldsymbol{h}|$, are

$$|\psi_+\rangle = \begin{pmatrix} \mathrm{e}^{-\mathrm{i}\frac{\varphi}{2}}\cos\frac{\theta}{2} \\ \mathrm{e}^{\mathrm{i}\frac{\varphi}{2}}\sin\frac{\theta}{2} \end{pmatrix},\ |\psi_-\rangle = \begin{pmatrix} \mathrm{e}^{-\mathrm{i}\frac{\varphi}{2}}\sin\frac{\theta}{2} \\ -\mathrm{e}^{\mathrm{i}\frac{\varphi}{2}}\cos\frac{\theta}{2} \end{pmatrix}, \tag{2.9}$$

respectively. We are free to multiply an arbitrary phase to these eigenstates. The choice shown in equation (2.9) is very symmetric, but the problem exists. If you consider that the Hamiltonian adiabatically precesses about the z-axis (a full circle in

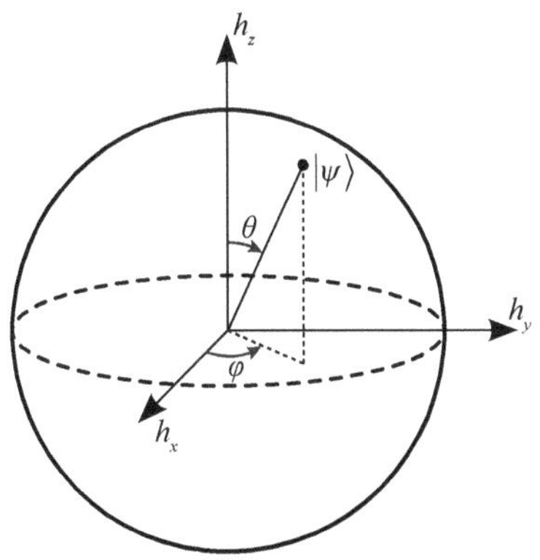

Figure 2.1. A practical graphical representation of a two-level Hamiltonian H is a Bloch sphere. The eigenenergies can be viewed as the distance of the point from the origin. The eigenstates depend on the direction of $\boldsymbol{h}$, i.e., on the polar angle $\theta \in [0, \pi]$ and azimuthal angle $\varphi \in [0, 2\pi)$, as in equation (2.9).

parameter space by keeping θ constant and let φ runs from 0 to 2π), we pick up a phase of π. This indicates that the current representation is not single-valued.

To fix this problem, it is possible to choose a different gauge simply by multiplying equation (2.9) by a factor, $e^{\pm i\frac{\varphi}{2}}$. Thus, we have

$$\left|\psi_+^{\mathrm{N}}\right\rangle = \begin{pmatrix} \cos\frac{\theta}{2} \\ e^{i\varphi}\sin\frac{\theta}{2} \end{pmatrix}, \left|\psi_-^{\mathrm{N}}\right\rangle = \begin{pmatrix} e^{-i\varphi}\sin\frac{\theta}{2} \\ -\cos\frac{\theta}{2} \end{pmatrix}. \tag{2.10}$$

The phase prefactor ± 1 seems could give us a single-valued representation. However, this choice of gauge makes the eigenstates not well-defined at the South Pole. At the South Pole, the two eigenstates read ${}^t(0, e^{i\varphi})$ and ${}^t(e^{-i\varphi}, 0)$ respectively, with undefined φ. In contrast, at the North Pole, they read ${}^t(1, 0)$ and ${}^t(0, -1)$ respectively, with no problem. This gauge is only problematic at the South Pole; hence, we add the superscript 'N'.

Our second attempt is to multiply equation (2.9) by another factor $e^{\mp i\frac{\varphi}{2}}$, and we have

$$\left|\psi_+^{\mathrm{S}}\right\rangle = \begin{pmatrix} e^{-i\varphi}\cos\frac{\theta}{2} \\ \sin\frac{\theta}{2} \end{pmatrix}, \left|\psi_-^{\mathrm{S}}\right\rangle = \begin{pmatrix} \sin\frac{\theta}{2} \\ -e^{i\varphi}\cos\frac{\theta}{2} \end{pmatrix}. \tag{2.11}$$

Compared with equation (2.10), the phase factor $e^{i\varphi}$ is now multiplied by $\cos\frac{\theta}{2}$ instead of $\sin\frac{\theta}{2}$. One can see that this representation is not well-defined at the North Pole, hence we use the superscript 'S'.

It is impossible to find a gauge that yields a smooth, nonsingular connection over the whole Bloch sphere, the singularity is unavoidable. Such singularity, can be moved but not removed.

2.2.2 Gauge-dependent Berry connections, gauge-independent Berry curvature and Chern number

We now calculate the Berry connection[2], which is a gauge-dependent object. e.g., for $|\psi_+\rangle$, we have

$$\nabla|\psi_+\rangle = \frac{1}{2h}\begin{pmatrix} -\,\mathrm{e}^{-\mathrm{i}\frac{\varphi}{2}}\sin\dfrac{\theta}{2} \\ \mathrm{e}^{\mathrm{i}\frac{\varphi}{2}}\cos\dfrac{\theta}{2} \end{pmatrix}\hat{\boldsymbol{\theta}} + \frac{\mathrm{i}}{2h\sin\theta}\begin{pmatrix} -\,\mathrm{e}^{-\mathrm{i}\frac{\varphi}{2}}\cos\dfrac{\theta}{2} \\ \mathrm{e}^{\mathrm{i}\frac{\varphi}{2}}\sin\dfrac{\theta}{2} \end{pmatrix}\hat{\boldsymbol{\varphi}}. \tag{2.13}$$

Hence the Berry connection

$$\mathscr{A}_+ = \mathrm{i}\langle\psi_+|\nabla|\psi_+\rangle = -\frac{1}{2h\sin\theta}\left(-\cos^2\frac{\theta}{2} + \sin^2\frac{\theta}{2}\right)\hat{\varphi} = \frac{\cos\theta}{2h\sin\theta}\hat{\varphi} = \frac{\cot\theta}{2h}\hat{\varphi}. \tag{2.14}$$

Similarly, the Berry connections for $|\psi_+^{\mathrm{N}}\rangle$ and $|\psi_+^{\mathrm{S}}\rangle$ read

$$\mathscr{A}_+^{\mathrm{N}} = \frac{\cos\theta - 1}{2h\sin\theta}\hat{\varphi} = -\frac{\tan\dfrac{\theta}{2}}{2h}\hat{\varphi},\ \mathscr{A}_+^{\mathrm{S}} = \frac{\cos\theta + 1}{2h\sin\theta}\hat{\varphi} = \frac{\cot\dfrac{\theta}{2}}{2h}\hat{\varphi}, \tag{2.15}$$

respectively. Next, the Berry curvature is obtained by taking the curl of the Berry connection[3]. We find it is a gauge-invariant quantity (thus we drop the subscripts 'N' and 'S') and reads

$$\boldsymbol{\Omega}_+ = \nabla\times\mathscr{A}_+ = \nabla\times\mathscr{A}_+^{\mathrm{N(S)}} = \frac{1}{h\sin\theta}\frac{\partial}{\partial\theta}\left(\sin\theta\frac{\cos\theta}{2h\sin\theta}\right)\frac{\boldsymbol{h}}{|\boldsymbol{h}|} = -\frac{\boldsymbol{h}}{2h^3}. \tag{2.17}$$

The Berry phase and curvature for the $|\psi_-\rangle$ state can also be derived likewise, as

$$\mathscr{A}_- = -\frac{\cot\theta}{2h}\hat{\varphi}, \tag{2.18}$$

$$\boldsymbol{\Omega}_- = \frac{\boldsymbol{h}}{2h^3}, \tag{2.19}$$

respectively. We recognize that Berry curvature $\boldsymbol{\Omega}_n$ is a field of a point-like monopole in the origin $\boldsymbol{h} = 0$, where the two eigenenergies become degenerate. Therefore, the origin acts as a source or drain of the Berry curvature flux. According to the definition

[2] Recall that in spherical coordinates, we have

$$\nabla f = \frac{\partial f}{\partial r}\hat{\boldsymbol{r}} + \frac{1}{r}\frac{\partial f}{\partial\theta}\hat{\boldsymbol{\theta}} + \frac{1}{r\sin\theta}\frac{\partial f}{\partial\varphi}\hat{\boldsymbol{\varphi}}. \tag{2.12}$$

[3] Also, in spherical coordinates, we have

$$\nabla\times\boldsymbol{A} = \frac{1}{r\sin\theta}\left[\frac{\partial}{\partial\theta}(A_\varphi\sin\theta) - \frac{\partial A_\theta}{\partial\varphi}\right]\hat{\boldsymbol{r}} + \frac{1}{r}\left[\frac{1}{\sin\theta}\frac{\partial A_r}{\partial\varphi} - \frac{\partial}{\partial r}(rA_\varphi)\right]\hat{\boldsymbol{\theta}} + \frac{1}{r}\left[\frac{\partial}{\partial r}(rA_\theta) - \frac{\partial A_r}{\partial\theta}\right]\hat{\boldsymbol{\varphi}}. \tag{2.16}$$

of the Berry phase, the Berry phase of a closed loop C is the flux of Berry curvature through the surface S whose boundary is C. Mathematically, it reads

$$\gamma_{\pm}(C) = \int_{S(C)} \boldsymbol{\Omega}_{\pm} \cdot \mathrm{d}\boldsymbol{S} = \mp\frac{1}{2}\int_{S(C)} \frac{\boldsymbol{h} \cdot \mathrm{d}\boldsymbol{S}}{h^3}. \tag{2.20}$$

From the definition of solid angle [36], it is straightforward to convince yourself that the Berry phase is half of the solid angle subtended by the curve C, Ω_C. Such a relationship reads

$$\gamma_{\pm}(C) = \mp\frac{\Omega_C}{2}. \tag{2.21}$$

Alternatively, we can use one-form expression for the Berry connection

$$\mathscr{A}_\theta = \mathrm{i}\langle\psi|\partial_\theta\psi\rangle,\ \mathscr{A}_\varphi = \mathrm{i}\langle\psi|\partial_\varphi\psi\rangle, \tag{2.22}$$

where $\partial_\theta = \frac{\partial}{\partial\theta}$ and $\partial_\varphi = \frac{\partial}{\partial\varphi}$. $\mathscr{A}_\theta$ and $\mathscr{A}_\varphi$ are real quantities, because $\partial_\theta(\langle\psi|\psi\rangle) = 0 = \langle\partial_\theta\psi|\psi\rangle = \langle\psi|\partial_\theta\psi\rangle$ and $(\langle\partial_\theta\psi|\psi\rangle)^* = \langle\psi|\partial_\theta\psi\rangle = -\langle\partial_\theta\psi|\psi\rangle$ result in $\langle\psi|\partial_\theta\psi\rangle$ and $\langle\psi|\partial_\varphi\psi\rangle$ being purely imaginary. From equation (2.22), the gauge-dependent Berry connection is given by

$$\begin{cases} \mathscr{A}_{-,\theta} = 0,\quad \mathscr{A}_{-,\varphi} = -\dfrac{\cos\theta}{2} & \text{for } |\psi_-\rangle, \\ \mathscr{A}^{\mathrm{N}}_{-,\theta} = 0,\quad \mathscr{A}^{\mathrm{N}}_{-,\varphi} = \sin^2\dfrac{\theta}{2} = \dfrac{1-\cos\theta}{2} & \text{for } \left|\psi^{\mathrm{N}}_-\right\rangle, \\ \mathscr{A}^{\mathrm{S}}_{-,\theta} = 0,\quad \mathscr{A}^{\mathrm{S}}_{-,\varphi} = -\cos^2\dfrac{\theta}{2} = -\dfrac{1+\cos\theta}{2} & \text{for } \left|\psi^{\mathrm{S}}_-\right\rangle. \end{cases} \tag{2.23}$$

The corresponding gauge-invariant Berry curvature is

$$\Omega_-^{\theta\varphi} = \partial_\theta\mathscr{A}_{-,\varphi} - \partial_\varphi\mathscr{A}_{-,\theta} = \partial_\theta\mathscr{A}^{\mathrm{N(S)}}_{-,\varphi} - \partial_\varphi\mathscr{A}^{\mathrm{N(S)}}_{-,\theta} = \frac{1}{2}\sin\theta. \tag{2.24}$$

The total flux integrated over the entire parameter space should be

$$\int_{S^2} \Omega_-^{\theta\varphi}\mathrm{d}\varphi\mathrm{d}\theta = \frac{1}{2}\int_0^{2\pi}\mathrm{d}\varphi\int_0^{\pi}\sin\theta\mathrm{d}\theta = 2\pi. \tag{2.25}$$

This can be seen as the integral of $\frac{1}{2}\sin\theta$ over the parameters (θ, φ), as we did not include the additional factor $\sin\theta$ for a surface element[4]. This is because the integration was not over the Bloch sphere but over a square $[0, \pi] \times [0, 2\pi)$. Now we use the Stokes' theorem to prove the results in equation (2.25). We define two subdomains in the Bloch sphere in figure 2.2: The caps N and S, representing

[4] Surface element in spherical coordinates:

$$\mathrm{d}S = r^2\sin\theta\mathrm{d}\theta\mathrm{d}\varphi. \tag{2.26}$$

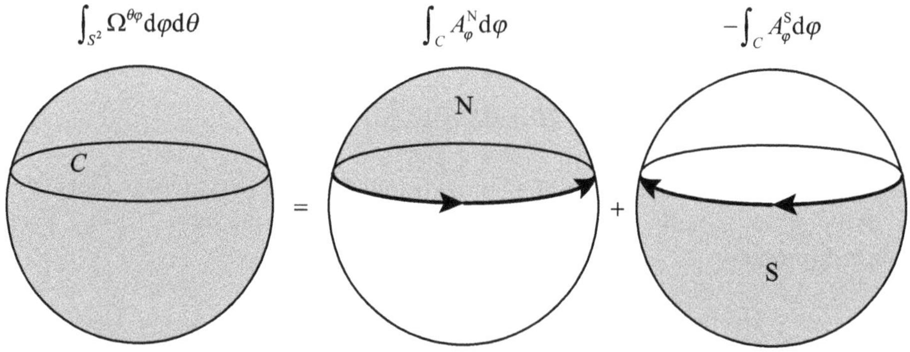

Figure 2.2. The total flux of the Berry curvature $\Omega^{\theta\varphi}$ through the whole Bloch sphere S^2 (left) is equal to the difference in circulations of the Berry connection $\mathscr{A}_\varphi^N$ and $\mathscr{A}_\varphi^S$ along a parallel C, i.e., $\int_{S^2} \Omega^{\theta\varphi} d\varphi d\theta = \oint_C \mathscr{A}_\varphi^N d\varphi - \oint_C \mathscr{A}_\varphi^S d\varphi$.

regions to the north and south of the parallel C with fixed θ, respectively. Within the north cap N, because $\mathscr{A}_{-,\varphi}^N$ is well-defined over N, we can apply Stokes' theorem,

$$\int_N \Omega_-^{\theta\varphi} d\varphi d\theta = \oint_C \mathscr{A}_{-,\varphi}^N d\varphi = 2\pi \sin^2 \frac{\theta}{2}. \tag{2.27}$$

Within the south cap S, we can similarly utilise Stokes' theorem,

$$\int_S \Omega_-^{\theta\varphi} d\varphi d\theta = -\oint_C \mathscr{A}_{-,\varphi}^S d\varphi = 2\pi \cos^2 \frac{\theta}{2}. \tag{2.28}$$

Notice that the minus sign is because of the orientation of the circle C. Finally, the total flux through the north and south caps is

$$\int_{S^2} \Omega_-^{\theta\varphi} d\varphi d\theta = \int_N \Omega_-^{\theta\varphi} d\varphi d\theta + \int_S \Omega_-^{\theta\varphi} d\varphi d\theta = 2\pi. \tag{2.29}$$

This is consistent with the evaluation given in equation (2.25). We can also verify that the total flux of Berry curvature $\Omega_+^{\theta\varphi}$ associated with the $|\psi_+\rangle$ integrated over the entire parameter space is -2π. In general, it shows that the total flux of the Berry curvature $\Omega_\pm^{\theta\varphi}$ through the whole Bloch sphere S^2 is equal to the difference of circulations of the Berry connection of two distinct gauges along a closed path, which equals a multiple of 2π. This allows us to define a physical quantity, which is the total flux of Berry curvature in units of 2π. We call it Chern number $\mathcal{C}$, in our case,

$$\mathcal{C}_\pm = \frac{1}{2\pi} \int_{S^2} \Omega_\pm^{\theta\varphi} d\varphi d\theta = \mp 1. \tag{2.30}$$

The value and sign of the Chern number are the fingerprints of different physical behaviours associated with topological materials.

2.3 Electron dynamics in an electric field

2.3.1 Anomalous velocity

The dynamics of Bloch electrons in the presence of an electric field is one of the oldest problems in solid-state physics [37]. The semiclassical model links eigenenergy $\mathcal{E}_n(\boldsymbol{k})$ to electron properties like position $\boldsymbol{r}$, wave vector $\boldsymbol{k}$, and band index n. In the presence of electric field $\boldsymbol{E}$, the time evolution of the position of an electron (electron velocity) with band index n is determined by the equation of motion [37]

$$v_n(\boldsymbol{k}) = \frac{1}{\hbar}\partial_k \mathcal{E}_n(\boldsymbol{k}). \tag{2.31}$$

Recent progress on the semiclassical dynamics of Bloch electrons shows that this description is incomplete since the presence of Berry curvature (magnetic field in momentum space) can modify the electron dynamics. In particular, in the presence of an electric field, an electron can acquire an anomalous velocity v_a proportional to the Berry curvature of the band [21, 38–40]. The corrected velocity becomes

$$v_n(\boldsymbol{k}) = \frac{1}{\hbar}\partial_k \mathcal{E}_n(\boldsymbol{k}) + v_\mathrm{a} = \frac{1}{\hbar}\partial_k \mathcal{E}_n(\boldsymbol{k}) - \frac{e}{\hbar}\boldsymbol{E} \times \boldsymbol{\Omega}_n(\boldsymbol{k}), \tag{2.32}$$

where $\boldsymbol{\Omega}_n(\boldsymbol{k})$ is the Berry curvature of the nth band[5]

$$\begin{aligned} \boldsymbol{\Omega}_n(\boldsymbol{k}) &= \nabla_k \times \langle \psi_n(\boldsymbol{k}) | \mathrm{i}\nabla_k | \psi_n(\boldsymbol{k}) \rangle \\ &= \mathrm{i}\langle \nabla_k \psi_n(\boldsymbol{k}) | \times | \nabla_k \psi_n(\boldsymbol{k}) \rangle. \end{aligned} \tag{2.33}$$

In 2D systems with parameter space (k_x, k_y), where this book focuses, the only surviving component of the Berry curvature is the z-component [41] [6]

$$\begin{aligned} \Omega_n(\boldsymbol{k}) &= \mathrm{i}\left[\left\langle \partial_{k_x}\psi_n(\boldsymbol{k}) \middle| \partial_{k_y}\psi_n(\boldsymbol{k}) \right\rangle - \left\langle \partial_{k_y}\psi_n(\boldsymbol{k}) \middle| \partial_{k_x}\psi_n(\boldsymbol{k}) \right\rangle \right] \\ &= -2\mathrm{Im}\left\langle \partial_{k_x}\psi_n(\boldsymbol{k}) \middle| \partial_{k_y}\psi_n(\boldsymbol{k}) \right\rangle. \end{aligned} \tag{2.34}$$

We can see that, in equation (2.32), the first term describes the usual band dispersion contribution, and the second term, in analogy with the magnetic field $\boldsymbol{B}$, which changes the motion of an electron in real space [39],

$$\dot{\boldsymbol{k}} = -\frac{e}{\hbar}\boldsymbol{E} - \frac{e}{\hbar}\dot{\boldsymbol{r}} \times \boldsymbol{B}. \tag{2.35}$$

Here, a single dot over a variable indicates the first derivative with respect to time. Therefore, we can view the Berry curvature in equation (2.32) as a momentum-space magnetic field [21, 39], which also changes the motion of an electron. Therefore, Berry curvature can be treated as a pseudo-magnetic field in momentum space [21]. The direction of Berry curvature is always transverse to the electric field $\boldsymbol{E}$, thus

---

[5] We use the vector identities $\nabla \times (f\boldsymbol{A}) = f(\nabla \times \boldsymbol{A}) + (\nabla f) \times \boldsymbol{A}$ and $\nabla \times (\nabla f) = 0$.

[6] We use the fact that if $(\boldsymbol{i}, \boldsymbol{j}, \boldsymbol{k})$ is a positively oriented orthonormal basis, the basis vectors satisfy $\boldsymbol{i} \times \boldsymbol{i} = \boldsymbol{j} \times \boldsymbol{j} = 0$, $\boldsymbol{i} \times \boldsymbol{j} = \boldsymbol{k}$, and the anticommutativity of the cross product, $\boldsymbol{j} \times \boldsymbol{i} = -\boldsymbol{k}$.

leading to Hall current, with Hall conductivity given as an integral of the Berry curvature [41–44],

$$\sigma_{\mathrm{H}} = g_{\mathrm{s}}\frac{e^2}{\hbar}\int \frac{\mathrm{d}^2k}{(2\pi)^2} f_{\mathrm{FD}}(\boldsymbol{k})\Omega_n(\boldsymbol{k}), \tag{2.36}$$

where g_s is the spin degeneracy (usually $g_s = 2$) and $f_{\mathrm{FD}}(\boldsymbol{k})$ is the Fermi–Dirac distribution function. Historically, even though the anomalous velocity has been obtained in the 1950s [45, 46], its relation to the Berry phase came 40 years later [38–40].

2.3.2 Berry curvature: symmetry considerations

Now we will discuss the conditions under which we cannot neglect the Berry curvature term. Following symmetry analysis, we can have a few useful properties of the Berry curvature [47].

1. If the system has TRS, flipping the signs of $v_n(\boldsymbol{k})$ and $\boldsymbol{k}$ while fixing $\boldsymbol{E}$ in equation (2.32) requires

$$\Omega_n(\boldsymbol{k}) = -\Omega_n(-\boldsymbol{k}). \tag{2.37}$$

2. If the system has spatial IS, $v_n(\boldsymbol{k})$, $\boldsymbol{k}$ and $\boldsymbol{E}$ change sign. In this case, we find

$$\Omega_n(\boldsymbol{k}) = \Omega_n(-\boldsymbol{k}). \tag{2.38}$$

Therefore, in the presence of both TRS and IS, $\Omega_n(\boldsymbol{k}) \equiv 0$ throughout the BZ, and equation (2.32) reduces to (2.31). However, in the system where both symmetries do not coexist, equation (2.32) is the proper description of motion due to nonzero Berry curvature. For broken TRS, one can find it in crystals with ferromagnetic ordering. Such as ferromagnetic bcc Fe [48] and $SrRuO_3$ [49]. However, one of the most interesting possibilities of having a Hall effect, is the chance which does not rely on magnetic field or magnetic ordering, but simply broken IS.

2.4 VHE in graphene

2.4.1 TB model

Let us examine the case of graphene with broken IS [50] by introducing staggered sublattice potential Semenoff mass [25], which modifies the Bloch Hamiltonian matrix in equation (1.15) to [51],

$$H_{\mathrm{S}}(\boldsymbol{k}) = \begin{pmatrix} \dfrac{\Delta_z}{2} & -\gamma_0 f(\boldsymbol{k}) \\ -\gamma_0 f(\boldsymbol{k})^* & -\dfrac{\Delta_z}{2} \end{pmatrix}, \tag{2.39}$$

with $f(\boldsymbol{k})$ being defined by equation (1.13). Notice that the diagonal terms indicate site energy difference Δ_z between sublattices and explicitly break the IS. This

Hamiltonian can also be utilized to describe hBN, which has boron on the A atom and nitrogen on the B atom [52–57]. The energy spectrum can be obtained by diagonalizing Hamiltonian (2.39),

$$\mathcal{E}_\xi(\boldsymbol{k}) = \xi\sqrt{\gamma_0^2\,|f(\boldsymbol{k})|^2 + \left(\frac{\Delta_z}{2}\right)^2}, \tag{2.40}$$

where $\xi = 1\,(-1)$ corresponds to the conduction (valence) band. As we can see, the Semenoff mass generates a gap, Δ_z, at the Dirac points. The observed gap, which typically ranges from 10 to several tens of meV [58], is present in a graphene monolayer on top of SiC ($\sim$0.26 eV [59]), graphite [60, 61], and hBN with long-wavelength moiré superlattice forms ($\sim$53 meV [62, 63]) [64–66]. A monolayer TMDC is also naturally analog to IS-broken graphene [67, 68]. We can readily compute the quantities of interest, Berry curvature [69], by employing the azimuthal $\varphi(\boldsymbol{k})$ and polar angles $\theta(\boldsymbol{k})$ on the Bloch sphere, where

$$\begin{aligned} \cos\varphi(\boldsymbol{k}) &= \frac{\Delta_z/2}{|\mathcal{E}_\xi(\boldsymbol{k})|}, \\ \sin\varphi(\boldsymbol{k}) &= -\frac{\gamma_0|f(\boldsymbol{k})|}{|\mathcal{E}_\xi(\boldsymbol{k})|}, \\ \theta(\boldsymbol{k}) &= \mathrm{Arg}[f(\boldsymbol{k})]. \end{aligned} \tag{2.41}$$

Thus, we can write the Hamiltonian (2.39) as

$$H_S(\boldsymbol{k}) = |\mathcal{E}_\xi(\boldsymbol{k})|\begin{pmatrix} \cos\varphi(\boldsymbol{k}) & \sin\varphi(\boldsymbol{k})\mathrm{e}^{\mathrm{i}\theta(\boldsymbol{k})} \\ \sin\varphi(\boldsymbol{k})\mathrm{e}^{-\mathrm{i}\theta(\boldsymbol{k})} & -\cos\varphi(\boldsymbol{k}) \end{pmatrix}. \tag{2.42}$$

The two eigenstates $|\psi_\xi\rangle$ of energies $\mathcal{E}_\xi(\boldsymbol{k}) = \xi|\mathcal{E}_\xi(\boldsymbol{k})|$ with $\xi = +1$ and -1 are

$$|\psi_+(\boldsymbol{k})\rangle = \begin{pmatrix} \cos\dfrac{\varphi(\boldsymbol{k})}{2} \\ \mathrm{e}^{-\mathrm{i}\theta(\boldsymbol{k})}\sin\dfrac{\varphi(\boldsymbol{k})}{2} \end{pmatrix}, \tag{2.43}$$

$$|\psi_-(\boldsymbol{k})\rangle = \begin{pmatrix} -\mathrm{e}^{\mathrm{i}\theta(\boldsymbol{k})}\sin\dfrac{\varphi(\boldsymbol{k})}{2} \\ \cos\dfrac{\varphi(\boldsymbol{k})}{2} \end{pmatrix}. \tag{2.44}$$

The Berry connection in band ξ is given by

$$\mathscr{A}_\xi(\boldsymbol{k}) = \mathrm{i}\langle\psi_\xi(\boldsymbol{k})|\nabla_{\boldsymbol{k}}\psi_\xi(\boldsymbol{k})\rangle = \xi\sin^2\frac{\varphi(\boldsymbol{k})}{2}\nabla_{\boldsymbol{k}}\theta(\boldsymbol{k}), \tag{2.45}$$

and the corresponding Berry curvature reads

$$\begin{aligned}\boldsymbol{\Omega}_\xi(\boldsymbol{k}) &= \nabla_{\boldsymbol{k}} \times \mathscr{A}_\xi(\boldsymbol{k}) \\ &= [\partial_{k_x}\mathscr{A}_y(\boldsymbol{k}) - \partial_{k_y}\mathscr{A}_x(\boldsymbol{k})]\hat{\boldsymbol{z}} \\ &= \frac{\xi}{2}\sin\varphi(\boldsymbol{k})[\partial_{k_x}\varphi(\boldsymbol{k})\partial_{k_y}\theta(\boldsymbol{k}) - \partial_{k_x}\theta(\boldsymbol{k})\partial_{k_y}\varphi(\boldsymbol{k})]\hat{\boldsymbol{z}}.\end{aligned} \tag{2.46}$$

After substituting equation (2.41) into (2.46) and doing some algebra, we obtain

$$\begin{aligned}\Omega_\xi(\boldsymbol{k}) &= \frac{\xi a^2\gamma_0^2\Delta_z}{4\sqrt{3}\ |\mathcal{E}_\xi(\boldsymbol{k})|^3}\sin\frac{k_y a}{2}\left(\cos\frac{\sqrt{3}k_x a}{2} - \cos\frac{k_y a}{2}\right) \\ &= \frac{\xi a^2\gamma_0^2\Delta_z}{2\sqrt{3}\ |\mathcal{E}_\xi(\boldsymbol{k})|^3}\sin\left(\boldsymbol{k}\cdot\frac{-\boldsymbol{a}_1}{2}\right)\sin\left(\boldsymbol{k}\cdot\frac{\boldsymbol{a}_2}{2}\right)\sin\left(\boldsymbol{k}\cdot\frac{\boldsymbol{a}_1-\boldsymbol{a}_2}{2}\right).\end{aligned} \tag{2.47}$$

Notice that the Berry curvature has both the C_3 rotational symmetry and translational symmetry.

In figure 2.3, we plot the Berry curvature $\Omega_+(\boldsymbol{k})$ (in units of a^2) in the conduction band as a function of the wavevector $\boldsymbol{k}$ (in units of $2\pi/a$). As we can see, the TRS ensures the satisfaction of equation (2.37) and the Berry curvatures in the two valleys have opposite signs. The integral of Berry curvature over the entire BZ vanishes. Once we reveal the structure of the Berry curvature, we can easily see the opposite sign of anomalous velocity in two valleys in response to the in-plane electric field $\boldsymbol{E}$. The opposite flow of carriers belonging to different valleys, perpendicular to $\boldsymbol{E}$, therefore emerges. This phenomenon is called the VHE [70].

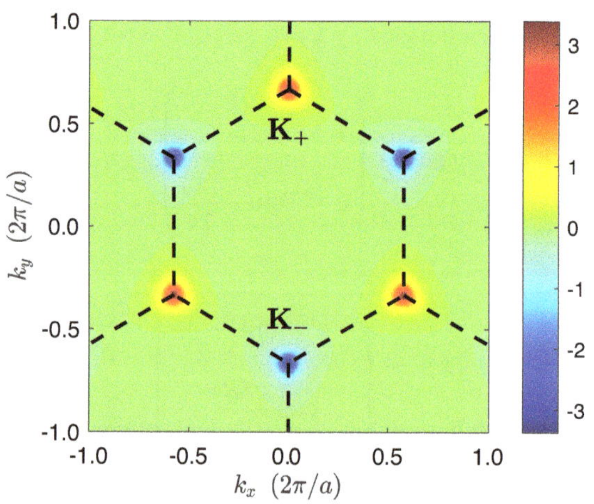

Figure 2.3. Berry curvature $\Omega_+(\boldsymbol{k})$ (in units of a^2) in the conduction band as a function of the wavevector $\boldsymbol{k} = (k_x, k_y)$ (in units of $2\pi/a$) over the hexagonal BZ (black dashed line) for $\gamma_0 = 3$ eV and $\Delta_z = 2.5$ eV. Δ_z is set to be much larger than experimental values to give a clearer view of the plot since the Berry curvature has a sharper peak at $\boldsymbol{K}_\pm$ as Δ_z decreases [50].

2.4.2 Low-energy model (massive 2D Dirac fermions)

Now, for pedagogical purposes and physical transparency, we present the effective Hamiltonian close to the $\boldsymbol{K}_{\pm}$ point derived from Hamiltonian (2.39). It is the simplest Hamiltonian which describes the IS broken system and was first proposed by Xiao *et al* [71] to manifest the valley-contrasting Berry curvature. The Hamiltonian reads

$$H_{\mathrm{S}}^{\kappa} = \frac{\sqrt{3}}{2}\gamma_0 a(q_x\sigma_x - \kappa q_y\sigma_y) + \frac{\Delta_z}{2}\sigma_z, \tag{2.48}$$

where the first term is the Hamiltonian of pristine graphene as given in equation (1.25), with the valley index κ, i.e., $\kappa = -1$ for $\boldsymbol{K}_{-}$ and $\kappa = 1$ for $\boldsymbol{K}_{+}$. The second term is the IS-breaking term, the Semenoff mass [25]. This Hamiltonian of gapped graphene is akin to the effective Hamiltonian of monolayer TMDC without SOC shown in equation (1.48) because they share similar symmetry properties.

Since equation (2.48) is a Hamiltonian for a two-level system, it is practicable to rewrite it as [20, 72]

$$H_{\mathrm{S}}^{\kappa} = \frac{\sqrt{3}}{2}\gamma_0 a\boldsymbol{h}\cdot\boldsymbol{\sigma}, \tag{2.49}$$

with

$$\boldsymbol{h} = \left(q_x, -\kappa q_y, \frac{\Delta_z}{\sqrt{3}\gamma_0 a}\right). \tag{2.50}$$

We can treat it as a Bloch sphere of a pseudospin defined by the occupation of the two graphene sublattices, and we can apply the results in section 2.2.2 to understand qualitatively the behaviour of the Berry curvature and the Berry phase in graphene. The Berry phase can be visualised in figure 2.4. When an electron moves around a circle in momentum space around $\boldsymbol{q} = 0$ for a fixed $\frac{\Delta_z}{\sqrt{3}\gamma_0 a}$, the Berry phase accumulated by the two-component wavefunction is equal to half of the solid angle

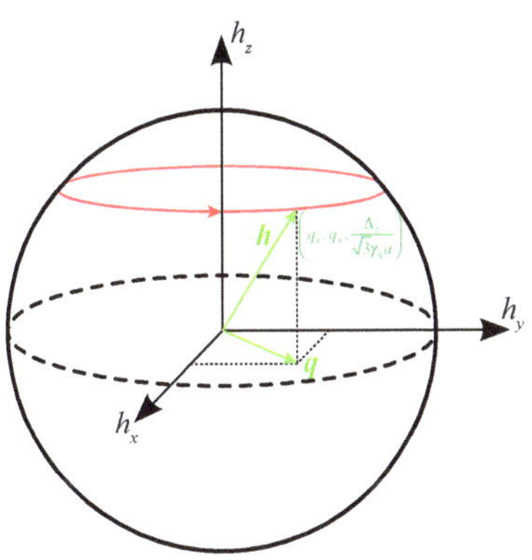

Figure 2.4. Fictitious field $\boldsymbol{h}$, as defined in equation (2.50), couples to sublattice pseudospin. A Berry phase is acquired when an electron undergoes an adiabatic rotation along the red circle.

subtended by the circle which the electron traces. In addition, because of the opposite sign of κ in two valleys in the y-component of $\boldsymbol{h}$, it is easy to see that the Berry phase acquires an opposite sign at the two valleys, rendering them distinguishable using the Berry phase [70, 72].

We emphasise that for $\Delta_z = 0$, the circle traced out by electron would be the equator of the Bloch sphere, and because of equation (2.21), the Berry phases at the two valleys would be $\pm\pi$, which are indistinguishable because they differ by modulo 2π. On the other hand, for $\Delta_z \neq 0$, the Berry phases at the two valleys become distinguishable, which is consistent with our investigation in the symmetry analysis.

Now we explicitly calculate the Berry curvature in the present system. The spectrum is

$$\mathcal{E}_{\pm}(\boldsymbol{q}) = \pm\sqrt{\left(\frac{\sqrt{3}}{2}\gamma_0 qa\right)^2 + \left(\frac{\Delta_z}{2}\right)^2}, \tag{2.51}$$

with a bandgap $|\Delta_z|$ at the $\boldsymbol{K}_{\pm}$ point; and the wave function of the conduction band is given by

$$\psi_{+}(\boldsymbol{q}) = \frac{1}{\sqrt{Q(\boldsymbol{q})}}\begin{pmatrix} \frac{\sqrt{3}}{2}\gamma_0 a(q_x + \mathrm{i}\kappa q_y) \\ |\mathcal{E}_{\pm}(\boldsymbol{q})| - \frac{\Delta_z}{2} \end{pmatrix}, \tag{2.52}$$

with $Q(\boldsymbol{q}) = \left(\frac{\sqrt{3}}{2}\gamma_0 qa\right)^2 + \left[|\mathcal{E}_{\pm}(\boldsymbol{q})| - \frac{\Delta_z}{2}\right]^2$. The Berry connection can be calculated as

$$\begin{aligned} \mathscr{A}_x(\boldsymbol{q}) &= \mathrm{i}\langle \psi_{+}(\boldsymbol{q})|\partial_{k_x}|\psi_{+}(\boldsymbol{q})\rangle = \frac{\kappa q_y}{2} g(\boldsymbol{q}), \\ \mathscr{A}_y(\boldsymbol{q}) &= \mathrm{i}\langle \psi_{+}(\boldsymbol{q})|\partial_{k_y}|\psi_{+}(\boldsymbol{q})\rangle = -\frac{\kappa q_x}{2} g(\boldsymbol{q}), \end{aligned} \tag{2.53}$$

with

$$g(\boldsymbol{q}) = \frac{\left(\frac{\sqrt{3}}{2}\gamma_0 a\right)^2}{\left(\frac{\sqrt{3}}{2}\gamma_0 qa\right)^2 - \frac{\Delta_z}{2}|\mathcal{E}_{\pm}(\boldsymbol{q})| + \left(\frac{\Delta_z}{2}\right)^2}. \tag{2.54}$$

Consequently, it is straightforward to obtain Berry curvature for the conduction band [71],

$$\Omega_{+}(\boldsymbol{q}) = (\partial_{k_x}\mathscr{A}_y - \partial_{k_y}\mathscr{A}_x)\hat{\boldsymbol{z}} = \kappa\frac{3\gamma_0^2 a^2 \Delta_z}{16\,|\mathcal{E}_{\pm}(\boldsymbol{q})|^3}\hat{\boldsymbol{z}} = \kappa\frac{3\gamma_0^2 a^2 \Delta_z}{2\left(3\gamma_0^2 q^2 a^2 + \Delta_z^2\right)^{3/2}}\hat{\boldsymbol{z}}. \tag{2.55}$$

The result is consistent with the one shown in figure 2.3, in which the sign of Berry curvature is valley-dependent. So, the total integral of Berry curvature is zero.

Indeed, one can easily prove that equation (2.55) can be derived by expanding (2.47) in the vicinity of $\boldsymbol{K}_{\pm}$. As a result of equations (2.32) and (2.55), electrons from the two valleys will acquire opposite anomalous velocity v_{a} perpendicular to $\boldsymbol{E}$ and sail in opposite directions, giving rise to VHE, as shown in figure 2.5. Equation (2.55) also shows that $\Omega_{+}(\boldsymbol{q})$ decays rapidly with increasing energy [64]. In the limit $\Delta_z \to 0$, we recover that the vanishing Berry curvature everywhere except at $\boldsymbol{q} = 0$. When $\boldsymbol{q} = 0$ and $\Delta_z \to 0$, the sign of Berry curvature (even for a given valley) is not well defined [73]. Similarly, we can find that the Berry curvature in the valence band has the same magnitude as the conduction band, but with an opposite sign [74, 75]. It reads

$$\boldsymbol{\Omega}_{-}(\boldsymbol{q}) = -\kappa\frac{3\gamma_0^2a^2\Delta_z}{16\,|\mathcal{E}_{\pm}(\boldsymbol{q})|^3}\hat{\boldsymbol{z}} = -\kappa\frac{3\gamma_0^2a^2\Delta_z}{2\left(3\gamma_0^2q^2a^2 + \Delta_z^2\right)^{3/2}}\hat{\boldsymbol{z}}. \tag{2.56}$$

Ignoring skew scattering, intervalley scattering [76], transforming coordinates from q to $\mathcal{E}$ and noticing $\mathcal{E}_{\pm}(\boldsymbol{q})\mathrm{d}\mathcal{E}_{\pm}(\boldsymbol{q}) = \frac{3}{4}\gamma_0^2a^2q\mathrm{d}q$, we find a valley-dependent Hall conductivity based on equation (2.36) at zero kelvin [77–79],

$$\sigma_{\mathrm{H}}(\kappa, \mathcal{E}_{\mathrm{F}}) = \begin{cases} -\kappa\dfrac{e^2}{h}\dfrac{\Delta_z}{2|\mathcal{E}_{\mathrm{F}}|} & \text{if } |\mathcal{E}_{\mathrm{F}}| > \dfrac{\Delta_z}{2} \\ -\kappa\dfrac{e^2}{h} & \text{if } |\mathcal{E}_{\mathrm{F}}| \leqslant \dfrac{\Delta_z}{2}. \end{cases} \tag{2.57}$$

We note that the $\sigma_{\mathrm{H}}(\kappa, \mathcal{E}_{\mathrm{F}})$ for $|\mathcal{E}_{\mathrm{F}}| > \frac{\Delta_z}{2}$ and $|\mathcal{E}_{\mathrm{F}}| \leqslant \frac{\Delta_z}{2}$ is continuous at $|\mathcal{E}_{\mathrm{F}}| = \frac{\Delta_z}{2}$, although the derivative is discontinuous [42]. The two opposite valleys, $\boldsymbol{K}_{+}$ and $\boldsymbol{K}_{-}$, have opposite but the same magnitude of Hall conductivity. Therefore, the summation of the Hall conductivity of both valleys vanishes. In addition, when the Fermi energy $\mathcal{E}_{\mathrm{F}}$ is situated inside the bandgap, such that the Berry curvature peak is totally covered by occupied states, the valley-dependent Hall conductivity becomes maximum and appears as a quantised quantity.

Once the valley-dependent Hall conductivity emerges, we can define the valley Hall conductivity [42, 70],

$$\sigma_{\mathrm{v}}(\mathcal{E}_{\mathrm{F}}) \equiv \sigma_{\mathrm{H}}(\kappa = 1, \mathcal{E}_{\mathrm{F}}) - \sigma_{\mathrm{H}}(\kappa = -1, \mathcal{E}_{\mathrm{F}}) \tag{2.58}$$

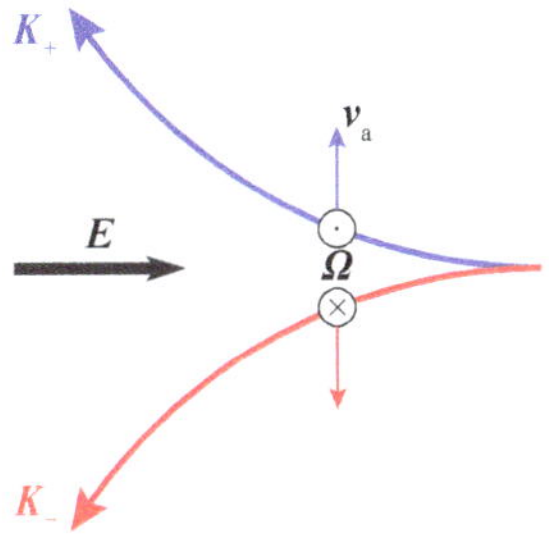

Figure 2.5. VHE. Electrons from two opposite valleys, $\boldsymbol{K}_{+}$ and $\boldsymbol{K}_{-}$, sail in opposite directions, because they acquire opposite anomalous velocity v_{a} perpendicular to the electric field $\boldsymbol{E}$ due to opposite Berry curvature $\boldsymbol{\Omega}$.

and we find

$$\sigma_{\mathrm{v}}(\mathcal{E}_{\mathrm{F}}) = \begin{cases} -\dfrac{e^2}{h}\dfrac{\Delta_z}{|\mathcal{E}_{\mathrm{F}}|} & \text{if } |\mathcal{E}_{\mathrm{F}}| > \dfrac{\Delta_z}{2} \\ -\dfrac{2e^2}{h} & \text{if } |\mathcal{E}_{\mathrm{F}}| \leqslant \dfrac{\Delta_z}{2}. \end{cases} \tag{2.59}$$

The VHE allows the nondissipative transfer of a voltage signal via the pure valley current since the valley current is driven without the flow of charge. Utilising valley current in logic calculation has the advantage of very low power consumption.

VHE has been predicted [67, 71] and also detected in various materials such as monolayer and bilayer graphene [80, 81], MoS_2 [82–84], with both transport measurements and optical methods.

2.4.3 Non-local measurement for detection of VHE

VHE arises from the finite Berry curvature arising from the lack of IS in systems possessing valley DOF [71] and can be regarded as an intrinsic effect without extrinsic contributions [64]. The prediction of VHE [71] was inspired by weak localization experiments showing suppression of intervalley scattering [85] and demonstration of the opening of a gap in graphene due to a substrate potential [59].

VHE could be understood to be similar to spin Hall effect through an analogy between valley and spin DOF. Once we have both the electrical conductivity σ_{xx} and valley Hall conductivity σ_{v}, we can formulate the charge and valley current in the presence of an in-plane electric field semi-classically [42, 70, 80, 86]. The following rank-two conductivity tensor allows us to examine the conversion between the electric field $\boldsymbol{E}$ (in the y-direction) and the electric current density $\boldsymbol{j}$, as well as between $\boldsymbol{E}$ and valley current density $\boldsymbol{j}_{\mathrm{v}}$ (in the x-direction), on a homogenous sheet:

$$\begin{pmatrix} j \\ j_{\mathrm{v}} \end{pmatrix} = \begin{pmatrix} \sigma_{xx} & -\sigma_{\mathrm{v}} \\ \sigma_{\mathrm{v}} & \sigma_{xx} \end{pmatrix} \begin{pmatrix} E \\ E_{\mathrm{v}} \end{pmatrix}, \tag{2.60}$$

where

$$E_{\mathrm{v}} = \frac{1}{2e}\boldsymbol{\nabla}\delta\mu_{\mathrm{v}} = \frac{1}{2e}\partial_x\delta\mu_{\mathrm{v}}, \tag{2.61}$$

is the 'field' induced by the chemical potential difference $\delta\mu_{\mathrm{v}}$ between the two valleys referred to as the 'valley voltage' because the imbalance in valley population. Taking $E_{\mathrm{v}} = 0$, the VHE reads

$$j_{\mathrm{v}} = \sigma_{\mathrm{v}} E, \tag{2.62}$$

as the description of the conversion of the electric field into the valley current. On the other hand, by taking the inverse of the conductivity tensor in equation (2.60), we have

$$\begin{pmatrix} E \\ E_{\mathrm{v}} \end{pmatrix} = \frac{1}{\sigma_{xx}^2 + \sigma_{\mathrm{v}}^2} \begin{pmatrix} \sigma_{xx} & \sigma_{\mathrm{v}} \\ -\sigma_{\mathrm{v}} & \sigma_{xx} \end{pmatrix} \begin{pmatrix} j \\ j_{\mathrm{v}} \end{pmatrix}. \tag{2.63}$$

Thus, when $j = 0$, the conversion of the valley current into the electric field is described by

$$E = \frac{\sigma_v}{\sigma_{xx}^2 + \sigma_v^2} j_v . \tag{2.64}$$

We employ a diffusion equation to take into account the inter-valley scattering, which occurs at graphene edges and/or at atomic-scale defects in experiments [64]. The diffusion equation reads [80]

$$\frac{\partial^2}{\partial x^2}\delta\mu_v = \frac{1}{\ell_v^2}\delta\mu_v, \tag{2.65}$$

where ℓ_v is the inter-valley scattering length.

Figure 2.6 shows the measurement setup for valley transport at zero magnetic field by several groups. Similar Hall bar devices have also been used for measurements of the spin Hall effect [87–92]. Experimentalists do not directly measure the valley currents flowing through the sample, they interpret the nonlocal resistance R_{NL} from the inverse effect of VHE as a signature of the presence of valley currents. The nonlocal resistance R_{NL} is defined to be the measured voltage V_{NL} at $x = L$ divided by the charge current I injected aside at $x = 0$ [80, 93]. When the valley Hall angle $\alpha_v = \sigma_v/\sigma_{xx}$ fulfils the condition $\alpha_v \ll 1$ [80], the valley transport can be understood by a consecutive series of conversions. Firstly, the electric current j on the left is converted into the in-plane electric field $\boldsymbol{E}$ with ratio $\rho_d = \sigma_{xx}^{-1}$, where ρ_d is the resistivity of the device; secondly, the in-plane electric field $\boldsymbol{E}$ gives rise to VHE described by equation (2.62) with ratio σ_v, VHE signifies that a charge neutral valley current j_v, which is transverse to $\boldsymbol{E}$ flows through the system; finally, the valley current j_v from VHE is converted back into nonlocal voltage V_{NL} on the right via

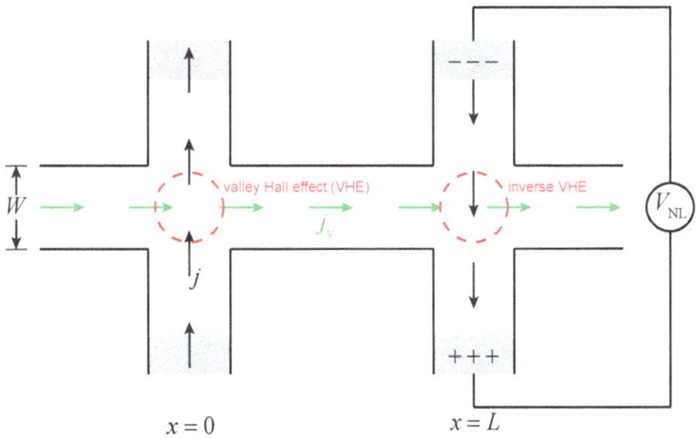

Figure 2.6. Schematic drawing of an experimental Hall bar setup for measuring nonlocal resistance due to the VHE. The current j on the left generates a pure valley current j_v in its transverse direction via the VHE. Subsequently, the valley current is converted into a voltage V_{NL} on the right via the inverse VHE. The nonlocal resistance R_{NL} is defined as V_{NL}/I.

inverse VHE described by equation (2.64) with ratio $\approx \frac{\sigma_v}{\sigma_{xx}^2}$. Incorporating the conversion ratio of all the processes, we find that

$$R_{\text{NL}} \propto \sigma_{xx}^{-1} \cdot \sigma_{\text{v}} \cdot \frac{\sigma_{\text{v}}}{\sigma_{xx}^2} = \sigma_{\text{v}}^2 \rho_{\text{d}}^3. \tag{2.66}$$

Using equations (2.60) and (2.65) self-consistently under an appropriate boundary condition, $\delta\mu_{\text{v}}(x = \pm\infty) = 0$, which we assume the valley voltage is zero at the infinitely distant points, we recover the conventional formula known for spin Hall systems [94],

$$R_{\text{NL}} = \frac{W}{2\ell_{\text{v}}} \frac{\sigma_{\text{v}}^2}{\sigma_{xx}^3} \exp\left(-\frac{L}{\ell_{\text{v}}}\right), \tag{2.67}$$

where W is the horizontal width of the Hall bar and L is the lateral distance between the two contacts as shown in figure 2.6. On the other hand, for $\alpha_{\text{v}} \gg 1$, we obtain

$$R_{\text{NL}} = \frac{W}{2\ell_{\text{v}}} \frac{1}{\sigma_{xx}} \exp\left(-\frac{L}{\ell_{\text{v}}}\right), \tag{2.68}$$

implying $R_{\text{NL}} \propto \rho_{\text{d}}$, in contrast with equation (2.66).

To provide experimental evidence of the valley current, one needs to consider contribution from trivial Ohmic resistance, which is due to classical diffusive charge transport to the nonlocal resistance. The Ohmic contribution can be derived from the van der Pauw formula [64, 91, 95–97]

$$R_{\Omega} = \frac{\rho_{\text{d}}}{\pi} \exp\left(-\pi\frac{L}{W}\right), \tag{2.69}$$

which is linearly scaled with ρ_{d}. Therefore, for $\alpha_{\text{v}} \ll 1$, compared with the $R_{\text{NL}} \propto \rho_{\text{d}}^3$ described by equation (2.66), we can safely exclude the Ohmic contribution as the origin of the experimentally observed R_{NL} [80]. It is noted that the situation for $\alpha_{\text{v}} \gg 1$ remains largely unexplored. This is because in conventional spin Hall systems, the counterpart of α_{v}, spin Hall angle $\alpha_{\text{s}} = \sigma_{\text{s}}/\sigma_{xx} \gg 1$ with σ_{s} as spin Hall conductivity, is inaccessible [70].

2.4.4 Experimental detection of VHE

Although the theoretical prediction of the VHE in monolayer graphene with broken IS was made in 2007 [71], its experimental observation was challenged due to the lack of routes to break the IS. Thanks to the advancement of sophisticated layer transfer techniques [99] used to construct vdW heterostructures, it is possible to align atomic layers with an angular precision of $\pm 1^{\circ}$. The early work [99] showed a huge improvement in the electronic performance of graphene because of the ultra-flat and low-defect nature of the hBN substrate. In 2014, Gorbachev *et al* [64] adjusted the alignment between a monolayer graphene sample with an ultra-flat hBN substrate for breaking the sublattice symmetry to induce VHE. Since hBN also has a hexagonal lattice with its lattice constant (2.50 Å [100]) close to that of graphene (mismatch ~1.7% [99, 101, 102]) and low defect nature, graphene aligned on hBN acquires periodic potential fluctuation with its period observed as a moiré pattern [65, 74, 103–106], because of the different sublattice potentials of graphene result from the binding of carbon to boron and nitrogen atoms. Regions of commensurate graphene/hBN regions form with the same A–B sublattice asymmetry sign, with surrounding strain boundaries. The size of these commensurate regions is about 10 nm, which is ten times smaller than the typical electron wavelength [87]. This indicates that this periodic potential produces local minibands and bandgap in the energy bands of graphene [62, 65, 98, 107–116]. The left inset of figure 2.7 schematically illustrates the band structure of the monolayer graphene aligned on an hBN. Even if currents can flow through the interface strained areas, the bandgap prevents the observation of thermally activated conductance.

In the experiment [64], a high-quality monolayer graphene shaped into a Hall bar is aligned on single-crystal hBN, as depicted in the top right inset of figure 2.7. At low charge density [87], a notable nonlocal signal was detected far away from the current injection region due to the valley Hall and its inverse effect discussed in the previous section [74]. The result obtained in figure 2.7 presents the typical R_{NL} and ρ_{d} measured as functions of the back-gate voltage V_{g}. A notable peak of R_{NL} is observed when the Fermi energy passes the energy of Berry curvature hot spots, near the neutrality point and minigaps. Several aspects seem to indicate that the peak is related to a VHE.

1. For typical aspect ratio $L/W \approx 4$, the van der Pauw formula in equation (2.69) only yields $R_{\Omega} \approx 0.01\ \Omega$ for the ρ_{d} measured in figure 2.7 [80].

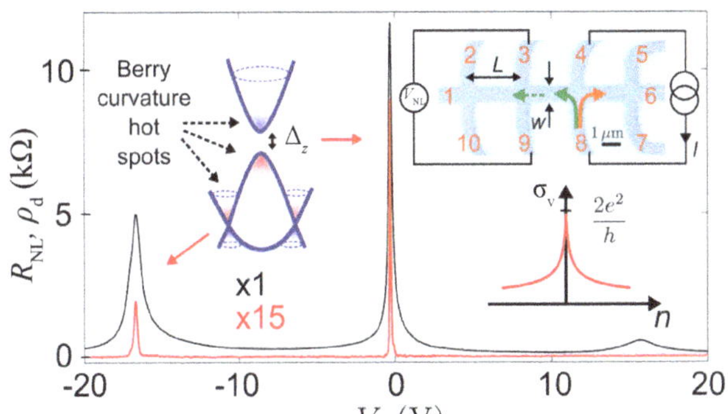

Figure 2.7. Nonlocal resistance R_{NL} multiplied by 15 (red curve) and longitudinal resistance ρ_{d} (black curve) measured in monolayer graphene aligned on hBN at 20 K. The back-gate voltage V_{g} was applied through a ≈130 nm thick dielectric (SiO_2 plus hBN). Top right inset: Optical micrograph of a typical device and the nonlocal measurement geometry, in which $L \approx 3.5$ μm, $W \approx 1$ μm. The valley $K_{\pm}$ currents are also schematically shown by the green and orange arrows. Left inset: Schematic band structure of graphene superlattices with Berry curvature hot spots arising near the finite bandgap of Δ_z and avoided band crossing regions [98]. Bottom right inset: Valley Hall conductivity σ_{V} modelled for gapped Dirac fermions as a function of carrier density n. From [64]. Reprinted with permission from the AAAS.

2. Nonaligned graphene on hBN, which is not subjected to the periodic potential modulation from the hBN, has a vanishing bandgap [87]. It does not show such a large R_{NL}; this excludes any possible spin Hall effect.
3. Any transport along edges is excluded, and the phenomenon has a purely bulk nature [87].
4. The nonlocal resistance is observed to decay exponentially with increasing L. The decay parameter $\approx 1\ \mu\mathrm{m}$ is compatible with the intervalley scattering due to edges or disorder [64, 87].

Therefore, the observed large R_{NL} is attributed to valley-mediated transport. The sharper response of R_{NL} to the gate voltage (which modulates the carrier density) than the resistivity ρ_{d} can be attributed to the cubic dependence of R_{NL} corresponds to ρ_{d} in equation (2.66).

The nonlocal voltage can also be observed in bilayer graphene if an out-of-plane electric field breaks the IS, introducing a bandgap and valley contrasting Berry curvature [74, 80, 81].

2.5 Wavepacket and its orbital moment

The wavepacket, unlike a classical point particle, has a finite spread in real space. Intuitively, the electrons also acquire a magnetic moment due to the self-rotation of the electron wavepacket known as the valley orbital magnetic moment [21, 40, 51, 83],

$$\boldsymbol{M}_n(\boldsymbol{k}) = -\mathrm{i}\frac{e}{2\hbar}\langle \nabla_{\boldsymbol{k}}\psi_n(\boldsymbol{k})| \times [\mathcal{E}_n(\boldsymbol{k}) - H(\boldsymbol{k})]|\nabla_{\boldsymbol{k}}\psi_n(\boldsymbol{k})\rangle. \tag{2.70}$$

This can be rewritten in a more tractable form for calculation,

$$\begin{aligned}\boldsymbol{M}_n(\boldsymbol{k}) &= -\mathrm{i}\frac{e}{2\hbar}\left[\langle \partial_{k_x}\psi_n(\boldsymbol{k})|H(\boldsymbol{k}) - \mathcal{E}_n(\boldsymbol{k})|\partial_{k_y}\psi_n(\boldsymbol{k})\rangle - \mathrm{c.c.}\right] \\ &= -\mathrm{i}\frac{e}{2\hbar}\sum_{n'\neq n}\left[\frac{\langle \psi_n(\boldsymbol{k})|\partial_{k_x}H(\boldsymbol{k})|\psi_{n'}(\boldsymbol{k})\rangle\langle \psi_{n'}(\boldsymbol{k})|\partial_{k_y}H(\boldsymbol{k})|\psi_n(\boldsymbol{k})\rangle}{\mathcal{E}_{n'}(\boldsymbol{k}) - \mathcal{E}_n(\boldsymbol{k})} - \mathrm{c.c.}\right].\end{aligned} \tag{2.71}$$

The orbital moment is an intrinsic property of the band. Its final expression, equation (2.70), does not depend on the actual shape and size of the wavepacket and only depends on the Bloch functions. Under symmetry operations, the orbital moment transforms exactly like the Berry curvature. Therefore, unless the system has both TRS and IS, $\boldsymbol{M}_n(\boldsymbol{k})$ is generally nonzero. In the particular case of a two-band model with electron–hole symmetry, the orbital magnetic moment has a simple relationship with the Berry curvature [51, 71]:

$$\boldsymbol{M}_n(\boldsymbol{k}) = \frac{e}{\hbar}\mathcal{E}_n(\boldsymbol{k})\boldsymbol{\Omega}_n(\boldsymbol{k}). \tag{2.72}$$

The explicit form of the valley-dependent orbital magnetic moment in monolayer graphene with broken IS in the vicinity of the $\boldsymbol{K}_{\pm}$ points for the simple two-band $\boldsymbol{k}\cdot\boldsymbol{p}$ model (equation (2.48)) can be derived as [71]

$$\boldsymbol{M}(\boldsymbol{q}) = \kappa\frac{3e}{4\hbar}\frac{\gamma_0^2 a^2\Delta_z}{3\gamma_0^2 q^2 a^2 + \Delta_z^2}\hat{\boldsymbol{z}}. \tag{2.73}$$

Note that the sign is switched between valleys to satisfy TRS. Since the motion of the wavepacket is in the 2D plane, the orbital magnetic moment is out of the plane. In contrast to the Berry curvature, the quantity is identical between the conduction and valence bands in the same valley. Since the energy of a carrier is quadratically dependent on its momentum near a CBM or a VBM, close to the valley centre ($\boldsymbol{q} = 0$), i.e., the $\boldsymbol{K}_\kappa$ point, the effective mass is [37, 42, 71, 117]

$$m^* = \hbar^2\left[\frac{\partial^2\mathcal{E}(\boldsymbol{q})}{\partial k^2}\right]^{-1} = \frac{2\hbar^2\Delta_z}{3\gamma_0^2 a^2}. \tag{2.74}$$

Equation (2.73) has an advised form for the magnetic moment at the $\boldsymbol{K}_\kappa$ point:

$$\boldsymbol{M}(\boldsymbol{q}) = \kappa\frac{3e}{4\hbar}\frac{\gamma_0^2 a^2}{\Delta_z}\hat{\boldsymbol{z}} = \kappa\boldsymbol{\mu}_{\mathrm{B}}^* \tag{2.75}$$

with $\boldsymbol{\mu}_{\mathrm{B}}^* = \frac{e\hbar}{2m^*}\hat{\boldsymbol{z}}$ as the effective Bohr magneton for the orbital magnetic moment. The valley magnetic moment can couple to an out-of-plane external magnetic field as $-\boldsymbol{\mu}_{\mathrm{B}}^*\cdot\boldsymbol{B}$ [72, 118]. As the orbital magnetic moments in the $\boldsymbol{K}_-$ and $\boldsymbol{K}_+$ valleys are opposite in sign, B_z causes the opposite energy shifts for the $\boldsymbol{K}_-$ and $\boldsymbol{K}_+$ valleys. This lifts the valley energy degeneracy and gives rise to valley Zeeman splitting [83, 119–124] and we will discuss this in section 7.6.

2.6 Valley and spin Hall effect in monolayer TMDCs

Monolayer TMDCs consist of a staggered hexagonal lattice with a broken IS. Thus, they inherit the valley-dependent Berry curvature in graphene with staggered sublattice potential [71]. Similar to the previous section, for massive Dirac fermions described by the effective Hamiltonian in equation (1.49), we can calculate the Berry curvature for conduction electrons as [67]

$$\boldsymbol{\Omega}_+(\boldsymbol{q}) = -\kappa\frac{a^2\gamma^2(\Delta_z - \kappa s\lambda_{\mathrm{SO}})}{4\left|\mathcal{E}_\xi^\kappa(\boldsymbol{q}) - \kappa s\dfrac{\lambda_{\mathrm{SO}}}{2}\right|^3}\hat{\boldsymbol{z}} = -\kappa\frac{2a^2\gamma^2(\Delta_z - \kappa s\lambda_{\mathrm{SO}})}{[4a^2\gamma^2 q^2 + (\Delta_z - \kappa s\lambda_{\mathrm{SO}})^2]^{3/2}}\hat{\boldsymbol{z}}, \tag{2.76}$$

where $\mathcal{E}_\xi^\kappa(\boldsymbol{q})$ is given in equation (1.50). Note that the Berry curvatures have opposite signs in opposite valleys. Similarly, we can find that the Berry curvature in the valence band has the same magnitude as the conduction band, but with an opposite sign, i.e., $\boldsymbol{\Omega}_+(\boldsymbol{k}) = -\boldsymbol{\Omega}_-(\boldsymbol{k})$. Theoretical modelling and DFT calculations in the entire BZ qualitatively support this result, showing that a sizeable Berry curvature is found in the neighbourhood of the $\boldsymbol{K}_{\pm}$ points [67, 125, 126], and the peak of Berry curvature is located exactly at $\boldsymbol{K}_{\pm}$ [83, 127]. Based on equation (2.36) and noticing $\left[\mathcal{E}_{\xi}^{\kappa}(\boldsymbol{q}) - \kappa s\frac{\lambda_{\rm SO}}{2}\right]\mathrm{d}\left[\mathcal{E}_{\xi}^{\kappa}(\boldsymbol{q}) - \kappa s\frac{\lambda_{\rm SO}}{2}\right] = a^2\gamma^2 q\mathrm{d}q$, we can find a valley-dependent Hall conductivity at zero kelvin,

$$\sigma_{\rm H}(\kappa,\, s,\, \mathcal{E}_{\rm F}) = \begin{cases} \kappa\dfrac{e^2}{h}\dfrac{\Delta_z - \kappa s\lambda_{\rm SO}}{4\left|\mathcal{E}_{\rm F} - \kappa s\dfrac{\lambda_{\rm SO}}{2}\right|} & \text{if } \mathcal{E}_{\rm F} < -\dfrac{\Delta_z}{2} + \kappa s\lambda_{\rm SO} \text{ or } \mathcal{E}_{\rm F} > \dfrac{\Delta_z}{2} \\ \dfrac{\kappa}{2}\dfrac{e^2}{h} & \text{if } -\dfrac{\Delta_z}{2} + \kappa s\lambda_{\rm SO} \leqslant \mathcal{E}_{\rm F} \leqslant \dfrac{\Delta_z}{2}. \end{cases} \tag{2.77}$$

Figure 2.8(a) schematically displays the low-energy dispersion around the $\boldsymbol{K}_{\pm}$ points for monolayer TMDC. We mark two Fermi energies for the electron and hole-doped systems, by solid and dashed horizontal lines, respectively, and schematically present the corresponding VHE driven by the Berry curvature in figure 2.8(b). Because of the finite Berry curvature with opposite signs in the two valleys, when an in-plane longitudinal electric field $\boldsymbol{E}$ is applied through the system, the carriers at $\boldsymbol{K}_{+}$ and $\boldsymbol{K}_{-}$ valleys have opposite anomalous velocities along the direction perpendicular to the longitudinal electric field $\boldsymbol{E}$. As a result, the carriers accumulate to the opposite transverse edges. In addition, the spin Hall effect will appear simultaneously with the VHE owing to the spin–valley locking relationship.

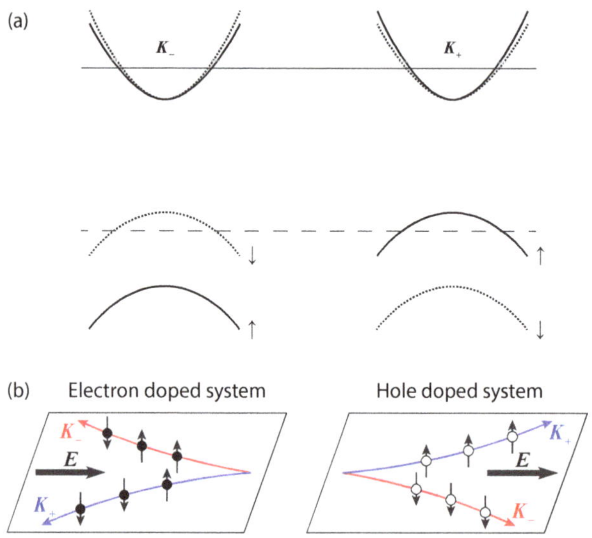

Figure 2.8. (a) Schematic drawing of the band structure at the band edges located at the $\boldsymbol{K}_{\pm}$ points of monolayer TMDCs. Spin up (down) bands are denoted by solid (dotted) curves. For two Fermi energies, denoted by solid and dashed horizontal lines, representing electron and hole doped systems, respectively, we show the corresponding VHE schematically in (b). The quantitative results are also discussed in the main text. In (b), electrons and holes are denoted by black and white dots, respectively. The spin Hall conductivity in the hole doped system is larger than the electron doped system, owing to the larger spin splitting in the valence band. Part (b) reprinted with permission from [67]. Copyright (2012) by the American Physical Society.

Now we explain figure 2.8(b) quantitatively. For moderate hole doping with $\mathcal{E}_F$ lying between the two split valence band edges, i.e., $-\frac{\Delta_z}{2} - \lambda_{SO} < \mathcal{E}_F < -\frac{\Delta_z}{2} + \lambda_{SO}$, as illustrated by the dashed horizontal line in figure 2.8(a). For $\mu \ll \Delta_z - \lambda_{SO}$, where μ is the Fermi energy measured from the VBM, i.e., $\mathcal{E}_F = -\frac{\Delta_z}{2} + \lambda_{SO} + \mu$, the valley Hall conductivity of holes is given by

$$\sigma_v(\mathcal{E}_F) \equiv \sum_s \sigma_H(\kappa = 1, s, \mathcal{E}_F) - \sum_s \sigma_H(\kappa = -1, s, \mathcal{E}_F) = \frac{2e^2}{h}\left(1 + \frac{\mu}{\Delta_z - \lambda_{SO}}\right). \tag{2.78}$$

On the other hand, the spin Hall conductivity of holes is [67]

$$\sigma_s(\mathcal{E}_F) \equiv \sum_\kappa \sigma_H(\kappa, s = 1, \mathcal{E}_F) - \sum_\kappa \sigma_H(\kappa, s = -1, \mathcal{E}_F) = \frac{2e^2}{h}\frac{\mu}{\Delta_z - \lambda_{SO}}. \tag{2.79}$$

As we can see, when $\mu = 0$, the spin Hall conductivity becomes 0, while the valley Hall conductivity becomes $\frac{2e^2}{h}$, which reproduces the results for graphene with broken IS [equation (2.59)].

When electron doping occurs, the Fermi energy is illustrated by the solid horizontal line in figure 2.8(a), the valley and spin Hall conductivities of electrons are given by

$$\sigma_v(\mathcal{E}_F) = \frac{2e^2}{h}\left(1 + \frac{2\Delta_z}{\Delta_z^2 - \lambda_{SO}^2}\mu\right), \tag{2.80}$$

$$\sigma_s(\mathcal{E}_F) = -\frac{4e^2}{h}\frac{\lambda_{SO}}{\Delta_z^2 - \lambda_{SO}^2}\mu, \tag{2.81}$$

respectively, where μ is the Fermi energy measured from the CBM, i.e., $\mathcal{E}_F = \frac{\Delta_z}{2} + \mu$. σ_s in the valence band is higher than that in the conduction band owing to the larger spin splitting in the valence band [127, 128]. The coexistence of the VHE and spin Hall effect makes 2D materials possible for the interdisciplinary study of spintronics and valleytronics.

2.7 Chern number and valley Chern number

The previous sections state that the Hall conductance can be found directly from the band structure and that it is an internal property of the bands. The physical quantity physicists use to identify this topological property of the band structure is called the Chern number [129], which is the integral of the Berry curvature of the band over the entire BZ. Similar to the definition in equation (2.30), the Chern number of the n-th band reads

$$\mathcal{C}_n = \frac{1}{2\pi}\int \mathrm{d}k_x \mathrm{d}k_y \Omega_n(\boldsymbol{k}). \tag{2.82}$$

In this case, one can calculate the Hall conductivity of the nth band as proportional to the Chern number, which is quantised in units of $\frac{e^2}{h}$ and denoted as [44, 130–134]

$$\sigma_{\mathrm{H},n} = \frac{e^2}{h} \cdot \mathcal{C}_n. \tag{2.83}$$

Therefore, for the conduction band of the system described by equation (2.48), the Chern number of spin s in the Dirac valley $\boldsymbol{K}_\kappa$ is [130]

$$\begin{aligned} \mathcal{C}_s^\kappa &= \frac{1}{2\pi}\int_0^\infty 2\pi q \mathrm{d}q \Omega_+(\boldsymbol{q}) \\ &= -\frac{\kappa}{2}\int_0^\infty \frac{3\gamma_0^2 a^2 \Delta_z q \mathrm{d}q}{\left(3\gamma_0^2 q^2 a^2 + \Delta_z^2\right)^{3/2}} \\ &= -\frac{\kappa}{2}\mathrm{sgn}(\Delta_z). \end{aligned} \tag{2.84}$$

This result is in line with equations (2.57) and (2.83) if spin degeneracy $g_\mathrm{s} = 2$ is considered. The Chern number is quantised as $\mathcal{C}_s^\kappa = -\frac{\kappa}{2}$ at the Dirac point $\boldsymbol{K}_\kappa$. It is insensitive to a deformation of the band structure provided the gap is open [50]. On the other hand, it changes its sign when Δ_z changes its sign. Such a quantity is a topological number [135]. This valley-dependent half-integer Chern number from the integration of Berry curvature not only induces an anomalous velocity for bulk wavepacket [21] in the previous sections, but also leads to the topological surface or kink states [136] that will be discussed in the next section.

The expressions derived for the Chern numbers above allow us to predict the physical implications of the signs and magnitude of the topological invariants. To this end, an insulator phase is indexed by a set of four topological invariants, i.e., the total Chern numbers, the valley Chern number, the spin Chern number and the spin–valley Chern number [50, 77, 135, 137]. Here, to focus on the scope of valleytronics, we only define the former two of them, the total Chern number $\mathcal{C}$ and the valley Chern number $\mathcal{C}_\mathrm{v}$. They read [7, 138–141]

$$\mathcal{C} = \mathcal{C}_\uparrow^+ + \mathcal{C}_\downarrow^+ + \mathcal{C}_\uparrow^- + \mathcal{C}_\downarrow^-, \tag{2.85}$$

$$\mathcal{C}_\mathrm{v} = \mathcal{C}_\uparrow^+ + \mathcal{C}_\downarrow^+ - \mathcal{C}_\uparrow^- - \mathcal{C}_\downarrow^-. \tag{2.86}$$

Note that the valley Chern number is well-defined only in the Dirac theory. Namely, they are ill-defined in the TB model since they are defined by the difference of the Berry curvatures at the $\boldsymbol{K}_+$ and $\boldsymbol{K}_-$ points [142]. On the other hand, we call $\mathcal{C}$ the genuine Chern number.

2.8 Topological kink states

In 2008, Semenoff *et al* [143] considered the simplest linelike domain-wall defects in the mass pattern in monolayer graphene, which is gapped by staggered sublattice potential. In figure 2.9, we present the schematic diagram of monolayer graphene

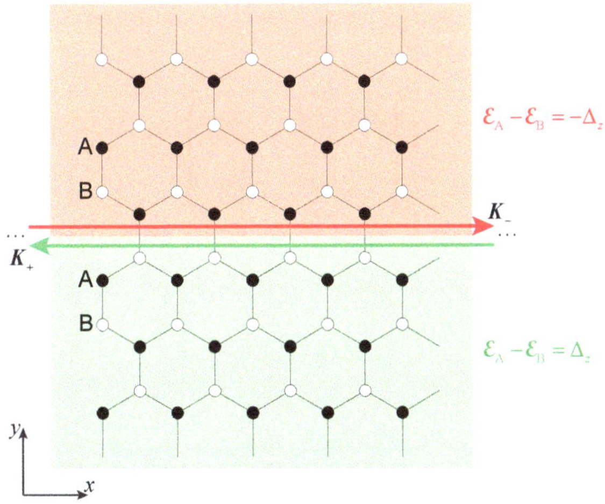

Figure 2.9. Schematics of a domain wall constructed in graphene between $y < 0$ (green area) and $y > 0$ (red area). The potential difference between the sublattices A and B is $\mathcal{E}_A - \mathcal{E}_B = -\text{sgn}(y)\Delta_z$. A pure topological valley current flows along the domain wall at $y = 0$ since the $\boldsymbol{K}_+$ and $\boldsymbol{K}_-$ valley electrons have opposite group velocities due to the definite chirality.

with a zigzag domain wall at $y = 0$, denoted by the boundary between the red and green background. The sublattices that have higher or lower potentials are interchanged at the domain wall [63]. Specifically, in the area with the green (red) background, the potential difference between the sublattices A and B, $\mathcal{E}_A - \mathcal{E}_B = \Delta_z$ $(-\Delta_z)$ with $\Delta_z > 0$. This implies that in the vicinity of $\boldsymbol{K}_\pm$ valley, the Chern number in the conduction band is

$$\mathcal{C}_+ = \begin{cases} \pm\dfrac{1}{2}, & \text{if } y > 0 \\ \mp\dfrac{1}{2}, & \text{if } y < 0 \end{cases}, \tag{2.87}$$

based on equation (2.84). Thus, the valley Chern number is $\mathcal{C}_v = 1\ (-1)$ for $y > 0$ (<0). The domain wall acts as a 1D metal embedded in a semiconductor, and it could potentially serve as a single-channel quantum wire [144].

For a graphene system with staggered sublattice potential, and its zigzag direction positioned along the x-axis, the effective Hamiltonian can be written as [145]

$$H_\kappa(\boldsymbol{q}) = \hbar v_F \begin{pmatrix} \dfrac{\Delta_z}{2} & \kappa q_x - \mathrm{i}q_y \\ \kappa q_x + \mathrm{i}q_y & -\dfrac{\Delta_z}{2} \end{pmatrix}. \tag{2.88}$$

The wavefunctions in the energy gap can be obtained by solving the above Hamiltonian [146],

$$\Psi_{\rm I}(y>0) = C_1\begin{pmatrix} \hbar v_{\rm F}(\kappa q_x - {\rm i}q_y) \\ \mathcal{E}(\boldsymbol{q}) - \dfrac{\Delta_z}{2} \end{pmatrix} {\rm e}^{{\rm i}(q_x x + q_y y)}, \tag{2.89}$$

$$\Psi_{\rm II}(y<0) = C_2\begin{pmatrix} \hbar v_{\rm F}(\kappa q_x + {\rm i}q_y) \\ \mathcal{E}(\boldsymbol{q}) + \dfrac{\Delta_z}{2} \end{pmatrix} {\rm e}^{{\rm i}(q_x x - q_y y)}, \tag{2.90}$$

with C_1 and C_2 as undetermined coefficients, $\hbar v_{\rm F} q_y = {\rm i}\sqrt{(\frac{\Delta_z}{2})^2 + (\hbar v_{\rm F} q_x)^2 - \mathcal{E}(\boldsymbol{q})^2}$. q_x is assumed to be conserved when the wavefunctions are scattered at the domain wall, $y = 0$. Note the wavefunctions in the energy gap are evanescent and different in the two areas because of the opposite staggered potential $\pm\Delta_z$ and location relative to the domain wall ${\rm sgn}(y)$. The wavefunctions are continuous at the domain wall, $\Psi_{\rm I}(y=0) = \Psi_{\rm II}(y=0)$. One can obtain the meaningful solution to the secular equation of its coefficient matrix,

$$\mathcal{E}(\boldsymbol{q}) = -\kappa\hbar v_{\rm F} q_x \tag{2.91}$$

by neglecting the other solution $\mathcal{E} = \kappa\frac{\Delta_z}{2}$ since we only focus on the energies within the bandgap, $\mathcal{E}(\boldsymbol{q}) < \frac{\Delta_z}{2}$. It is evident that such interface states possess a definite chirality, with the $\boldsymbol{K}_+$ and $\boldsymbol{K}_-$ valleys have opposite ones. Therefore, we can designate them as kink states, wherein pure valley current flows along the domain wall at $y = 0$. Such behaviour is also illustrated in figure 2.9. The current is topologically protected and robust against valley-conservation scattering.

To further clarify the valley kink states, in figure 2.10, we plot the band structure of the system shown in figure 2.9 with a total of 100 zigzag lines. In our calculations, parameters are taken as $\gamma_0 = 3$ eV and $\Delta_z = 1$ eV. One can see that in addition to the bulk states with bandgap at $|\mathcal{E}(\boldsymbol{k})| \geqslant \frac{\Delta_z}{2}$, two subbands with linear dispersion relation appear in the bandgap $|\mathcal{E}(\boldsymbol{k})| < \frac{\Delta_z}{2}$ and cross the Fermi energy $\mathcal{E}_{\rm F} = 0$ at the $\boldsymbol{K}_+$ and $\boldsymbol{K}_-$ points [147]. This means the bandgap closes at the domain wall since it connects the areas where $\mathcal{C}_{\rm v} = 1$ and -1. Further, at $\boldsymbol{K}_+$ ($\boldsymbol{K}_-$) valley, the band of states intersects the Fermi energy with a negative (positive) group velocity and defines the left-moving (right-moving) chiral modes. One can also regard the topological kink states as the bulk–boundary correspondence [148–150], which states that the difference between the number of right-moving and left-moving modes, $N_{\rm R} - N_{\rm L}$, is the difference in the Chern number across the interface $\Delta\mathcal{C}$ [151, 152]. Mathematically,

$$N_{\rm R} - N_{\rm L} = \Delta\mathcal{C}, \tag{2.92}$$

which can be easily verified distinctively at $\boldsymbol{K}_+$ and $\boldsymbol{K}_-$ valleys from equation (2.87).

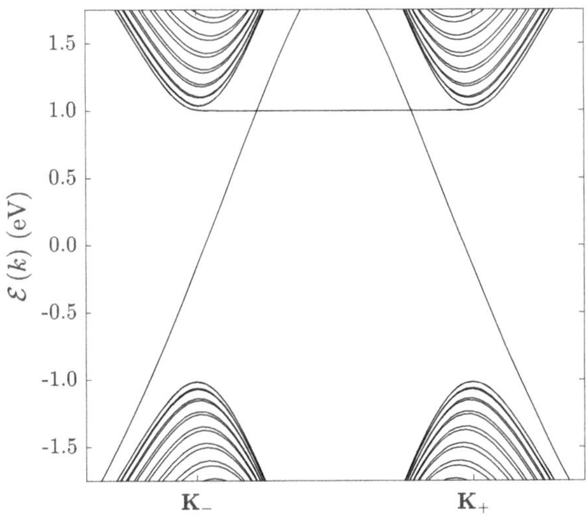

Figure 2.10. The band structure of the system presented in figure 2.9 with 100 zigzag lines. The parameters are $\gamma_0 = 3$ eV and $\Delta_z = 1$ eV.

These kink states cannot be observed experimentally in monolayer graphene because of the impossibility of assigning enormous enough sublattice-specific potentials. The value of this system is that it can be used to simulate different applications using kink states, while the computational effort is significantly lower compared with simulations using bilayer graphene.

Similarly, topological kink states can also appear along the line defect of monolayer graphene theoretically and the staggered sublattice potential should have a mirror symmetry with respect to the line defect [153]. The topological kink states can also be realized in silicene, where the buckled structure enables the generation of a staggered sublattice potential through an external perpendicular electric field [146, 154]. It is also possible to extend to the systems where spin degeneracy is broken, generating spin and valley-polarized current in the domain wall [155].

At last, we briefly discuss the utilization of the studied topological kink states. As we have shown in figure 2.9, the direction of the propagation is determined by the valley. For the given potential, the $\boldsymbol{K}_-$ ($\boldsymbol{K}_+$) valley electrons can enter the state only from the left (right) side. Therefore, applying a voltage difference along the domain wall can lead to the realization of a valley filter. In this case, only electrons from one of the valleys will contribute to the electric current, while the other valley is filtered out. The potential determines which valley can be transported [156]. A valley valve can also be possible by connecting two filters in series that are separated by an area without any potential applied to it. After the current passes through the first valley filter, the electrons are either fully $\boldsymbol{K}_+$ or $\boldsymbol{K}_-$ valley-polarized. If the polarity of the second filter has the same (opposite) sign as the first one, the electrons can pass through (be blocked) [157]. In the valley Y-junction, the valley filters are in the two arms of the Y-structure. The electrons enter from the base of the Y-junction. The polarities of the two arms are opposite. This means the $\boldsymbol{K}_+$ and $\boldsymbol{K}_-$ valley electrons enter different arms of the Y-junction. When we ignore the intervalley scattering, electrons always transport ballistically along the edges of the domain wall. Furthermore, earlier proposals [158–160] for valley filters and valves based on graphene have stringent requirements on perfect zigzag edges, making their practical realization very challenging. On the other hand, the topological kink states shown in this section are far less demanding since they appear away from the edges of the graphene. Some technologies, such as breaking the IS in bilayer graphene by gate voltage, have already been demonstrated experimentally [161], as we will discuss in section 6.1.

References

[1] Asbóth J K, Oroszlány L and Pályi A 2016 *A Short Course on Topological Insulators: Band Structure and Edge States in One and Two Dimensions* Lecture Notes in Physics **vol 919** (Cham: Springer)

[2] Qi X-L and Zhang S-C 2011 Topological insulators and superconductors *Rev. Mod. Phys.* **83** 1057–10

[3] Zanardi P and Lloyd S 2003 Topological protection and quantum noiseless subsystems *Phys. Rev. Lett.* **90** 067902

[4] Ren Y, Qiao Z and Niu Q 2016 Topological phases in two-dimensional materials: a review *Rep. Prog. Phys.* **79** 066501

[5] Pollmann F, Berg E, Turner A M and Oshikawa M 2012 Symmetry protection of topological phases in one-dimensional quantum spin systems *Phys. Rev.* B **85** 075125

[6] Klitzing K v, Dorda G and Pepper M 1980 New method for high-accuracy determination of the fine-structure constant based on quantized Hall resistance *Phys. Rev. Lett.* **45** 494–7

[7] Brooks M 2019 Spin qubits in two-dimensional semiconductors *PhD Thesis* University of Konstanz, Konstanz

[8] Hasan M Z and Kane C L 2010 Colloquium: topological insulators *Rev. Mod. Phys.* **82** 3045–67

[9] Laughlin R B 1983 Anomalous quantum Hall effect: an incompressible quantum fluid with fractionally charged excitations *Phys. Rev. Lett.* **50** 1395–8

[10] Tsui D C, Stormer H L and Gossard A C 1982 Two-dimensional magnetotransport in the extreme quantum limit *Phys. Rev. Lett.* **48** 1559–62

[11] Jotzu G, Messer M, Desbuquois R, Lebrat M, Uehlinger T, Greif D and Esslinger T 2014 Experimental realization of the topological Haldane model with ultracold fermions *Nature* **515** 237–40

[12] Khanikaev A B and Shvets G 2017 Two-dimensional topological photonics *Nat. Photon.* **11** 763–73

[13] Lu L, Joannopoulos J D and Soljačić M 2014 Topological photonics *Nat. Photon.* **8** 821–9

[14] Bernevig B A and Zhang S-C 2006 Quantum spin Hall effect *Phys. Rev. Lett.* **96** 106802

[15] Pan H, Li Z, Liu C-C, Zhu G, Qiao Z and Yao Y 2014 Valley-polarized quantum anomalous Hall effect in silicene *Phys. Rev. Lett.* **112** 106802

[16] Imhof S *et al* 2018 Topolectrical-circuit realization of topological corner modes *Nat. Phys.* **14** 925–9

[17] Kitaev A Yu 2001 Unpaired Majorana fermions in quantum wires *Phys.-Usp.* **44** 131

[18] Chen Y L *et al* 2009 Experimental realization of a three-dimensional topological insulator, Bi_2Te_3 *Science* **325** 178–81
[19] Berry M V 1984 Quantal phase factors accompanying adiabatic changes *Proc. R. Soc. A: Math. Phys. Eng. Sci.* **392** 45–57
[20] Wang Z 2019 Spin- and valley-dependent excitons in atomically thin transition metal dichalcogenides *PhD Thesis* The Pennsylvania State University, University Park, PA
[21] Xiao D, Chang M-C and Niu Q 2010 Berry phase effects on electronic properties *Rev. Mod. Phys.* **82** 1959–2007
[22] Culcer D, MacDonald A and Niu Q 2003 Anomalous Hall effect in paramagnetic two-dimensional systems *Phys. Rev.* B **68** 045327
[23] Liu C-X, Qi X-L, Dai X, Fang Z and Zhang S-C 2008 Quantum anomalous Hall effect in $Hg_{1-y}Mn_yTe$ quantum wells *Phys. Rev. Lett.* **101** 146802
[24] De Chiara G and Palma G M 2003 Berry phase for a spin 1/2 particle in a classical fluctuating field *Phys. Rev. Lett.* **91** 090404
[25] Semenoff G W 1984 Condensed-matter simulation of a three-dimensional anomaly *Phys. Rev. Lett.* **53** 2449–52
[26] Haldane F D M 1988 Model for a quantum Hall effect without Landau levels: condensed-matter realization of the "parity anomaly" *Phys. Rev. Lett.* **61** 2015–8
[27] Su W P, Schrieffer J R and Heeger A J 1979 Solitons in polyacetylene *Phys. Rev. Lett.* **42** 1698–701
[28] Rice M J and Mele E J 1982 Elementary excitations of a linearly conjugated diatomic polymer *Phys. Rev. Lett.* **49** 1455–9
[29] Vanderbilt D and King-Smith R D 1993 Electric polarization as a bulk quantity and its relation to surface charge *Phys. Rev.* B **48** 4442–55
[30] Onoda S, Murakami S and Nagaosa N 2004 Topological nature of polarization and charge pumping in ferroelectrics *Phys. Rev. Lett.* **93** 167602
[31] Zhang C, Dudarev A M and Niu Q 2006 Berry phase effects on the dynamics of quasiparticles in a superfluid with a vortex *Phys. Rev. Lett.* **97** 040401
[32] Wang S, Ukhtary M S and Saito R 2020 Strain effect on circularly polarized electro-luminescence in transition metal dichalcogenides *Phys. Rev. Res.* **2** 033340
[33] Band Y B and Avishai Y 2013 *Quantum Mechanics with Applications to Nanotechnology and Information Science* (Oxford: Academic Press)
[34] El-Batanouny M 2020 *Advanced Quantum Condensed Matter Physics: One-Body, Many-Body, and Topological Perspectives* (Cambridge: Cambridge University Press)
[35] Araújo M and Sacramento P 2021 *Topology in Condensed Matter: An Introduction* (Singapore: World Scientifc)
[36] Hassani S 2009 *Mathematical Methods: For Students of Physics and Related Fields* 2nd edn (New York: Springer)
[37] Ashcroft N W and Mermin N D 1976 *Solid State Physics* (Boston, MA: Cengage Learning)
[38] Chang M-C and Niu Q 1995 Berry phase, hyperorbits, and the Hofstadter spectrum *Phys. Rev. Lett.* **75** 1348–51
[39] Chang M-C and Niu Q 1996 Berry phase, hyperorbits, and the Hofstadter spectrum: semiclassical dynamics in magnetic Bloch bands *Phys. Rev.* B **53** 7010–23
[40] Sundaram G and Niu Q 1999 Wave-packet dynamics in slowly perturbed crystals: gradient corrections and Berry-phase effects *Phys. Rev.* B **59** 14915–25

[41] Jungwirth T, Niu Q and MacDonald A H 2002 Anomalous Hall effect in ferromagnetic semiconductors *Phys. Rev. Lett.* **88** 207208
[42] Enoki T and Ando T (ed) 2020 *Physics and Chemistry of Graphene: Graphene to Nanographene* 2nd edn (Singapore: Jenny Stanford Publishing)
[43] Nagaosa N, Sinova J, Onoda S, MacDonald A H and Ong N P 2010 Anomalous Hall effect *Rev. Mod. Phys.* **82** 1539–92
[44] Lado J L and Fernández-Rossier J 2015 Quantum anomalous Hall effect in graphene coupled to skyrmions *Phys. Rev.* B **92** 115433
[45] Karplus R and Luttinger J M 1954 Hall effect in ferromagnetics *Phys. Rev.* **95** 1154–60
[46] Adams E N and Blount E I 1959 Energy bands in the presence of an external force field—II: anomalous velocities *J. Phys. Chem. Solids* **10** 286–303
[47] Barré E 2021 Exploring optical and optoelectronic properties in transition metal dichalcogenides *PhD Thesis* Stanford University, Stanford
[48] Yao Y, Kleinman L, MacDonald A H, Sinova J, Jungwirth T, Wang D-S, Wang E and Niu Q 2004 First principles calculation of anomalous Hall conductivity in ferromagnetic bcc Fe *Phys. Rev. Lett.* **92** 037204
[49] Fang Z, Nagaosa N, Takahashi K S, Asamitsu A, Mathieu R, Ogasawara T, Yamada H, Kawasaki M, Tokura Y and Terakura K 2003 The anomalous Hall effect and magnetic monopoles in momentum space *Science* **302** 92–5
[50] Rao C N R and Waghmare U V (ed) 2018 *2D Inorganic Materials beyond Graphene* (London: World Scientifc)
[51] Fuchs J N, Piéchon F, Goerbig M O and Montambaux G 2010 Topological Berry phase and semiclassical quantization of cyclotron orbits for two dimensional electrons in coupled band models *Eur. Phys. J.* B **77** 351–62
[52] Wang S, Pratama F R, Ukhtary M S and Saito R 2020 Independent degrees of freedom in two-dimensional materials *Phys. Rev.* B **101** 081414(R)
[53] Jin C, Lin F, Suenaga K and Iijima S 2009 Fabrication of a freestanding boron nitride single layer and its defect assignments *Phys. Rev. Lett.* **102** 195505
[54] Glavin N R *et al* 2016 Amorphous boron nitride: a universal, ultrathin dielectric for 2D nanoelectronics *Adv. Funct. Mater.* **26** 2640–7
[55] Novoselov K S, Jiang D, Schedin F, Booth T J, Khotkevich V V, Morozov S V and Geim A K 2005 Two-dimensional atomic crystals *Proc. Natl. Acad. Sci. USA* **102** 10451–3
[56] Sławińska J, Zasada I and Klusek Z 2010 Energy gap tuning in graphene on hexagonal boron nitride bilayer system *Phys. Rev.* B **81** 155433
[57] Ghalamkari K, Tatsumi Y and Saito R 2018 Energy band gap dependence of valley polarization of the hexagonal lattice *J. Phys. Soc. Jpn.* **87** 024710
[58] Yin L-J, Bai K-K, Wang W-X, Li S-Y, Zhang Y and He L 2017 Landau quantization of Dirac fermions in graphene and its multilayers *Front. Phys.* **12** 127208
[59] Zhou S Y, Gweon G-H, Fedorov A V, First P N, de Heer W A, Lee D-H, Guinea F, Castro Neto A H and Lanzara A 2007 Substrate-induced bandgap opening in epitaxial graphene *Nat. Mater.* **6** 770–5
[60] Li G, Luican A and Andrei E Y 2009 Scanning tunneling spectroscopy of graphene on graphite *Phys. Rev. Lett.* **102** 176804
[61] Li G, Luican-Mayer A, Abanin D, Levitov L and Andrei E Y 2013 Evolution of Landau levels into edge states in graphene *Nat. Commun.* **4** 1744

[62] Giovannetti G, Khomyakov P A, Brocks G, Kelly P J and van den Brink J 2007 Substrate-induced band gap in graphene on hexagonal boron nitride: Ab initio density functional calculations *Phys. Rev.* B **76** 073103
[63] Zarenia M, Leenaerts O, Partoens B and Peeters F M 2012 Substrate-induced chiral states in graphene *Phys. Rev.* B **86** 085451
[64] Gorbachev R V *et al* 2014 Detecting topological currents in graphene superlattices *Science* **346** 448
[65] Hunt B *et al* 2013 Massive Dirac fermions and Hofstadter butterfly in a van der Waals heterostructure *Science* **340** 1427–30
[66] Sachs B, Wehling T O, Katsnelson M I and Lichtenstein A I 2011 Adhesion and electronic structure of graphene on hexagonal boron nitride substrates *Phys. Rev.* B **84** 195414
[67] Xiao D, Liu G-B, Feng W, Xu X and Yao W 2012 Coupled spin and valley physics in monolayers of MoS_2 and other group-VI dichalcogenides *Phys. Rev. Lett.* **108** 196802
[68] Wu S 2016 Device physics of two-dimensional crystalline materials *PhD Thesis* University of Washington, Seattle, WA
[69] Gosselin P, Bérard A, Mohrbach H and Ghosh S 2009 Berry curvature in graphene: a new approach *Eur. Phys. J.* C **59** 883–9
[70] Yamamoto M, Shimazaki Y, Borzenets I V and Tarucha S 2015 Valley Hall effect in two-dimensional hexagonal lattices *J. Phys. Soc. Jpn.* **84** 121006
[71] Xiao D, Yao W and Niu Q 2007 Valley-contrasting physics in graphene: magnetic moment and topological transport *Phys. Rev. Lett.* **99** 236809
[72] Mak K F, Xiao D and Shan J 2018 Light–valley interactions in 2D semiconductors *Nat. Photon.* **12** 451–60
[73] Cayssol J and Fuchs J N 2021 Topological and geometrical aspects of band theory *J. Phys.: Mater.* **4** 034007
[74] Chi D, Johnson Goh K E and Wee A T S (ed) 2019 *2D Semiconductor Materials and Devices* Materials Today (Amsterdam: Elsevier)
[75] Cai T, Yang S A, Li X, Zhang F, Shi J, Yao W and Niu Q 2013 Magnetic control of the valley degree of freedom of massive Dirac fermions with application to transition metal dichalcogenides *Phys. Rev.* B **88** 115140
[76] Park J, Kim K J, Kim S and Yoo J-W 2021 Geulaepin-eseo seupin mich baelli susong-e gwanhan yeongu donghyang [The spin and valley transport in graphene] *J. Korean Magn. Soc.* **31** 277–86
[77] Spencer M J S and Morishita T (ed) 2016 *Silicene: Structure, Properties and Applications* Springer Series in Materials Science **vol 235** (Cham: Springer International Publishing)
[78] Ishikawa K 1984 Chiral anomaly and quantized Hall effect *Phys. Rev. Lett.* **53** 1615–8
[79] Narikiyo O and Kuboki K 1993 Parity anomaly of two-dimensional Dirac fermion in a magnetic field *J. Phys. Soc. Jpn.* **62** 1812–3
[80] Shimazaki Y, Yamamoto M, Borzenets I V, Watanabe K, Taniguchi T and Tarucha S 2015 Generation and detection of pure valley current by electrically induced Berry curvature in bilayer graphene *Nat. Phys.* **11** 1032–6
[81] Sui M *et al* 2015 Gate-tunable topological valley transport in bilayer graphene *Nat. Phys.* **11** 1027–31
[82] Lee J, Mak K F and Shan J 2016 Electrical control of the valley Hall effect in bilayer MoS_2 transistors *Nat. Nanotechnol* **11** 421–5

[83] Xu X, Yao W, Xiao D and Heinz T F 2014 Spin and pseudospins in layered transition metal dichalcogenides *Nat. Phys.* **10** 343–50
[84] Mak K F, McGill K L, Park J and McEuen P L 2014 The valley Hall effect in MoS_2 transistors *Science* **344** 1489
[85] Morozov S V, Novoselov K S, Katsnelson M I, Schedin F, Ponomarenko L A, Jiang D and Geim A K 2006 Strong suppression of weak localization in graphene *Phys. Rev. Lett.* **97** 016801
[86] Osada T 2016 Genshi-sō ni okeru toporojī butsuri [Topology in the physics of atomic layers] *Hyomen Kagaku* **37** 535–40
[87] Cresti A, Nikolić B K, García J H and Roche S 2016 Charge, spin and valley Hall effects in disordered graphene *Riv. Nuovo Cimento* **39** 587–667
[88] Nikolić B K, Zârbo L P and Souma S 2006 Imaging mesoscopic spin Hall flow: spatial distribution of local spin currents and spin densities in and out of multiterminal spin-orbit coupled semiconductor nanostructures *Phys. Rev.* B **73** 075303
[89] Wang Z, Liu H, Jiang H and Xie X C 2016 Numerical study of the giant nonlocal resistance in spin-orbit coupled graphene *Phys. Rev.* B **94** 035409
[90] Van Tuan D, Marmolejo-Tejada J M, Waintal X, Nikolić B K, Valenzuela S O and Roche S 2016 Spin Hall effect and origins of nonlocal resistance in adatom-decorated graphene *Phys. Rev. Lett.* **117** 176602
[91] Abanin D A *et al* 2011 Giant nonlocality near the Dirac point in graphene *Science* **332** 328–30
[92] Maekawa S, Adachi H, Uchida K-i, Ieda J and Saitoh E 2013 Spin current: experimental and theoretical aspects *J. Phys. Soc. Jpn.* **82** 102002
[93] Azari M 2020 Valleytronics of quantum dots of topological materials *PhD Thesis* Simon Fraser University, Burnaby
[94] Abanin D A, Shytov A V, Levitov L S and Halperin B I 2009 Nonlocal charge transport mediated by spin diffusion in the spin Hall effect regime *Phys. Rev.* B **79** 035304
[95] van der Pauw L J 1958 A method of measuring specific resistivity and Hall effect of discs of arbitrary shape *Philips Res. Rep.* **13** 1–9
[96] Balakrishnan J, Kok Wai Koon G, Jaiswal M, Castro Neto A H and Özyilmaz B 2013 Colossal enhancement of spin–orbit coupling in weakly hydrogenated graphene *Nat. Phys.* **9** 284–7
[97] Ozdemir S 2021 *Electronic Properties of Rhombohedral Graphite* Springer Theses (Cham: Springer)
[98] Song J C W, Samutpraphoot P and Levitov L S 2015 Topological Bloch bands in graphene superlattices *Proc. Natl. Acad. Sci. USA* **112** 10879–83
[99] Dean C R *et al* 2010 Boron nitride substrates for high-quality graphene electronics *Nat. Nanotechnol.* **5** 722–6
[100] Wang J, Ma F and Sun M 2017 Graphene, hexagonal boron nitride, and their heterostructures: properties and applications *RSC Adv.* **7** 16801–22
[101] Decker R, Wang Y, Brar V W, Regan W, Tsai H-Z, Wu Q, Gannett W, Zettl A and Crommie M F 2011 Local electronic properties of graphene on a BN substrate via scanning tunneling microscopy *Nano Lett.* **11** 2291–5
[102] Jain N, Bansal T, Durcan C A, Xu Y and Yu B 2013 Monolayer graphene/hexagonal boron nitride heterostructure *Carbon* **54** 396–402
[103] Woods C R *et al* 2014 Commensurate–incommensurate transition in graphene on hexagonal boron nitride *Nat. Phys.* **10** 451–6

[104] Ponomarenko L A *et al* 2013 Cloning of Dirac fermions in graphene superlattices *Nature* **497** 594–7
[105] Dean C R *et al* 2013 Hofstadter's butterfly and the fractal quantum Hall effect in moiré superlattices *Nature* **497** 598–602
[106] Chen Z-G, Shi Z, Yang W, Lu X, Lai Y, Yan H, Wang F, Zhang G and Li Z 2014 Observation of an intrinsic bandgap and Landau level renormalization in graphene/boron-nitride heterostructures *Nat. Commun.* **5** 4461
[107] Park C-H, Yang L, Son Y-W, Cohen M L and Louie S G 2008 New generation of massless Dirac fermions in graphene under external periodic potentials *Phys. Rev. Lett.* **101** 126804
[108] Kindermann M, Uchoa B and Miller D L 2012 Zero-energy modes and gate-tunable gap in graphene on hexagonal boron nitride *Phys. Rev.* B **86** 115415
[109] Wallbank J R, Patel A A, Mucha-Kruczyński M, Geim A K and Fal'ko V I 2013 Generic miniband structure of graphene on a hexagonal substrate *Phys. Rev.* B **87** 245408
[110] Song J C W, Shytov A V and Levitov L S 2013 Electron interactions and gap opening in graphene superlattices *Phys. Rev. Lett.* **111** 266801
[111] Lee M, Wallbank J R, Gallagher P, Watanabe K, Taniguchi T, Fal'ko V I and Goldhaber-Gordon D 2016 Ballistic miniband conduction in a graphene superlattice *Science* **353** 1526–9
[112] Wang J, Ma F, Liang W, Wang R and Sun M 2017 Optical, photonic and optoelectronic properties of graphene, h-BN and their hybrid materials *Nanophotonics* **6** 943–76
[113] Park C-H, Yang L, Son Y-W, Cohen M L and Louie S G 2008 Anisotropic behaviours of massless Dirac fermions in graphene under periodic potentials *Nat. Phys.* **4** 213–7
[114] Utama M I B *et al* 2021 Visualization of the flat electronic band in twisted bilayer graphene near the magic angle twist *Nat. Phys.* **17** 184–8
[115] Berdyugin A I *et al* 2020 Minibands in twisted bilayer graphene probed by magnetic focusing *Sci. Adv.* **6** eaay7838
[116] Ulstrup S *et al* 2020 Direct observation of minibands in a twisted graphene/WS_2 bilayer *Sci. Adv.* **6** eaay6104
[117] Liu J-M and Lin I-T 2018 *Graphene Photonics* (Cambridge: Cambridge University Press)
[118] Xie H 2019 Probing excitonic mechanics in suspended and strained transition metal dichalcogenides monolayers *PhD Thesis* The Pennsylvania State University, University Park, PA
[119] MacNeill D, Heikes C, Mak K F, Anderson Z, Kormányos A, Zólyomi V, Park J and Ralph D C 2015 Breaking of valley degeneracy by magnetic field in monolayer $MoSe_2$ *Phys. Rev. Lett.* **114** 037401
[120] Li Y *et al* 2014 Valley splitting and polarization by the Zeeman effect in monolayer $MoSe_2$ *Phys. Rev. Lett.* **113** 266804
[121] Aivazian G, Gong Z, Jones A M, Chu R-L, Yan J, Mandrus D G, Zhang C, Cobden D, Yao W and Xu X 2015 Magnetic control of valley pseudospin in monolayer WSe_2 *Nat. Phys.* **11** 148–52
[122] Srivastava A, Sidler M, Allain A V, Lembke D S, Kis A and Imamoğlu A 2015 Valley Zeeman effect in elementary optical excitations of monolayer WSe_2 *Nat. Phys.* **11** 141–7
[123] Wang G, Bouet L, Glazov M M, Amand T, Ivchenko E L, Palleau E, Marie X and Urbaszek B 2015 Magneto-optics in transition metal diselenide monolayers *2D Mater.* **2** 034002
[124] Stier A V, McCreary K M, Jonker B T, Kono J and Crooker S A 2016 Exciton diamagnetic shifts and valley Zeeman effects in monolayer WS_2 and MoS_2 to 65 Tesla *Nat. Commun.* **7** 10643

[125] Cao T *et al* 2012 Valley-selective circular dichroism of monolayer molybdenum disulphide *Nat. Commun.* **3** 887
[126] Feng W, Yao Y, Zhu W, Zhou J, Yao W and Xiao D 2012 Intrinsic spin Hall effect in monolayers of group-VI dichalcogenides: a first-principles study *Phys. Rev.* B **86** 165108
[127] Tang C S, Yin X and Wee A T S (ed) 2023 *Two-Dimensional Transition-Metal Dichalcogenides: Phase Engineering and Applications in Electronics and Optoelectronics* (Weinheim: Wiley-VCH)
[128] Conder K 2018 Optimisation of transition metal dichalcogenide devices for measurement of spin-Hall voltages generated by optical spin orientation *PhD Thesis* University of Exeter, Exeter
[129] Hatsugai Y 1993 Chern number and edge states in the integer quantum Hall effect *Phys. Rev. Lett.* **71** 3697–700
[130] Kotetes P 2019 *Topological Insulators* (San Rafael, CA: Morgan & Claypool Publishers)
[131] Thouless D J, Kohmoto M, Nightingale M P and den Nijs M 1982 Quantized Hall conductance in a two-dimensional periodic potential *Phys. Rev. Lett.* **49** 405–8
[132] Marino E C 2017 *Quantum Field Theory Approach to Condensed Matter Physics* (Cambridge: Cambridge University Press)
[133] Ferraz A, Gupta K S, Semenoff G W and Sodano P (ed) 2020 *Strongly Coupled Field Theories for Condensed Matter and Quantum Information Theory: Proceedings, International Institute of Physics, Natal, Rn, Brazil, 2–21 August 2015* Springer Proceedings in Physics **vol 239** (Cham: Springer)
[134] Larson J, Sjöqvist E and Öhberg P 2020 *Conical Intersections in Physics: An Introduction to Synthetic Gauge Theories* Lecture Notes in Physics **vol 965** (Cham: Springer)
[135] Ezawa M 2015 Monolayer topological insulators: silicene, germanene, and stanene *J. Phys. Soc. Jpn.* **84** 121003
[136] Akitsu T (ed) 2018 *Symmetry (Group Theory) and Mathematical Treatment in Chemistry* (Rijeka: IntechOpen)
[137] Vogt P and Le Lay G (ed) 2018 *Silicene: Prediction, Synthesis, Application* NanoScience and Technology (Cham: Springer)
[138] Zhang F, MacDonald A H and Mele E J 2013 Valley Chern numbers and boundary modes in gapped bilayer graphene *Proc. Natl. Acad. Sci. USA* **110** 10546–51
[139] Ezawa M 2013 Topological Kirchhoff law and bulk-edge correspondence for valley Chern and spin-valley Chern numbers *Phys. Rev.* B **88** 161406
[140] Tse W-K, Qiao Z, Yao Y, MacDonald A H and Niu Q 2011 Quantum anomalous Hall effect in single-layer and bilayer graphene *Phys. Rev.* B **83** 155447
[141] Ezawa M 2014 Symmetry protected topological charge in symmetry broken phase: spin-Chern, spin-valley-Chern and mirror-Chern numbers *Phys. Lett.* A **378** 1180–4
[142] Ezawa M 2014 From Graphene to Silicene: A New 2D Topological Insulator *JPS Conf. Proc.* **1** 012003
[143] Semenoff G W, Semenoff V and Zhou F 2008 Domain walls in gapped graphene *Phys. Rev. Lett.* **101** 087204
[144] Wang Z, Cheng S, Liu X and Jiang H 2021 Topological kink states in graphene *Nanotechnology* **32** 402001
[145] Raza H (ed) 2012 *Graphene Nanoelectronics: Metrology, Synthesis, Properties and Applications* NanoScience and Technology (Heidelberg: Springer)

[146] Wang S K, Wang J and Chan K S 2014 Multiple topological interface states in silicene *New J. Phys.* **16** 045015

[147] Yao W, Yang S A and Niu Q 2009 Edge states in graphene: from gapped flat-band to gapless chiral modes *Phys. Rev. Lett.* **102** 096801

[148] Franz M and Molenkamp L (ed) 2013 *Topological Insulators* Contemporary Concepts of Condensed Matter Science **vol 6** (Oxford: Elsevier)

[149] Fukui T, Shiozaki K, Fujiwara T and Fujimoto S 2012 Bulk-edge correspondence for Chern topological phases: a viewpoint from a generalized index theorem *J. Phys. Soc. Jpn.* **81** 114602

[150] Mong R S K and Shivamoggi V 2011 Edge states and the bulk-boundary correspondence in Dirac Hamiltonians *Phys. Rev.* B **83** 125109

[151] Li J, Morpurgo A F, Büttiker M and Martin I 2010 Marginality of bulk-edge correspondence for single-valley Hamiltonians *Phys. Rev.* B **82** 245404

[152] Li J, Martin I, Büttiker M and Morpurgo A F 2012 Marginal topological properties of graphene: a comparison with topological insulators *Phys. Scr.* **2012** 014021

[153] Song J, Liu H, Jiang H, Sun Q-f and Xie X C 2012 One-dimensional quantum channel in a graphene line defect *Phys. Rev.* B **86** 085437

[154] Yang J-E, Lü X-L, Zhang C-X and Xie H 2020 Topological spin–valley filtering effects based on hybrid silicene-like nanoribbons *New J. Phys.* **22** 053034

[155] Sun Y, Zhao H, Yu Z-M and Pan H 2019 Valley current and spin-valley filter in topological domain wall *J. Appl. Phys.* **125** 123904

[156] Cheng S-g, Zhou J, Jiang H and Sun Q-F 2016 The valley filter efficiency of monolayer graphene and bilayer graphene line defect model *New J. Phys.* **18** 103024

[157] Prikoszovich K 2022 Valley-filters using kink states in bilayer graphene *PhD Thesis* Technische Universität Wien, Wien

[158] Recher P, Trauzettel B, Rycerz A, Blanter Y M, Beenakker C W J and Morpurgo A F 2007 Aharonov-Bohm effect and broken valley degeneracy in graphene rings *Phys. Rev.* B **76** 235404

[159] Rycerz A, Tworzydło J and Beenakker C W J 2007 Valley filter and valley valve in graphene *Nat. Phys.* **3** 172–5

[160] Rycerz A 2008 Nonequilibrium valley polarization in graphene nanoconstrictions *Phys. Stat. Sol. (a)* **205** 1281–9

[161] Oostinga J B, Heersche H B, Liu X, Morpurgo A F and Vandersypen L M K 2008 Gate-induced insulating state in bilayer graphene devices *Nat. Mater.* **7** 151–7

Chapter 3

Strain effect on valley polarisation (VP)

2D nanosheets are usually deposited on substrates, where the interfacial lattice mismatch can induce the deformation of the 2D nanosheets [1, 2]. Therefore, it is essential to investigate how strain due to these deformations affects the physical properties of graphene [3–5]. Strain engineering emerges as a powerful tool to manipulate valley properties, as the strain on the bond length can affect the hopping parameters between neighbouring atoms, and further introduces asymmetrics in energy landscapes for two inequivalent valleys as required by TRS [6–10], resulting in the break of valley degeneracy. Fujita *et al* [11] first proposed that uniform uniaxial strain can be an effective and robust approach to separating $\boldsymbol{K}_+$ and $\boldsymbol{K}_-$ valley electrons in the space of incident angle. Other proposals of strain-induced VP include mechanical resonator [12], nanobubble [13, 14], snakelike nanoribbon [15] and strained fold-like out-of-plane deformation [16].

This chapter investigates the influence of mechanical strain on VP in 2D materials. It explores theoretical models of strain-induced VP, detailing how different types of strain—uniaxial, biaxial, and shear—affect valley dynamics. Emphasis is placed on the practical implications of strain tuning for valleytronic devices, including enhanced valley filtering, and the development of various formulations of strain. Through this examination, the chapter highlights the potential of strain engineering as a versatile method for advancing valley-based technologies.

3.1 Effect of uniaxial strain

In this section, we refocus on the strain effect on VP and give a detailed picture of the valley-filtering effect as a function of Fermi energy, the width of the strained region, the smoothness of the strain barrier and the incident angle.

3.1.1 Model

Let us start with the graphene device shown in figure 3.1, which comprises a strained region (region II) denoted by grey background with width D in pristine graphene. The x-axis is oriented along the armchair direction and the strained region extends infinitely along the y-axis (zigzag direction). In the strained region, the graphene sheet is uniformly stretched (or compressed) along a tunable direction, which makes an angle α with the positive x-axis. The strain tensor resulting from stress can be written in terms of strain e_0 and Poisson's ratio ν [18–23], as

$$\varepsilon_0 = \begin{pmatrix} \varepsilon_{xx} & \varepsilon_{xy} \\ \varepsilon_{yx} & \varepsilon_{yy} \end{pmatrix}, \tag{3.1}$$

where normal or extensional strain components

$$\begin{aligned} \varepsilon_{xx} &= e_0(\cos^2\alpha - \nu\sin^2\alpha), \\ \varepsilon_{yy} &= e_0(\sin^2\alpha - \nu\cos^2\alpha), \end{aligned} \tag{3.2}$$

and shear strain components

$$\varepsilon_{xy} = \varepsilon_{yx} = e_0[(1+\nu)\sin\alpha\cos\alpha]. \tag{3.3}$$

For Poisson's ratio, we use the value for the basal plane of graphite, $\nu = 0.165$ [20, 24–26].

3.1.2 Deformation-induced gauge fields in terms of the strain tensor

If a strain is introduced into graphene owing to the presence of stress, the NN hopping parameters γ_0, in general, can be anisotropic. We now repeat the derivation of the effective Hamiltonian [27] in section 1.2.1 with unequal hopping parameters. The Hamiltonian of the strained graphene reads

$$H_s(\boldsymbol{k}) = \begin{pmatrix} 0 & H_{s,AB}(\boldsymbol{k}) \\ H^*_{s,AB}(\boldsymbol{k}) & 0 \end{pmatrix}, \tag{3.4}$$

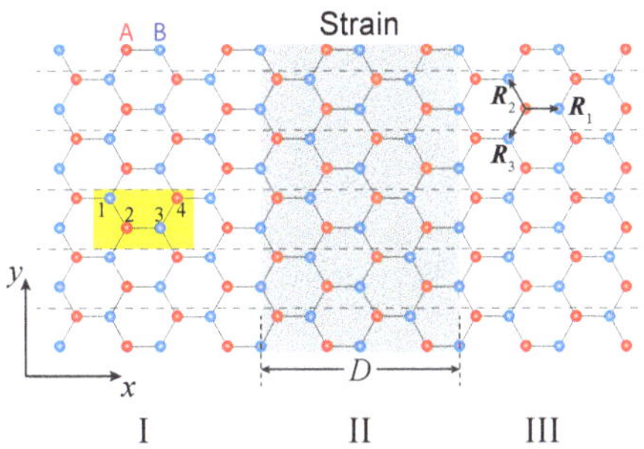

Figure 3.1. The proposed valley filter comprises a strained region (region II) denoted by grey background with width D in pristine graphene. The structure extends infinitely along the y-axis with a period a as indicated by the grey dashed lines, where a is the lattice constant of pristine graphene. The vectors $\boldsymbol{R}_i$ ($i = 1, 2, 3$) connect A (red) atoms to their B (blue) atom neighbours. The yellow rectangle denotes a supercell containing four atoms. Reproduced from [17].

where

$$H_{\mathrm{s,AB}}(\boldsymbol{k}) = -\sum_{i=1}^{3}\gamma_i e^{i\boldsymbol{k}\cdot\boldsymbol{R}_i} \tag{3.5}$$

$$= -\sum_{i=1}^{3}(\gamma_0 + \Delta\gamma_i)e^{i\boldsymbol{k}\cdot\boldsymbol{R}_i} \tag{3.6}$$

$$= -\sum_{i=1}^{3}\gamma_0 e^{i\boldsymbol{k}\cdot\boldsymbol{R}_i} - \sum_{i=1}^{3}\Delta\gamma_i e^{i\boldsymbol{k}\cdot\boldsymbol{R}_i} \tag{3.7}$$

with γ_i ($i = 1, 2, 3$) as the TB hopping parameters under deformation corresponds to $|\boldsymbol{R}_i|$ (equation (1.2)) and $\Delta\gamma_i$ is the correction to γ_0. For equation (3.7), we can see that the first term is the same as that in pristine graphene [27–29], which is shown in equation (1.13) and the second term is the correction due to strain. We expand the Hamiltonian (3.4) in the vicinity of the two inequivalent Dirac points $\boldsymbol{K}_\kappa = \kappa\frac{4\pi}{3a}\hat{\boldsymbol{y}}$ to the first order,

$$H_\kappa(\boldsymbol{p}_\kappa) = v_{\mathrm{F}}\begin{pmatrix} 0 & \chi[(p_{x,\kappa} + \kappa A_x) + i\kappa(p_{y,\kappa} + \kappa A_y)] \\ \chi^*[(p_{x,\kappa} + \kappa A_x) - i\kappa(p_{y,\kappa} + \kappa A_y)] & 0 \end{pmatrix}, \tag{3.8}$$

where momentum $\boldsymbol{p}_\kappa = \hbar\boldsymbol{q}_\kappa$ with $\boldsymbol{q}_\kappa = \boldsymbol{k} - \boldsymbol{K}_\kappa$, which measures the deviation from $\boldsymbol{K}_\kappa$, and the x- and y-components of deformation-induced gauge field $\boldsymbol{A}$ read

$$\begin{aligned} A_x &= \frac{\sqrt{3}}{2v_F}(\Delta\gamma_2 - \Delta\gamma_3), \\ A_y &= \frac{1}{2v_F}(\Delta\gamma_2 + \Delta\gamma_3 - 2\Delta\gamma_1). \end{aligned} \tag{3.9}$$

The phase $\chi = e^{-i\pi/2}$ can be excluded by a unitary transformation of the basis functions [28, 30]. Thus, the effective Hamiltonian (3.8) in the vicinity of $\boldsymbol{K}_\kappa$ can be simplified as

$$H_\kappa(\boldsymbol{p}_\kappa) = v_{\mathrm{F}}\begin{pmatrix} 0 & (p_{x,\kappa} + \kappa A_x) + i\kappa(p_{y,\kappa} + \kappa A_y) \\ (p_{x,\kappa} + \kappa A_x) - i\kappa(p_{y,\kappa} + \kappa A_y) & 0 \end{pmatrix}. \tag{3.10}$$

Now we show the relationship between the deformation-induced gauge field $\boldsymbol{A}$ and strain tensor ε [31].

Under deformation, the TB hopping parameters γ_i can be parameterized as [32, 33]

$$\gamma_i = \gamma_0 \exp\left[-\beta\left(\frac{d_i}{a_{\mathrm{C}}} - 1\right)\right], \tag{3.11}$$

where $a_{\mathrm{C}} = a/\sqrt{3}$ is the C–C bond length in pristine graphene, and the factor β gives the exponential decay of hopping energy with the increase of C–C bond length

d_i corresponds to $|\boldsymbol{R}_i|$. The validity of such parameterization has been verified by DFT calculations [26]. Here we set $\beta = 3.37$ [20, 33, 34].

According to finite strain theory [35], if we only keep the terms up to the first order of strain, the strained C–C bond length d_i can be expressed, in terms of Green–St. Venant or Green–Lagrange strain tensor $\varepsilon = \frac{1}{2}[\nabla \boldsymbol{u} +^{\mathrm{t}}(\nabla \boldsymbol{u}) +^{\mathrm{t}}(\nabla \boldsymbol{u})\nabla \boldsymbol{u}]$ with displacement field $\boldsymbol{u}$, as [35–37]

$$d_i \approx \sqrt{\boldsymbol{R}_i \cdot \boldsymbol{R}_i + 2\boldsymbol{R}_i \cdot \varepsilon \cdot \boldsymbol{R}_i} \approx a_{\mathrm{C}} + \frac{1}{a_{\mathrm{C}}}\boldsymbol{R}_i \cdot \varepsilon \cdot \boldsymbol{R}_i. \tag{3.12}$$

The TB hopping parameters in equation (3.11) can be further written as

$$\gamma_i = \gamma_0 \exp\left(-\frac{\beta}{a_{\mathrm{C}}^2}\boldsymbol{R}_i \cdot \varepsilon \cdot \boldsymbol{R}_i\right) \approx \gamma_0\left(1 - \frac{\beta}{a_{\mathrm{C}}^2}\boldsymbol{R}_i \cdot \varepsilon \cdot \boldsymbol{R}_i\right). \tag{3.13}$$

Thus, the changes in the hopping energy $\Delta\gamma_i = \gamma_i - \gamma_0 = -\beta\gamma_0 \frac{\boldsymbol{R}_i \cdot \varepsilon \cdot \boldsymbol{R}_i}{a_{\mathrm{C}}^2}$ can be further derived by substituting equations (1.2) and (3.1), as

$$\begin{aligned} \Delta\gamma_1 &= -\beta\gamma_0\varepsilon_{xx} \\ \Delta\gamma_2 &= -\beta\gamma_0\left(\frac{1}{4}\varepsilon_{xx} - \frac{\sqrt{3}}{2}\varepsilon_{xy} + \frac{3}{4}\varepsilon_{yy}\right) \\ \Delta\gamma_3 &= -\beta\gamma_0\left(\frac{1}{4}\varepsilon_{xx} + \frac{\sqrt{3}}{2}\varepsilon_{xy} + \frac{3}{4}\varepsilon_{yy}\right). \end{aligned} \tag{3.14}$$

At last, by substituting equation (3.14) into (3.9), the deformation-induced gauge field $\boldsymbol{A}$ can be described in terms of strain tensor. The x- and y-components of $\boldsymbol{A}$ read

$$\begin{aligned} A_x &= \frac{\sqrt{3}\hbar\beta}{a}\varepsilon_{xy}, \\ A_y &= \frac{\sqrt{3}\hbar\beta}{2a}(\varepsilon_{xx} - \varepsilon_{yy}). \end{aligned} \tag{3.15}$$

It is noted that the effects of normal strain components on $|\boldsymbol{R}_2|$ and $|\boldsymbol{R}_3|$ are the same. Together with equation (3.9), we can find that the normal strain components cannot affect A_x. Therefore, the shear and normal strain components give rise to the x- and y-components of $\boldsymbol{A}$, respectively.

The energy band dispersion and the eigenfunction [29] can be obtained from the eigenvalues and eigenvectors of the Hamiltonian (3.10), as

$$\begin{aligned} \mathcal{E}_{\mathrm{F}}(\boldsymbol{q}_\kappa) &= \xi\hbar v_{\mathrm{F}}\sqrt{\left(q_{x,\kappa} + \kappa\frac{A_x}{\hbar}\right)^2 + \left(q_{y,\kappa} + \kappa\frac{A_y}{\hbar}\right)^2}, \\ \psi_\xi^\kappa(\boldsymbol{r}) &= \frac{\mathrm{e}^{\mathrm{i}\boldsymbol{k}\cdot\boldsymbol{r}}}{\sqrt{2}}\begin{pmatrix} \xi\mathrm{e}^{\mathrm{i}\kappa\theta_\kappa} \\ 1 \end{pmatrix}, \end{aligned} \tag{3.16}$$

where $\xi = +1\ (-1)$ denotes the conduction (valence) band, and

$$e^{i\kappa\theta_\kappa} = \frac{\left(q_{x,\kappa} + \kappa\frac{A_x}{\hbar}\right) + i\kappa\left(q_{y,\kappa} + \kappa\frac{A_y}{\hbar}\right)}{|\mathcal{E}_F(\boldsymbol{q}_\kappa)|/(\hbar v_F)}. \tag{3.17}$$

According to equations (3.15) and (3.16), we can see that after the application of strain, the Dirac cone originally at $\boldsymbol{K}_\kappa$ shifts to $\boldsymbol{K}_\kappa - \kappa\frac{\boldsymbol{A}}{\hbar}$ while keeping its shape unchanged [38]. e.g., when tensile strain along the x-axis ($\varepsilon_{xx} > 0$) and negative shear strain ($\varepsilon_{xy} < 0$) [22, 39–42] are applied, the Dirac cone originally at $\boldsymbol{K}_\pm$ shifts along $\pm\hat{\boldsymbol{k}}_x$ direction by a distance $|A_x|/\hbar$ and $\mp\hat{\boldsymbol{k}}_y$ direction by a distance $|A_y|/\hbar$. Figure 3.2(a) illustrates such a situation, which shows the relationship between the original Fermi circle (lighter red/blue dashed circle) and the new Fermi circle (darker red/blue solid circle).

In the pristine region, where $\boldsymbol{A}$ vanishes, the energy band dispersion and the eigenfunction reduce to [28, 30, 43]

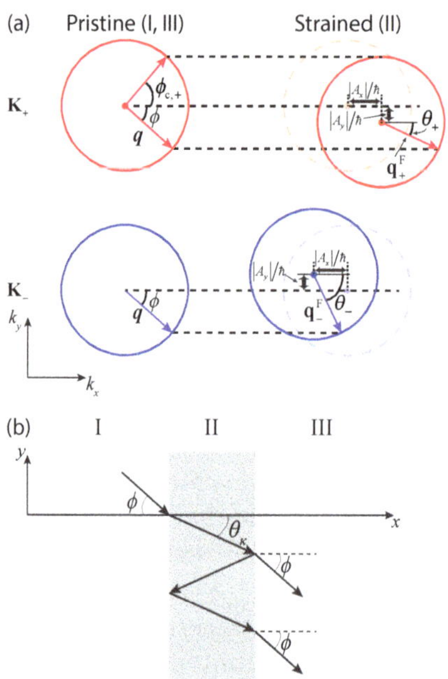

Figure 3.2. (a) Refraction of momentum due to the conservation of energy (equi-energy surfaces denoted by identical Fermi circles) and momentum k_y during the transmission from pristine to strained region (Region II). Here we take $\varepsilon_{xx} > 0$ ($A_y > 0$) and $\varepsilon_{xy} < 0$ ($A_x < 0$) as an example. The red (blue) Fermi circles correspond to the $\boldsymbol{K}_+$ ($\boldsymbol{K}_-$) valley with incident angle ϕ in the pristine region and refraction angle θ_+ (θ_-) in the strained region. (b) Scattering of Dirac electrons in the system with multiple internal reflections inside the strained region (Region II). Region I: incident angle ϕ, region II: refraction angle θ_κ, region III: refraction angle ϕ. Reproduced from [17].

$$\mathcal{E}_{\mathrm{F}}(\boldsymbol{q}) = \xi\hbar v_{\mathrm{F}}|\boldsymbol{q}| = \xi\hbar v_{\mathrm{F}}\sqrt{q_x^2 + q_y^2},$$
$$\psi_{\xi}(\boldsymbol{r}) = \frac{\mathrm{e}^{\mathrm{i}\boldsymbol{k}\cdot\boldsymbol{r}}}{\sqrt{2}}\begin{pmatrix}\xi\mathrm{e}^{\mathrm{i}\kappa\phi}\\ 1\end{pmatrix}, \tag{3.18}$$

with

$$\mathrm{e}^{\mathrm{i}\kappa\phi} = \frac{q_x + \mathrm{i}\kappa q_y}{\sqrt{q_x^2 + q_y^2}}, \tag{3.19}$$

which reproduces equations (1.22) and (1.27).

3.1.3 Valley-polarized transmission across a sharp barrier

Now we analyse the transmission characteristics across the strained graphene structure presented in figure 3.1. Let us consider a de Broglie wave of an electron [38, 44, 45] with energy $\mathcal{E}_{\mathrm{F}}$ and wavevector $\boldsymbol{q} = (q_x, q_y)$ incident from the left (region I), with incident angle ϕ. The electron observes two conservation laws [34, 46]:

1. Energy: from time translational invariance;
2. y-component of momentum p_y or wavevector k_y: from translational invariance along the y-axis.

Figure 3.2(a) shows the diagram illustrating the conservation of y-component momentum in the vicinity of two nonequivalent valleys. When the electron approaches the left boundary of the strained region (region II), the wave is partly reflected to the state with momentum $(-q_x, q_y)$ and partly transmitted to the state with momentum $\boldsymbol{K}_\kappa + (q_{x,\kappa}, q_y) = \boldsymbol{K}_\kappa + \boldsymbol{q}_\kappa^{\mathrm{F}} - \kappa\frac{\boldsymbol{A}}{\hbar}$ with Fermi wavevector $\boldsymbol{q}_\kappa^{\mathrm{F}} = (q_{x,\kappa}^{\mathrm{F}}, q_{y,\kappa}^{\mathrm{F}}) = (q_{x,\kappa}^{\mathrm{F}}, q_y + \kappa\frac{A_y}{\hbar})$ [46–48]. Because of the energy conservation and the unchanged Dirac cone shape under the effect of small strain as can be perceived from equations (3.16) and (3.18), the Fermi circles in the pristine and strained regions are identical. This will result in an unchanged magnitude of the Fermi wavevector in the whole process, i.e., $q = |\boldsymbol{q}| = |\boldsymbol{q}_\kappa^{\mathrm{F}}|$. As a result, the incident angle $\phi = \arctan\frac{q_y}{q_x}$ and the refraction angle $\theta_\kappa = \arctan\frac{q_{y,\kappa}^{\mathrm{F}}}{q_{x,\kappa}^{\mathrm{F}}} = \arctan\frac{q_y + \kappa A_y/\hbar}{q_{x,\kappa} + \kappa A_x/\hbar}$ in the strained region, as illustrated in figures 3.2(a) and (b), should satisfy

$$\sin\phi = \sin\theta_\kappa - \kappa\frac{A_y}{\hbar q}. \tag{3.20}$$

Equation (3.20) states that when $|\sin\phi + \kappa\frac{A_y}{\hbar q}| > 1$, corresponding to the situation in which, for $\boldsymbol{K}_\kappa$ valley, q_y in pristine graphene does not have a corresponding state in strained graphene. Under the circumstances, no angle θ_κ satisfies equation (3.20), thus the interface between regions I and II is entirely reflective for $\boldsymbol{K}_\kappa$ valley electron with incident angle ϕ. This situation is illustrated in figure 3.2(a), where we define critical incident angle $\phi_{c,+}$ for $\boldsymbol{K}_+$ valley electron,

$$\phi_{c,+} = \arcsin\left(1 - \kappa\frac{A_y}{\hbar q}\right). \tag{3.21}$$

When $\phi > \phi_{c,+}$, the system is completely reflective. We can easily find that $\phi_{c,\kappa} = -\phi_{c,-\kappa}$, as guaranteed by the TRS [49].

The wavefunction for $\boldsymbol{K}_\kappa$ valley in the three regions of figure 3.1 can then be written in terms of incident and reflected waves as [12, 43, 47, 50]

$$\begin{aligned}
\Psi_{\text{I}} &= \frac{1}{\sqrt{2}}\left\{\begin{pmatrix} \xi e^{i\kappa\phi} \\ 1 \end{pmatrix} e^{iq_x x} + r_\kappa \begin{pmatrix} \xi e^{i\kappa(\pi-\phi)} \\ 1 \end{pmatrix} e^{-iq_x x}\right\} e^{iq_y y}, \\
\Psi_{\text{II}} &= \frac{1}{\sqrt{2}}\left\{a\begin{pmatrix} \xi e^{i\kappa\theta_\kappa} \\ 1 \end{pmatrix} e^{iq_{x,\kappa}^{\text{F}} x} + b\begin{pmatrix} \xi e^{i\kappa(\pi-\theta_\kappa)} \\ 1 \end{pmatrix} e^{-iq_{x,\kappa}^{\text{F}} x}\right\} e^{iq_y y}, \\
\Psi_{\text{III}} &= \frac{t_\kappa}{\sqrt{2}}\begin{pmatrix} \xi e^{i\kappa\phi} \\ 1 \end{pmatrix} e^{iq_x x} e^{iq_y y}.
\end{aligned} \tag{3.22}$$

Requiring the continuity of the wavefunction at $x = 0$ (interface between Regions I and II) and $x = D$ (interface between Regions II and III) [51, 52], we obtain the following expressions for the coefficients

$$\begin{aligned}
r_\kappa &= \kappa\frac{e^{i\kappa\phi}\sin\left(q_{x,\kappa}^{\text{F}} D\right)(\sin\phi - \sin\theta_\kappa)}{\cos\phi\cos\theta_\kappa\cos\left(q_{x,\kappa}^{\text{F}} D\right) - i\sin\left(q_{x,\kappa}^{\text{F}} D\right)(1 - \sin\phi\sin\theta_\kappa)}, \\
t_\kappa &= \frac{e^{-iq_x D}\cos\phi\cos\theta_\kappa}{\cos\phi\cos\theta_\kappa\cos\left(q_{x,\kappa}^{\text{F}} D\right) - i\sin\left(q_{x,\kappa}^{\text{F}} D\right)(1 - \sin\phi\sin\theta_\kappa)}.
\end{aligned} \tag{3.23}$$

The transmission probability for $\boldsymbol{K}_\kappa$ valley electron can then be obtained straightforwardly from $T_\kappa = t_\kappa t_\kappa^* = 1 - r_\kappa r_\kappa^*$:

$$T_\kappa = \frac{\cos^2\phi\cos^2\theta_\kappa}{\cos^2\phi\cos^2\theta_\kappa + \sin^2\left(q_{x,\kappa}^{\text{F}} D\right)(\sin\phi - \sin\theta_\kappa)^2}. \tag{3.24}$$

Equation (3.24) indicates that under resonance conditions

$$q_{x,\kappa}^{\text{F}} D = n\pi \tag{3.25}$$

with $n \in \mathbf{N}$, the strained region becomes perfectly transparent for $\boldsymbol{K}_\kappa$ electrons ($T_\kappa = 1$). This behaviour can be modelled by multiple internal reflections, as in Fabry–Pérot etalon [53, 54]. In figure 3.2(b), we see two trajectories, the first transmitted through both interfaces and the second reflected at the second interface, which proceeds with two further traversals of the length D to interfere with the first trajectory [55, 56]. The phase $\Delta\varphi$ gained by an electron bouncing between the interfaces is [25, 57, 58]

$$\Delta\varphi = 2\varphi_{\mathrm{WKB}} + \Delta\varphi_1 + \Delta\varphi_2, \tag{3.26}$$

where $\Delta\varphi_1$ ($\Delta\varphi_2$) is the back-reflection phase for interface 1 (2) and

$$\varphi_{\mathrm{WKB}} = \frac{1}{\hbar}\int_0^D p^{\mathrm{F}}_{x,\,\kappa}\,\mathrm{d}x = q^{\mathrm{F}}_{x,\,\kappa}D, \tag{3.27}$$

where $p^{\mathrm{F}}_{x,\,\kappa} = \hbar q^{\mathrm{F}}_{x,\,\kappa}$ is the x-component of Fermi momentum [59] in the strained region. In our system, the incident angles at interfaces 1 and 2 have opposite signs, and thus the jumps in $\Delta\varphi_{1(2)}$ cancel [55]. To exhibit constructive interference, which requires $\Delta\varphi = 2n\pi$, from equations (3.26) and (3.27) we can see that $q^{\mathrm{F}}_{x,\,\kappa}D$ should be multiples of π. This is consistent with the resonance condition shown in equation (3.25).

Similar to equation (1.106), VP can be defined as

$$P = \frac{T_+ - T_-}{T_+ + T_-}. \tag{3.28}$$

When $P = 1$ (-1), the transmission only happens for $\boldsymbol{K}_+$ ($\boldsymbol{K}_-$) valley electrons. It is noted that another major insight into equation (3.24) is that the x-component of the deformation-induced gauge field $\boldsymbol{A}$ does not affect transmission [47] and the VP. Therefore, according to equation (3.15), we will only discuss the nonzero extensional strain component in the strained region, i.e., $\varepsilon_{xx} - \varepsilon_{yy} \neq 0$.

In figure 3.3, we plot the transmission for (a) $\boldsymbol{K}_+$ valley electron T_+ and (b) $\boldsymbol{K}_-$ valley electron T_- as functions of incident angle ϕ and Fermi energy $\mathcal{E}_{\mathrm{F}}$ at $\varepsilon_{xx} - \varepsilon_{yy} = 1\%$. The transmission T_+ at $\varepsilon_{xx} - \varepsilon_{yy} = 2\%$ is plotted in figure (c). The geometrical parameter $D = 100\,\mathrm{nm}$. The plot contains several stripes of transmission peak ($T_+ = 1$) when resonance conditions in equation (3.25) are satisfied. We can see that the transmission T_- can be obtained by taking the mirror symmetry of T_+ about ϕ, i.e., $T_\kappa(\phi) = T_{-\kappa}(-\phi)$, as guaranteed by TRS [60]. In figures 3.3(a) and (b), the white dashed line represents the critical incident angle $\phi_{\mathrm{c},+}$ and $\phi_{\mathrm{c},-}$ as a function of energy $\mathcal{E}_{\mathrm{F}}$, respectively, obtained from equations (3.15), (3.18) and (3.21). For a particular Fermi energy $\mathcal{E}_{\mathrm{F}}$, the transmission T_+ (T_-) is absent when $\phi > \phi_{\mathrm{c},+}$ ($\phi < \phi_{\mathrm{c},-}$).

Now we explain the transmission T_+ shown in figure 3.3(a). In figure 3.3(d), we plot the Fermi circles for $\boldsymbol{K}_+$ valley in pristine and strained graphene, as well as the resonance conditions (vertical lines), equation (3.25), in $\boldsymbol{k}$-space. The Fermi circles represent Fermi energies from 20 to 160 meV with an increment of 20 meV. Because the electron propagates rightward, therefore, each intersection of the resonance condition ($q^{\mathrm{F}}_{x,\,+}D = n\pi$ with $n \in \mathbb{Z}^{0+}$) and the Fermi circle for a given electron energy $\mathcal{E}_{\mathrm{F}}$ in overlapping k_y-space gives $T_+ = 1$.

When $\mathcal{E}_{\mathrm{F}} < 34.1$ meV, as illustrated by the Fermi circle of $\mathcal{E}_{\mathrm{F}} = 20$ meV in figure 3.3(d), the transmission in the whole ϕ-space is absent. This is because the

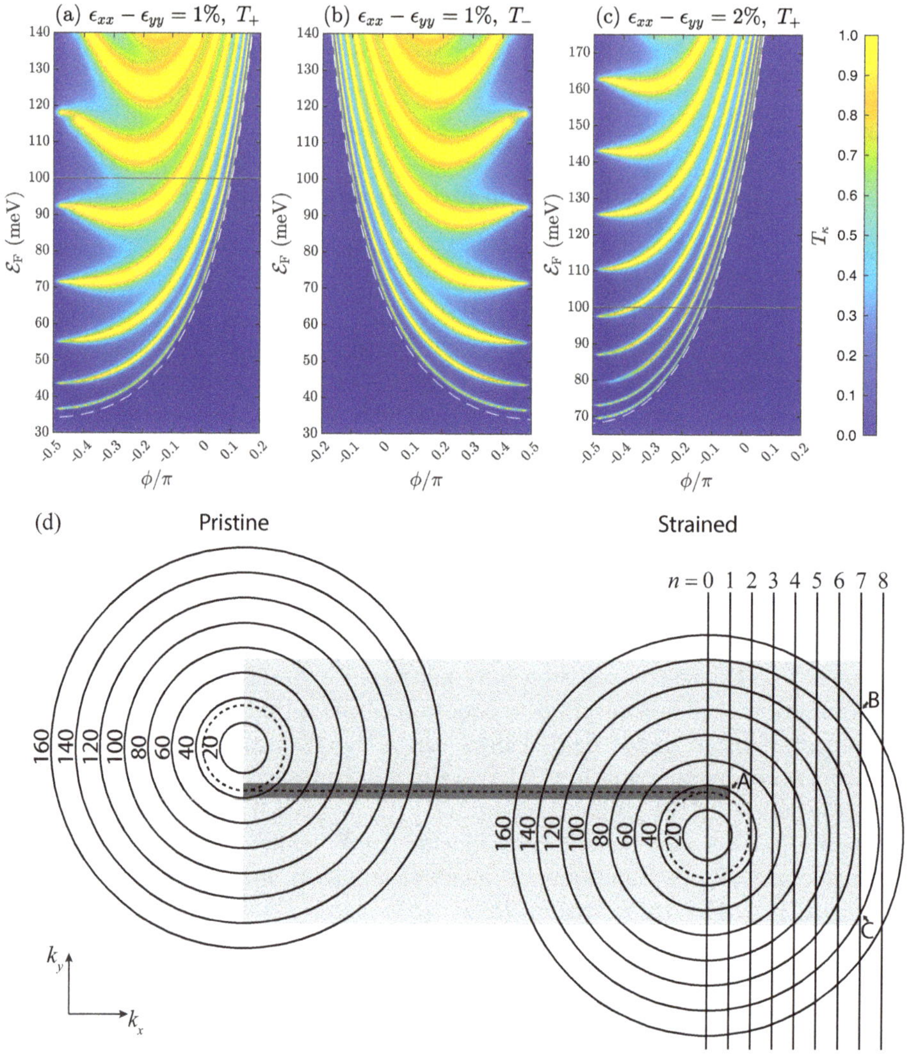

Figure 3.3. Transmission for (a) $\boldsymbol{K}_+$ valley electron T_+ and (b) $\boldsymbol{K}_-$ valley electron T_- as functions of incident angle ϕ and Fermi energy $\mathcal{E}_F$ when $\varepsilon_{xx} - \varepsilon_{yy} = 1\%$, as well as (c) T_+ when $\varepsilon_{xx} - \varepsilon_{yy} = 2\%$. The white dashed line represents the critical incident angle (a) $\phi_{c,+}$, (b) $\phi_{c,-}$ and (c) $\phi_{c,+}$ as a function of Fermi energy. The horizontal lines at $\mathcal{E}_F = 100$ meV in (a) and (c) is plotted to guide the comparison with the results in figure 3.7 at $\ell = 0$. (d) The relative positions of the Fermi circles of pristine and strained graphene with $\varepsilon_{xx} - \varepsilon_{yy} = 1\%$ for $\boldsymbol{K}_+$ valley, as well as the resonance condition, $q^{\mathrm{F}}_{x,+}D = n\pi$ with $n \in \mathbb{Z}^{0+}$, illustrated by an array of vertical lines. The solid Fermi circles represent Fermi energies from 20 to 160 meV with an increment of 20 meV, as labelled beside the Fermi circles. The dashed Fermi circles represent Fermi energy $\mathcal{E}^{\mathrm{i}}_{\mathrm{F}} = 34.1$ meV, which corresponds to the onset of nonzero transmission. The overlapping k_y between the Fermi circles of pristine and strained graphene for $\mathcal{E}_F = 20$ and 140 meV is denoted by dark and light grey, respectively. The geometric parameter $D = 100$ nm. Reproduced from [17]. © 2023 IOP Publishing Ltd. All rights reserved.

diameter of the Fermi circle $2q$ [61] is smaller than the strain-induced shift of the Fermi circle $\frac{|A_y|}{\hbar}$ in the k_y direction. In such a situation, there is no overlap between the Fermi surface of pristine and strained graphene in k_y-space. Notice that in the extreme case, $\mathcal{E}_{\mathrm{F}}^{\mathrm{i}} \approx 34.1$ meV, as shown in the dashed Fermi circle in figure 3.3(d), the diameter of the Fermi circle equals the strain-induced shift of the Fermi circle. From equation (3.15), we have

$$\mathcal{E}_{\mathrm{F}}^{\mathrm{i}} = \frac{v_{\mathrm{F}}|A_y|}{2} = \frac{3}{8}\beta\gamma_0(\varepsilon_{xx} - \varepsilon_{yy}). \tag{3.29}$$

The resonance condition in equation (3.25) can be satisfied with $q_{x,+}^{\mathrm{F}} = 0$, $\phi = -\frac{\pi}{2}$ and $\theta_+ = \frac{\pi}{2}$. However, in this case, T_+ is meaningless since the Fermi wavevector in the strained region $\boldsymbol{q}_+^{\mathrm{F}}$ is parallel to the interface. $\mathcal{E}_{\mathrm{F}}^{\mathrm{i}} \approx 34.1$ meV can be defined as the onset of nonzero transmission $T_\pm$, when $\mathcal{E}_{\mathrm{F}}$ becomes larger, the overlap between the Fermi surface of pristine and strained graphene in k_y-space emerges and critical incident angle $\phi_{\mathrm{c},+}$ increases. The nonzero T_+ can be observed at $\phi \in (-\frac{\pi}{2}, \phi_{\mathrm{c},+})$. This is consistent with reference [11]. When $\mathcal{E}_{\mathrm{F}}$ reaches 36.5 meV, we can observe resonance $T_+ = 1$ at $\phi \approx -\frac{\pi}{2}$, this corresponds to $q_{x,+}^{\mathrm{F}} D = \pi$. After the further increase of $\mathcal{E}_{\mathrm{F}}$, ϕ increases to retain the resonance condition $q_{x,+}^{\mathrm{F}} D = \pi$, as illustrated by the intersection A in figure 3.3(d). At $\phi \approx -\frac{\pi}{2}$, the resonance conditions $q_{x,+}^{\mathrm{F}} D = 2\pi$, 3π and 4π can be observed at $\mathcal{E}_{\mathrm{F}} = 43.5$, 55.1 and 71.5 meV respectively. Each stripe of transmission peak ($T_+ = 1$) denotes the situation of the same $q_{x,+}^{\mathrm{F}}$ which satisfies the resonance condition, for the same n in equation (3.25). When $\mathcal{E}_{\mathrm{F}}$ reaches critical Fermi energy, $\mathcal{E}_{\mathrm{F}}^{\mathrm{c}} \approx 68.2$ meV, $\phi_{\mathrm{c},+} = 0$, this indicates that we can find the total separation of T_+ and T_- in ϕ-space when $\mathcal{E}_{\mathrm{F}} \leqslant 68.2$ meV, i.e., VP $P = 1$ (-1) when $\phi < \phi_{\mathrm{c},+}$ ($\phi > -\phi_{\mathrm{c},+}$), as can be seen from figures 3.3(a) and (b). In this situation, the magnitude of the wavevector q corresponding to the critical Fermi energy $\mathcal{E}_{\mathrm{F}}^{\mathrm{c}}$ equals the strain-induced shift of the Fermi circle, $\frac{|A_y|}{\hbar}$. Therefore, similar to equation (3.29), we have

$$\mathcal{E}_{\mathrm{F}}^{\mathrm{c}} = v_{\mathrm{F}}|A_y| = \frac{3}{4}\beta\gamma_0(\varepsilon_{xx} - \varepsilon_{yy}). \tag{3.30}$$

We can see that $\mathcal{E}_{\mathrm{F}}^{\mathrm{c}} = 2\mathcal{E}_{\mathrm{F}}^{\mathrm{i}}$ from equation (3.29).

Another interesting phenomenon can be found when $\mathcal{E}_{\mathrm{F}} > 80$ meV, the newly emerged resonance condition can be observed twice for a particular $\mathcal{E}_{\mathrm{F}}$. This situation is illustrated for Fermi energy $\mathcal{E}_{\mathrm{F}} = 140$ meV in figure 3.3(d). We can see that the resonance condition $q_{x,+}^{\mathrm{F}} D = 7\pi$ have two intersections with the Fermi circle of $\mathcal{E}_{\mathrm{F}} = 140$ meV in the strained region, labelled as B and C. Both the intersections are in the overlapping k_y-space (lighter grey), and each intersection corresponds to an incident angle ϕ, which represents resonance.

In figure 3.3(c), where we plot T_+ at $\varepsilon_{xx} - \varepsilon_{yy} = 2\%$, the trends of the resonance condition and $\phi_{\mathrm{c},+}$ are similar to the case of $\varepsilon_{xx} - \varepsilon_{yy} = 1\%$ in figure 3.3(a).

However, the Fermi energy corresponding to the onset of nonzero transmission $\mathcal{E}_F^i = 68.2$ meV and the critical Fermi energy $\mathcal{E}_F^c = 136.5$ meV, below which the total separation of T_+ and T_- in ϕ-space can occur, both doubled from the case of $\varepsilon_{xx} - \varepsilon_{yy} = 1\%$. This is an effect of

1. The linear dispersion relation of graphene;
2. The linear relationship between the deformation-induced gauge field $\boldsymbol{A}$ and strain $\varepsilon_{xx} - \varepsilon_{yy}$ as shown in equation (3.15);
3. Both $\mathcal{E}_F^i$ and $\mathcal{E}_F^c$ are proportional to the shift of Fermi surface $\frac{|A_y|}{\hbar}$ due to strain, as can be seen from equations (3.29) and (3.30).

In figure 3.4, we further plot the Fermi energy $\mathcal{E}_F^i$ and the critical Fermi energy $\mathcal{E}_F^c$ as functions of strain $\varepsilon_{xx} - \varepsilon_{yy}$. We can see that $\mathcal{E}_F^i$ and $\mathcal{E}_F^c$ are linearly proportional to $\varepsilon_{xx} - \varepsilon_{yy}$ with slope $3\beta\gamma_0/8$ and $3\beta\gamma_0/4$, respectively, according to equations (3.29) and (3.30). The increase of $\varepsilon_{xx} - \varepsilon_{yy}$ results in the total separation of T_+ and T_- in a larger Fermi energy range. Experimentally, the Fermi energy in graphene can be modulated by various strategies [62], including heteroatom substitution [63, 64], physisorption [65], and dynamic doping modulation using external stimuli [66, 67].

In figure 3.5, by doubling the width of the strained region, $D = 200$ nm, we plot the transmission for $\boldsymbol{K}_+$ valley electron T_+ as a function of incident angle ϕ and energy $\mathcal{E}_F$ at $\varepsilon_{xx} - \varepsilon_{yy} = 1\%$. Compared with figure 3.3(a), after the increase of D, the resonance conditions illustrated in figure 3.3(d) become denser, and more situations in $\mathcal{E}_F$–ϕ space will satisfy the resonance condition. This can be witnessed by more yellow stripes in figure 3.5 compared with figure 3.3(a). It is noted that the resonance conditions in figure 3.3(a) can still be spotted in figure 3.5, since the set of $q_{x,\kappa}^F$'s which satisfy equation (3.25) for $D = 100$ nm is a subset of $D = 200$ nm. Therefore, D should be increased to have larger transmission due to more resonance.

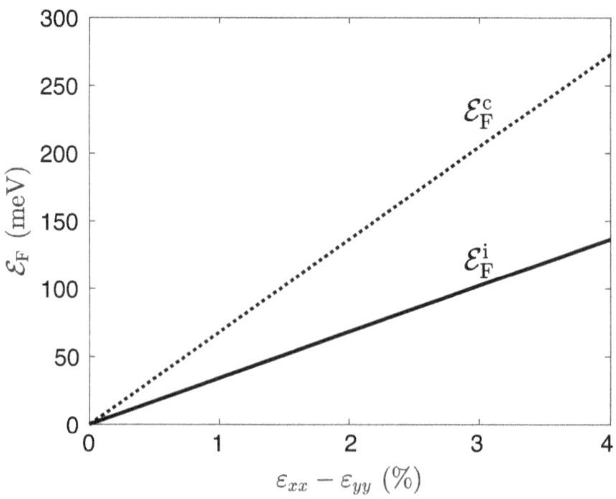

Figure 3.4. (Colour online) The Fermi energy $\mathcal{E}_F^i$ (solid line) corresponds to the onset of nonzero transmission $T_\pm$ and the critical Fermi energy $\mathcal{E}_F^c$ (dotted line) below which the total separation of T_+ and T_- in ϕ-space can occur, as a function of strain $\varepsilon_{xx} - \varepsilon_{yy}$. Reproduced from [17].

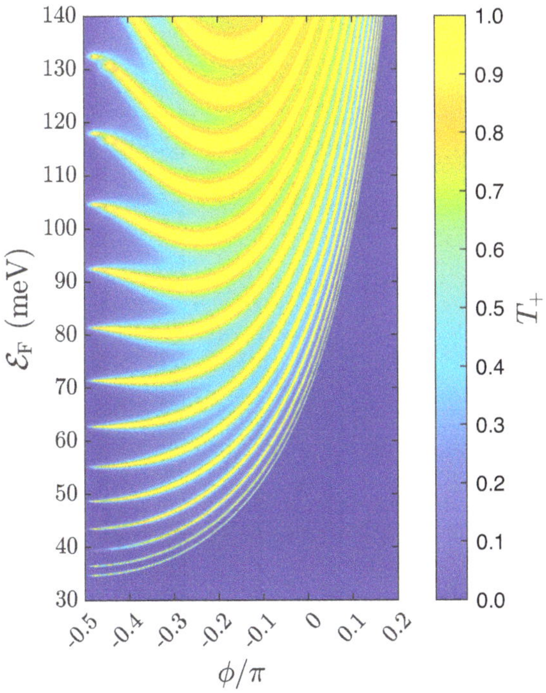

Figure 3.5. (Colour online) Transmission for K_+ valley electron T_+ as functions of incident angle ϕ and Fermi energy $\mathcal{E}_F$ when $\varepsilon_{xx} - \varepsilon_{yy} = 1\%$. The geometric parameter $D = 200$ nm. Reproduced from [17]. © 2023 IOP Publishing Ltd. All rights reserved.

3.1.4 Valley-polarized transmission across a smooth strain barrier

Although one can gain substantial insight from the analytical solutions in equation (3.24) to the problem of sharp strain barriers, from realistic and experimental points of view [68], because of the lattice mismatch between the pristine and strained graphene, the strain will always develop incrementally. Such a strain profile will result in a continuous Fermi wavevector and further modify the valley-polarized transmission. Therefore, we consider a smooth strain profile [64, 69, 70] that mimics Wood–Saxon potential [71] with two symmetric steps [72], which reads [73, 74]

$$\varepsilon(x) = \frac{\varepsilon_0}{\tanh(D/4\ell)}\left[\frac{1}{1+\mathrm{e}^{-x/\ell}} - \frac{1}{1+\mathrm{e}^{-(x-D)/\ell}}\right]. \tag{3.31}$$

The prefactor $1/\tanh(D/4\ell)$ is introduced to ensure $\varepsilon(D/2) = \varepsilon_0$. We plot the shape of the strain profile as the solid line in figure 3.6. For such a strain profile, the strain $\varepsilon(x)$ monotonically changes from 0 to ε_0 over a distance of 4ℓ around the interface ($x = 0$ and $x = D$) [70]. In the limit $\ell \to 0$, the strain profile is a sharp barrier shown as the dashed line, such a case is investigated in the previous subsection.

For the smooth strain barrier, the wavefunction matching approach introduced in the previous section is no longer applicable. Therefore, here we introduce the non-equilibrium Green's function (NEGF) method for calculating the transmission probability in this system [75].

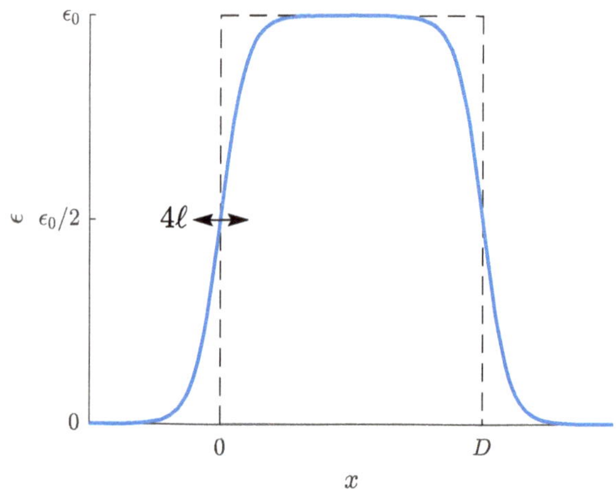

Figure 3.6. Solid line: smooth strain profile shown in equation (3.31) with two symmetric steps. Dashed line: sharp barrier by taking the limit, $\ell \to 0$. Reproduced from [17].

Let us start with the NN TB model of graphene with Hamiltonian [34, 76, 77]

$$H_{\mathrm{ss}} = \sum_{ij} \gamma_{ij}(x) c_i c_j^\dagger, \tag{3.32}$$

where c_i ($c_i^\dagger$) is the annihilation (creation) operator for an electron at site i, and $\gamma_{ij}(x)$ denotes the hopping energy between NN sites labelled by i and j. When the lattice suffers deformation, $\gamma_{ij}(x)$ can bc approximated as [13, 20, 78]

$$\gamma_{ij}(x) = \gamma_0 \exp\left\{-\beta\left[\frac{d_{ij}(x)}{a_{\mathrm{C}}} - 1\right]\right\}. \tag{3.33}$$

Here, $d_{ij}(x)$ is the interatomic distance between sites i and j under strain at coordinate x, which is defined from the midpoint of sites i and j. The relationship between $d_{ij}(x)$ and $\varepsilon(x)$ is [35–37]

$$d_{ij}(x) \approx \sqrt{\boldsymbol{\delta}_{ij} \cdot \boldsymbol{\delta}_{ij} + 2\boldsymbol{\delta}_{ij} \cdot \varepsilon(x) \cdot \boldsymbol{\delta}_{ij}}, \tag{3.34}$$

where $\boldsymbol{\delta}_{ij}$ is the vector which connects the NN sites i and j before the deformation. Notice that here we use the subscript 'ij' instead of 'i' in equations (3.11) and (3.12); this is because, for the sharp strain case, we only have three distinct hopping energies γ_i corresponding to three vectors $\boldsymbol{R}_i$ which connect NNs, while for the smooth strain case, we have infinite numbers of hopping energies from the combinations of sites i and j.

Since the y-component of the wavevector k_y is a conserved quantity because of the translational invariance along the y-axis, the creation and annihilation operators can be transformed in the momentum space according to the Fourier transformation [76]:

$$c_i^\dagger = \sum_{k_y} c_{k_y, i_x} \mathrm{e}^{-\mathrm{i}k_y i_y a} \text{ and } c_i = \sum_{k_y} c_{k_y, i_x} \mathrm{e}^{\mathrm{i}k_y i_y a}. \tag{3.35}$$

In the momentum space, the Hamiltonian matrix can be described as [49, 76]

$$H_{k_y} = -\sum_i \zeta_i^\dagger H_0(x)\zeta_i - \sum_i \zeta_i^\dagger H_1(x)\zeta_{i+1} + \text{h.c.}, \tag{3.36}$$

where $\zeta_i^\dagger = [c_{k_y,\,i,\,1}^\dagger, c_{k_y,\,i,\,2}^\dagger, c_{k_y,\,i,\,3}^\dagger, c_{k_y,\,i,\,4}^\dagger]$ with i representing the position of a supercell. $H_0(x)$ and $H_1(x)$ describe the Hamiltonian matrix of a supercell itself and the interaction between neighbouring supercells, respectively. According to the index of four atoms in a supercell shown in figure 3.1, in pristine graphene, they read

$$H_0 = \gamma_0 \begin{pmatrix} 0 & 2\cos\frac{k_y a}{2} & 0 & 0 \\ 2\cos\frac{k_y a}{2} & 0 & 1 & 0 \\ 0 & 1 & 0 & 2\cos\frac{k_y a}{2} \\ 0 & 0 & 2\cos\frac{k_y a}{2} & 0 \end{pmatrix}, \quad H_1 = \begin{pmatrix} 0 & 0 & 0 & 0 \\ 0 & 0 & 0 & 0 \\ 0 & 0 & 0 & 0 \\ \gamma_0 & 0 & 0 & 0 \end{pmatrix}, \tag{3.37}$$

and the Hamiltonian in the deformed region can be constructed by replacing the corresponding hopping parameter according to equation (3.33). Notice that H_0 has a similar form to Hamiltonian (1.30).

Once the Hamiltonian matrices are constructed, the transmission probability of an electron with energy $\mathcal{E}_\text{F}$ and momentum k_y from the left can be calculated by the NEGF method [64, 79–82],

$$T_{k_y}(\mathcal{E}_\text{F}) = \text{tr}[\Gamma_\text{L}(\mathcal{E}_\text{F})G^\text{r}(\mathcal{E}_\text{F})\Gamma_\text{R}(\mathcal{E}_\text{F})G^\text{a}(\mathcal{E}_\text{F})], \tag{3.38}$$

where $G^{\text{r},\,\text{a}}(\mathcal{E}_\text{F})$ is the retarded (r) or advanced (a) Green's function related to the Hamiltonian and $\Gamma_{\text{L,R}}(\mathcal{E}_\text{F}) = \text{i}[\Sigma_{\text{L, R}}^\text{r}(\mathcal{E}_\text{F}) - \Sigma_{\text{L, R}}^\text{a}(\mathcal{E}_\text{F})]$ is the line-width function with the retarded [advanced] self-energy $\Sigma_{\text{L, R}}^\text{r}(\mathcal{E}_\text{F})$ $[\Sigma_{\text{L, R}}^\text{a}(\mathcal{E}_\text{F})]$ of left or right lead. The left or right lead is represented by a semi-infinite quasi-one-dimensional graphene lattice. The self-energies as well as $G^\text{r}(\mathcal{E}_\text{F})$ can be obtained by the recursive algorithm [64, 83, 84]. For a specific electron energy $\mathcal{E}_\text{F}$ and incident angle ϕ, we can obtain q_y from equation (3.18). Further, in the vicinity of $\boldsymbol{K}_\kappa = (0, \kappa\frac{4\pi}{3a})$, $k_y = \kappa\frac{4\pi}{3a} + q_y$. Therefore, the transmission coefficient from $\boldsymbol{K}_\kappa$ valley can be calculated as a function of the electron incident angle ϕ.

In figure 3.7, we set the electron energy to $\mathcal{E}_\text{F} = 100$ meV, corresponding to Fermi wavelength $\lambda_\text{F} = \frac{\hbar v_\text{F}}{2\pi\mathcal{E}_\text{F}} \approx 0.9$ nm, and show our numerical results of T_κ as functions of incident angle ϕ for $\ell = 0$, 1, 3 and 6 nm at (a) $\varepsilon_{xx} = 1\%$ and (b) 2%. Other parameters are the same as those in figure 3.3. Notice that for $\varepsilon_{xx} = 1\%$, we only show T_+ for clarity, because of the overlapping of nonzero T_+ and T_- in ϕ-space. For $\ell = 0$, the sharp case, figures 3.7(a) and (b) reproduce figure 3.3(a) and (c) at $\mathcal{E}_\text{F} = 100$ meV (horizontal guideline), respectively, with five resonance peaks. After we further increase the smoothness of the strain profile, the transmission increases

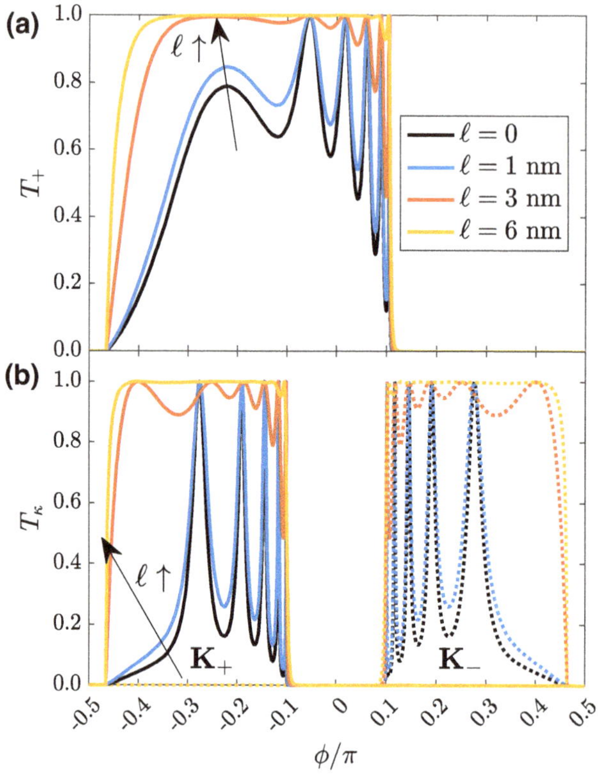

Figure 3.7. The transmission probability T_κ for $\boldsymbol{K}_+$ (solid lines) and $\boldsymbol{K}_-$ valley electrons (dotted lines) as functions of incident angle ϕ for various smooth strain profiles $\ell = 0$, 1, 3 and 6 nm at (a) $\varepsilon_{xx} = 1\%$ and (b) 2%. Reproduced from [17].

and the oscillation of the transmission becomes suppressed. At $\ell = 6$ nm, the transmission almost reaches unity. The range of incident angle ϕ for nonzero transmission is the same for all the values of ℓ; this is governed by the conservation of energy and the y-component of the wavevector depicted in figure 3.2. Such behaviours indicate stronger transmission but suppressed VP in the ϕ region where nonzero T_+ and T_- overlapped. We can also note that the characteristic width [43] of the smooth strain potential ℓ should be compared with the Fermi wavelength λ_F when we investigate transmission probability. Usually, the results from the sharp case in the previous section can be justified if the Fermi wavelength, λ_F, of the electron is much larger than the characteristic width of the smooth strain potential [43]. As we can see from figure 3.7, for $\ell \ll \lambda_F$, the potential profile can be approximated as a sharp potential. For $\ell = 1$ nm $\sim\lambda_F$, the transmission spectra are still similar to the sharp case. However, when $\ell > 1$nm, the detailed shape of the strain profile becomes important.

For practical purposes, it is necessary to measure the directional currents in figures 3.3, 3.5 and 3.7. Experimentally, such currents can be measured using multiple directional leads [38, 49, 85–96], which is schematically illustrated in figure 3.8. The directional currents partly flow to the right lead and partly flow to

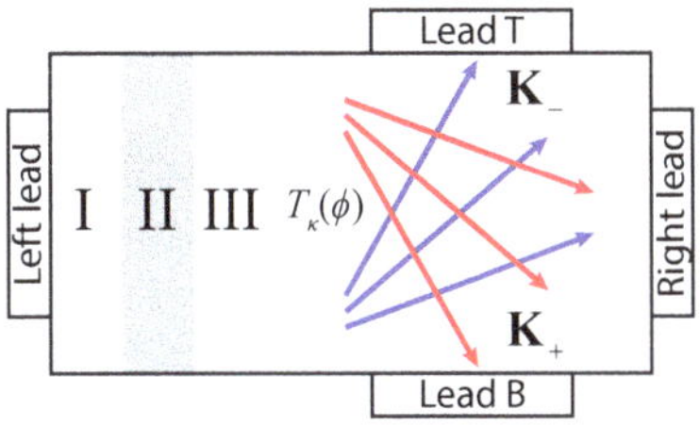

Figure 3.8. (Colour online) Schematics of a device with multiple leads for measuring directional currents. Reproduced from [17].

leads T and B. The currents directed to leads T and B are mainly valley-polarized. In chapter 7, we will introduce an optical method to detect VP.

Up to now, we have discussed the valley filtering behaviour when extensional strain components $\varepsilon_{xx} - \varepsilon_{yy} > 0$. At last, we would like to discuss the effect of the strain and crystal orientation on the polarity of the valley filter.

1. Reverse the extensional strain components $(\varepsilon_{xx} - \varepsilon_{yy})$, which flips the effect of ϕ, i.e., $T_\kappa(\phi) \to T_\kappa(-\phi)$;
2. Change the shear strain component (ε_{xy}), which does not affect the result, since T_κ is independent of A_x;
3. If the transmission is along the zigzag direction of graphene, only the shear component of strain can result in VP, since the extensional strain cannot give rise the deformation-induced gauge field in the transverse direction [8, 12, 31, 97].

3.2 Local Gaussian bump for valley filtering

The previous section has shown that the most important effect produced by deformation is the strain-induced gauge field in the continuum Hamiltonian of graphene. If the zigzag direction of graphene is positioned along the x-axis, in the $\{\boldsymbol{K}_+A, \boldsymbol{K}_+B, -\boldsymbol{K}_-B, \boldsymbol{K}_-A\}$ basis, the low-energy continuum model which includes the effect of deformation reads [31]

$$H_\kappa = v_F \boldsymbol{\sigma} \cdot (\boldsymbol{p} + \kappa \boldsymbol{A}), \tag{3.39}$$

where pseudovector

$$A_x = -\frac{\hbar\beta}{2a}(\varepsilon_{xx} - \varepsilon_{yy}), \tag{3.40}$$

$$A_y = \frac{\hbar\beta}{a}\varepsilon_{xy}, \tag{3.41}$$

with the strain tensor components given by [98]

$$\varepsilon_{ij} = \frac{1}{2}(\partial_j u_i + \partial_i u_j + \partial_i z \partial_j z), \tag{3.42}$$

where $u_{i,j}$ and z are in-plane and out-of-plane displacement fields, respectively. In this section, we present a model of out-of-plane deformation, a localized bubble-like shape with circular symmetry, which is naturally seen in experiments or can be easily engineered.

In figure 3.9, we present the schematics of such a bubble-like shape, which is called a Gaussian bump [99]. Effects of the Gaussian bump deformation on the valley-polarized transport properties of graphene have been initiated by several groups [13, 14]. Experimentally, the bump-like structure can be created when graphene membranes deposit on top of a locally rough substrate, or introducing pressure to a substrate with circular aperture [100, 101], or having an intercalated impurity cluster between graphene and a flat substrate [102–104]. In this case, the membrane bends out of plane to accommodate to the roughness of the underlying surface. Using scanning tunnel microscope [105–109] or atomic force microscope tips [110] to lift the graphene membrane from a substrate is also a useful technique.

A natural mathematical description of the Gaussian bump is given by the Monge representation, which is well suited to describe almost planar configurations of surfaces [111]. In Monge representation, a surface is described by a function $z(x, y)$, where x and y are the coordinates of the (flat) Euclidean plane. The Gaussian bump in our study is modelled by

$$z(r) = h_\mathrm{d} \exp\left(-\frac{r^2}{2\sigma_\mathrm{d}^2}\right), \tag{3.43}$$

where $h_\mathrm{d} = 3.5$ nm and $\sigma_\mathrm{d} = 5$ nm are the height and width of the deformation, corresponding to a maximum strain of approximately 8.5%. The results are robust and scalable for other deformation dimensions [99].

The centrosymmetric local structure of the Gaussian bump produces scalar potentials given by

$$A_x = -A_0 \cos 2\theta \tag{3.44}$$

$$A_y = A_0 \sin 2\theta, \tag{3.45}$$

with

$$A_0 = \frac{\hbar\beta}{a}\left(\frac{h_\mathrm{d} r}{2\sigma_\mathrm{d}^2}\right)^2 \exp\left(-\frac{r^2}{\sigma_\mathrm{d}^2}\right). \tag{3.46}$$

Figure 3.9. Schematics of the out-of-plane Gaussian bump of 2D graphene. Reprinted figures with permission from [99]. Copyright (2018) by the American Physical Society.

Since the scalar potential $\boldsymbol{A}$ is in the xOy plane, the only surviving component of the associated pseudomagnetic field (PMF) $\boldsymbol{B}_s = \nabla \times \boldsymbol{A}$ is its z-component

$$B_s^z = -\frac{\hbar\beta}{a}\frac{h_d^2}{2}\left(\frac{r}{\sigma_d^2}\right)^3 \exp\left(-\frac{r^2}{\sigma_d^2}\right)\sin 3\theta. \tag{3.47}$$

Figure 3.10(a) shows the PMF for $\boldsymbol{K}_+$ valley produced by the Gaussian bump, an equal magnitude field but opposite sign is experienced by $\boldsymbol{K}_-$ valley. One can see that the PMF exhibits threefold symmetry with alternating positive and negative regions, which can be inferred from the factor sin 3θ in equation (3.47). The classical circular trajectories expected for such PMF are also shown in figure 3.10(a) for the $\boldsymbol{K}_+$ (blue arrows) and $\boldsymbol{K}_-$ valleys (red arrows).

We suppose a plane wave is incident along the zigzag direction and we show, in figure 3.10(b), the resulting local currents at the lowest resonance energy of the Gaussian bump by employing the patched Green's function approach [112] with recursive routine [113] based on a NN TB model. The size and direction of the arrows indicate the magnitude and direction of the local current. To increase visibility, the arrows are averaged over several sites. We note that the local current is largest at the interface between PMF regions. The spatially resolved current density ($|j_d|$) relative to that without strain ($|j_0|$), evaluated at the rightmost edge of subfigure (b) is presented in figure 3.10(c), which indicates that the bump focuses the initially uniform current at this electron energy.

Comparing the classical circular trajectories in figure 3.10(a) and the direction of local currents in figure 3.10(b), we find that the direction of local currents match the circular trajectories of $\boldsymbol{K}_-$ valley. That is to say, electrons in the $\boldsymbol{K}_+$ valley 'see' only the vortex pattern indicated by blue arrows in figure 3.10(a), which tends to be backscattered. Conversely, $\boldsymbol{K}_-$ electrons 'see' the pattern shown by the red arrows, and are guided through the bump.

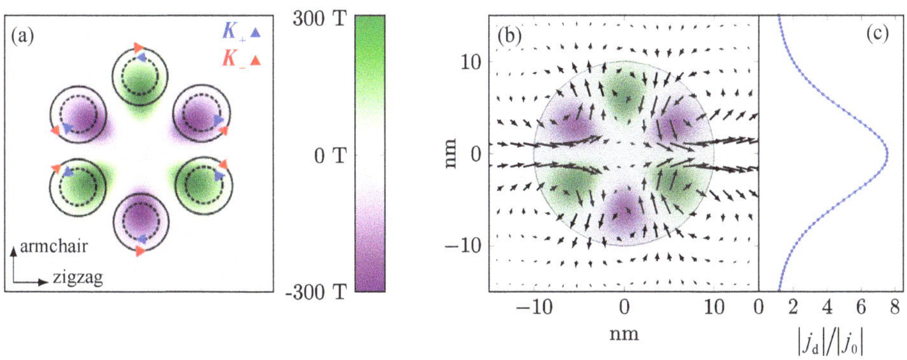

Figure 3.10. (a) The colour map indicates the threefold symmetric flower-like PMF induced by the circularly symmetric Gaussian bump shown in figure 3.9. The vortices show trajectories corresponding to the field experienced in the $\boldsymbol{K}_+$ (blue arrows) and $\boldsymbol{K}_-$ valleys (red arrows). (b) Calculation of the local current incident from the left along the zigzag direction at $\mathcal{E} = 0.01\gamma_0$. The arrows indicate the direction and magnitude of the current. The $r \leqslant 2\sigma_d$ region is shaded in grey. (c) Spatially resolved current density with strain ($|j_d|$), relative to that without strain ($|j_0|$), evaluated at the rightmost edge of subfigure (b), indicating that the bump focuses the initially uniform current. Reprinted figures with permission from [14]. Copyright (2016) by the American Physical Society.

More findings are presented in figure 3.11, in which we compare the local current flow and VP for different incident directions. The real space map of relative occupations of $\boldsymbol{K}_+$ and $\boldsymbol{K}_-$ presented in figures 3.11(d)–(f) is produced by computing the spectral density for each value of $\boldsymbol{k}$. The main finding is: The VP originates from the effect of the underlying PMF and strongly depends on the incident direction. For example, for a current incident from the right, the trajectories of $\boldsymbol{K}_+$ valley (Figure 3.10(a)) match the direction of the incoming wave (figure 3.11(b)). Thus, in figure 3.11(e), the corresponding valley occupations show that $\boldsymbol{K}_+$ valley is transmitted from the right, whereas the $\boldsymbol{K}_-$ valley is transmitted from the left. This is consistent with the previous discussion for figures 3.10(a) and (b). Finally, if we compare incidence along the zigzag direction (figures 3.11(a) and (d)) to incidence along the armchair direction (figures 3.11(c) and (f)), we witness a mixture of the two valleys, resulting in low VP for this case.

All the findings above and some further findings can be obtained via the scattering theory in the continuum formalism [31]. A recent study [99] has shown that electrons coming from the two valleys exhibit different scattering cross-sections in the second-order Born approximation. Figure 3.12 presents the differential cross section (in units of Å) for each valley in the low and high energy regimes in the left and right columns, respectively, for an incident electron from three representative directions. In condensed matter physics, the differential cross section characterizes the elastic scattering process. The numerical results show that

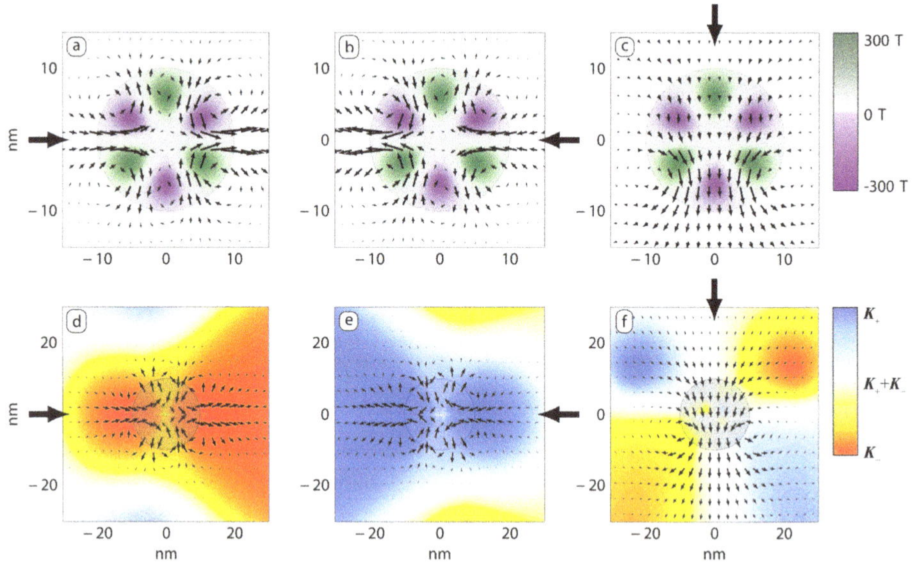

Figure 3.11. Comparing different incident directions for the Gaussian bump. An incident wave with $\mathcal{E} = 0.01\gamma_0$ from the (a) left, (b) right and (c) top as indicated by the large black arrows. Other elements in (a)–(c) are explained in figure 3.10(b). Panel (a) shows the same data as figure 3.10(b) and are reproduced here for convenience. (d)–(f) Real space map of the relative occupation of $\boldsymbol{K}_+$ and $\boldsymbol{K}_-$ corresponding to different incident directions in (a)–(c), respectively. Panels (d) and (e) show real space filtering of the $\boldsymbol{K}_-$ and $\boldsymbol{K}_+$ valleys, respectively. Reprinted figures with permission from [14]. Copyright (2016) by the American Physical Society.

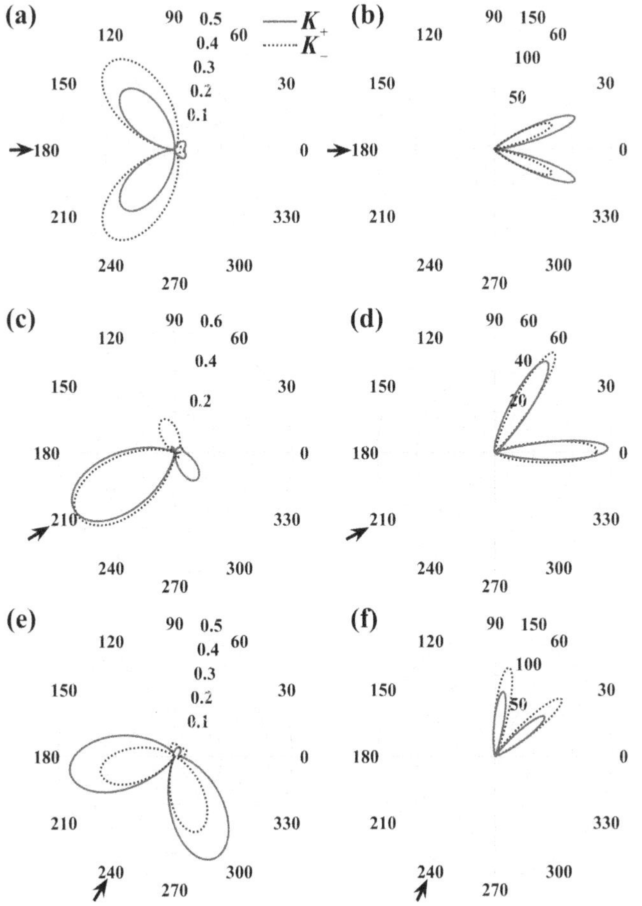

Figure 3.12. Polar plots of the differential cross section (in units of Å) for K_+ (solid line) and K_- valleys (dotted line) with different incident angles indicated by the arrows. Incident angle: first row, 0°, second row, 30°, and third row, 60°. The left (right) column corresponds to electron energy $\mathcal{E} = 20$ (300) meV. Other parameters: $\sigma_d = 10.6$ nm, $h_d = 1.5$ nm. Reprinted figures with permission from [99]. Copyright (2018) by the American Physical Society.

the degree of VP is highly dependent on the incident direction in both regimes. i.e., for particles incident along the armchair directions (odd multiples of 30°), the VP is the weakest, whereas for particles incident along the zigzag direction (even multiples of 30°), the VP is the strongest. In addition, comparing the first with the third row, the VP is reversed as the incident angle changes by 60°. This is because of the relative rotation of the pattern of PMF and the incident direction. All these observations are qualitatively consistent with TB results in figure 3.11. If we increase the energy of the electron, by comparing the first and the second row, the differential cross section also increases. This is because electrons with low (high) energy are more probable to be reflected (transmitted). Thus, valley filtering is more effective at higher energies.

Another important finding in reference [14] is that upon a change of the sign of the Fermi energy, one can switch between the two valleys in the filtered region. This observation allows us to change the valley selectivity by a simple back gate as the Fermi energy is shifted between positive and negative values. Further findings [114] claimed that the valley filtering effect can be enhanced by the scalar potential, which was disregarded in many previous studies.

3.3 Other proposals corresponding to strain

Following the original ideas on deformation-induced valley-polarised current, researchers designed other strain profiles to enrich the applications of deformation in valleytronics. Cavalcante *et al* [15] studied snake states transport in a GNR with a bended region as shown in figure 3.13(a). This particular kind of strain generates a pseudo-magnetic kink barrier along the ribbon width. They find that electrons from opposite valleys propagate in opposite directions. This shows an efficient valley filtering process and the efficiency can be further optimized by tuning Fermi energy.

Carrillo-Bastos *et al* [16] used a Gaussian fold-like out-of-plane deformation to obtain VP, as shown in figure 3.13(b). A fold-like out-of-plane deformation can be prepared through an atomic force microscope indentation approach [110]. Numerical calculations show that a current injected along the fold will be split into two currents: part of the current along the centre of the fold constituted by states from one valley, while the regions around sides of the fold are filled with states from the other valley. If disorder is presented at the edges, the edge states [115] as well as

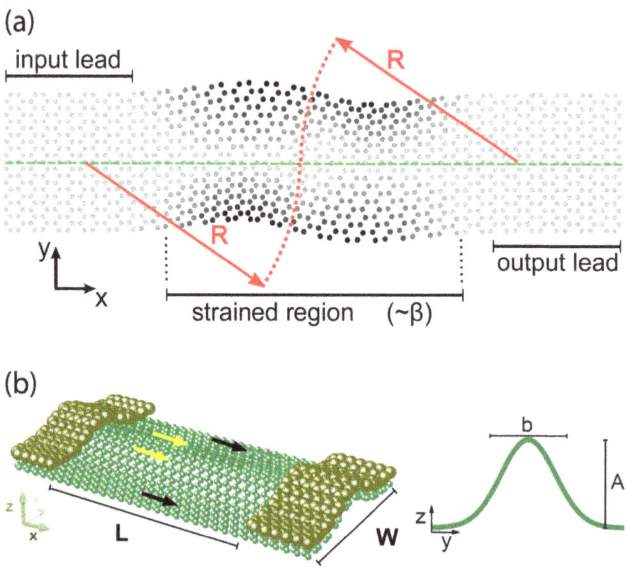

Figure 3.13. Sketch of (a) a particular strained GNR with a bended region. Electrons from opposite valleys propagate in opposite directions. (b) ZGNR with a Gaussian fold-like out-of-plane deformation. Arrows in different colours indicate currents from different valleys. Reprinted figures with permission from [15] and [16], respectively. Copyright (2016) by the American Physical Society.

states further away from the centre of the ribbon would be destroyed. Due to the separation of states in real space, at a given point, the current is expected to be composed mostly by states from one valley, results in VP currents.

References

[1] Ni Z H, Chen W, Fan X F, Kuo J L, Yu T, Wee A T S and Shen Z X 2008 Raman spectroscopy of epitaxial graphene on a SiC substrate *Phys. Rev.* B **77** 115416

[2] Si C, Sun Z and Liu F 2016 Strain engineering of graphene: a review *Nanoscale* **8** 3207–17

[3] Chen X, Deng Z-Y and Ban Y 2014 Delay time and Hartman effect in strain engineered graphene *J. Appl. Phys.* **115** 173703

[4] Wang S and Yu J 2018 Tuning electronic properties of silicane layers by tensile strain and external electric field: a first-principles study **654** 107–15

[5] Wang S and Yu J 2018 Bandgap modulation of partially chlorinated graphene (C_4Cl) nanosheets via biaxial strain and external electric field: a computational study *Appl. Phys.* A **124** 487

[6] Grujić M M, Tadić M and Peeters F M 2014 Spin-valley filtering in strained graphene structures with artificially induced carrier mass and spin-orbit coupling *Phys. Rev. Lett.* **113** 046601

[7] Sasaki K-i and Saito R 2008 Pseudospin and deformation-induced gauge field in graphene *Prog. Theor. Phys. Suppl.* **176** 253–78

[8] Vozmediano M A H, Katsnelson M I and Guinea F 2010 Gauge fields in graphene *Phys. Rep.* **496** 109–48

[9] K-i Sasaki Y, Kawazoe and Saito R 2005 Local energy gap in deformed carbon nanotubes *Prog. Theor. Phys.* **113** 463–80

[10] Castro Neto A H, Guinea F, Peres N M R, Novoselov K S and Geim A K 2009 The electronic properties of graphene *Rev. Mod. Phys.* **81** 109–62

[11] Fujita T, Jalil M B A and Tan S G 2010 Valley filter in strain engineered graphene *Appl. Phys. Lett.* **97** 043508

[12] Jiang Y, Low T, Chang K, Katsnelson M I and Guinea F 2013 Generation of pure bulk valley current in graphene *Phys. Rev. Lett.* **110** 046601

[13] Milovanović S P and Peeters F M 2016 Strain controlled valley filtering in multi-terminal graphene structures *Appl. Phys. Lett.* **109** 203108

[14] Settnes M, Power S R, Brandbyge M and Jauho A-P 2016 Graphene nanobubbles as valley filters and beam splitters *Phys. Rev. Lett.* **117** 276801

[15] Cavalcante L S, Chaves A, da Costa D R, Farias G A and Peeters F M 2016 All-strain based valley filter in graphene nanoribbons using snake states *Phys. Rev.* B **94** 075432

[16] Carrillo-Bastos R, León C, Faria D, Latgé A, Andrei E Y and Sandler N 2016 Strained fold-assisted transport in graphene systems *Phys. Rev.* B **94** 125422

[17] Wang S, Tian H and Sun M 2023 Valley-polarized and enhanced transmission in graphene with a smooth strain profile *J. Phys.: Condens. Matter* **35** 304002

[18] Nye J F 1985 *Physical Properties of Crystals: Their Representation by Tensors and Matrices* (Oxford: Clarendon Press)

[19] Atanackovic T M and Guran A 2000 *Theory of Elasticity for Scientists and Engineers* (New York: Springer Science)

[20] Pereira V M, Castro Neto A H and Peres N M R 2009 Tight-binding approach to uniaxial strain in graphene *Phys. Rev.* B **80** 045401

[21] Kaxiras E and Joannopoulos J D 2019 *Quantum Theory of Materials* (Cambridge: Cambridge University Press)

[22] Sadd M H 2020 *Elasticity: Theory, Applications, and Numerics* 4th edn (Oxford: Academic Press)

[23] Luo Y, Ren C, Xu Y, Yu J, Wang S and Sun M 2021 A first principles investigation on the structural, mechanical, electronic, and catalytic properties of biphenylene *Sci. Rep.* **11** 19008

[24] Blakslee O L, Proctor D G, Seldin E J, Spence G B and Weng T 1970 Elastic constants of compression-annealed pyrolytic graphite *J. Appl. Phys.* **41** 3373–82

[25] Wolf E L 2013 *Graphene: A New Paradigm in Condensed Matter and Device Physics* (Oxford: Oxford University Press)

[26] Ribeiro R M, Pereira V M, Peres N M R, Briddon P R and Castro Neto A H 2009 Strained graphene: tight-binding and density functional calculations *New J. Phys.* **11** 115002

[27] Saito R, Dresselhaus G and Dresselhaus M S 1998 *Physical Properties of Carbon Nanotubes* (London: World Scientific)

[28] Liu J-M and Lin I-T 2018 *Graphene Photonics* (Cambridge: Cambridge University Press)

[29] Foa Torres L E F, Roche S and Charlier J-C 2020 *Introduction to Graphene-Based Nanomaterials: From Electronic Structure to Quantum Transport* 2nd edn (Cambridge: Cambridge University Press)

[30] Katsnelson M I 2020 *The Physics of Graphene* 2nd edn (Cambridge: Cambridge University Press)

[31] Zhai D and Sandler N 2019 Electron dynamics in strained graphene *Mod. Phys. Lett.* B **33** 1930001

[32] Oliva-Leyva M and Naumis G G 2013 Understanding electron behavior in strained graphene as a reciprocal space distortion *Phys. Rev.* B **88** 085430

[33] Botello-Méndez A R, Obeso-Jureidini J C and Naumis G G 2018 Toward an accurate tight-binding model of graphene's electronic properties under strain *J. Phys. Chem.* C **122** 15753–60

[34] Du L, Ren C D, Cui L, Lu W T, Tian H Y and Wang S K 2022 Robust valley filter induced by quantum constructive interference in graphene with line defect and strain *Phys. Scr.* **97** 125825

[35] Kružík M and Roubíček T 2019 *Mathematical Methods in Continuum Mechanics of Solids* Interaction of Mechanics and Mathematics (Cham: Springer)

[36] Reddy J N 2015 *An Introduction to Nonlinear Finite Element Analysis: With Applications to Heat Transfer, Fluid Mechanics, and Solid Mechanics* 2nd edn (Oxford: Oxford University Press)

[37] Eck C, Garcke H and Knabner P 2017 *Mathematical Modeling* Springer Undergraduate Mathematics Series (Cham: Springer)

[38] Nguyen V H, Dechamps S, Dollfus P and Charlier J-C 2016 Valley filtering and electronic optics using polycrystalline graphene *Phys. Rev. Lett.* **117** 247702

[39] Dixit P M and Dixit U S 2008 *Modeling of Metal Forming and Machining Processes: By Finite Element and Soft Computing Methods* Engineering Materials and Processes (London: Springer)

[40] Muvdi B B and Elhouar S 2016 *Mechanics of Materials: With Applications in Excel* 1st edn (Boca Raton, FL: CRC Press)

[41] Kumar P 2022 *Mechanics of Materials: A Friendly Approach* (Singapore: World Scientific)

[42] Muvdi B B and McNabb J W 1991 *Engineering Mechanics of Materials* 3rd edn (New York: Springer)

[43] Katsnelson M I, Novoselov K S and Geim A K 2006 Chiral tunnelling and the Klein paradox in graphene *Nat. Phys.* **2** 620–5
[44] Chiu K-L and Xu Y 2017 Single-electron transport in graphene-like nanostructures *Phys. Rep.* **669** 1–42
[45] Kostadinova E G 2018 *Spectral Approach to Transport Problems in Two-Dimensional Disordered Lattices: Physical Interpretation and Applications* Springer Theses (Cham: Springer)
[46] Allain P E and Fuchs J N 2011 Klein tunneling in graphene: optics with massless electrons *Eur. Phys. J.* B **83** 301
[47] McRae A C, Wei G and Champagne A R 2019 Graphene quantum strain transistors *Phys. Rev. Appl.* **11** 054019
[48] Fogler M M, Guinea F and Katsnelson M I 2008 Pseudomagnetic fields and ballistic transport in a suspended graphene sheet *Phys. Rev. Lett.* **101** 226804
[49] Tian H, Ren C and Wang S 2022 Valleytronics in two-dimensional materials with line defect *Nanotechnology* **33** 212001
[50] Yokoyama T 2014 Spin and valley transports in junctions of Dirac fermions *New J. Phys.* **16** 085005
[51] Gasiorowicz S 2003 *Quantum Physics* 3rd edn (New York: Wiley)
[52] Griffiths D J and Schroeter D F 2018 *Introduction to Quantum Mechanics* 3rd edn (Cambridge: Cambridge University Press)
[53] Fabry C and Pérot A 1899 Théorie et applications d'une nouvelle méthode de spectroscopie interférentielle *Ann. Chim. Phys.* **16** 115–44
[54] Pérot A and Fabry C 1899 On the application of interference phenomena to the solution of various problems of spectroscopy and metrology *Astrophys. J.* **9** 87
[55] Shytov A V, Rudner M S and Levitov L S 2008 Klein backscattering and Fabry-Pérot interference in graphene heterojunctions *Phys. Rev. Lett.* **101** 156804
[56] Shytov A, Rudner M, Gu N, Katsnelson M and Levitov L 2009 Atomic collapse, Lorentz boosts, Klein scattering, and other quantum-relativistic phenomena in graphene *Solid State Commun.* **149** 1087–93
[57] Sokolnikov A U 2017 *Graphene for Defense and Security* (Boca Raton, FL: CRC Press)
[58] Rossi E, Bardarson J H, Brouwer P W and Das Sarma S 2010 Signatures of Klein tunneling in disordered graphene p-n-p junctions *Phys. Rev.* B **81** 121408(R)
[59] Mitin V V, Kochelap V A, Dutta M and Stroscio M A 2019 *Introduction to Optical and Optoelectronic Properties of Nanostructures* (Cambridge: Cambridge University Press)
[60] Wu Z, Zhai F, Peeters F M, Xu H Q and Chang K 2011 Valley-dependent Brewster angles and Goos-Hänchen effect in strained graphene *Phys. Rev. Lett.* **106** 176802
[61] Eisenstein J P, Gramila T J, Pfeiffer L N and West K W 1991 Probing a two-dimensional Fermi surface by tunneling *Phys. Rev.* B **44** 6511–4
[62] Johannsen J C *et al* 2015 Tunable carrier multiplication and cooling in graphene *Nano Lett.* **15** 326–31
[63] Luo Z, Lim S, Tian Z, Shang J, Lai L, MacDonald B, Fu C, Shen Z, Yu T and Lin J 2011 Pyridinic N doped graphene: synthesis, electronic structure, and electrocatalytic property *J. Mater. Chem.* **21** 8038–44
[64] Wang S, Hung N T, Tian H, Islam M S and Saito R 2021 Switching behavior of a heterostructure based on periodically doped graphene nanoribbon *Phys. Rev. Appl.* **16** 024030

[65] Jeong H K, Kim K-j, Kim S M and Lee Y H 2010 Modification of the electronic structures of graphene by viologen *Chem. Phys. Lett.* **498** 168–71
[66] Jang S K, Jang J-r, Choe W-S and Lee S 2015 Harnessing denatured protein for controllable bipolar doping of a monolayer graphene *ACS Appl. Mater. Interfaces* **7** 1250–6
[67] Lee H, Paeng K and Kim I S 2018 A review of doping modulation in graphene *Synth. Met.* **244** 36–47
[68] Zhang X *et al* 2022 Gate-tunable Veselago interference in a bipolar graphene microcavity *Nat. Commun.* **13** 6711
[69] Akhmerov A R, Bardarson J H, Rycerz A and Beenakker C W J 2008 Theory of the valley-valve effect in graphene nanoribbons *Phys. Rev.* B **77** 205416
[70] Wang J, Chen X, Zhu B-F and Zhang S-C 2012 Topological p-n junction *Phys. Rev.* B **85** 235131
[71] Flügge S 1999 *Practical Quantum Mechanics* Classics in Mathematics 1st edn (Berlin: Springer)
[72] Cayssol J, Huard B and Goldhaber-Gordon D 2009 Contact resistance and shot noise in graphene transistors *Phys. Rev.* B **79** 075428
[73] Pellegrino F M D, Angilella G G N and Pucci R 2011 Transport properties of graphene across strain-induced nonuniform velocity profiles *Phys. Rev.* B **84** 195404
[74] Pellegrino F M D, Angilella G G N and Pucci R 2012 Ballistic transport properties across nonuniform strain barriers in graphene *High Press. Res.* **32** 18–22
[75] Cui L, Liu H, Ren C, Yang L, Tian H and Wang S 2023 Influence of local deformation on valley transport properties in the line defect of graphene *Acta Phys. Sin.* **72** 166101
[76] Liu Y, Song J, Li Y, Liu Y and Sun Q-f 2013 Controllable valley polarization using graphene multiple topological line defects *Phys. Rev.* B **87** 195445
[77] Fomin V M (ed) 2025 *Physics of Quantum Rings* NanoScience and Technology 3rd edn (Cham: Springer)
[78] Bonča J and Kruchinin S (ed) 2018 *Nanostructured Materials for the Detection of CBRN* NATO Science for Peace and Security Series A: Chemistry and Biology (Dordrecht: Springer)
[79] Wang J J, Liu S, Wang J and Liu J-F 2018 Valley-coupled transport in graphene with Y-shaped Kekulé structure *Phys. Rev.* B **98** 195436
[80] Wang S K, Wang J and Chan K S 2014 Multiple topological interface states in silicene *New J. Phys.* **16** 045015
[81] Wang S K and Wang J 2015 Valley precession in graphene superlattices *Phys. Rev.* B **92** 075419
[82] Wang S-K, Wang J and Liu J-F 2016 Topological phase in one-dimensional Rashba wire *Chin. Phys.* B **25** 077305
[83] Lopez Sancho M P, Lopez Sancho J M and Rubio J 1984 Quick iterative scheme for the calculation of transfer matrices: application to Mo (100) *J. Phys. F: Met. Phys.* **14** 1205–15
[84] Thorgilsson G, Viktorsson G and Erlingsson S I 2014 Recursive Green's function method for multi-terminal nanostructures *J. Comput. Phys.* **261** 256–66
[85] Gorbachev R V *et al* 2014 Detecting topological currents in graphene superlattices *Science* **346** 448
[86] Sui M *et al* 2015 Gate-tunable topological valley transport in bilayer graphene *Nat. Phys.* **11** 1027–31
[87] Shimazaki Y, Yamamoto M, Borzenets I V, Watanabe K, Taniguchi T and Tarucha S 2015 Generation and detection of pure valley current by electrically induced Berry curvature in bilayer graphene *Nat. Phys.* **11** 1032–6

[88] Rickhaus P, Maurand R, Liu M-H, Weiss M, Richter K and Schönenberger C 2013 Ballistic interferences in suspended graphene *Nat. Commun.* **4** 2342

[89] Rickhaus P, Makk P, Liu M-H, Richter K and Schönenberger C 2015 Gate tuneable beamsplitter in ballistic graphene *Appl. Phys. Lett.* **107** 251901

[90] Rickhaus P, Liu M-H, Makk P, Maurand R, Hess S, Zihlmann S, Weiss M, Richter K and Schönenberger C 2015 Guiding of electrons in a few-mode ballistic graphene channel *Nano Lett.* **15** 5819–25

[91] Lee G-H, Park G-H and Lee H-J 2015 Observation of negative refraction of Dirac fermions in graphene *Nat. Phys.* **11** 925–9

[92] Chen S *et al* 2016 Electron optics with p-n junctions in ballistic graphene *Science* **353** 1522–5

[93] Yu Q *et al* 2011 Control and characterization of individual grains and grain boundaries in graphene grown by chemical vapour deposition *Nat. Mater.* **10** 443–9

[94] Tsen A W, Brown L, Levendorf M P, Ghahari F, Huang P Y, Havener R W, Ruiz-Vargas C S, Muller D A, Kim P and Park J 2012 Tailoring electrical transport across grain boundaries in polycrystalline graphene *Science* **336** 1143–6

[95] Yasaei P *et al* 2014 Chemical sensing with switchable transport channels in graphene grain boundaries *Nat. Commun.* **5** 4911

[96] Kochat V, Tiwary C S, Biswas T, Ramalingam G, Hsieh K, Chattopadhyay K, Raghavan S, Jain M and Ghosh A 2016 Magnitude and origin of electrical noise at individual grain boundaries in graphene *Nano Lett.* **16** 562–7

[97] Oliva-Leyva M and Naumis G G 2015 Generalizing the Fermi velocity of strained graphene from uniform to nonuniform strain *Phys. Lett.* A **379** 2645–51

[98] Landau L D, Lifshitz E M, Kosevich A M and Pitaevskii L P 1986 *Theory of Elasticity* Course of Theoretical Physics **vol 7** 3rd edn (Oxford: Elsevier)

[99] Zhai D and Sandler N 2018 Local versus extended deformed graphene geometries for valley filtering *Phys. Rev.* B **98** 165437

[100] Shin Y, Lozada-Hidalgo M, Sambricio J L, Grigorieva I V, Geim A K and Casiraghi C 2016 Raman spectroscopy of highly pressurized graphene membranes *Appl. Phys. Lett.* **108** 221907

[101] Smith A D *et al* 2016 Piezoresistive properties of suspended graphene membranes under uniaxial and biaxial strain in nanoelectromechanical pressure sensors *ACS Nano* **10** 9879–86

[102] Levy N, Burke S A, Meaker K L, Panlasigui M, Zettl A, Guinea F, Castro Neto A H and Crommie M F 2010 Strain-induced pseudo-magnetic fields greater than 300 tesla in graphene nanobubbles *Science* **329** 544–7

[103] Gao L, Guest J R and Guisinger N P 2010 Epitaxial graphene on Cu(111) *Nano Lett.* **10** 3512–6

[104] Khestanova E, Guinea F, Fumagalli L, Geim A K and Grigorieva I V 2016 Universal shape and pressure inside bubbles appearing in van der Waals heterostructures *Nat. Commun.* **7** 12587

[105] Georgi A *et al* 2017 Tuning the pseudospin polarization of graphene by a pseudomagnetic field *Nano Lett.* **17** 2240–5

[106] Klimov N N, Jung S, Zhu S, Li T, Wright C A, Solares S D, Newell D B, Zhitenev N B and Stroscio J A 2012 Electromechanical properties of graphene drumheads *Science* **336** 1557–61

[107] Mashoff T, Pratzer M, Geringer V, Echtermeyer T J, Lemme M C, Liebmann M and Morgenstern M 2010 Bistability and oscillatory motion of natural nanomembranes appearing within monolayer graphene on silicon dioxide *Nano Lett.* **10** 461–5

[108] Xu P *et al* 2012 Atomic control of strain in freestanding graphene *Phys. Rev.* B **85** 121406

[109] Neek-Amal M, Xu P, Schoelz J K, Ackerman M L, Barber S D, Thibado P M, Sadeghi A and Peeters F M 2014 Thermal mirror buckling in freestanding graphene locally controlled by scanning tunnelling microscopy *Nat. Commun.* **5** 4962

[110] Nemes-Incze P, Kukucska G, Koltai J, Kürti J, Hwang C, Tapasztó L and Biró L P 2017 Preparing local strain patterns in graphene by atomic force microscope based indentation *Sci. Rep.* **7** 3035

[111] Nelson D, Piran T and Weinberg S (ed) 2004 *Statistical Mechanics of Membranes and Surfaces* 2nd edn (Singapore: World Scientific)

[112] Power S R and Ferreira M S 2011 Electronic structure of graphene beyond the linear dispersion regime *Phys. Rev.* B **83** 155432

[113] Settnes M, Power S R, Lin J, Petersen D H and Jauho A-P 2015 Patched Green's function techniques for two-dimensional systems: electronic behavior of bubbles and perforations in graphene *Phys. Rev.* B **91** 125408

[114] Carrillo-Bastos R, Ochoa M, Zavala S A and Mireles F 2018 Enhanced asymmetric valley scattering by scalar fields in nonuniform out-of-plane deformations in graphene *Phys. Rev.* B **98** 165436

[115] Fujita M, Wakabayashi K, Nakada K and Kusakabe K 1996 Peculiar localized state at zigzag graphite edge *J. Phys. Soc. Jpn.* **65** 1920–3

IOP Publishing

Two-dimensional Valleytronic Materials
From principles to device applications
Sake Wang and Hongyu Tian

Chapter 4

Other valley filter and valve effects

This chapter serves as a supplement to the preceding discussions on valley filters and valve effects, offering a deeper exploration of lesser-known or emerging phenomena in valleytronics. Building on the foundational understanding of valleytronics, this chapter begins by analyzing the effects of trigonal warping (TW), which distort electronic bands and influence valley transport properties. Next, it examines the impact of spin–orbit coupling (SOC), particularly how it modulates valley filtering and spin–valley interactions.

The chapter also introduces ferrovalley materials, which exhibit spontaneous valley polarization (VP) due to intrinsic magnetic ordering, providing new opportunities for valleytronics applications. Finally, it discusses the detection of VP through superconducting contacts, a novel approach that offers high sensitivity and could pave the way for future device integration. Collectively, these effects present diverse methodologies for controlling and utilizing valley degrees of freedom (DOF), enhancing the versatility of valleytronic devices in next-generation electronics.

4.1 Trigonal warping effects

In the following year of the proposals to develop valleytronic devices [1, 2], there have been several works that investigated the VP of the transmitted electrons tunnelling through potential barriers [3–5]. These studies focused solely on the effect of TW [6] on the tunnelling probability. As we discussed in section 1.2.2, inherited from crystal symmetry, TW is a modification of the isotropic dispersion relation of energy bands as the energy scale increases. Their results [3, 4] show that valley-dependent transmission can be induced by considering TW.

If the zigzag direction of graphene is positioned along the x-axis, in the TB approximation of the low energy band structure for graphene, the dispersion relation

doi:10.1088/978-0-7503-5562-9ch4

including both the linear and TW terms can be obtained by a procedure similar to the one mentioned in section 1.2.2. It reads [4, 7]

$$\mathcal{E}_{\mathrm{F}}(\boldsymbol{q}) = \pm \hbar v_{\mathrm{F}} \sqrt{q^2 + \kappa \frac{a_{\mathrm{C}}}{2}\left(3q_x q_y^2 - q_y^3\right) + \frac{a_{\mathrm{C}}^2 q^4}{16}}. \tag{4.1}$$

Figure 4.1 illustrates the idea of the original proposal [3], which shows a field effect n–p$^+$ device of graphene. Figures 4.1(a) and (b) show the two limiting cases for the relative orientations between the gate and the n region, (a) zigzag and (b) armchair junctions. Figures 4.1(c) and (d) show the direction of group velocity determined by the conservation of momentum parallel to the junction interfaces (k_y) for the two types of barriers. In the p$^+$ region (grey region), the applied gate voltage brings the chemical potential to energies where the TW effect becomes evident. For the (c) zigzag interface, the currents associated to each valley travel at different angle ranges, but symmetric with respect to x-axis, leading to valley-polarized currents. For the (d) armchair interface, the current associated to $\boldsymbol{K}_-$ valley is more dispersed than $\boldsymbol{K}_+$ because the incident angle is almost perpendicular to the Fermi surface for $\boldsymbol{K}_+$ valley.

In the original proposal [3], the authors assumed that the transmission coefficient would not be significantly affected by the inclusion of TW. Thus, they considered only the dependence of the carrier group velocity on the gate voltage. Later, Pereira *et al* demonstrated that the transmission probability can be modified by TW. Their results show that TW introduces an anisotropy in the valley-dependent transmission, thus can be used as a way of generating valley-polarized current in graphene. We

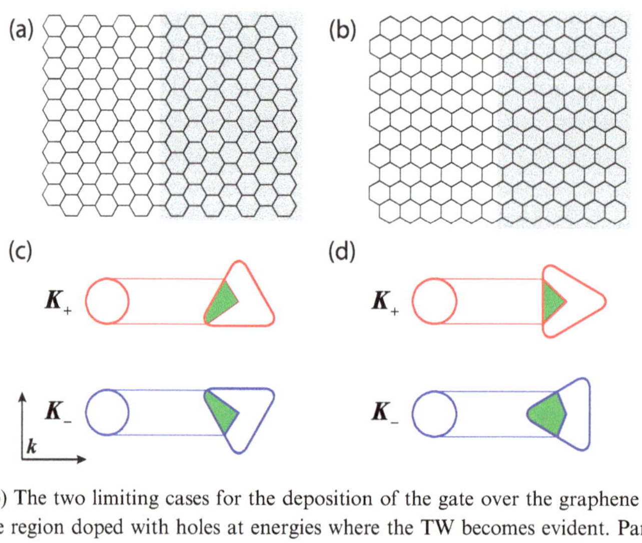

Figure 4.1. (a), (b) The two limiting cases for the deposition of the gate over the graphene sample. The grey region denotes the region doped with holes at energies where the TW becomes evident. Panel (a) zigzag and (b) armchair interface of the gate. Panels (c) and (d) represent the Fermi surface of the two valleys and the conservation of the wavevector (k_y) parallel to the zigzag and armchair interface, respectively. For (c) the zigzag interface, the refractions in the two valleys have different angle ranges. For (d) the armchair interface, the transmission in $\boldsymbol{K}_-$ valley is more dispersed than $\boldsymbol{K}_+$ valley.

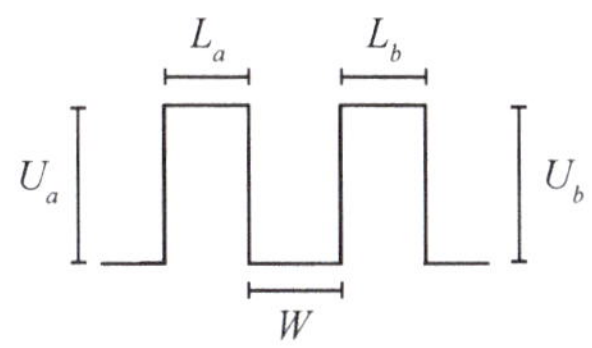

Figure 4.2. Schematic depiction of the double barrier system with potential parameters.

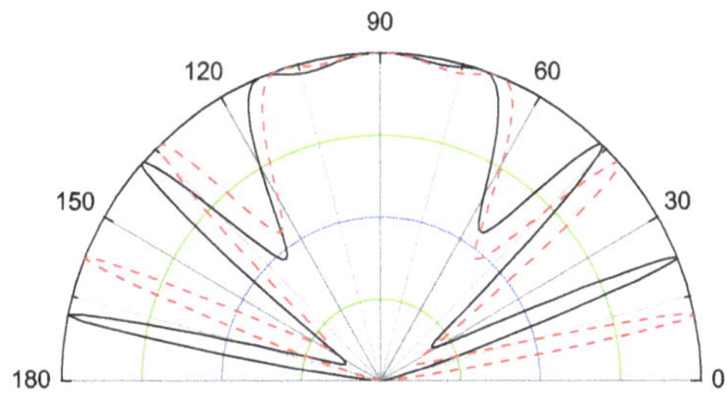

Figure 4.3. Incident angle dependence of the transmission coefficient for a single barrier of the zigzag case, for carriers associated with $\boldsymbol{K}_+$ valleys (red dashed line) and $\boldsymbol{K}_-$ valleys (black solid line). The incident electron has energy $\mathcal{E}_\mathrm{F} = 50$ meV and the potential barrier with width $L = 500$ nm has height $U = 200$ meV. Reproduced from [4].

consider a square potential barrier system and impose the continuity of the wavefunction solutions at the barrier interfaces. For the double barrier case, as can be seen from figure 4.2, the two potential barriers are assumed to have height U_a, U_b, and widths L_a, L_b, separated by a distance W. For the single barrier case, only U_a is nonzero. Depending on the orientation of the interface (zigzag or armchair), we consider two limiting cases: 1) The zigzag case, figure 4.1(a), with the potential interface aligned parallel to the zigzag direction; 2) The armchair case, figure 4.1(b), with the potential interface aligned parallel to the armchair direction.

4.1.1 The zigzag case

In this case, the solutions in different regions are matched along the zigzag direction of the graphene lattice. According to equation (1.29), the x-component of the wavevector is given by

$$q_x = \pm\sqrt{-q_y^2 - \kappa\frac{12q_y}{a_\mathrm{C}} - \frac{8}{a_\mathrm{C}^2} \pm \frac{4}{a_\mathrm{C}^2}\sqrt{2\kappa q_y^3 a_\mathrm{C}^3 + (3\kappa q_y a_\mathrm{C} + 2)^2 + \frac{a_\mathrm{C}^2}{\hbar v_\mathrm{F}}(\mathcal{E}_\mathrm{F} - U)^2}}\,. \tag{4.2}$$

This shows that TW creates an anisotropy of the dispersion, and the value of q_x is sensitive to the sign of q_y. This anisotropy should further affect the transmission properties of electrons through potential barriers. In figure 4.3, we plot the incident angle dependence of the transmission coefficient for a single barrier with $U = 200$ meV and $L = 500$ nm. The incident electron has energy $\mathcal{E}_\mathrm{F} = 50$ meV. We can see the anisotropy creates the asymmetry in the angular dependence of the transmission. Further, compared with the transmission for $\boldsymbol{K}_+$ valley (red dashed line), the range of nonzero transmission for $\boldsymbol{K}_-$ valley (black solid line) shifts to higher incident angles. This is consistent with the schematic drawing in figure 4.1(c). We should also notice that the transmission for $\boldsymbol{K}_+$ valley (red dashed line) is the mirror symmetry of the transmission for $\boldsymbol{K}_-$ valley (black solid line) with respect to incident angle 90°, i.e., $T_+(\mathcal{E}_F, \phi) = T_-(\mathcal{E}_F, \pi - \phi)$. This can be explained by equation (4.2), a change of sign for the valley index κ is equivalent to a change of sign of q_y.

4.1.2 The armchair case

For the armchair case, one has to find q_x by solving the equation

$$\frac{a_C^2}{16}q_x^4 - \kappa\frac{a_C}{2}q_x^3 + \left(\frac{q_y^2 a_C^2}{8} + 1\right)q_x^2 + \kappa\frac{3q_y^2 a}{2}q_x + \left[\frac{q_y^4 a_C^2}{16} + q_y^2 - \left(\frac{\mathcal{E}_F - U}{\hbar v_F}\right)^2\right] = 0. \quad (4.3)$$

For a given energy, the values of q_x are expected to differ for different κ. However, in contrast with the zigzag case, the transmission is expected to be invariant under a transformation $q_y \to -q_y$. The asymmetry in the transmission for two valleys in this case occurs because the transmission in $\boldsymbol{K}_-$ valley is more dispersed than $\boldsymbol{K}_+$ valley, as can be seen from the schematic drawing in figure 4.1(d). Therefore, for the armchair interface, the large potential steps created in p–n junctions can create a valley-polarized current by means of a valley-selective transmission. Another important feature one should notice is that a change of sign for the valley index κ is equivalent to change of sign of q_x. This means that the valley filtering effect for the $\boldsymbol{K}_+$ ($\boldsymbol{K}_-$) valley when carriers incident from the left is the same as the valley filtering effect for the $\boldsymbol{K}_-$ ($\boldsymbol{K}_+$) valley when carriers incident from the left. Figure 4.4 shows the transmission coefficient as a function of energy for a double barrier for $\boldsymbol{K}_+$ (red dashed line) and $\boldsymbol{K}_-$ (black solid line), where (a) $q_y = 0.05$ nm^{-1} and (b) $q_y = 0.01$ nm^{-1}. The red dashed line corresponds to rightward (leftward) and the black solid

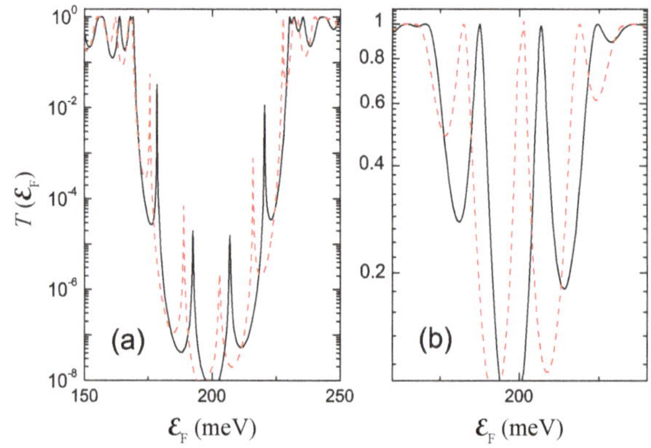

Figure 4.4. Transmission coefficient as a function of energy for a double barrier with armchair interfaces for $\boldsymbol{K}_+$ (red dashed line) and $\boldsymbol{K}_-$ (black solid line) valley, where (a) $q_y = 0.05$ nm^{-1} and (b) $q_y = 0.01$ nm^{-1}. Other parameters are $U_a = U_b = 200$ meV and $L_a = L_b = W = 100$ nm. Reproduced from [4].

line corresponds to leftward (rightward) carriers of the $\boldsymbol{K}_+$ ($\boldsymbol{K}_-$) valley. It shows that the position and distribution of the transmission peaks are momentum-dependent.

Reference [4] also shows that for potential barriers $U \gtrsim 200$ meV, the total transmission peaks of carriers through double barriers are shifted in energy depending on the valley index, which are within reach of current experiments. However, for lower barriers, the VP becomes negligible.

4.2 Spin–orbit coupling effects

In section 1.4.2, we find that the large value of SOC $\lambda_{\mathrm{SO}} = 3.9$ meV in silicene [8] couples the spin and valley DOFs. Besides, the exchange field has been proven to exist at the interface between a ferromagnetic insulator (FMI) and 2D materials, originating from an overlap of electronic wavefunctions [9–11]. Therefore, magnetization could be present [12, 13] in silicene by depositing Fe atoms to the silicene surface or depositing silicene on an appropriate FMI, such as EuO or $Y_3Fe_2(FeO_4)_3$ [14–19], typically EuS, without inducing external Rashba-type interactions [20], as has also been argued for graphene [10, 21–25]. Specially, the exchange field induced in silicene due to the magnetic proximity effect could be of the order of 1 meV for EuO [10]. Theoretical proposals predict that due to the exchange field of the ferromagnetic graphene which couples spin DOF, spin transport can be tuned by a gate voltage [10, 26, 27]. Therefore, one may control valley and spin DOFs independently by considering both SOC and the exchange field [28] (figure 4.5). In this section, we mainly study the valley transport in a normal/ferromagnetic/normal silicene junction, with a gate electrode attached on the ferromagnetic segment, which has length D. We set the x (y)-axis to be perpendicular (parallel) to the interface, and the interfaces are located at $x = 0$ and $x = D$. From sections 1.2.2, 1.4.2 and 2.4, we can have the effective Hamiltonian of the ferromagnetic silicene [8, 14, 29], and it reads

$$H_{\mathrm{Si}}(\boldsymbol{q}) = \hbar v_{\mathrm{F}}(q_x\sigma_x - \kappa q_y\sigma_y) - \Delta_{\mathrm{f}}\sigma_z - sh, \tag{4.4}$$

where $\Delta_{\mathrm{f}} = \kappa s\lambda_{\mathrm{SO}} - \Delta_z$. Δ_z ($-\Delta_z$) is the on-site potential of A (B) sublattices, which can be tuned by an electric field applied perpendicular to the silicene plane due to the buckled structure[1]. h is the exchange field in the ferromagnetic region. $s = 1$ (-1) denotes spin up (down) electrons. In the normal region, we have $\Delta_z = h = 0$. An electron observes energy conservation during the transport process. Also,

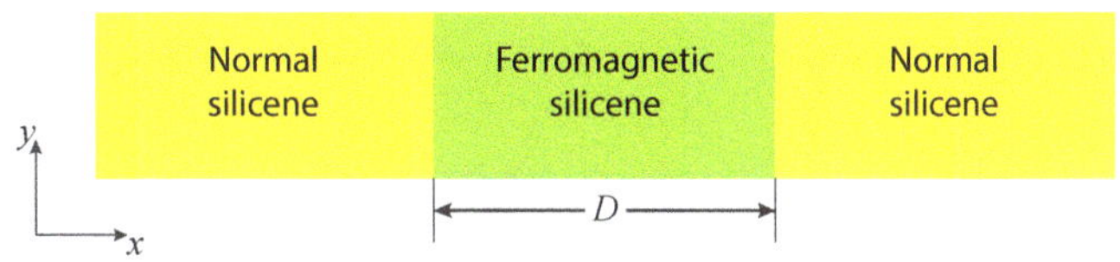

Figure 4.5. Schematic picture of a normal/ferromagnetic/normal silicene junction.

[1] In this section, the definition of Δ_z follows the convention used in reference [28], which is twice the value of the Δ_z used elsewhere in this book.

because of the translational invariance along the y-axis, the y-component of the momentum, k_y, is conservative.

By diagonalising the Hamiltonian, we obtain the eigenenergies in the normal (first line) and ferromagnetic silicene (second line) as follows

$$\begin{aligned}\mathcal{E}_{\xi,\kappa,s} &= \xi\sqrt{(\hbar v_{\mathrm{F}} q)^2 + \Delta_{\mathrm{n}}^2} \\ &= \xi\sqrt{(\hbar v_{\mathrm{F}} q')^2 + \Delta_{\mathrm{f}}^2} - sh,\end{aligned} \tag{4.5}$$

with $\Delta_{\mathrm{n}} = \kappa s \lambda_{\mathrm{SO}}$ and $\xi = 1\ (-1)$ corresponding to the conduction (valence) band. $\boldsymbol{q} = (q_x, q_y)$ and $\boldsymbol{q}' = (q_x', q_y)$ are the momenta in the normal and the ferromagnetic regions. In the ferromagnetic region, we can obtain the following relationship,

$$\mathcal{E}_{\xi,\kappa,s} = -\mathcal{E}_{-\xi,-\kappa,-s}. \tag{4.6}$$

Also, when $h = 0$, $\mathcal{E}_{\xi,\kappa,s} = \mathcal{E}_{\xi,-\kappa,-s}$, the system is both spin- and valley degenerate. When $\Delta_z = 0$, $\mathcal{E}_{\xi,\kappa,s} = \mathcal{E}_{\xi,-\kappa,s}$, the system is valley degenerate. In figure 4.6, we show the band structures near the $\boldsymbol{K}_{\pm}$ point in (a) normal silicene and (b), (c) ferromagnetic silicene. The horizontal line denotes Fermi energy $\mathcal{E}$. Spin up (down) bands are denoted by black solid (red dotted) lines. We set $h/\mathcal{E} = 0.3$ and $\lambda_{\mathrm{SO}}/\mathcal{E} = 0.5$. In (b) and (c), we set $\Delta_z/\mathcal{E} = 0.5$ and 1.5, respectively, since Δ_z can be tuned by the gate electrode connected to ferromagnetic silicene. As shown in figure 4.6(b), for small Δ_z, the bandgap is small and hence the Fermi level crosses the bands near both $\boldsymbol{K}_+$ and $\boldsymbol{K}_-$ points in the ferromagnetic region. Thus, the electrons near both the $\boldsymbol{K}_+$ and $\boldsymbol{K}_-$ points will contribute to the conductance. In figure 4.6(c), for large Δ_z, the Fermi level locates inside the bandgap near the $\boldsymbol{K}_-$ point, while crossing the spin up band near the $\boldsymbol{K}_+$ point. In this case, the current will entirely carried by the electrons near the $\boldsymbol{K}_+$ point.

Now we investigate the transport properties of the system. Since the energy is conserved in the transmission process, we will simply denote $\mathcal{E}_{\xi,\kappa,s}$ as $\mathcal{E}$. By solving equation (4.5), we have $q_x' = (\hbar v_{\mathrm{F}})^{-1}\sqrt{(\mathcal{E} + sh)^2 - (\hbar v_{\mathrm{F}} q_y)^2 - \Delta_{\mathrm{f}}^2}$. The wavefunctions corresponding to the eigenenergies read

$$\psi_{\mathrm{n}} = \frac{1}{\sqrt{2\mathcal{E}\mathcal{E}_{\mathrm{n}}}}\begin{pmatrix}\hbar v_{\mathrm{F}}(q_x + \mathrm{i}\kappa q_y) \\ \mathcal{E}_{\mathrm{n}}\end{pmatrix}, \tag{4.7}$$

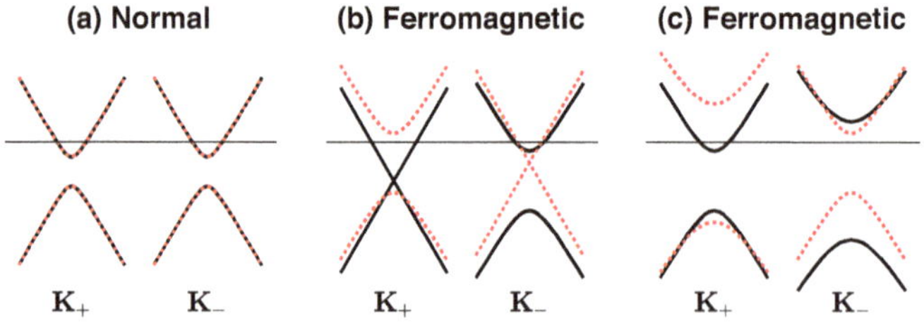

Figure 4.6. Band structures near the $\boldsymbol{K}_{\pm}$ point in (a) normal silicene and (b), (c) ferromagnetic silicene. The horizontal line denotes Fermi energy $\mathcal{E}$. Spin up (down) bands are denoted by black solid (red dotted) lines. We set $h/\mathcal{E} = 0.3$ and $\lambda_{\mathrm{SO}}/\mathcal{E} = 0.5$. In (b) and (c), we set $\Delta_z/\mathcal{E} = 0.5$ and 1.5, respectively.

$$\psi_{\mathrm{f}} = \frac{1}{\sqrt{2\mathcal{E}\mathcal{E}_{\mathrm{f}}}} \begin{pmatrix} \hbar v_{\mathrm{F}}\left(q_x' + \mathrm{i}\kappa q_y\right) \\ \mathcal{E}_{\mathrm{f}} \end{pmatrix}, \tag{4.8}$$

respectively, where $\mathcal{E}_{\mathrm{n}} = \mathcal{E} + \Delta_{\mathrm{n}}$ and $\mathcal{E}_{\mathrm{f}} = \mathcal{E} + sh + \Delta_{\mathrm{f}}$. The wavefunctions in each region can be written in terms of incident and reflected waves as

$$\begin{aligned} \Psi(x<0) &= \frac{1}{\sqrt{2\mathcal{E}\mathcal{E}_{\mathrm{n}}}} \begin{pmatrix} \hbar v_{\mathrm{F}}(q_x + \mathrm{i}\kappa q_y) \\ \mathcal{E}_{\mathrm{n}} \end{pmatrix} \mathrm{e}^{\mathrm{i}q_x x} + \frac{r_{\kappa s}}{\sqrt{2\mathcal{E}\mathcal{E}_{\mathrm{n}}}} \begin{pmatrix} \hbar v_{\mathrm{F}}(-q_x + \mathrm{i}\kappa q_y) \\ \mathcal{E}_{\mathrm{n}} \end{pmatrix} \mathrm{e}^{-\mathrm{i}q_x x}, \\ \Psi(0<x<D) &= a_{\kappa s} \begin{pmatrix} \hbar v_{\mathrm{F}}\left(q_x' + \mathrm{i}\kappa q_y\right) \\ \mathcal{E}_{\mathrm{f}} \end{pmatrix} \mathrm{e}^{\mathrm{i}q_x' x} + b_{\kappa s} \begin{pmatrix} \hbar v_{\mathrm{F}}\left(-q_x' + \mathrm{i}\kappa q_y\right) \\ \mathcal{E}_{\mathrm{f}} \end{pmatrix} \mathrm{e}^{-\mathrm{i}q_x' x}, \\ \Psi(x>D) &= \frac{t_{\kappa s}}{\sqrt{2\mathcal{E}\mathcal{E}_{\mathrm{n}}}} \begin{pmatrix} \hbar v_{\mathrm{F}}(q_x + \mathrm{i}\kappa q_y) \\ \mathcal{E}_{\mathrm{n}} \end{pmatrix} \mathrm{e}^{\mathrm{i}q_x x}. \end{aligned} \tag{4.9}$$

Here, $r_{\kappa s}$ and $t_{\kappa s}$ are reflection and transmission coefficients, respectively. By matching the wavefunctions at the interfaces, we obtain the transmission coefficient

$$t_{\kappa s} = \frac{2q_x q_x' \mathcal{E}_{\mathrm{n}} \mathcal{E}_{\mathrm{f}} \mathrm{e}^{-\mathrm{i}q_x D}}{2q_x q_x' \mathcal{E}_{\mathrm{n}} \mathcal{E}_{\mathrm{f}} \mathrm{e}^{-\mathrm{i}q_x' D} - \mathrm{i}\left\{[q_y(\mathcal{E}_{\mathrm{f}} - \mathcal{E}_{\mathrm{n}})]^2 + \left(q_x \mathcal{E}_{\mathrm{f}} - q_x' \mathcal{E}_{\mathrm{n}}\right)^2\right\} \sin\left(q_x' D\right)}. \tag{4.10}$$

The transmission can then be obtained straightforwardly from $T_{\kappa s} = t_{\kappa s} t_{\kappa s}^*$:

$$T_{\kappa s} = \frac{\left(2q_x q_x' \mathcal{E}_{\mathrm{n}} \mathcal{E}_{\mathrm{f}}\right)^2}{\left(2q_x q_x' \mathcal{E}_{\mathrm{n}} \mathcal{E}_{\mathrm{f}}\right)^2 + \left\{[q_y(\mathcal{E}_{\mathrm{f}} - \mathcal{E}_{\mathrm{n}})]^2 + \left(q_x \mathcal{E}_{\mathrm{f}} + q_x' \mathcal{E}_{\mathrm{n}}\right)^2\right\}\left\{[q_y(\mathcal{E}_{\mathrm{f}} - \mathcal{E}_{\mathrm{n}})]^2 + \left(q_x \mathcal{E}_{\mathrm{f}} - q_x' \mathcal{E}_{\mathrm{n}}\right)^2\right\} \sin^2\left(q_x' D\right)}. \tag{4.11}$$

It indicates that under resonance conditions $\sin^2(q_x' D) = 0$ or

$$q_x' D = n\pi \tag{4.12}$$

with $n \in \mathbf{N}$, the ferromagnetic region becomes totally transparent ($T_{\kappa s} = 1$). By setting $q_x = q\cos\phi$ and $q_y = q\sin\phi$, we define the normalized valley and spin resolved conductance:

$$G_{\kappa s} = \frac{1}{2}\int_{-\pi/2}^{\pi/2} T_{\kappa s} \cos\phi \, \mathrm{d}\phi. \tag{4.13}$$

The conductance for $\boldsymbol{K}_{\pm}$ valley is then defined as

$$G_{\pm} = \frac{1}{2}(G_{\pm\uparrow} + G_{\pm\downarrow}). \tag{4.14}$$

We also introduce charge conductance G_{c} and the VP P:

$$G_{\mathrm{c}} = G_{+} + G_{-}, \tag{4.15}$$

$$P = \frac{G_{+} - G_{-}}{G_{\mathrm{c}}}. \tag{4.16}$$

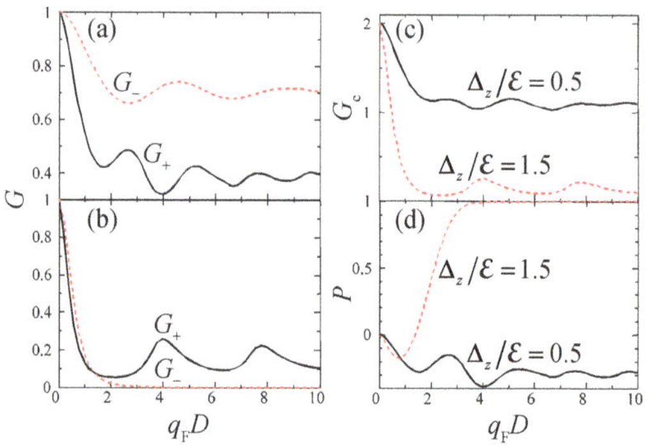

Figure 4.7. Valley-resolved conductance $G_\pm$ for (a) $\Delta_z/\mathcal{E} = 0.5$ and (b) $\Delta_z/\mathcal{E} = 1.5$ as a function of $q_F D$. (c) Charge conductance G_c and (d) VP P as functions of $q_F D$. Other parameters are the same as that in figure 4.6. Reprinted figures with permission from [28]. Copyright (2013) by the American Physical Society.

Figure 4.7 displays valley resolved conductance $G_\pm$ for (a) $\Delta_z/\mathcal{E} = 0.5$ and (b) $\Delta_z/\mathcal{E} = 1.5$ as a function of $q_F D$, as well as the corresponding (c) charge conductance G_c and (d) VP P, where q_F is defined as $q_F = \mathcal{E}/(\hbar v_F)$. In figure 4.7(a), both G_+ and G_- decay with D in an oscillatory way. This oscillation should attribute to the $\sin^2(q_x' D)$ term in equation (4.11), and can be analogous with the tunnelling of massless Dirac fermion (Klein tunneling) [30]. Further, for a given Fermi energy, electrons experience different 'potentials', thus different q_x'. Therefore, near the $\boldsymbol{K}_+$ and $\boldsymbol{K}_-$ points, the periods of the oscillations of G_+ and G_- are different. This will further result in the oscillatory behaviour shown in figures 4.7(c) and (d). The conductance $G_+ < G_-$ is because around the $\boldsymbol{K}_+$ point, the Fermi level crosses one band, only spin-up electron contributes to the conductance; while near the $\boldsymbol{K}_-$ point, the Fermi level crosses two bands, both spin species contribute to the conductance. In figure 4.7(b), G_+ behaves similarly to figure (a), while G_- is strongly suppressed and becomes almost zero. This can be explained by figure 4.6(c), where the Fermi level is located inside the bandgap near the $\boldsymbol{K}_-$ point. The current is entirely carried by the electrons near the $\boldsymbol{K}_+$ point. This indicates the charged current is fully valley-polarized, as shown by the red dotted line in figure 4.7(d) with $P = 1$. It should be noted that when the exchange field is absent, $h = 0$, the charge conductances contributed from $\boldsymbol{K}_+$ and $\boldsymbol{K}_-$ are equal to each other due to TRS [31].

Now we obtain the condition of fully valley-polarized conductance. As we can see from equation (4.6), the bandgaps in $\boldsymbol{K}_+$ and $\boldsymbol{K}_-$ valleys are the same, and the energy ranges of bandgaps are opposite. Therefore, fully valley-polarized conductance can occur when Condition 1: valley is not degenerate, and Condition 2: Fermi energy is located within the bandgap. Condition 1 requires that both Δ_z and h are nonzero. For Condition 2, we need to find the condition for the existence of bandgap in either $\boldsymbol{K}_+$ or $\boldsymbol{K}_-$. According to equation (4.6), if bandgap exists in $\boldsymbol{K}_+$ valley, bandgap will also exists in $\boldsymbol{K}_-$ valley, and vice versa. Therefore, in the following, we only consider $\boldsymbol{K}_+$ point. The CBM for spin up and spin down electrons are, $|\lambda_{SO} - \Delta_z| - h$ and $|-\lambda_{SO} - \Delta_z| + h$, respectively. The VBM for spin up and spin down electrons are, $-|\lambda_{SO} - \Delta_z| - h$ and $-|-\lambda_{SO} - \Delta_z| + h$, respectively. Therefore, for nonoverlapping conduction and valence bands, the following two inequalities should be satisfied,

$$\begin{aligned} |\lambda_{\mathrm{SO}} - \Delta_z| - h &> -|-\lambda_{\mathrm{SO}} - \Delta_z| + h, \\ |-\lambda_{\mathrm{SO}} - \Delta_z| + h &> -|\lambda_{\mathrm{SO}} - \Delta_z| - h. \end{aligned} \tag{4.17}$$

The solution to the above inequalities is $|h| < \max(\lambda_{\mathrm{SO}}, |\Delta_z|)$. Therefore, together with Condition 1, we obtain the condition necessary for the fully valley-polarized transport as

$$|h| < \max(\lambda_{\mathrm{SO}}, |\Delta_z|) \text{ and } \lambda_{\mathrm{SO}} \neq |\Delta_z|. \tag{4.18}$$

Later, Soodchomshom [17] extended this finding by controlling the chemical potential in the barrier.

4.3 Ferrovalley materials with spontaneous VP

Parts of section 4.3 have been reprinted from [40], copyright (2020), with permission from Elsevier.

For nonmagnetic monolayer TMDCs, the degeneracy at $\boldsymbol{K}_+$ and $\boldsymbol{K}_+$ valleys is lifted due to the breaking of spatial IS and strong SOC. The VP can be obtained by various external means, e.g., a circularly polarised optical excitation [32] or an external magnetic field [33], as discussed in the previous section. However, the VP cannot be maintained when the external field is removed. Therefore, it is imperative to search for materials with spontaneous VP. In this section, we introduce the investigation progress of ferrovalley materials from forming VP of the orbit of monolayer TMDC in 2H phase [34, 35].

The concept of room-temperature ferrovalley as a new member of ferroic family was first proposed by Tong *et al* [36] in the *d*-orbital VSe_2 system, as a peculiar ferromagnetic semiconductor among TMDCs [37–39]. Its structure is the same as that of MoS_2 in 2H phase, except for the presence of unpaired electrons in the 3*d* orbital of the V atom, which leads to the existence of intrinsic magnetism and causes the monolayer VSe_2 to be ferromagnetic.

4.3.1 Two-band effective model including exchange interaction

The mechanism of spontaneous VP induced by ferromagnetism and SOC effects can be represented using a two-band $\boldsymbol{k} \cdot \boldsymbol{p}$ Hamiltonian,

$$H_{\mathrm{Fe}}(\boldsymbol{k}) = I_2 \otimes H_0(\boldsymbol{k}) + H_{\mathrm{SO}}(\boldsymbol{k}) + H_{\mathrm{x}}(\boldsymbol{k}). \tag{4.19}$$

The $H_{\mathrm{SO}}(\boldsymbol{k})$ term represents the SOC effect, which is equivalent to a momentum space-dependent PMF. Specifically, as discussed in section 1.5.3, at $\boldsymbol{K}_+$ valley, the SOC effect shifts the spin-up (down) valence band upward (downward), while the spin splitting at $\boldsymbol{K}_-$ valley is opposite. The additional term $H_{\mathrm{x}}(\boldsymbol{k})$ is utilised to describe the valley-independent intrinsic magnetic exchange interaction, which is equivalent to an intrinsic magnetic field, moving the spin-up and down states oppositely. Combining the SOC effect with the valley-independent exchange interaction, the degeneracy of the two valleys is broken, leading to the VP without external field. Therefore, according to Hamiltonian (4.19), the schematic band structure near the valleys $\boldsymbol{K}_{\pm}$ can be easily deduced as shown in figure 4.8. The figure also shows the valley-dependent optical selection rule, which states that the electrons at $\boldsymbol{K}_{+}$ and $\boldsymbol{K}_{-}$ valley could only be excited by the left-handed circularly polarized (LCP) and right-handed circularly polarised (RCP) light, respectively. At this stage, the readers can only memorise this rule as it will be deduced in section 7.3.3. The Fermi level lies between the upper band (UB) and lower band (LB). The energy band in the absence of SOC effect with broken IS is shown in figure 4.8(a). At present, the system is described by Hamiltonian (2.48). According to equation (2.51), the optical bandgaps, $\mathcal{E}_{\mathrm{g}}^{\mathrm{opt}}(\boldsymbol{K}_{\kappa})$, of the two valleys are identical and can be demonstrated by

$$\mathcal{E}_{\mathrm{g}}^{\mathrm{opt}}(\boldsymbol{K}_{+}) = \mathcal{E}_{\mathrm{g}}^{\mathrm{opt}}(\boldsymbol{K}_{-}) = \Delta_{z}. \tag{4.20}$$

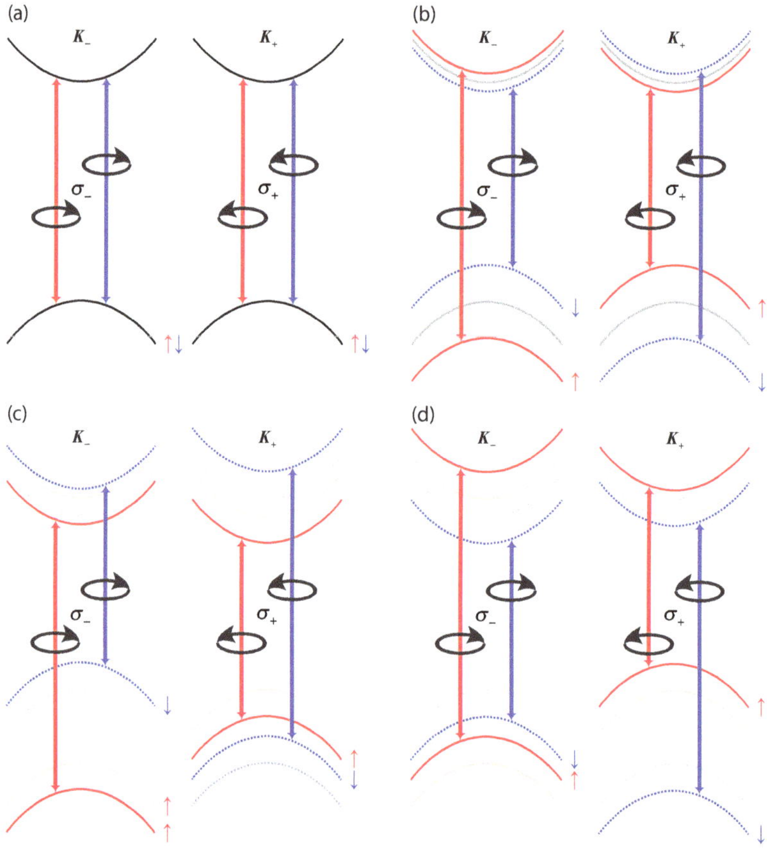

Figure 4.8. The schematic evolution of band structures at $\boldsymbol{K}_{-}$ and $\boldsymbol{K}_{+}$ valleys of representative monolayer TMDCs. Black solid, blue solid and red dotted lines represent spin degenerated, spin-up and spin-down bands, respectively. (a) Without SOC effect, (b) With SOC effect, (c) with SOC effect and a positive exchange field, which is the valley-polarised case. (d) is the same as (c) but with a negative exchange field. σ_{+} and σ_{-} represent LCP and RCP radiation, respectively. Note that the band structures in (b) is referenced to (a), while the band structures in (c) and (d) are both referenced to (b). The references are shown in lighter colour [36].

When the SOC effect is considered in such a representative monolayer TMDC in figure 4.8(b), the spin splitting of UB induced by SOC effect has the opposite sign to the one of LB. Note that the handedness of light (LCP/RCP) is locked in each valley. Especially at the $\boldsymbol{K}_+$ valley, both spin-up and spin-down components can only be excited by LCP light, while the $\boldsymbol{K}_-$ valley only corresponds to RCP light. The valley-dependent SOC effect splits the previously degenerated bands, and makes the optical bandgaps in $\boldsymbol{K}_+$ and $\boldsymbol{K}_-$ valleys stemming from different spin states. The spin splitting energy at the bottom of UB (the top of LB) is labelled as $2\lambda_U$ ($2\lambda_L$), which is defined by the energy difference $\mathcal{E}_{U\uparrow} - \mathcal{E}_{U\downarrow}$ ($\mathcal{E}_{L\uparrow} - \mathcal{E}_{L\downarrow}$) at the $\boldsymbol{K}_+$ points. Although the electron transitions of $\boldsymbol{K}_-$ and $\boldsymbol{K}_+$ correspond to different spin states, $\mathcal{E}_g^{opt}(\boldsymbol{K}_\kappa)$ of the two valleys are the same, which means the energy levels are still degenerate:

$$\mathcal{E}_g^{opt}(\boldsymbol{K}_+) = \mathcal{E}_g^{opt}(\boldsymbol{K}_-) = \Delta_z - \lambda_L + \lambda_U. \tag{4.21}$$

Once the intrinsic exchange field exists in figures 4.8(c) and (d), the TRS of the system is broken, which decouples the energetically degenerated valleys and results in the occurrence of VP. Taking the positive exchange field as an example (figure 4.8 (c)), the optical bandgap of $\boldsymbol{K}_+$ ($\boldsymbol{K}_-$) valley corresponds to the transition of spin-up (spin-down) electrons excited by LCP (RCP) radiation, is given by

$$\mathcal{E}_g^{opt}(\boldsymbol{K}_\kappa) = \Delta_z - \lambda_L + \lambda_U - \kappa(m_L - m_U), \tag{4.22}$$

where m_U (m_L) represents the effective exchange splitting in the band edge of UB (LB), which defines the energy difference between the spin-up and spin-down bands, i.e., $\mathcal{E}_{U\uparrow} - \mathcal{E}_{U\downarrow}$ ($\mathcal{E}_{L\uparrow} - \mathcal{E}_{L\downarrow}$). The effect of the intrinsic exchange field leads to the difference in the optical bandgap between the two valleys with the magnitude of

$$\left|\mathcal{E}_g^{opt}(\boldsymbol{K}_+) - \mathcal{E}_g^{opt}(\boldsymbol{K}_-)\right| = 2|m_L - m_U|. \tag{4.23}$$

For such a valley-polarized system, once the polarity of circularly polarized light is reversed, different optical bandgaps can be detected directly, indicating the possibility of judging the VP by utilizing noncontact and nondestructive circularly polarized optical means [40].

The above two-band $\boldsymbol{k} \cdot \boldsymbol{p}$ theory analysis clearly demonstrates the general guideline to hunt for ferrovalley materials with spontaneous VP—the coexistence of SOC effects with intrinsic exchange interaction. Fortunately, monolayer VSe_2 in 2H phase [37–39, 41] has been predicted to be such a ferrovalley material [36]. As a peculiar ferromagnetic semiconductor among TMDCs, the unpaired 3*d* electrons of the V atom lead to the existence of an intrinsic magnetic moment with the magnitude of $1.01\mu_B$ [36], which implies a remarkable exchange interaction and causes significant spontaneous VP.

4.3.2 Band structures from density functional theory (DFT)

Figure 4.9 presents the detailed evolution of band structures of monolayer VSe_2 from *ab initio* calculations. When the ferromagnetism in monolayer VSe_2 is not considered, as shown in figure 4.9(a), its band structure is essentially similar to the

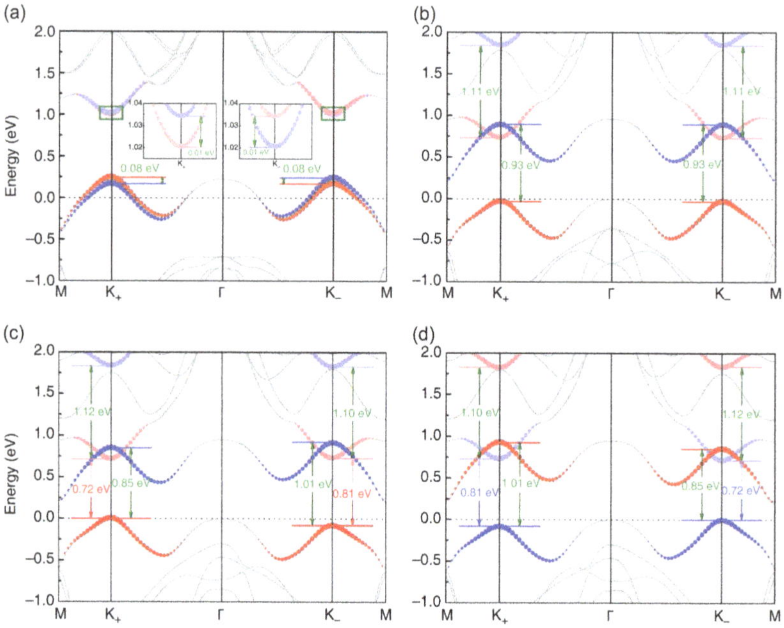

Figure 4.9. The band structures of monolayer VSe_2. (a) With SOC effect but without ferromagnetism, and (b) with magnetic moment but without SOC effect. (c) The real case including both the magnetism and SOC effect. (d) is the same as (c) but with opposite magnetic moment. The insets in (a) zoom in on the splitting splitting at the bottom of the upper band. The radius of dots is proportional to its population in the corresponding state near valleys: red (blue) ones for spin-up (spin-down) components of $d_{x^2-y^2}$ and d_{xy} orbitals on cation-V, while light red (light blue) symbols represent spin-up (spin-down) states for d_{z^2} characters. The Fermi level is set to zero in all cases. Reproduced from [36]. CC BY 4.0.

representative one for TMDCs (figure 4.8(b)). The system is metallic with the Fermi level passing through the states predominantly comprising $d_{x^2-y^2}$ and d_{xy} orbitals of cation-V. On the other hand, as shown in figure 4.9(b), when we start from the gapped graphene described in figure 4.8(a) and only consider the intrinsic exchange interaction of unpaired d electrons, which corresponds to a massive PMF, the band structure shows complete splitting of the degenerated band near the Fermi level into spin-up and spin-down states. Inspiringly, the spin-splitting energy of the valence band is up to 0.93 eV. In this case, the system exhibits ferromagnetic semiconductor properties with a limited indirect bandgap. Although the valence band top is situated at the Γ point at the centre of the BZ, the direct bandgap remains at the two valleys.

Finally, if we superpose the relatively small but non-negligible SOC effect (0.08 eV in valence band at $\boldsymbol{K}_{\pm}$ valleys) and the strong exchange interaction from the intrinsic magnetic moment of the V-d electrons, the spontaneous VP can be induced. When the magnetic moment is positive as shown in figure 4.9(c), the spin splitting of states predominantly occupied by $d_{x^2-y^2}$ and d_{xy} orbitals equals to $|2m_L + 2\lambda_L| \approx 0.85$ eV at $\boldsymbol{K}_+$ valley, which is much smaller than $|2m_L - 2\lambda_L| \approx$ 1.01 eV at $\boldsymbol{K}_-$ valley. Conversely, the spin splitting of states predominantly occupied by the d_{z^2} orbital is relatively larger at $\boldsymbol{K}_+$ point ($|2m_{\mathrm{U}} + 2\lambda_{\mathrm{U}}| \approx 1.12$ eV) than that at $\boldsymbol{K}_-$ ($|2m_{\mathrm{U}} - 2\lambda_{\mathrm{U}}| \approx 1.10$ eV) because of the opposite contribution from SOC effect between 1) $\boldsymbol{K}_-$ and $\boldsymbol{K}_+$ valleys; 2) UB and LB. Upon reversing the magnetic moment, as illustrated in figure 4.9, the direction of VP undergoes an inverse polarity [42].

4.3.3 Chirality-dependent optical bandgap and Berry curvatures

Importantly, optical calculations demonstrate that the bandgap with energy difference

$$\left| \mathcal{E}_{\mathrm{g}}^{\mathrm{opt}}(\boldsymbol{K}_+) - \mathcal{E}_{\mathrm{g}}^{\mathrm{opt}}(\boldsymbol{K}_-) \right| = |2\lambda_{\mathrm{L}} - 2\lambda_{\mathrm{U}}| \approx 0.09 \text{ eV}, \tag{4.24}$$

will directly reflect in the optical properties excited by circularly polarized light. As we know, $\boldsymbol{K}_+$ valleys can only be excited by LCP light, while $\boldsymbol{K}_-$ valleys correspond to RCP light. As shown in figure 4.10(a), compared with the absorption peak related to the LCP light, the RCP light experiences a blueshift. When the magnetic moment is reversed, VP possesses reversed polarity according to figure 4.9(d). Therefore, as shown in figure 4.10(b), in comparison with the LCP one, the redshift of the optical bandgap excited by RCP light happens. The above chirality-dependent optical bandgap indicates that circularly polarized light can be used to characterize the occurrence of VP and its chiral characteristics [40].

It is also interesting to examine the Berry curvature, which has a crucial influence on the electronic transport properties and is the key parameter to various Hall effects. Here, we first consider the spin-resolved nonzero z-component Berry curvature derived from equation (2.34) [43]. Then, we sum the Berry curvatures of spin-up and spin-down

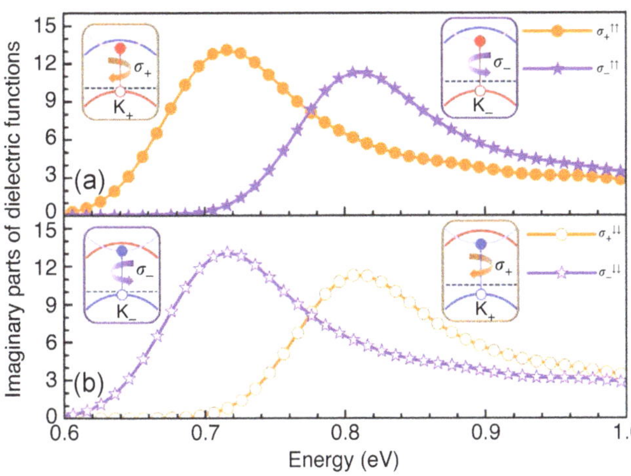

Figure 4.10. Schematic diagram of interband transitions dependent on optical bandgap. The imaginary parts of complex dielectric function ε_2 for monolayer VSe_2. Cases excited by LCP radiation σ_+ and RCP radiation σ_- (a) with positive magnetic moment (associated to figures 4.9(c)), and (b) with negative one (associated to figure 4.9(d)) are presented. Insets are the schematic interband transitions related to certain optical bandgap. Reproduced from [36]. CC BY 4.0.

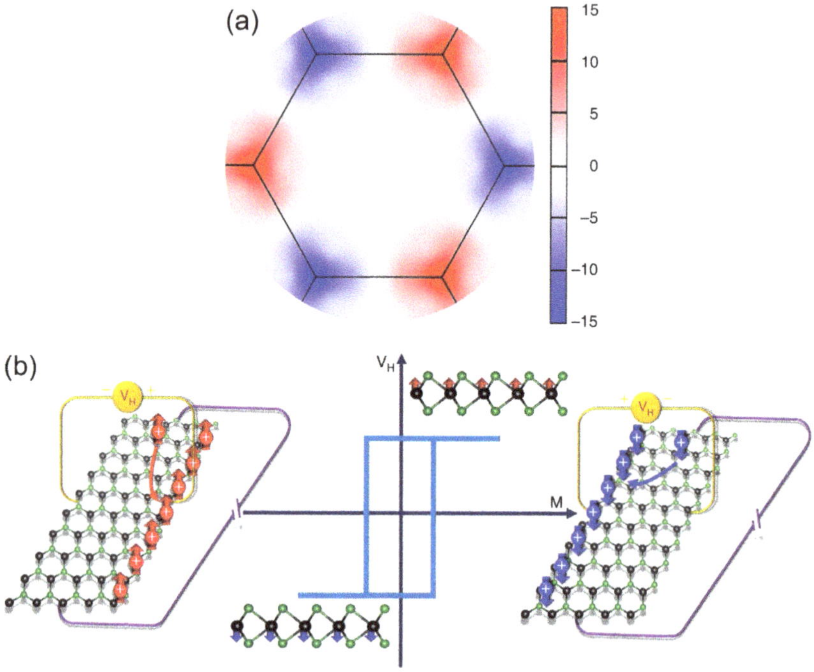

Figure 4.11. (a) Contour map of Berry curvature (in units of Å^2) in the $\boldsymbol{k}$-space for bands mainly occupied by $d_{x^2-y^2}$ and d_{xy} orbitals for monolayer VSe_2. (b) Sketch of data storage or memory device utilizing hole-doped ferrovalley materials based on AVHE. The carriers denoted by white '+' symbols are holes. Upward arrows in red and downward arrows in blue represent spin-up and spin-down carriers, respectively. Reproduced from [36]. CC BY 4.0.

bands, which have a major contribution from the $d_{x^2-y^2}$ and d_{xy} orbitals of V atoms, i.e., the red and blue LBs in figure 4.9. For a system with equilibrium valleys, as presented in equation (2.37), the Berry curvature is an odd function in the momentum space because of TRS and broken space IS. However, in the ferrovalley material, as shown in figure 4.11(a), the absolute values in opposite valleys are no longer identical, although Berry curvatures still have opposite signs [36, 40].

4.3.4 Discussion

As we learned from section 2.4, the sign change of Berry curvature in opposite valleys will lead to the VHE, which has been widely found in 2D honeycomb lattices [14, 44–47]. The accumulation of spin and valley information at the boundary of the sample has brought about a series of interesting phenomena such as the emission of photons with opposite circularly polarized light on the two boundaries.

In ferrovalley materials, the VHE exhibits a more interesting feature, which is the presence of additional charge Hall current originating from the spontaneous VP. Analogous to the anomalous Hall effect in ferromagnetic materials [48, 49], this effect in ferrovalley materials is named anomalous VHE (AVHE). Since the charge Hall current is more straightforward to measure experimentally, the AVHE offers a possible route to realize data storage utilizing ferrovalley materials. Figure 4.11(b) illustrates an example for moderate p-type VSe_2 achieved by slight hole doping with Fermi energy lying between the VBMs of $\boldsymbol{K}_+$ and $\boldsymbol{K}_-$ valleys. The p-type VSe_2 possesses 100% spin polarization around the Fermi level. In the presence of an external electric field, because of the almost zero Berry curvature in the central region of BZ, the carriers from the Γ point and its neighbouring $\boldsymbol{k}$-points travel through the ribbon directly without experiencing transverse deflection. When the p-type VSe_2 possesses positive VP, as illustrated in the right panel, the majority carriers, which are spin-down holes from the $\boldsymbol{K}_+$ valley, gain transverse velocities towards the left edge of the sample. The accumulation of holes in the left boundary of the ribbon generates a charge Hall current, which is measurable as a positive Hall voltage. When the polarity of VP is reversed, as schematically drawn in the left panel, spin-up holes from the $\boldsymbol{K}_-$ valley will accumulate on the right edge of the sample because of the negative Berry curvature, resulting in measurable transverse voltage with an opposite sign. Electrons with a particular spin or valley can also be accumulated to one edge with n-type doping [50].

After the proposal of ferrovalley materials such as VSe_2 and GeSe [51], many intrinsic ferrovalley materials have been proposed. Here, we categorize them according to orbitals [34].

1. *p*-orbital: monolayer XY (X= K, Rb, Cs; Y= N, P, As, Sb, Bi) [52] and PbO [53];
2. *d*-orbital: MBr_2 (M= Ru, Os) [54], VS_2 [55], $RuCl_2$ [56], h-MNX (M= Ti, Zr, Hf; X= Cl, Br) [57], monolayer Janus structure $VSiXN_4$ (X= C, Si, Ge, Sn, Pb) [58], VSSe [59–61], VClBr [62], etc [34];
3. *f*-orbital: XI_2 (X= Ce [63], La, Pr [64]), $LaBr_2$ [65], and Janus GdXY (X, Y= Cl, Br, I; $X \neq Y$) [66].

Among them, the *d*-orbital ferrovalley materials have been most extensively explored. In addition, the heterostructure systems based on polarized materials and 2D channels may provide many interests for the future study of valleytronics. The interlayer of heterostructure is bonded by a weak vdW interaction, which facilitates tuning VP by the multi-field [34], such as electrical [67–69], interlayer stacking mode [70, 71], strain [54] and interface neighbouring effect [72]. Even though 2D ferrovalley materials have been predicted early by the DFT calculation, the experimental progress has been limited, partly related to the lack of powerful sample characterization and preparation techniques. Recently, Guan *et al* [73] first reported the successful experimental preparation of intrinsic bulk ferrovalley material. For 2D ferrovalleys, so far, there has been no experimental progress.

The key to electronic device design is to construct a binary logic switching state. Based on the AVHE in ferrovalley materials, the electrically reading and magnetically writing memory devices are arising. In AVHE, because of the strong locking between valley and spin in ferrovalley materials, there is exclusively one kind of carrier of spin or charge, originating from a particular valley, leading to the emergence of an extra Hall current of charge. Further, the binary logic information is stored by the charge, spin polarisation and VP, which can be manipulated by the magnetic moment of the applied external magnetic field. It can be then handily read out by measuring the sign of the transverse Hall voltage. Furthermore, in contrast to conventional valleytronic materials, due to the intrinsic VP of ferrovalley materials, the memory applications are nonvolatile, which is of great significance to the innovation of storage technology and the development of next-generation data storage. At this point, the experimental efforts on V-group dichalcogenides with 2H phase are strongly advocated, in order to achieve the related valley physics and valleytronic devices. In addition, the combination of charge, spin and valley DOFs endows the new members of the Hall family with appeal in the fields of electronics, spintronics, valleytronics and even their interdisciplinary subjects [42, 74]. Typically, the spin based on the valley is more stable, and the spin control and detection can be easily realised with the help of the valley, which significantly advances and expands the application prospect of the device.

4.4 Detection of VP with a superconducting contact

The Andreev reflection (AR) is a unique transport process where an electron incident from the conductor is reflected as a hole at the interface of a conductor–superconductor. Simultaneously, a Cooper pair is formed in the superconductor [75, 76]. In graphene, the reflected holes and incident electrons are from different valleys due to the conservation of TRS. This makes it possible to achieve the VP and detection through the AR effect. In this part, we will introduce the detection of the VP of edge states produced by a magnetic field through the AR effect [77].

4.4.1 The physical origin of VP

In the presence of perpendicular magnetic fields, the quantized Hall conductance in graphene exhibits the half-integer quantization characteristic, $G_{\mathrm{H}} = \frac{4e^2}{h}(n + \frac{1}{2})$, which is determined by the number of edge state bands crossing the Fermi level [78]. For the lowest LL (LLL), the conductance $G_{\mathrm{H}} = \frac{2e^2}{h}$ and the factor 2 stems from twofold spin degeneracy while the valley degeneracy is removed. However, the components of the edge states of the LLL are dependent on the crystallographic orientation of the edge; for the zigzag termination, the edge state lies fully in a single valley while it admixes states from different valleys for the armchair termination [79]. When a superconducting electrode is applied, an electron approaching in one valley will be converted to a hole leaving in the other valley, as shown in figure 4.12. Therefore, the subgap conductance through the normal–superconducting (NS) interface is sensitive to nanoribbon configuration, which is related to the VP. For some configurations the conductance equals zero when the edge states at the Fermi level lie exclusively in a single valley while it can be e^2/h for other configurations indicating the VP.

For the the two-terminal device, the conductance measured between the superconductor and a normal–metal contact is given by [76]

$$G_{\mathrm{NS}} = \frac{2e^2}{h}(1 - \cos\Theta), \tag{4.25}$$

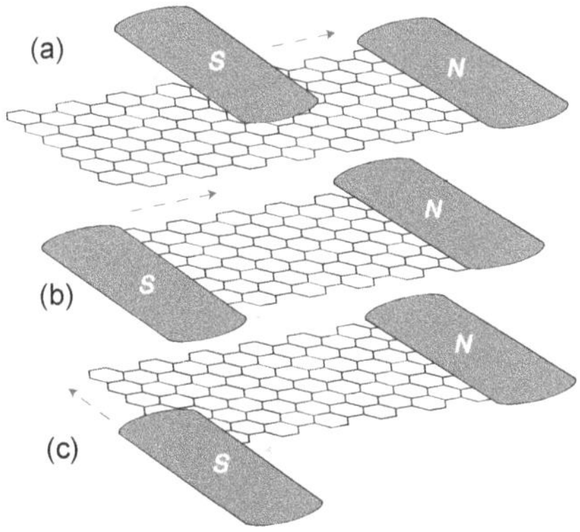

Figure 4.12. Three diagrams of a graphene sheet contacted by one normal metal (N) and one superconducting (S) electrode, where the S electrode covers a single boundary in (a), covers the opposite boundaries in (b) and covers the adjacent boundaries in (c). The green (isospin ν_1) and blue (isospin ν_2) solid arrows represent the edge states approaching and leaving the S electrode, and the red dashed line represents a hole state (isospin $-\nu_2$). Reprinted figures with permission from [77]. Copyright (2007) by the American Physical Society.

when the Hall conductance $G_H = 2e^2/h$ is on the LLL. In equation (4.25), Θ is the angle difference between the valley isospins of the states entering (ν_1) and leaving (ν_2) the superconductor which is dependent only on the device geometry. For instance, the angle $\Theta = 0$ when the two states flow in the same directions along the boundary [see figure 4.12(a)] while $\Theta = \pi$ for the antiparallel situation [see figure 4.12(b)]. For other angels of Θ, G_{NS} varies from 0 to $4e^2/h$ when the orientation of the valley isospins along the two edges changes from 0 to π, dependent on the relative orientation of the valley isospins along the two edges. One can directly determine the component of the current, from one valley or two valleys, through the changes in conductance.

4.4.2 The Dirac–Bogoliubov–de Gennes (DBdG) equation

The DBdG equation is adopted for the calculations [76], which can describe the superconducting correlation between massless Dirac fermions with different valley indices,

$$\begin{pmatrix} H - \mu & \Delta \\ \Delta^* & \mu - \mathcal{T}H\mathcal{T}^{-1} \end{pmatrix} \Psi = \mathcal{E}\Psi, \tag{4.26}$$

where H is the Dirac Hamiltonian, Δ is the superconducting pair potential, $\mathcal{T}$ is the time-reversal operator, and $\mathcal{E}$ is the excitation energy measured relative to the Fermi energy μ. $\Psi = (\Psi_e, \Psi_h)$ is the wavefunction and Ψ_e/Ψ_h represents the electron/hole excitation. The 2D electron spinor then has components $\Psi_e = (\Psi_{A+}, \Psi_{B+})$ while the hole spinor $\Psi_h = \mathcal{T}\Psi_e$ has the components $\Psi_h = (\Psi^*_{A-}, \Psi^*_{B-})$. The electron excitations in one valley are coupled by the superconductor to hole excitations in the other valley.

In equation (4.26), the Hamiltonian H under a perpendicular magnetic field can be given by [80]

$$H = v_F\begin{pmatrix} (\boldsymbol{p} + e\boldsymbol{A})\cdot\boldsymbol{\sigma} & 0 \\ 0 & (\boldsymbol{p} + e\boldsymbol{A})\cdot\boldsymbol{\sigma} \end{pmatrix} = v_F\kappa_0 \otimes (\boldsymbol{p} + e\boldsymbol{A})\cdot\boldsymbol{\sigma}, \tag{4.27}$$

where v_F, $\boldsymbol{p}$ and $\boldsymbol{A}$ are the Fermi velocity, the momentum operator and the vector potential under a perpendicular magnetic field B_z, respectively. σ_i and κ_i are the Pauli matrices in the sublattices pseudospin and valley DOFs, and σ_0 (κ_0) represents the 2×2 unit matrix. The time-reversal operator in the valley isotropic basis reads

$$\mathcal{T} = \begin{pmatrix} 0 & \mathrm{i}\sigma_y \\ -\mathrm{i}\sigma_y & 0 \end{pmatrix}\mathscr{C} = -(\kappa_y \otimes \sigma_y)\mathscr{C}, \tag{4.28}$$

with $\mathscr{C}$ the operator of complex conjugation.

Substitution of equations (4.27) and (4.28) into equation (4.26) gives the DBdG equation in the valley isotropic form,

$$\begin{pmatrix} H_+ - \mu & \Delta \\ \Delta^* & \mu - H_- \end{pmatrix}\Psi = \mathcal{E}\Psi, \tag{4.29}$$

with

$$H_\pm = v_F\kappa_0 \otimes (\boldsymbol{p} \pm e\boldsymbol{A})\cdot\boldsymbol{\sigma}. \tag{4.30}$$

To gain the AR effect, the incident energies of electrons should be below the excitation gap Δ_0 in the superconductor. Electron and hole excitations cannot propagate into the superconductor at subgap energies. Under a magnetic field, the electrons are confined in the normal region to within a magnetic length $l_B = \sqrt{\hbar/eB}$ (equation (1.69)) due to the Lorentz force. The edges are assumed to be smooth on the scale of l_B ($\approx$25 nm at $B = 1$ T), so that the electron trajectory can be treated locally as a straight line with a homogeneous boundary condition. The magnetic field should be less than the critical field of the superconductor. For Nb [81], the critical field is 2.6 T, below which the superconductivity is well maintained in the quantum Hall effect regime.

4.4.3 Effective boundary conditions

Actually, what people care about are the edge states along the insulating edge of the graphene layer and along the interface with the superconductor. The two edge states are different because of the different boundary conditions. The effective boundary conditions for the envelope function Ψ can be expressed in general terms as [82]

$$\Psi = \mathsf{M}\Psi, \tag{4.31}$$

where M is a Hermitian, unitary, 4×4 matrix and $\mathsf{M}^2 = 1$. Because no current can occur at the direction normal to the boundary, the following relation should be satisfied,

$$\langle\Psi|\hat{\boldsymbol{n}}\cdot\boldsymbol{J}|\Psi\rangle = 0, \tag{4.32}$$

where $\hat{\boldsymbol{n}}$ is a unit vector in the xOy plane while normal to the boundary and pointing outward, $\boldsymbol{J}$ is the particle current operator and $\hat{\boldsymbol{n}} \cdot \boldsymbol{J}$ represents the particle current operator in the direction $\hat{\boldsymbol{n}}$ with $\hat{\boldsymbol{n}} \cdot \boldsymbol{J} = v_F \kappa_0 \otimes (\boldsymbol{\sigma} \cdot \hat{\boldsymbol{n}})$. This is also equivalent to the requirement of anticommutation of the matrix M with the current operator,

$$\{\mathsf{M}, \hat{\boldsymbol{n}} \cdot \boldsymbol{J}\} = 0. \tag{4.33}$$

At the NS interface, the matrix M is expressed as

$$\mathsf{M} = \begin{pmatrix} 0 & M_{\mathrm{NS}} \\ M_{\mathrm{NS}}^{\dagger} & 0 \end{pmatrix}, \tag{4.34}$$

with $M_{\mathrm{NS}} = \kappa_0 \otimes \mathrm{e}^{\mathrm{i}\phi + \mathrm{i}\beta \hat{\boldsymbol{n}} \cdot \boldsymbol{\sigma}}$ and $\beta = \arccos(\mathcal{E}/\Delta_0) \in (0, \pi)$ which is determined by the order parameter $\Delta = \Delta_0 \mathrm{e}^{\mathrm{i}\phi}$ in the superconductor.

In the presence of a perpendicular magnetic field, the incident electrons and the reflected holes form the cyclotron orbits along the boundary of the device because of the Lorentz force. The helical direction of them can be contrary or identical when the retroreflected holes and the incident electrons come from the same or different bands [83]. For instance, when $\mu > \mathcal{E}$, the electrons and holes are in the same band, bringing about the Andreev retroreflection at the NS interface, and the electrons and reflected holes have the same helical direction and move in the same direction. When $\mu < \mathcal{E}$ the electrons and holes are from different bands and the specular AR occurs at the NS interface, the electrons and holes have an opposite helical direction and move in the opposite direction. However, the electrons cannot accomplish a complete circular motion in a magnetic field, then the TRS is still preserved.

For the insulating (I) edge, it does not mix electrons and holes and M is block-diagonal with electron block M_{I} and hole block $\mathcal{T} M_{\mathrm{I}} \mathcal{T}^{-1}$. M_{I} commutes with $\mathcal{T}$ because the TRS is well preserved. The general matrix M is given by [82]

$$\mathsf{M} = \begin{pmatrix} M_{\mathrm{I}} & 0 \\ 0 & M_{\mathrm{I}} \end{pmatrix}, \; M_{\mathrm{I}} = (\boldsymbol{\nu} \cdot \boldsymbol{\kappa}) \otimes (\hat{\boldsymbol{n}}_{\perp} \cdot \boldsymbol{\sigma}), \tag{4.35}$$

where $\boldsymbol{\nu}$ and $\hat{\boldsymbol{n}}_{\perp}$ are 3D unit vectors. The vector $\hat{\boldsymbol{n}}_{\perp}$ is orthogonal to $\hat{\boldsymbol{n}}$ while $\boldsymbol{\nu}$ is not so constrained. Electron–hole symmetry restricts the boundary matrix M_{I} to two classes: zigzag edge ($\boldsymbol{\nu} = \pm\hat{\boldsymbol{z}}$, $\hat{\boldsymbol{n}}_{\perp} = \hat{\boldsymbol{z}}$) and armchair edge ($\boldsymbol{\nu} \cdot \hat{\boldsymbol{z}} = 0$, $\hat{\boldsymbol{n}}_{\perp} \cdot \hat{\boldsymbol{z}} = 0$).

For the boundary along the y-axis ($\hat{\boldsymbol{n}} = -\hat{\boldsymbol{x}}$), the wave number q along the boundary is a good quantum number and the eigenstates of equation (4.29) in the $x > 0$ ($\Delta \equiv 0$) region can be given by (the local gauge $\boldsymbol{A} = Bx\hat{\boldsymbol{y}}$)

$$\Psi(x, y) = \mathrm{e}^{\mathrm{i}qy} \begin{pmatrix} C_{\mathrm{e}} \otimes \Phi_{\mathrm{e}}(x + q) \\ C_{\mathrm{h}} \otimes \Phi_{\mathrm{h}}(x - q) \end{pmatrix}, \tag{4.36}$$

with

$$\Phi_{\rm e}(\zeta) = {\rm e}^{-\zeta^2/2}\begin{pmatrix} -{\rm i}(\mu + \mathcal{E})\mathcal{H}_{(\mu+\mathcal{E})^2/2-1}(\zeta) \\ \mathcal{H}_{(\mu+\mathcal{E})^2/2}(\zeta) \end{pmatrix},$$
$$\Phi_{\rm h}(\zeta) = {\rm e}^{-\zeta^2/2}\begin{pmatrix} \mathcal{H}_{(\mu+\mathcal{E})^2/2}(\zeta) \\ -{\rm i}(\mu - \mathcal{E})\mathcal{H}_{(\mu-\mathcal{E})^2/2-1}(\zeta) \end{pmatrix}. \quad (4.37)$$

In the above equations, the energies and lengths are in units of $\hbar v_{\rm F}/l_B$ and l_B. $\mathcal{H}_{|n|}(\zeta)$ is the Hermite function and the two-component spinors $C_{\rm e}$ and $C_{\rm h}$ can determine the valley isospin of the electron and hole components.

4.4.4 The dispersion relation

The dispersion relation between energy and momentum q can be obtained by substitution of the state equation (4.36) into the boundary condition equation (4.31). At the NS interface, when the equation (4.34) is token as the boundary condition, one can obtain

$$f_{\mu+\mathcal{E}}(q) - f_{\mu-\mathcal{E}}(-q) = \frac{\mathcal{E}[f_{\mu+\mathcal{E}}(q)f_{\mu-\mathcal{E}}(-q) + 1]}{\sqrt{\Delta_0^2 - \mathcal{E}^2}},$$
$$f_\alpha(q) = \frac{\mathcal{H}_{\alpha^2/2}(q)}{\alpha\mathcal{H}_{\alpha^2/2-1}(q)}. \quad (4.38)$$

The solutions $\mathcal{E}_n(q)$ as a function of q are plotted in figure 4.13. As is shown, the dispersion relation has the IS $\mathcal{E}(q) - -\mathcal{E}(-q)$. Each mode has a twofold valley degeneracy, because the boundary condition is isotropic in the valley isospin $\boldsymbol{\nu}$. The two degenerate eigenstates (labelled $\pm$) have $C_{\rm e}^{\pm} = c_{\rm e}|\pm\boldsymbol{\nu}\rangle$, $C_{\rm h}^{\pm} = c_{\rm h}|\pm\boldsymbol{\nu}\rangle$, with $|\pm\boldsymbol{\nu}\rangle$ being eigenstates of $\boldsymbol{\nu}\cdot\boldsymbol{\tau}$.

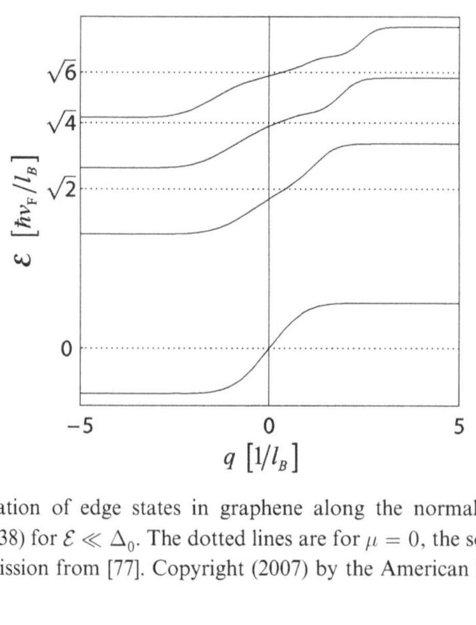

Figure 4.13. Dispersion relation of edge states in graphene along the normal–superconducting interface, calculated from equation (4.38) for $\mathcal{E} \ll \Delta_0$. The dotted lines are for $\mu = 0$, the solid lines for $\mu = 0.4\hbar v_{\rm F}/l_B$. Reprinted figures with permission from [77]. Copyright (2007) by the American Physical Society.

The expectation value of the velocity $v_n = \hbar^{-1}\frac{d\mathcal{E}_n}{dq}$ along the boundary in the nth mode is determined by the derivative of the dispersion relation. We see from figure 4.13 that the edge states all propagate in the same direction, dictated by the sign of B_z and μ. The velocity vanishes for $|q| \to \infty$, as the NS edge states evolve into the usual dispersionless LLs deep in the normal region. For $q \to -\infty$ the LLs contain electron excitations at energy $\mathcal{E}_n = \sqrt{2}(\hbar v_F/l_B)\text{sgn}(n)\sqrt{|n|} - \mu$, while for $q \to \infty$ they contain hole excitations with $\mathcal{E}_n = \sqrt{2}(\hbar v_F/l_B)\text{sgn}(n)\sqrt{|n|} + \mu$. For $\mu = 0$ the NS edge states have zero velocity at any q for $|\mathcal{E}| \ll \Delta_0$. However, the localisation of the edge states can happen when $\mu \to 0$ because the electron and hole excitations will move in opposite directions along the boundary for $|\mathcal{E}| > |\mu|$, while they move in the same direction for $|\mathcal{E}| < |\mu|$ due to the AR effect. The boundary condition equation (4.35) is adopted to discuss the insulating edge. For an edge along the y-axis we have $\hat{\boldsymbol{n}}_\perp = (0, \sin\vartheta, \cos\vartheta)$, where ϑ is the chiral angle of GNR [82]. The valley degeneracy is broken in general, with different dispersion relations for the two eigenstates $|\pm\boldsymbol{\nu}\rangle$ of $\boldsymbol{\nu}\cdot\boldsymbol{\kappa}$. The dispersion relations for electrons and holes are related by $\mathcal{E}_\text{h}^{\pm}(q) = -\mathcal{E}_\text{e}^{\mp}(-q)$. For sufficiently small μ there is one electron and one hole state at the Fermi level, of opposite isospins. (Note that electrons and holes from the same valley have opposite isospins). We fix the sign of $\boldsymbol{\nu}$ such that $|+\boldsymbol{\nu}\rangle$ is the electron eigenstate and $|-\boldsymbol{\nu}\rangle$ is the hole eigenstate. Then $\mathcal{E}_\text{e}^{+}(q)$ is determined by the equation

$$f_{\mu+\mathcal{E}}(q) = \tan\frac{\vartheta}{2}, \tag{4.39}$$

while $\mathcal{E}_\text{e}^{-}(q)$ is determined by

$$f_{\mu+\mathcal{E}}(q) = -\cot\frac{\vartheta}{2}. \tag{4.40}$$

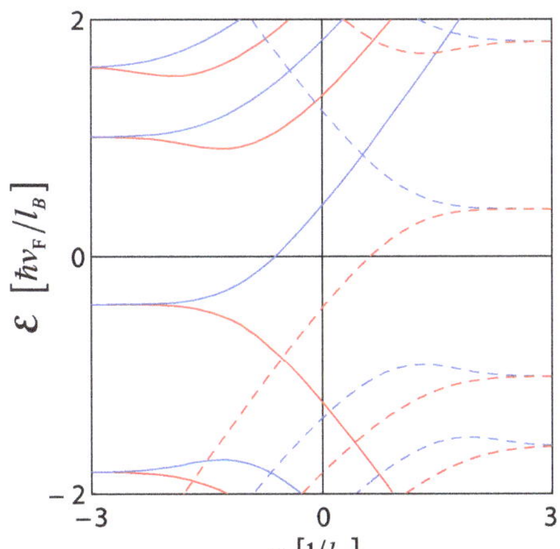

Figure 4.14. (a) Dispersion relation of states along the insulating edge, calculated from equation (4.39) and equation (4.40) for $\mu = 0.4\hbar v_F/l_m$ and $\vartheta = \pi/2$. The solid lines are the electron states (blue $\mathcal{E}_\text{e}^{+}$, red $\mathcal{E}_\text{e}^{-}$), the dashed lines are the hole states (blue $\mathcal{E}_\text{h}^{+}$, red $\mathcal{E}_\text{h}^{-}$). Reprinted figures with permission from [77]. Copyright (2007) by the American Physical Society.

The dispersion relations plotted in figure 4.14 are for the case $\vartheta = \pi/2$ of an armchair edge. The case $\vartheta = 0$ of a zigzag edge contains additional dispersionless states away from the Fermi level [79], but they play no role in the electrical conduction.

4.4.5 The conductance

To determine the conductance G_{NS}, one needs to calculate the transmission matrix t of the edge states at the Fermi level. Edge states approach the superconductor along the insulating edge I_1 (with parameters $\boldsymbol{\nu}_1$, ϑ_1), then propagate along the NS interface, and finally return along the insulating edge I_2 (with parameters $\boldsymbol{\nu}_2$, ϑ_2). At sufficiently small μ each insulating edge I_p supports only two propagating modes, one electron mode $\propto|\boldsymbol{\nu}_p\rangle$ and one hole mode $\propto|-\boldsymbol{\nu}_p\rangle$. The NS interface also supports two propagating modes at small μ, of mixed electron–hole character and valley degeneration. The conductance is given by

$$G_{\mathrm{NS}} = \frac{2e^2}{h}(1 - T_{\mathrm{ee}} + T_{\mathrm{he}}) = \frac{4e^2}{h}T_{\mathrm{he}}, \tag{4.41}$$

with $T_{\mathrm{ee}} = |t_{++}|^2$ as the probability that an electron incident along I_1 returns along I_2 as an electron and $T_{\mathrm{he}} = |t_{-+}|^2$ as the probability that the electron returns as a hole.

Since the unidirectional motion of the edge states prevents reflections, the transmission matrix t from I_1 to I_2 is the product of the transmission matrices t_1 from I_1 to NS and t_2 from NS to I_2. Each of the matrices t_p is a 2×2 unitary matrix, diagonal in the basis $|\pm\boldsymbol{\nu}_p\rangle$:

$$t_p = \mathrm{e}^{\mathrm{i}\phi_p}|+\boldsymbol{\nu}_p\rangle\langle+\boldsymbol{\nu}_p| + \mathrm{e}^{\mathrm{i}\phi_p'}|-\boldsymbol{\nu}_p\rangle\langle-\boldsymbol{\nu}_p|. \tag{4.42}$$

The phase shifts ϕ_p and ϕ_p' need not be determined. Using $|\langle\boldsymbol{\nu}_1| \pm \boldsymbol{\nu}_2\rangle|^2 = \frac{1}{2}(1 \pm \boldsymbol{\nu}_1 \cdot \boldsymbol{\nu}_2)$, one can obtain the required transmission probabilities according to $t = t_2 t_1$:

$$T_{\mathrm{he}} = 1 - T_{\mathrm{ee}} = \frac{1}{2}(1 - \boldsymbol{\nu}_1 \cdot \boldsymbol{\nu}_2). \tag{4.43}$$

Substituting equation (4.43) into equation (4.41) gives the central result, equation (4.25).

According to figure 4.12, it can be found that $G_{\mathrm{NS}} = 0$ in the case of (a) a superconducting contact to a single edge ($\boldsymbol{\nu}_1 = \boldsymbol{\nu}_2$)—regardless of whether the edge is zigzag or armchair. In the case of (c) a contact between a zigzag and an armchair edge, $\boldsymbol{\nu}_1 \cdot \boldsymbol{\nu}_2 = 0$ and $G_{\mathrm{NS}} = 2e^2/h$. The case (b) of a contact between two opposite edges has $\boldsymbol{\nu}_1 = -\boldsymbol{\nu}_2$ and $G_{\mathrm{NS}} = 4e^2/h$ if both edges are zigzag; the same holds if both edges are armchair separated by a multiple of three hexagons (as in the figure); if the number of hexagons separating the two armchair edges is not a multiple of three, then $\boldsymbol{\nu}_1 \cdot \boldsymbol{\nu}_2 = 1/2$ and $G_{\mathrm{NS}} = e^2/h$.

When the intervalley relaxation is considered, the conductance formula equation (4.25) will be rewritten as

$$G_{\mathrm{NS}} = \frac{2e^2}{h}(1 - \mathrm{e}^{-\Gamma_{\mathrm{r}}L/v_0}\cos\Theta), \tag{4.44}$$

where L is the length of the NS interface, $v_0 = \hbar^{-1}\frac{d\mathcal{E}_0}{dq} \simeq \min(v_F/2, \sqrt{2}\,\mu l_B/\hbar)$ is the velocity along the interface. Γ_r denotes the intervalley relaxation rate, which tends to equalize the populations of the two degenerate modes propagating along the NS interface. The intervalley relaxation becomes appreciable when $\Gamma_r L/v_0 \gtrsim 1$. The density matrix $\rho = \rho_0(1 - e^{-\Gamma_r L/v_0}) + \rho_1 e^{-\Gamma_r L/v_0}$ then contains a valley isotropic part $\rho_0 \propto \kappa_0$ with $T_{ee} = T_{eh} = 1/2$ and a nonequilibrium part $\rho_0 \propto |\boldsymbol{\nu}_1\rangle\langle\boldsymbol{\nu}_1|$ with T_{ee}, T_{eh} given by equation (4.43). A nonzero conductance when the supercurrent covers a single edge ($\Theta = 0$) is the direct measure of the intervalley relaxation.

References

[1] Rycerz A, Tworzydło J and Beenakker C W J 2007 Valley filter and valley valve in graphene *Nat. Phys.* **3** 172–5

[2] Xiao D, Yao W and Niu Q 2007 Valley-contrasting physics in graphene: magnetic moment and topological transport *Phys. Rev. Lett.* **99** 236809

[3] Garcia-Pomar J L, Cortijo A and Nieto-Vesperinas M 2008 Fully valley-polarized electron beams in graphene *Phys. Rev. Lett.* **100** 236801

[4] Pereira J M Jr, Peeters F M, Costa Filho R N and Farias G A 2008 Valley polarization due to trigonal warping on tunneling electrons in graphene *J. Phys.: Condens. Matter* **21** 045301

[5] Pereira J M, Peeters F M, Chaves A and Farias G A 2010 Klein tunneling in single and multiple barriers in graphene *Semicond. Sci. Technol.* **25** 033002

[6] Gold C, Knothe A, Kurzmann A, Garcia-Ruiz A, Watanabe K, Taniguchi T, Fal'ko V, Ensslin K and Ihn T 2021 Coherent jetting from a gate-defined channel in bilayer graphene *Phys. Rev. Lett.* **127** 046801

[7] Raza H (ed) 2012 *Graphene Nanoelectronics: Metrology, Synthesis, Properties and Applications* NanoScience and Technology (Heidelberg: Springer)

[8] Liu C-C, Jiang H and Yao Y 2011 Low-energy effective Hamiltonian involving spin-orbit coupling in silicene and two-dimensional germanium and tin *Phys. Rev.* B **84** 195430

[9] Swartz A G, Odenthal P M, Hao Y, Ruoff R S and Kawakami R K 2012 Integration of the ferromagnetic insulator EuO onto graphene *ACS Nano* **6** 10063–9

[10] Haugen H, Huertas-Hernando D and Brataas A 2008 Spin transport in proximity-induced ferromagnetic graphene *Phys. Rev.* B **77** 115406

[11] Semenov Y G, Kim K W and Zavada J M 2007 Spin field effect transistor with a graphene channel *Appl. Phys. Lett.* **91** 153105

[12] Wang S and Yu J 2018 Magnetic behaviors of 3d transition metal-doped silicane: a first-principle study *J. Supercond. Nov. Magn.* **31** 2789–95

[13] He W, Zhang S, Luo Y and Wang S 2024 Exploring monolayer GaN doped with transition metals: insights from first-principles studies *J. Supercond. Nov. Magn.* **37** 157–63

[14] Ezawa M 2012 Valley-polarized metals and quantum anomalous Hall effect in silicene *Phys. Rev. Lett.* **109** 055502

[15] Ezawa M 2013 Spin valleytronics in silicene: quantum spin Hall–quantum anomalous Hall insulators and single-valley semimetals *Phys. Rev.* B **87** 155415

[16] Wang S-K, Tian H-Y, Yang Y-H and Wang J 2014 Spin and valley half metal induced by staggered potential and magnetization in silicene *Chin. Phys.* B **23** 017203

[17] Soodchomshom B 2014 Perfect spin-valley filter controlled by electric field in ferromagnetic silicene *J. Appl. Phys.* **115** 023706

[18] Prarokijjak W and Soodchomshom B 2018 Large magnetoresistance dips and perfect spin-valley filter induced by topological phase transitions in silicene *J. Magn. Magn. Mater.* **452** 407–14

[19] Jatiyanon K and Soodchomshom B 2018 Spin-valley and layer polarizations induced by topological phase transitions in bilayer silicene *Superlattices Microstruct.* **120** 540–52

[20] Yarmohammadi M 2017 The effect of Rashba spin–orbit coupling on the spin- and valley-dependent electronic heat capacity of silicene *RSC Adv.* **7** 10650–9

[21] Wei P *et al* 2016 Strong interfacial exchange field in the graphene/EuS heterostructure *Nat. Mater.* **15** 711

[22] Qiao Z, Yang S A, Feng W, Tse W-K, Ding J, Yao Y, Wang J and Niu Q 2010 Quantum anomalous Hall effect in graphene from Rashba and exchange effects *Phys. Rev.* B **82** 161414

[23] Tse W-K, Qiao Z, Yao Y, MacDonald A H and Niu Q 2011 Quantum anomalous Hall effect in single-layer and bilayer graphene *Phys. Rev.* B **83** 155447

[24] Uchida K *et al* 2010 Spin Seebeck insulator *Nat. Mater.* **9** 894

[25] Wang S, Zhang P, Ren C, Tian H, Pang J, Song C and Sun M 2019 Valley Hall effect and magnetic moment in magnetized silicene *J. Supercond. Nov. Magn.* **32** 2947–7

[26] Yokoyama T 2008 Controllable spin transport in ferromagnetic graphene junctions *Phys. Rev.* B **77** 073413

[27] Yokoyama T and Linder J 2011 Anomalous magnetic transport in ferromagnetic graphene junctions *Phys. Rev.* B **83** 081418

[28] Yokoyama T 2013 Controllable valley and spin transport in ferromagnetic silicene junctions *Phys. Rev.* B **87** 241409

[29] Ezawa M 2012 A topological insulator and helical zero mode in silicene under an inhomogeneous electric field *New J. Phys.* **14** 033003

[30] Katsnelson M I, Novoselov K S and Geim A K 2006 Chiral tunnelling and the Klein paradox in graphene *Nat. Phys.* **2** 620–5

[31] Yamakage A, Ezawa M, Tanaka Y and Nagaosa N 2013 Charge transport in pn and npn junctions of silicene *Phys. Rev.* B **88** 085322

[32] Zeng H, Dai J, Yao W, Xiao D and Cui X 2012 Valley polarization in MoS_2 monolayers by optical pumping *Nat. Nanotechnol.* **7** 490–3

[33] Srivastava A, Sidler M, Allain A V, Lembke D S, Kis A and Imamoğlu A 2015 Valley Zeeman effect in elementary optical excitations of monolayer WSe_2 *Nat. Phys.* **11** 141–7

[34] Li P, Liu B, Chen S, Zhang W-X and Guo Z-X 2024 Progress on two-dimensional ferrovalley materials *Chin. Phys.* B **33** 017505

[35] Chu J *et al* 2021 2D polarized materials: ferromagnetic, ferrovalley, ferroelectric materials, and related heterostructures *Adv. Mater.* **33** 2004469

[36] Tong W-Y, Gong S-J, Wan X and Duan C-G 2016 Concepts of ferrovalley material and anomalous valley Hall effect *Nat. Commun.* **7** 13612

[37] Li F, Tu K and Chen Z 2014 Versatile electronic properties of VSe_2 bulk, few-layers, monolayer, nanoribbons, and nanotubes: a computational exploration *J. Phys. Chem.* C **118** 21264–74

[38] Pan H 2014 Electronic and magnetic properties of vanadium dichalcogenides monolayers tuned by hydrogenation *J. Phys. Chem.* C **118** 13248–53

[39] Manchanda P and Skomski R 2016 2D transition-metal diselenides: phase segregation, electronic structure, and magnetism *J. Phys.: Condens. Matter* **28** 064002

[40] Liu W and Xu Y (ed) 2020 *Spintronic 2D Materials: Fundamentals and Applications* Materials Today (Amsterdam: Elsevier)

[41] Dai Y, Wei W, Ma Y and Niu C 2022 *Calculations and Simulations of Low-Dimensional Materials: Tailoring Properties for Applications* (Weinheim: Wiley-VCH)

[42] Wang F, Zhang Y, Yang W, Zhang H and Xu X 2024 Recent progress on valley polarization and valley-polarized topological states in two-dimensional materials *Chin. Phys.* B **33** 017306

[43] Thouless D J, Kohmoto M, Nightingale M P and den Nijs M 1982 Quantized Hall conductance in a two-dimensional periodic potential *Phys. Rev. Lett.* **49** 405–8

[44] Xiao D, Liu G-B, Feng W, Xu X and Yao W 2012 Coupled spin and valley physics in monolayers of MoS_2 and other group-VI dichalcogenides *Phys. Rev. Lett.* **108** 196802

[45] Mak K F, McGill K L, Park J and McEuen P L 2014 The valley Hall effect in MoS_2 transistors *Science* **344** 1489

[46] Pan H, Li Z, Liu C-C, Zhu G, Qiao Z and Yao Y 2014 Valley-polarized quantum anomalous Hall effect in silicene *Phys. Rev. Lett.* **112** 106802

[47] Feng W, Yao Y, Zhu W, Zhou J, Yao W and Xiao D 2012 Intrinsic spin Hall effect in monolayers of group-VI dichalcogenides: a first-principles study *Phys. Rev.* B **86** 165108

[48] Hall E H 1881 XVIII. On the "rotational coefficient" in nickel and cobalt *Lond. Edinb. Dubl. Phil. Mag.* **12** 157–72

[49] Nagaosa N, Sinova J, Onoda S, MacDonald A H and Ong N P 2010 Anomalous Hall effect *Rev. Mod. Phys.* **82** 1539–92

[50] Jiang J and Mi W 2023 Two-dimensional magnetic Janus monolayers and their van der Waals heterostructures: a review on recent progress *Mater. Horiz.* **10** 788–807

[51] Shen X-W, Tong W-Y, Gong S-J and Duan C-G 2018 Electrically tunable polarizer based on 2D orthorhombic ferrovalley materials *2D Mater.* **5** 011001

[52] Wang K, Li Y, Mei H, Li P and Guo Z-X 2022 Quantum anomalous Hall and valley quantum anomalous Hall effects in two-dimensional d^0 orbital XY monolayers *Phys. Rev. Mater.* **6** 044202

[53] Jia Y, Luo F, Hao X, Meng Q, Dou W, Zhang L, Wu J, Zhai S and Zhou M 2021 Intrinsic valley polarization and high-temperature ferroelectricity in two-dimensional orthorhombic lead oxide *ACS Appl. Mater. Interfaces* **13** 6480–8

[54] Huan H, Xue Y, Zhao B, Gao G, Bao H and Yang Z 2021 Strain-induced half-valley metals and topological phase transitions in MBr_2 monolayers (M=Ru, Os) *Phys. Rev.* B **104** 165427

[55] Shen C, Wang G, Wang T, Xia C and Li J 2020 Spin orientation and strain tuning valley polarization with magneto-optic Kerr effects in ferrovalley VS_2 monolayer *Appl. Phys. Lett.* **117** 042406

[56] Sheng K, Zhang B, Yuan H-K and Wang Z-Y 2022 Strain-engineered topological phase transitions in ferrovalley 2H-$RuCl_2$ monolayer *Phys. Rev.* B **105** 195312

[57] Zhao P, Liang Y, Ma Y and Frauenheim T 2023 Valley physics and anomalous valley Hall effect in single-layer h-MNX (M = Ti, Zr, Hf; X = Cl, Br) *Phys. Rev.* B **107** 035416

[58] Li P, Yang X, Jiang Q-S, Wu Y-Z and Xun W 2023 Built-in electric field and strain tunable valley-related multiple topological phase transitions in $VSiXN_4$ (X = C, Si, Ge, Sn, Pb) monolayers *Phys. Rev. Mater.* **7** 064002

[59] Zhang C, Nie Y, Sanvito S and Du A 2019 First-principles prediction of a room-temperature ferromagnetic Janus VSSe monolayer with piezoelectricity, ferroelasticity, and large valley polarization *Nano Lett.* **19** 1366–70

[60] Luo C, Peng X, Qu J and Zhong J 2020 Valley degree of freedom in ferromagnetic Janus monolayer H-VSSe and the asymmetry-based tuning of the valleytronic properties *Phys. Rev.* B **101** 245416

[61] Li C and An Y 2022 Two-dimensional intrinsic ferrovalley Janus 2H-VSeS monolayer with high Curie temperature and robust valley polarization *Phys. Rev. Mater.* **6** 094012

[62] Zhao Y-F, Shen Y-H, Hu H, Tong W-Y and Duan C-G 2021 Combined piezoelectricity and ferrovalley properties in Janus monolayer VClBr *Phys. Rev.* B **103** 115124

[63] Sheng K, Chen Q, Yuan H-K and Wang Z-Y 2022 Monolayer CeI_2: an intrinsic room-temperature ferrovalley semiconductor *Phys. Rev.* B **105** 075304

[64] Sharan A and Singh N 2022 Intrinsic valley polarization in computationally discovered two-dimensional ferrovalley materials: LaI_2 and PrI_2 monolayers *Adv. Theory Simul.* **5** 2100476

[65] Zhao P, Ma Y, Lei C, Wang H, Huang B and Dai Y 2019 Single-layer $LaBr_2$: two-dimensional valleytronic semiconductor with spontaneous spin and valley polarizations *Appl. Phys. Lett.* **115** 261605

[66] Li C and An Y 2023 Two-dimensional rare-earth Janus $2H$-GdXY (X, Y = Cl, Br, I; $X \neq Y$) monolayers: bipolar ferromagnetic semiconductors with high Curie temperature and large valley polarization *Phys. Rev.* B **107** 115428

[67] Liang L, Yang Y, Wang X and Li X 2023 Tunable valley and spin splittings in VSi_2N_4 bilayers *Nano Lett.* **23** 858–62

[68] Hu H, Tong W-Y, Shen Y-H and Duan C-G 2020 Electrical control of the valley degree of freedom in 2D ferroelectric/antiferromagnetic heterostructures *J. Mater. Chem.* C **8** 8098–106

[69] Du W, Peng R, He Z, Dai Y, Huang B and Ma Y 2022 Anomalous valley Hall effect in antiferromagnetic monolayers *npj 2D Mater. Appl.* **6** 11

[70] Zhang T, Xu X, Huang B, Dai Y, Kou L and Ma Y 2023 Layer-polarized anomalous Hall effects in valleytronic van der Waals bilayers *Mater. Horiz.* **10** 483–90

[71] Peng R, Zhang T, He Z, Wu Q, Dai Y, Huang B and Ma Y 2023 Intrinsic layer-polarized anomalous Hall effect in bilayer $MnBi_2Te_4$ *Phys. Rev.* B **107** 085411

[72] Ma X, Yin L, Zou J, Mi W and Wang X 2019 Strain-tailored valley polarization and magnetic anisotropy in two-dimensional 2H-VS_2/Cr_2C heterostructures *J. Phys. Chem.* C **123** 17440–8

[73] Guan Y *et al* 2023 Layered semiconductor $Cr_{0.32}Ga_{0.68}Te_{2.33}$ with concurrent broken inversion symmetry and ferromagnetism: a bulk ferrovalley material candidate *J. Am. Chem. Soc.* **145** 4683–90

[74] Wang S, Pratama F R, Ukhtary M S and Saito R 2020 Independent degrees of freedom in two-dimensional materials *Phys. Rev.* B **101** 081414(R)

[75] Wang J and Liu S 2012 Crossed Andreev reflection in a zigzag graphene nanoribbon-superconductor junction *Phys. Rev.* B **85** 035402

[76] Beenakker C W J 2006 Specular Andreev reflection in graphene *Phys. Rev. Lett.* **97** 067007

[77] Akhmerov A R and Beenakker C W J 2007 Detection of valley polarization in graphene by a superconducting contact *Phys. Rev. Lett.* **98** 157003

[78] Novoselov K S, Geim A K, Morozov S V, Jiang D, Katsnelson M I, Grigorieva I V, Dubonos S V and Firsov A A 2005 Two-dimensional gas of massless Dirac fermions in graphene *Nature* **438** 197–200

[79] Brey L and Fertig H A 2006 Edge states and the quantized Hall effect in graphene *Phys. Rev.* B **73** 195408

[80] Castro Neto A H, Guinea F, Peres N M R, Novoselov K S and Geim A K 2009 The electronic properties of graphene *Rev. Mod. Phys.* **81** 109–62

[81] Eroms J, Weiss D, Boeck J D, Borghs G and Zülicke U 2005 Andreev reflection at high magnetic fields: evidence for electron and hole transport in edge states *Phys. Rev. Lett.* **95** 107001

[82] McCann E and Fal'ko V I 2004 Symmetry of boundary conditions of the Dirac equation for electrons in carbon nanotubes *J. Phys.: Condens. Matter* **16** 2371

[83] Hou Z, Xing Y and Guo A-M 2016 and Q-F Sun. Crossed Andreev effects in two-dimensional quantum Hall systems *Phys. Rev.* B **94** 064516

Chapter 5

Valley-polarized current in polycrystalline systems

The large-area 2D materials (graphene, silicene, monolayer group-VI TMDCs, etc) synthesized by CVD technique are always found to be polycrystalline in nature [1–11], consisting of two domains with different crystallographic orientations [12] and the grain boundaries (GBs) stitching them together. Polycrystalline 2D materials exhibit many excellent properties, including magnetic [13–18], thermal transport [19–23], and electrical properties [24–32], etc. At the same time, they are also ideal candidates in valleytronics due to the broken spatial IS. In this chapter, we will introduce the physical mechanism of valley scattering in polycrystalline systems and some measurements to regulate valley transport, most of which are based on the line defects [33].

5.1 Line defect

5.1.1 Model

In polycrystalline graphene, one of the most studied configurations in valleytronics is the line defect, containing the octagonal and pentagonal sp^2-hybridized carbon ring, which is also termed as 585 ring; see figure 5.1(a). In experiments, this kind of line defect has been fabricated on specific substrates [1, 2, 34], such as Ni(111) [34]. In graphene with a line defect, the positions of the two Dirac valleys are no longer located at their original positions due to the change of the unit vectors when the defect atoms are introduced, as shown in figure 5.1(b). For pristine graphene, the two unit vectors $\boldsymbol{a}_1$ and $\boldsymbol{a}_2$ are given in equation (1.1). The coordinates of the two unequivalent Dirac valleys are given in equations (1.19) and (1.20), which are located at the six corners of the hexagonal BZ, as shown in figure 5.1(b). However, for the line defect the unit vectors are changed to [35]

$$\boldsymbol{a}_1^{\mathrm{d}} = 2a\hat{\boldsymbol{y}},\ \boldsymbol{a}_2^{\mathrm{d}} = \sqrt{3}a\hat{\boldsymbol{x}}, \tag{5.1}$$

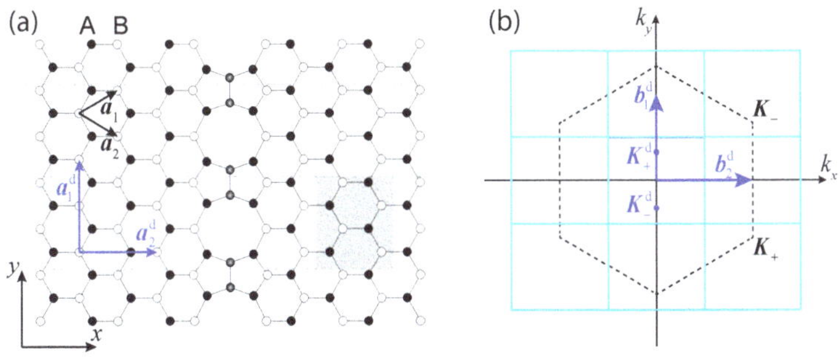

Figure 5.1. (a) Schematic drawing of a 585 line defect in graphene. The structure has a period $2a$ along the line defect as indicated by the gray dashed lines, where a is the lattice constant of pristine graphene. $\boldsymbol{a}_i$ and $\boldsymbol{a}_i^{\mathrm{d}}$ $(i = 1, 2)$ are the unit vectors of pristine graphene and graphene with line defect, respectively. (b) The first BZ of the pristine graphene (dashed lines) and graphene with 585 line defect (cyan lines). $\boldsymbol{b}_i^{\mathrm{d}}$ is the reciprocal lattice vector of graphene with line defect. The Dirac valleys $\boldsymbol{K}_-$ and $\boldsymbol{K}_+$ for the pristine graphene are folded to $\boldsymbol{K}_-^{\mathrm{d}}$ and $\boldsymbol{K}_+^{\mathrm{d}}$ for the graphene with line defect, respectively.

by considering the period $2a$ along the line defect (y direction) [36], see figure 5.1(a). Following the scheme in section 1.2.1, if we set $\boldsymbol{a}_3^{\mathrm{d}}$ as the unit vector that points in the direction perpendicular to the 2D graphene plane, we can derive the primitive vectors of the reciprocal lattice [33, 37]

$$\begin{aligned}\boldsymbol{b}_1^{\mathrm{d}} &= 2\pi\frac{\boldsymbol{a}_2^{\mathrm{d}} \times \boldsymbol{a}_3^{\mathrm{d}}}{\boldsymbol{a}_1^{\mathrm{d}} \cdot (\boldsymbol{a}_2^{\mathrm{d}} \times \boldsymbol{a}_3^{\mathrm{d}})} = \frac{\pi}{a}\hat{\boldsymbol{y}}, \\ \boldsymbol{b}_2^{\mathrm{d}} &= 2\pi\frac{\boldsymbol{a}_3^{\mathrm{d}} \times \boldsymbol{a}_1^{\mathrm{d}}}{\boldsymbol{a}_2^{\mathrm{d}} \cdot (\boldsymbol{a}_3^{\mathrm{d}} \times \boldsymbol{a}_1^{\mathrm{d}})} = \frac{2\pi}{\sqrt{3}a}\hat{\boldsymbol{x}}.\end{aligned} \tag{5.2}$$

The corresponding first BZ of the lattice is indicated by the blue rectangles shown in figure 5.1(b), bounded by $k_x \in [-\frac{\pi}{\sqrt{3}a}, \frac{\pi}{\sqrt{3}a}]$ and $k_y \in [-\frac{\pi}{2a}, \frac{\pi}{2a}]$. The energy bands are preferably plotted within the first BZ, as it is easy to describe the physical properties using the reduced-zone scheme. Thus, we can translate (fold [38]) any wavevector into the first BZ through reciprocal lattice vectors [37]. In the line defect, the Dirac valleys $\boldsymbol{K}_-$ and $\boldsymbol{K}_+$ for the pristine graphene are folded to

$$\boldsymbol{K}_-^{\mathrm{d}} = -\frac{\pi}{3a}\hat{\boldsymbol{y}},\ \boldsymbol{K}_+^{\mathrm{d}} = \frac{\pi}{3a}\hat{\boldsymbol{y}}, \tag{5.3}$$

in the first BZ of graphene with a line defect, respectively, because of the translational symmetry given by the following equations [39]

$$\begin{aligned}\boldsymbol{K}_+^{\mathrm{d}} &= \boldsymbol{K}_+ + m_1\boldsymbol{b}_1^{\mathrm{d}} + n_1\boldsymbol{b}_2^{\mathrm{d}}, \\ \boldsymbol{K}_-^{\mathrm{d}} &= \boldsymbol{K}_- + m_2\boldsymbol{b}_1^{\mathrm{d}} + n_2\boldsymbol{b}_2^{\mathrm{d}},\end{aligned} \tag{5.4}$$

where m_i and n_i $(i = 1, 2)$ are integers.

5.1.2 The physical mechanism

The valley transmission mechanism across the line defect was first put forward by Gunlycke *et al* [32] according to the symmetry of the lattice structure. In the low-energy regime, the NN TB Hamiltonian of graphene is given in section 1.2.2 [40, 41], as

$$H_\kappa = \hbar v_{\rm F}(q_x\sigma_y + \kappa q_y\sigma_x), \tag{5.5}$$

where $v_{\rm F}$ is the Fermi velocity, σ_x and σ_y are Pauli matrices, $\kappa = \pm 1$ are for the two valleys. Here, $\boldsymbol{q} = (q_x, q_y)$ is the wavevector measured from the centre of the occupied valley located at $\boldsymbol{K}_\kappa = \kappa\frac{4\pi}{3a}\hat{\boldsymbol{y}}$ with $q_x = q\cos\phi$ and $q_y = q\sin\phi$. According to equation (5.5), the energy eigenvalues can be written as $\mathcal{E} = \xi\hbar v_{\rm F} q$ with $\xi = 1\ (-1)$ denoting the quasiparticle is an electron (hole). Due to the linear dispersion relation in the low-energy limit, the quasiparticle group velocity is $v_{\rm F}\hat{\boldsymbol{q}}$ where $\hat{\boldsymbol{q}}$ is the unit vector in the direction of $\boldsymbol{q}$. The eigenstate of the above graphene Hamiltonian for a given valley κ and scattering angle ϕ can be expressed as

$$|\Phi_\kappa\rangle = \frac{1}{\sqrt{2}}(|A\rangle + \mathrm{i}e^{-\mathrm{i}\kappa\phi}|B\rangle), \tag{5.6}$$

where $|A\rangle$ and $|B\rangle$ are the two sublattices in graphene.

The lattice structure of the line defect is mirror-symmetric with respect to the line defect, as shown in figure 5.1(a). Therefore, the reflection operator commutes with the graphene translation operator perpendicular to the line defect as $q \to 0$. One can construct the symmetry-adapted states $|\pm\rangle$ according to the eigenstates of the graphene Hamiltonian and the reflection operator. Because the reflection operator maps A sites onto B sites and vice versa, it is convenient to adopt the operator σ_x acting on the two sublattices to describe the reflection operator. According to the eigenstates of σ_x, the symmetry-adapted states can be obtained as

$$|\pm\rangle = \frac{1}{\sqrt{2}}(|A\rangle \pm |B\rangle). \tag{5.7}$$

The graphene state equation (5.6) can be rewritten with the symmetry-adapted basis as

$$|\Phi_\kappa\rangle = \frac{1 + \mathrm{i}e^{-\mathrm{i}\kappa\phi}}{\sqrt{2}}|+\rangle + \frac{1 - \mathrm{i}e^{-\mathrm{i}\kappa\phi}}{\sqrt{2}}|-\rangle. \tag{5.8}$$

The total system contains three parts: the pristine graphene, the isolated line defect and the interaction between the pristine graphene and the line defect. Therefore, the Hamiltonian of the full system can be represented by $H_\kappa^{\rm d} = H_\kappa + H_{\rm i} + H_{\rm d}$, where the first term denotes the pristine graphene, the second term is the interaction between the pristine graphene and the line defect and the third term corresponds to the line defect. Each of the three terms commutes with the reflection operator σ_x, hence the full Hamiltonian must commute with the reflection operator, the eigenstates of H_κ^{d} in the symmetry-adapted basis are either symmetric or antisymmetric about the line defect. However, because the antisymmetric states have a node at the line defect, there are no matrix elements within the NN model coupling the left and right sides. Therefore, antisymmetric states will not contribute to any transmission across the line defect. There are two symmetric states at the Fermi level without a node on the line defect, which are extended eigenstates of the full Hamiltonian; they carry quasiparticles across the line defect and contribute to the transmission. Thus, the transmission probability of the quasiparticle across the line defect is

$$T_\kappa = |\langle + | \Phi_\kappa \rangle|^2 = \frac{1}{2}(1 + \kappa \sin \phi). \tag{5.9}$$

According to equation (5.9), it can be easily concluded that the line defect is semitransparent for valley states because $\sum_\kappa T_\kappa = 1$. The semitransparent here means that for a quasiparticle from $\boldsymbol{K}_+$ valley ($\kappa = +1$), it can transmit along the $\phi > 0$ direction while the transmission along the $\phi < 0$ direction is suppressed. However, the situation is exactly the opposite for $\boldsymbol{K}_-$ valley ($\kappa = -1$). Figure 5.2(d) illustrates the transmission process of the incident state Φ_κ across the line defect. At a large angle ϕ, the symmetric component of $\boldsymbol{K}_+$ valley can pass through the line defect with a high transmission probabilities while that of $\boldsymbol{K}_-$ valley ($\kappa = -1$) is mostly reflected in this direction. This can also be clearly seen from figures 5.3 and 5.4, where only one valley state can transmit through the line defect at a large angle $\phi \approx \pm\pi/2$ while the other valley is almost reflected in the same direction [42].

The transmission probability of a quasiparticle changes slowly from 1 to 0 as ϕ varies, as shown in figure 5.3. This transmission property indicates that the perfect

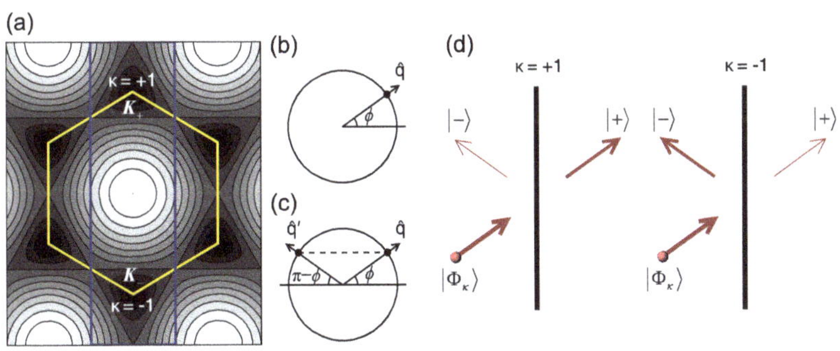

Figure 5.2. (a) The constant energy surface of graphene where the dark contours indicate the Fermi level and are located at the corners of the first BZ contained within the gold hexagon. The two valleys, $\boldsymbol{K}_+$ ($\kappa = 1$) and $\boldsymbol{K}_-$ ($\kappa = -1$), are identified by the two disjoint low-energy regions in the reciprocal primitive cell enclosed by the blue rectangle. (b) An incident quasiparticle state is defined by the wavevector $\boldsymbol{q}$ pointed in the direction $\hat{\boldsymbol{q}}$ with incidence angle $\phi = \arctan(q_y/q_x)$. (c) Owing to energy and momentum conservation along the line defect, the quasiparticle state from $\boldsymbol{K}_+$ valley can transmit along $\hat{\boldsymbol{q}}$ direction (with incidence angle ϕ) while the quasiparticle state from $\boldsymbol{K}_-$ valley can transmit along $\hat{\boldsymbol{q}}'$ direction (with incidence angle $\pi - \phi$). (d) Schematic diagram of the scattering process for the incident state $|\Phi_\kappa\rangle$ composed of the symmetric $|+\rangle$ and antisymmetric $|-\rangle$ components, the transmission/reflection probability is indicated by the thickness of the arrows. Reprinted figures with permission from [32]. Copyright (2011) by the American Physical Society.

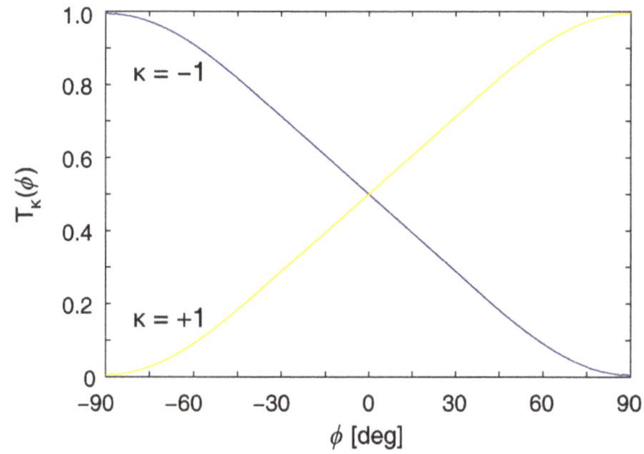

Figure 5.3. The transmission probability of T_κ as a function of incidence angle ϕ across the line defect. Reprinted figures with permission from [32]. Copyright (2011) by the American Physical Society.

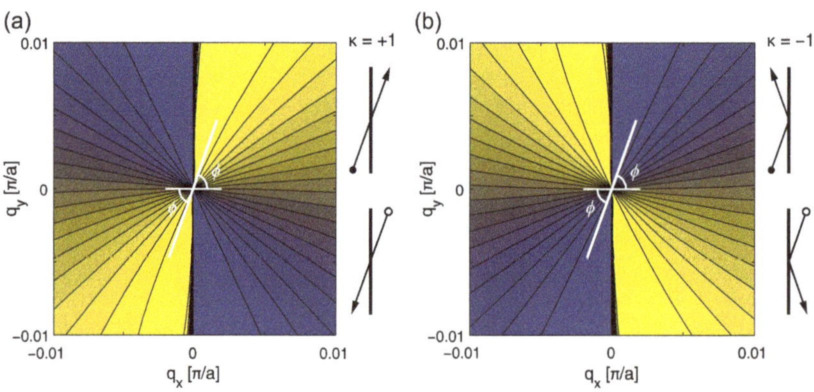

Figure 5.4. Transmission probability (indicated by the brightness) across the line defect with a finite energy. (a) The transmission probability for a quasiparticle from $\boldsymbol{K}_+$ ($\kappa = +1$) valley for different incidence angle. For the angle shown, the transmission is near one for $\boldsymbol{K}_+$ ($\kappa = +1$) valley quasiparticle, as illustrated on the right panel where the filled (open) circle represents an electron (hole) quasiparticle. (b) The transmission probability for the corresponding quasiparticle from $\boldsymbol{K}_-$ ($\kappa = -1$) valley. Reprinted figures with permission from [32]. Copyright (2011) by the American Physical Society.

VP can only exist for large incidence angles. According to the definition of VP in equation (3.28), one can obtain the VP from equation (5.9) quantitatively,

$$P = \sin \phi. \tag{5.10}$$

From equation (5.10) it can be found that the VP P varies from -1 to 1 as ϕ changes from $-\pi/2$ to $\pi/2$. However, at small incidence angles, the VP is unsatisfactory and reaches zero for normal incidence.

The above discussions about the transmission of the valley states across the line defect and the VP are based on the symmetry arguments at low-energy limit (the quasiparticle energy $\mathcal{E} \to 0$). However, it should be noted that the results always hold to an excellent approximation as long as $|\mathcal{E}| \ll \hbar v_F/a \approx 2.3$ eV. In figure 5.4, the transmission probability is plotted as a function of the components of the wavevector $\boldsymbol{q}$ within the NN TB model. Due to the linear dispersion relation, the quasiparticle energy and the quasiparticle wavevector $\boldsymbol{q}$ obey $|\mathcal{E}| = \hbar v_F q = \hbar v_F \sqrt{q_x^2 + q_y^2}$. It can be seen that as q_x (q_y) varies from 0 to $0.01\pi/a$, the transmission probability changes slightly for a definite angle ϕ (the black line) and does not change at all for high angle. Therefore, the VP stays unchanged as long as $qa \ll 1$ or $|\mathcal{E}| \ll \hbar v_F/a$.

Because the structure in figure 5.1(a) exhibits a symmetry plane through the centre of the primitive cell, there is TRS in the direction along the line defect. This TRS implies that the transmission probability of a quasiparticle with valley index $-\kappa$ can be obtained from that of a quasiparticle with valley index κ by letting $q_y \to -q_y$ or $T_\kappa(\phi) = T_{-\kappa}(-\phi)$. This relationship between $T_\kappa(\phi)$ and $T_{-\kappa}(\phi)$ can be seen in figure 5.4. However, the transmission probabilities in figure 5.4 are not symmetric about q_y, for instance, $T_\kappa(q_x, q_y) \neq T_\kappa(q_x, -q_y)$ ($T_\kappa(\phi) \neq T_\kappa(-\phi)$), thus one can conclude that the scattering of a quasiparticle depends on the valley index, which is a necessary requirement for a valley filter. As both graphene and the line defect have Hamiltonians exhibiting electron–hole symmetry, one might expect that $T_\kappa(\boldsymbol{q}) = T_\kappa(-\boldsymbol{q})$. However, the scattering does not obey electron–hole symmetry according to the numerical results obtained in figure 5.4 which accounts for the evanescent waves, for instance, $T_\kappa(0.005\pi/a, 0.01\pi/a) \neq T_\kappa(-0.005\pi/a, -0.01\pi/a)$, see figure 5.4. This is different from the condition depicted in equation (5.9) which neglects the evanescent waves. The lack of electron–hole symmetry in the combined graphene–line defect system can be explained by the lattice structure of the line defect, see figure 5.1(a). For the line defect, it is not biparticle as graphene because the sites participating in the pentagons at the line defect cannot be divided into two types where one type has only NNs of the other type.

The NEGF method can also be adopted to investigate the transport issues across the line defect of graphene in the low-energy limit ($q \to 0$). When dealing with the Green's function method, it requires the scattering region, the electrodes and the interaction between the scattering region and the electrodes, so as to obtain the retarded Green's function. The latter two parts are mainly used to calculate the retarded self energy Σ which accounts for the coupling between the scattering region and the electrodes. As shown in figure 5.1(a), the line defect has translational symmetry along the y direction, one can focus on those atoms within a primitive cell (the region between the two parallel dashed lines). For the line defect, the two defect atoms can act as the scattering region, the semi-infinite portion of graphene on each side of the line defect corresponds to the electrodes, and the interaction between the scattering region and the electrodes can be depicted with the coupling between the two defect atoms and their neighboring atoms.

The Hamiltonian of the scattering region (the isolated line defect) is given by

$$H_d = \gamma_0 \begin{pmatrix} 0 & 1 \\ 1 & 0 \end{pmatrix}, \tag{5.11}$$

with γ_0 as the NN hopping energy in graphene. Now we should derive the expression of Σ. Expressed in the basis of the atoms parallel to the line defect, the graphene state with valley index κ can be written as

$$|\kappa\rangle = \frac{1}{\sqrt{2}}\begin{pmatrix} 1 \\ e^{-2\pi i\kappa/3} \end{pmatrix}. \tag{5.12}$$

Requiring Σ to be retarded fixes the relative phase between the atoms on the line defect and their neighbours, resulting in the relation $\Sigma\langle B|\Phi_\kappa\rangle = \gamma_0\langle A|\Phi_\kappa\rangle|\kappa\rangle\langle\kappa|$. Then one can obtain

$$\Sigma = -\frac{i\gamma_0}{2}e^{i\kappa\phi}\begin{pmatrix} 1 & e^{2\pi i\kappa/3} \\ e^{-2\pi i\kappa/3} & 1 \end{pmatrix}. \tag{5.13}$$

Because the left and right semi-infinite graphene electrodes are symmetrical with respect to the central two defect atoms, the retarded self energy of the two electrodes should be identical. Once H_d and Σ are given, one can calculate the retarded Green's function on the line defect, $G^r = (\mathcal{E}I - H_d - 2\Sigma)^{-1}$, where I is the unit matrix and the factor 2 originates from the two semi-infinite graphene electrodes on both sides of the line defect. To zeroth order in q, one can obtain

$$G^r = -\frac{\gamma_0^{-1}}{1 + ie^{i\kappa\phi}}\begin{pmatrix} ie^{i\kappa\phi} & 1 - ie^{i\kappa(\phi+2\pi/3)} \\ 1 - ie^{i\kappa(\phi-2\pi/3)} & ie^{i\kappa\phi} \end{pmatrix}. \tag{5.14}$$

The transmission probability across the line defect can be given by

$$T_\kappa = \langle\kappa|\Gamma G^r\Gamma(G^r)^\dagger|\kappa\rangle, \tag{5.15}$$

where $\Gamma = i(\Sigma - \Sigma^\dagger)$ is the line-width function. Inserting equations (5.12–5.14) into equation (5.15), one recovers equation (5.9) exactly. A more detailed derivation can be found in references [35, 43, 44].

5.1.3 Other viewpoints on VP

The VP in the line defect is derived from the symmetry arguments [32]. However, the first-principles calculations indicate that the VP property across the line defect may be related to the electronic states localized at the line defect rather than from the symmetry of the lattice structure, as shown in figure 5.5. The band structure of the 585 line defect is given in figure 5.5(a). It is found that there are several bands crossing the Dirac cone of the projected 2D band structure of pristine graphene (shaded area). These bands correspond to the electronic states localized at the line defect, where two of them are in the vicinity of the Dirac point ($\mathcal{E} = 0$) and one at significantly higher energies (0.5 eV $< \mathcal{E} <$ 1 eV). The transmission probability across the line defect is shown in figure 5.5(b) and a noteworthy feature emerges where the transmission probability is strongly suppressed at the Dirac point. This property is caused by the resonant backscattering of charge carriers by the states localized at the line defect [45]. The author is convinced that the valley-filtering properties are dominated by this suppression of the charge-carrier transmission rather than by

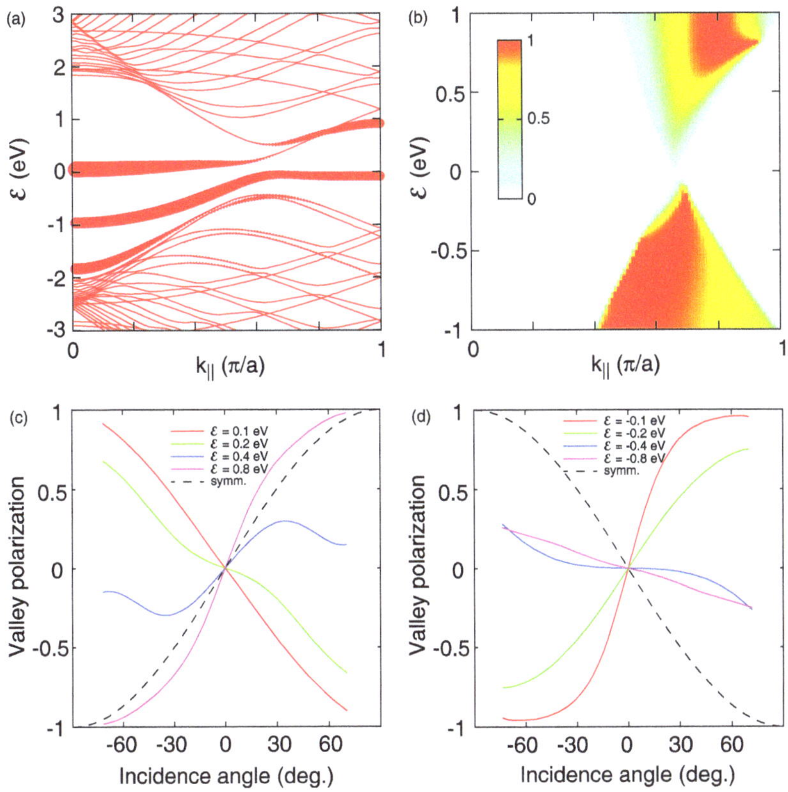

Figure 5.5. Electronic transport properties and VP of the line defect obtained from the first-principles calculations. (a) The electronic band structure of the line defect along the $k_{\|}$ direction (momentum along the line defect) at $k_{\perp} = 0$ (momentum perpendicular to the defect). The circles indicate the degree of localization of electronic states at the line defect. The shaded area corresponds to the continuum of bulk graphene states projected onto the 1D BZ of the line defect. (b) Transmission probability across the line defect as a function of momentum $k_{\|}$ and energy $\mathcal{E}$. (c), (d) VPs calculated for electrons and holes as a function of the incident angle of charge carriers for different energies. The dashed lines correspond to the VP according to the symmetry-based model in reference [32]. Reprinted figures with permission from [45]. Copyright (2014) by the American Physical Society.

symmetry-based consideration. Their reasons are as follows. In figures 5.5(c) and (d), the VP P as a function of incidence angle ϕ for different energies calculated for electron and hole charge carriers, respectively, are plotted. It is found that the dependences of P on ϕ are practically opposite to the ones predicted by the symmetry considerations (dashed lines) for the energies -0.2 eV $\leqslant \mathcal{E} \leqslant 0.2$ eV. For instance, when $\mathcal{E} = 0.1$ eV and $\phi = -\pi/2$, the first-principles calculations show that perfect VP with $P = 1$ occurs while the symmetry-based model [32] gives $P = -1$ (dashed line), see figure 5.5(c). However, for the high-energies when charge carriers are not affected by the resonance backscattering, the behaviour P predicted from symmetry arguments [32] is mostly recovered, e.g., the VP at $\mathcal{E} = 0.8$ eV agrees well with the dashed line in figure 5.5(c). Therefore, it can be concluded that the valley filter property across the line defect is dominated by the resonance backscattering.

5.1.4 Multiple line defects

Anyway, the perfect VP in line defects can only appear for the electrons with large incident angles, and the efficiency is reduced or even completely disappears when an electron is transmitted through the line defect perpendicularly. This effect indicates that the electron must always follow the direction of the line defect to maintain a relatively large VP, bringing about major challenges for the experimental studies of this phenomenon.

Liu *et al* [35] constructed multiple parallel line defects to filter one valley state, as shown in figure 5.6. In figure 5.6(a), the transmission probabilities of the two valleys for a line defect are mapped as a function of incident angle ϕ for different energies. It is found that at low energy ($\mathcal{E} = 0.001$ in unit of γ_0), the transmission probability overlaps exactly with $(1 \pm \sin\phi)/2$ (the dotted line), which reproduces the result represented in equation (5.9) and the high VP can only occur for large incident angles. However, some interesting phenomena emerge in the presence of two parallel line defects with distance $N_\mathrm{d}|\boldsymbol{a}_2^\mathrm{d}|$. For instance, the transmission probabilities are almost unchanged at small N_d while some transmission peaks are observed obviously at small incident angles for relative large N_d, as shown in figure 5.6(b). When $N_\mathrm{d} = 100$, the new peak emerges at $\phi = -0.1\pi$ and it appears at $\phi = 0.18\pi$ as $N_\mathrm{d} = 150$. To explore the physical origin of the new peak, the transmission

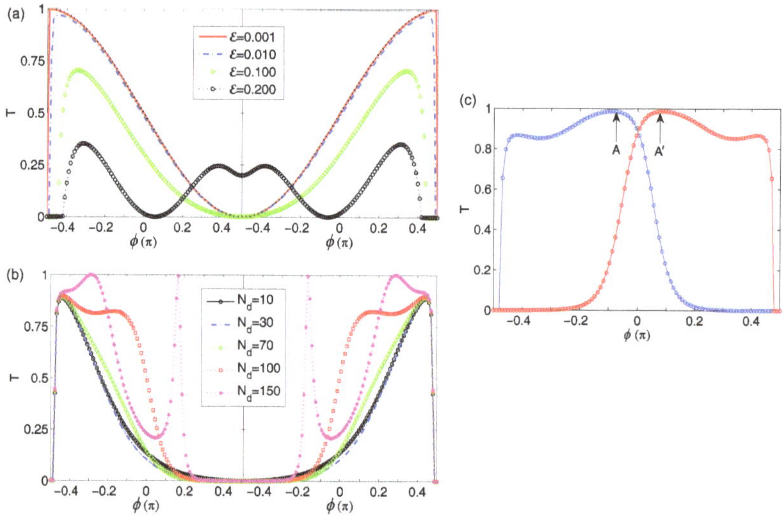

Figure 5.6. (a) The transmission probabilities T_+ (left panel) and T_- (right panel) across a 585 line defect as a function of incident angle ϕ for different energies (in units of γ_0). (b) The transmission probabilities T_+ (left panel) and T_- (right panel) as a function of incident angle ϕ for two parallel line defects at $\mathcal{E} = 0.01$, where N_d (in unit of $\sqrt{3}a$) is the distance between the two line defects. (c) The transmission probabilities of T_+ (blue line) and T_- (red line) as a function of ϕ with a fixed momentum $q = 0.017/a$ and $N_\mathrm{d} = 74$. Reprinted figures with permission from [35]. Copyright (2013) by the American Physical Society.

probabilities of the two valleys at a fixed momentum $q = 0.017/a$ are plotted in figure 5.6(c). It is found that the new peaks appear at $\phi \approx \pm 0.05\pi$ (the A/A' points). According to the parameters given in figure 5.6(c), the electron's wavelength $\lambda = 1/\mathcal{E} = 2/\sqrt{3}qa$ is roughly equivalent to the length it travels between the two line defects along the transmission direction $l = N_d/\cos\phi$. This indicates that these peaks should be induced by the quantum interferences effect in the presence of two line defects, e.g., the electron's resonant tunnelling phenomena, which is also discussed in sections 3.1 and 4.2. Compared with a single line defect, the perfect VP can be expanded to $\phi \in [-0.5\pi, -0.2\pi]$ ($[0.2\pi, 0.5\pi]$) although it is still unsatisfactory for small incident angles (A and A' points). Nevertheless, the results verify that the high valley transmission at a small incident angle is feasible by scattering off multiple line defects. In fact, for more parallel line defects, the VP can be promoted where the perfect VP can be realized in the angle region $\phi \in [-0.5\pi, 0]$ ($[0, 0.5\pi]$) for five parallel line defects at a definite N_d.

5.1.5 Line defect superlattice

Besides the line defect, the line-defect graphene superlattice (LDGSL) where the line defect is arranged along x direction, is also an ideal candidate to achieve the VP in graphene [46]. In figure 5.7(a), the pristine graphene/LDGSL/pristine graphene structure is constructed to filter one valley state. The special lattice structure brings about the unusual band structures. The lowest conduction subband is analogous to the flat-bottomed subband of a zigzag nanoribbon near the Fermi energy, except for two small dips around the Dirac points. In contrast to other bands of the LDGSL, the velocity direction of an electron in the lowest conduction band is valley dependent. For instance, only the rightward (leftward) motion is allowed when an electron is in the $\boldsymbol{K}_+$ ($\boldsymbol{K}_-$) valley, as shown in the middle panel of figure 5.7(b). The unique band structure plays the key role to realize VP in the structure presented in figure 5.7(a). When a low-energy electron in the conduction band of a pristine graphene is injected from the left electrode to the LDGSL, as shown in figure 5.7(b), the transmission is allowed only if such an electron belongs to the $\boldsymbol{K}_+$ valley. However, it will be strongly reflected by the LDGSL when it belongs to the $\boldsymbol{K}_-$ valley. As a result, an effective valley-filtering effect can be realized in such a simple electron transport structure.

The numerical calculations about the electronic wavevector-dependent transmission spectrum through the LDGSL are given in figures 5.7(c)–(e). In the calculations, the notation (N_x, N_z)-LDGSL is adopted to denote a structure shown in figure 5.7(a). N_x and N_z are the number of line defects along the x direction and the zigzag chains along the y direction, respectively. When the incident electron energy is confined in the range of $0 < \mathcal{E} < \Delta$, there are two rightward propagating modes in the electrodes that belong to $\boldsymbol{K}_+$ and $\boldsymbol{K}_-$ valleys. We define $T_{\kappa\kappa'}$ as the probability of an incident electron in $\boldsymbol{K}_\kappa$ valley been transmitted as an electron in the $\boldsymbol{K}_{\kappa'}$ valley. Therefore, there are totally four distinct kinds of transmission probabilities, which are represented by $T_{++}(\mathcal{E})$, $T_{--}(\mathcal{E})$, $T_{+-}(\mathcal{E})$, and $T_{-+}(\mathcal{E})$. The former two correspond to the intravalley transmissions and the latter two are the intervalley

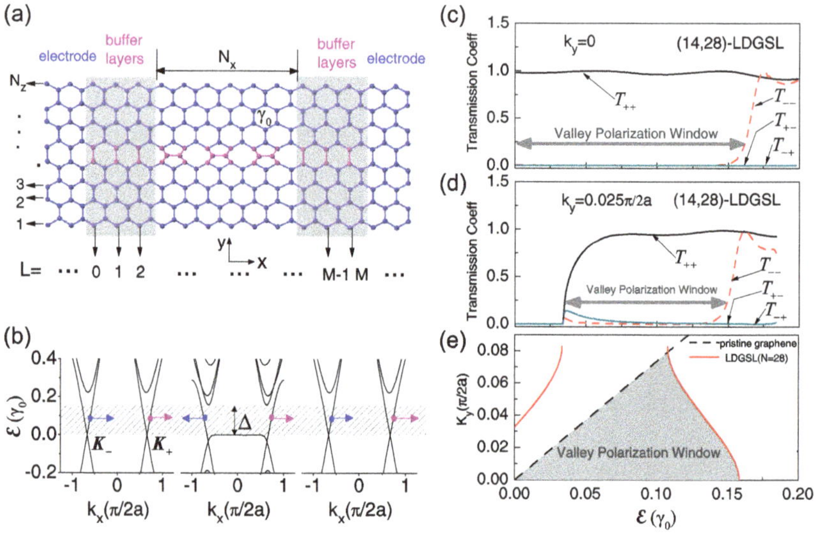

Figure 5.7. (a) Schematic of valley-polarized electronic transport device which consists of a finite length (N_x, N_z)-LDGSL sandwiched between two pieces of semi-infinite pristine graphene. The scattering region is composed of LDGSL and two buffer layers. N_x and N_z are the number of line defects along x direction and the zigzag chains along y direction, respectively. (b) The band structures of the electrodes and the LDGSL, where Δ is energy difference between the second conduction subband minimum and the flat bottom of the first conduction subband. (c), (d) The transmission probability of (14, 28)-LDGSL as a function of incident electronic energy $\mathcal{E}$, for the cases of (c) $k_y = 0$ and (d) $k_y = 0.025\pi/2a$. (e) The dashed line represents the dispersion relation of pristine graphene while the solid lines denote the lowest and second conduction subband bottoms of LDGSL as functions of electronic wavevector k_y with $k_x = \pi/3a$. The shaded region indicates the VP window. Reprinted figures with permission from [46]. Copyright (2012) by the American Physical Society.

transmissions. In figure 5.7(c), the electronic transmission probability of (14, 28)-LDGSL as a function of the incident electron energy $\mathcal{E}$ is mapped. It is found that in the case of $k_y = 0$, $T_{++}(\mathcal{E})$ is close to unity whereas the others are negligibly small in the energy range of $0 < \mathcal{E} < \Delta$. This demonstrates the VP effect in this structure and also verifies the theoretical analysis above. This means that when the $\boldsymbol{K}_+$ ($\boldsymbol{K}_-$) electron is injected into the LDGSL from the left electrode, it can pass through the device completely (will be reflected totally). When the electron energy is higher than the bottom of the second conduction subband of the LDGSL, it provides new propagating modes and the $\boldsymbol{K}_-$ valley electron will participate in the transport process. Therefore, $T_{--}(\mathcal{E})$ increases rapidly and the VP disappears in this energy region.

For a larger k_y, the VP effect still exists while the energy window to realize VP becomes narrow. In figure 5.7(d), the transmission spectra are shown for a fixed electronic wavevector $k_y = 0.025\pi/2a$. It is found that the energy range for $T_{++} \gg T_{--}$ becomes smaller in contrast to the case of $k_y = 0$. This can be understood by the fact that the bottom of the second conduction subband of the LDGSL descends as k_y goes away from zero. Meanwhile, the conduction-band bottom of pristine graphene (electrode) ascends because of the nonzero bandgap in the case of $k_y \neq 0$. These two characteristic energies determine the VP window together. In figure 5.7(e), the calculated VP window as a function of the electronic wavevector k_y is plotted.

5.2 Other polycrystalline graphene

Not confined to the line defect, various polycrystalline graphene has also been widely studied theoretically and experimentally [1–3, 12, 47–51]. Before further introducing the VP in polycrystalline graphene, we need to introduce a topological invariant that describes polycrystalline graphene, the Burgers vector. The Burgers vector $\boldsymbol{d}$ reflects the magnitude and direction of the crystalline lattice distortion produced by a dislocation [12]. The polycrystalline 2D materials can be deemed as embedding the periodic strip of width $\boldsymbol{d}$ into perfect 2D crystalline lattice. Therefore, the polycrystalline graphene are composed of different single-crystal grains separated by tilt GBs, which are the 1D interfaces between two domains of material with different crystallographic orientations. The periodicity of the translational vector $\boldsymbol{d}$ can be defined by the commensurability condition of two translation vectors $(n_{\rm L}, m_{\rm L})$ and $(n_{\rm R}, m_{\rm R})$ containing integers belonging to the left and right crystalline domains, by matching two translational vectors in the 'left' and 'right' domains separated by the GBs [12],

$$\boldsymbol{d} = n_{\rm L}\boldsymbol{a}_1^{\rm L} + m_{\rm L}\boldsymbol{a}_2^{\rm L} = n_{\rm R}\boldsymbol{a}_1^{\rm R} + m_{\rm R}\boldsymbol{a}_2^{\rm R}. \tag{5.16}$$

Here, $(\boldsymbol{a}_1^{\rm L}, \boldsymbol{a}_2^{\rm L})$ and $(\boldsymbol{a}_1^{\rm R}, \boldsymbol{a}_2^{\rm R})$ are the lattice vectors of the 'left' and 'right' domains. The construction is illustrated by a representative example of a GB structure shown in figure 5.8(a). The repeat vector $\boldsymbol{d}$ in figure 5.8(a) is defined by the (5, 3) and (7, 0) matching vectors in the left and right domains, respectively; the length of both vectors is $d = |n\boldsymbol{a}_1 + m\boldsymbol{a}_2| = a\sqrt{n^2 + nm + m^2}$ ($a = 0.246$ nm is the length of unit vectors $\boldsymbol{a}_1$ and $\boldsymbol{a}_2$ of the graphene lattice). Another outstanding characteristic of polycrystalline graphene is the rotation of the hexagonal BZ on each side of the GBs, defined by the crystallographic orientations, $\theta_{\rm L}$ and $\theta_{\rm R}$, in the left and right domains (with respect to the normal of the boundary line, see figure 5.8(b)) [12]. Therefore, the $\boldsymbol{K}_+$ (or $\boldsymbol{K}_-$) valley is no longer located at the same position in the momentum space on both sides of the GBs. Such a feature renders the polycrystalline graphene with tilt GBs to be an ideal candidate in valleytronics.

In 2016, Nguyen *et al* [52] put forward a practical scheme to manipulate highly valley-polarized currents, and explore the optical-like behaviours of charge carriers in polycrystalline graphene with tilt GBs under a strain. The studied polycrystalline configurations are shown in figures 5.9(a) and (b), in which a uniaxial strain with magnitude e_0 and direction α is applied globally. Here, we ignore Poisson's ratio. According to the discussions in section 3.1 (equations (3.15) and (3.16)), when strain is applied, Dirac cones of graphene are displaced and no longer located at the original $\boldsymbol{K}_+$ and $\boldsymbol{K}_-$ points [53]. Further, because of the different crystallographic orientations in the left and right domains, for a definite $\boldsymbol{K}_+$ valley, it will move in different directions on both sides of the GB, resulting in a separation of Dirac cones in momentum space, labelled as $\boldsymbol{K}_+^{\rm L}$ and $\boldsymbol{K}_+^{\rm R}$ in figure 5.9(c). The misalignment of Dirac cones can result in a finite transport gap, which is dependent on strain (e_0 and α), lattice symmetry [54], and the incident energy. This is inevitable because at a lower energy ($\mathcal{E} < \mathcal{E}_1$), the two Dirac valleys ($\boldsymbol{K}_+^{\rm L}$ and $\boldsymbol{K}_+^{\rm R}$) are separated in the

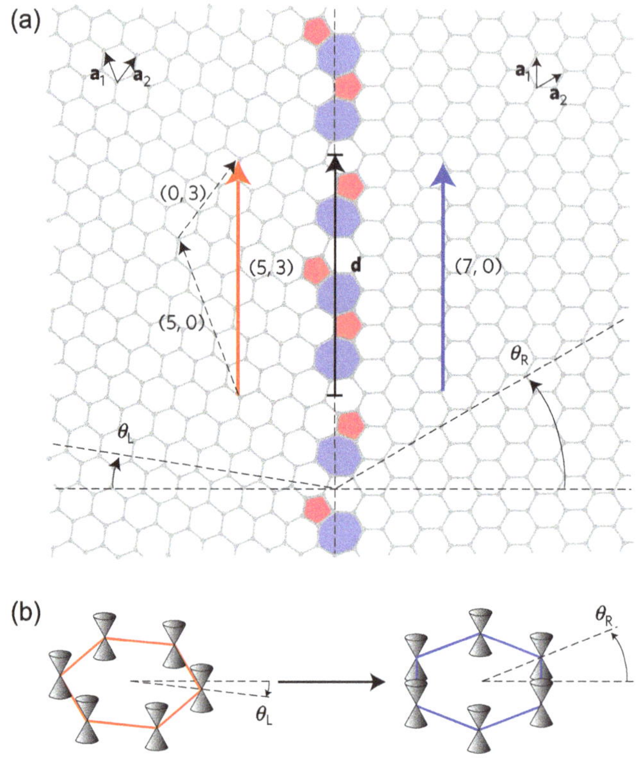

Figure 5.8. Structure of GBs in graphene. (a) An example of a tilt GB in graphene separating two crystalline domains rotated by $\theta_M = \theta_L + \theta_R = 8.2° + 30.0° = 38.2°$ with respect to each other [47]. The repeat vector $\boldsymbol{d}$ of the GB structure is defined by the matching vectors (5, 3) and (7, 0) in the left and right domains, respectively. A possible atomic structure of the interface region involves three elementary dislocation dipoles (pentagon–heptagon pairs) per repeat cell. (b) Effective rotation of the hexagonal BZ of graphene experienced by charge carriers crossing the (5, 3)|(7, 0) GB. Reproduced from [12], with permission from Springer Nature.

momentum space along the k_y direction and the two corresponding Fermi circles do not intersect, thus leading to the perfect reflection at arbitrary incident angle.

The transmission probability of the valley state across the GBs can be obtained according to the conservation of energy and momentum k_y. Suppose a de Broglie wave of electron with velocity $\boldsymbol{v}_L = v_L(\cos\phi, \sin\phi)$ and momentum $\boldsymbol{k}_L = \boldsymbol{K}_+^L + \boldsymbol{q}_L$ approaches the GB from the left domain, it will be partly reflected to the state with $\boldsymbol{v}_L^r = v_L^r(-\cos\phi, \sin\phi)$ and partly transmitted to the state with $\boldsymbol{v}_R = v_R(\cos\theta_+, \sin\theta_+)$ and $\boldsymbol{k}_R = \boldsymbol{K}_+^R + \boldsymbol{q}_R$ in the right domain, as shown in figure 5.9(c). Here, ϕ and θ_κ are the incident and outgoing angle for electrons in $\boldsymbol{K}_\kappa$ valley, respectively. Also, $\boldsymbol{q}_L = q_L(\cos\phi, \sin\phi)$, $\boldsymbol{q}_R = q_R(\cos\theta_+, \sin\theta_+)$. According to the conservation of k_y, one can obtain the following equation,

$$q_L \sin\phi - q_R \sin\theta_+ = K_{+,y}^R - K_{+,y}^L, \tag{5.17}$$

where $K_{\kappa,y}^{\Xi}$ denotes the y-component of $\boldsymbol{K}_\kappa^{\Xi}$. For the $\boldsymbol{K}_-$ valley, we have a similar relation but $K_{-,y}^R - K_{-,y}^L = -(K_{+,y}^R - K_{+,y}^L)$. Under small strain, the difference of

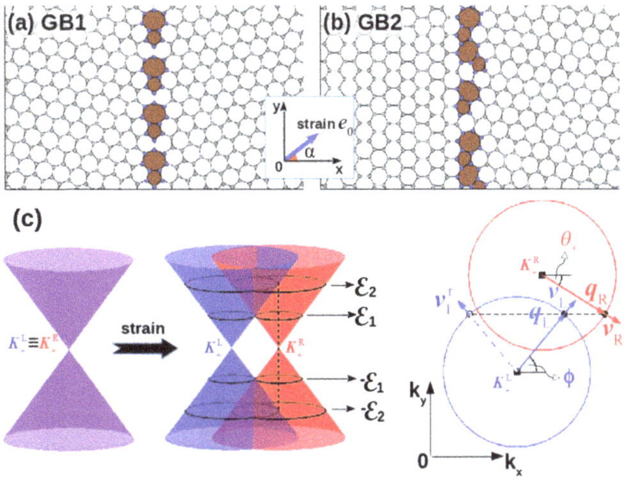

Figure 5.9. Structure of two types of GBs (a) (2, 1)|(1, 2) and (b) (0, 7)|(3, 5), termed as GB1 and GB2, respectively. A uniaxial strain with magnitude e_0 and direction α is applied. (c) Diagrams illustrating the strain effects on the band structure of graphene domains (left panel) and the momentum conservation rule (right panel). $\boldsymbol{K}_+^{\mathrm{L}}$ and $\boldsymbol{K}_+^{\mathrm{R}}$ indicate the new $\boldsymbol{K}_+$ valley result from the strain effect on the left and right side of the GB, respectively. ϕ and θ_+ are the incidence angle from the left side of GB and the outgoing angle in the right side of the GB, respectively. $\mathcal{E}_1$ is a critical energy at which the two Fermi circles from $\boldsymbol{K}_+^{\mathrm{L}}$ and $\boldsymbol{K}_+^{\mathrm{R}}$ valleys contact. Reprinted figures with permission from [52]. Copyright (2016) by the American Physical Society.

energy dispersion around the Dirac cones between the two graphene domains is negligible, leading to $|\boldsymbol{q}_{\mathrm{L}}| = |\boldsymbol{q}_{\mathrm{R}}| = |\boldsymbol{q}|$. Therefore, the relationship between ϕ and θ_κ can be rewritten as

$$\sin\phi - \sin\theta_\kappa = \kappa\beta(q). \tag{5.18}$$

where $\beta(q) = (K_{+,\,y}^{\mathrm{R}} - K_{+,\,y}^{\mathrm{L}})/q$. Because the group velocity and momentum are parallel (antiparallel) for electrons (holes), the above equation is also valid for holes when the sign of the right-hand side is changed. In addition, under small strain, the energy dispersion around Dirac points of graphene is still linear and can be described by equation (1.22) [55]. Hence, equation (5.18) can apply to both electrons and holes by using $\beta(q) = \hbar v_{\mathrm{F}}(K_{+,\,y}^{\mathrm{R}} - K_{+,\,y}^{\mathrm{L}})/\mathcal{E}$. Equation (5.18) indicates that the transmission in the two valleys can be modulated differently by strain, because the sign on the right side of the equation is opposite for the two valleys because of κ. Simultaneously, equation (5.18) also reveals a new refraction rule different from that of ordinary graphene, $\sin\theta_\kappa = \pm(q_{\mathrm{L}}/q_{\mathrm{R}})\sin\phi$ [56, 57]. In the following, we will expatiate the VP and optical-like behaviours of charge carriers in polycrystalline graphene according to equation (5.18).

5.2.1 The transmission characteristics and VP

In figure 5.10, the transmission probability $T_\pm$ in the two valleys and VP are plotted as a function of incident angle ϕ for different applied strains, where figures 5.10(a), (c) are for GB (2, 1)|(1, 2) and figures 5.10(b), (d) are for GB (0, 7)|(3, 5). It is

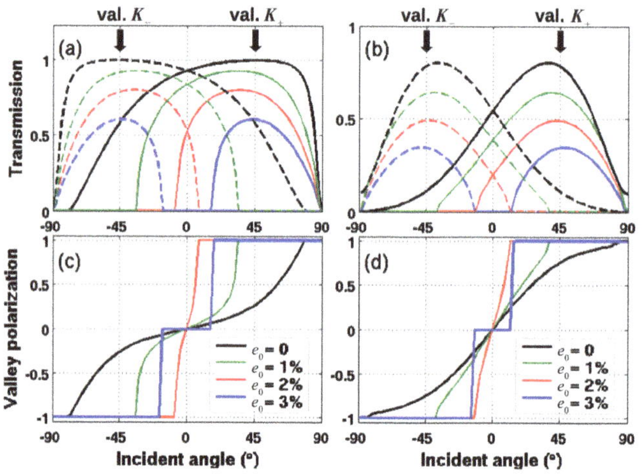

Figure 5.10. (a), (b) Transmission probability in the two valleys and (c), (d) VP as a function of incident angle for energy $\mathcal{E} = 0.3$ eV and various strains. (a), (c) and (b), (d) are for (2, 1)|(1, 2) (strain direction $\alpha = 45°$) and (0, 7)|(3, 5) (for $\alpha = 22°$), respectively. Reprinted figures with permission from [52]. Copyright (2016) by the American Physical Society.

shown that a similar transmission phenomenon as that in the line defect of graphene [32] can occur in GBs, where $T_\pm$ are angle dependent and the high VP P only exists at large angles without strain ($e_0 = 0$). From this perspective, the VP in the line defect of graphene should originate from the localized states at the line defect, the lattice structure of the two domains on both sides of the GB are not exactly symmetrical with respect to the GB while exhibiting the similar VP as the line defect. When strain is applied, the transmission gaps of $T_\pm$ ($T_\pm = 0$) will occur in different ranges of ϕ and hence P can be strongly enhanced. The transmission gap originates from the movement of the Dirac valleys. When strain strengthens, a perfect $P = \pm 1$ can be achieved in a wider range of ϕ in both systems.

The transmission properties can be explained according to equation (5.18), which is explored to distinguish different transport regimes as illustrated in figure 5.9(c). The k_y conservation implies that the transmission is nonzero only if the solution of equation (5.18) exists. In other words, for a specific ϕ, the absolute value of θ_κ should satisfy $|\theta_\kappa| \leqslant 90°$, which is dependent on the shift of the Dirac valleys. In this case, the angle ϕ has to satisfy

$$\max\left[\kappa\beta(q) - 1, -1\right] \leqslant \sin\phi \leqslant \min\left[\kappa\beta(q) + 1, 1\right]. \tag{5.19}$$

We can discuss the above equation in three regimes.

1. In the low-energy regime: $|\mathcal{E}| < \mathcal{E}_1$, i.e., $|\beta(q)| > 2$ with $\mathcal{E}_1 = \hbar v_F |K^R_{+,y} - K^L_{+,y}|/2$, no angle ϕ satisfies equation (5.19) and thus the system is totally reflective, corresponding to a transport gap.
2. In the second regime: $\mathcal{E}_1 \leqslant \mathcal{E} < \mathcal{E}_2$, i.e., $1 < |\beta(q)| \leqslant 2$ with $\mathcal{E}_2 = \hbar v_F |K^R_{+,y} - K^L_{+,y}|$, the transmission is found to be nonzero in finite ranges of ϕ, which satisfies the relation $\beta(q) - 1 \leqslant \sin\phi \leqslant 1$ and

$-1 \leqslant \sin\phi \leqslant 1 - \beta(q)$ for valleys $\boldsymbol{K}_+$ and $\boldsymbol{K}_-$, respectively. These two ranges are fully separated, e.g., when $\phi > 0$, T_+ adopts a finite magnitude and $T_- = 0$. The situation is exactly opposite when $\phi < 0$. In such a regime, the transmission is allowed for only one valley, leading to perfect VP $P = \pm 1$ as observed for $e_0 = 3\%$ in figure 5.10.

3. The third regime corresponds to $\mathcal{E} \geqslant \mathcal{E}_2$ with $\beta(q) \leqslant 1$. In this regime, the ranges of ϕ allowing for finite $T_\pm$ are larger than those observed in the second regime and could overlap. The perfect VP with $P = \pm 1$ can only occur outside the overlapped range of ϕ whereas P has a finite value inside the overlapped range, see $e_0 = 1\%$ and 2% in figure 5.10.

5.2.2 Optical-like behaviours

Equation (5.18) implies that electrons transmitted across the system have optical-like behaviours with new refraction rules. The relationship between ϕ and θ_κ extracted from the data in figure 5.10 is displayed in figures 5.11(a) and (b). Note that when $e_0 = 0$ and/or in systems containing domains of the same orientation (such as the line defect [32]), $\beta(q) = 0$ and $\phi = \theta_\kappa$, representing the linear transmission of electrons and the negative refraction does not exist. The refraction index can be easily modulated by strain and/or by varying carrier energy, i.e., $\sin\theta_\kappa = \sin\phi - \kappa\beta(q)$ from equation (5.18). For small strains ($e_0 = 1\%$ or 2% with $|\mathcal{E}| \geqslant \mathcal{E}_2$), the refraction indexes can be positive or negative. At a large strain

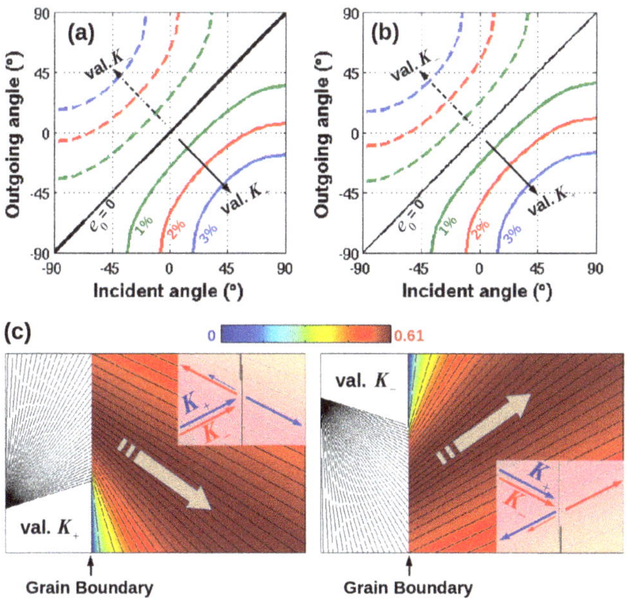

Figure 5.11. Strain-induced modulation of refraction index in the (a) GB1 and (b) GB2 systems extracted from the data in figure 5.10. (c) Diagrams illustrating the properties of electron beams with negative refraction when $e_0 = 3\%$ in (a) and colour maps showing the transmission function for different outgoing angles. Reprinted figures with permission from [52]. Copyright (2016) by the American Physical Society.

($e_0 = 3\%$ with $\mathcal{E}_1 \leqslant |\mathcal{E}| < \mathcal{E}_2$), the refraction index is always negative in the full ϕ and θ_κ ranges with finite transmission, in other words, the sign of ϕ and θ_κ are opposite at arbitrary angle.

In figure 5.11(c), it can be found that the maximum transmission probabilities T_κ are distributed around a specific angle satisfying $\theta_\kappa \cong -\phi = -\phi_\mathrm{h}$. Here, $\phi_\mathrm{h} = \kappa \arcsin[\beta(q)/2]$ can be tuned by changing carrier energy and/or strain. This property can be an important ingredient for controlling directional currents and highly focused beams. In graphene with p–n junctions, the high transmission occurs only for normal incidence ($\phi = 0$) while the transmission is low for large angles, especially when the transition length between highly doped regions is large [57]. This essentially limits the performance of corresponding electronic–optics devices, especially at high temperature [58, 59]. Moreover, when injected into the graphene–GB system, carriers in one (the other) valley are totally (partly) reflected [60], implying that both transmitted and reflected beams are highly valley polarized. In p–n junctions, these two beams can also be achieved but without VP [57, 61–63]. In a real experiment, these directional currents can be measured using multiple directional leads [49, 64] as shown in figure 3.8.

5.3 Line defects of MoS_2

The polycrystalline monolayer TMDCs are a hot research topic in electronics, and the most studied configurations are the line defects that have been observed experimentally [2, 66, 67]. For MoS_2, the line defects [10, 68] include the single vacancy line of sulphurs (SVL) [69], and the mirror-twin boundaries [69–72] with inversion domain boundaries (equivalent to 60° GBs), such as the 4|4P structure (IDB1) consisting of fourfold rings sharing a point at the chalcogen site and the 4|4E structure (IDB2) consisting of fourfold rings which share an edge [73], as shown in figure 5.12. The line defects of MoS_2 exhibit many unusual electronic transport properties, which is different from that of graphene and silicene due to the peculiar spin–valley coupling effect (figures 5.13(a) and (b)).

5.3.1 The transmission characteristics

Before further introducing the electronic transport properties across the line defects of MoS_2, it is necessary to discuss the lattice structure of the line defects. The line defects exhibit well-ordered periodic structures and the lattice structures on both

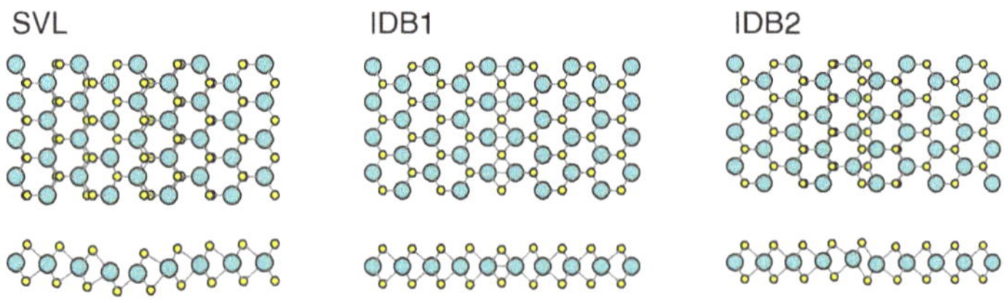

Figure 5.12. Top and side views of the three different kinds of extended line defects in monolayer MoS_2: SVL (left), IDB1 (centre) and IDB2 (right). Reprinted figures with permission from [65]. Copyright (2016) by the American Physical Society.

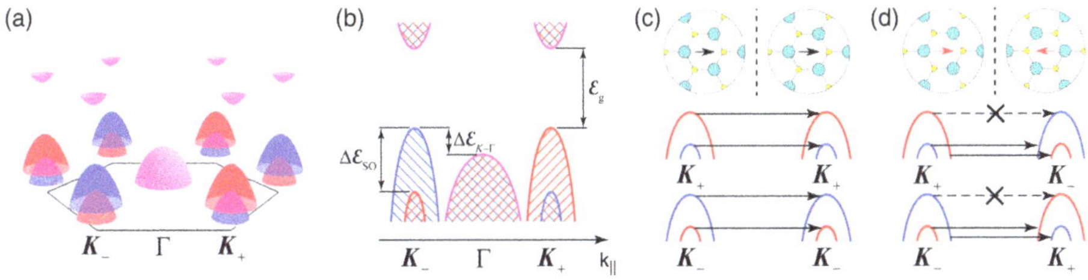

Figure 5.13. (a) The energy band structure of monolayer TMDC materials where the spin is split in the valence band at the points $\boldsymbol{K}_{\pm}$ of the BZ, while the spin is nearly degenerate at Γ and conduction bands at points $\boldsymbol{K}_{\pm}$. The red and blue bands correspond to the spin-up and spin-down bands, respectively. (b) The energy band structure of monolayer TMDC projected onto the direction of momentum parallel to the defect $k_{\|}$. The spin splitting at the $\boldsymbol{K}_{\pm}$ point $\Delta\mathcal{E}_{SO} = 0.15$ eV, the energy difference between the maximum of the valence band and the Γ point $\Delta\mathcal{E}_{K-\Gamma} = 0.055$ eV, and the bulk bandgap $\mathcal{E}_g = 1.61$ eV. The valley transmission processes for the hole charge carriers in the $\boldsymbol{K}_{\pm}$ valley across the line defects of TMDC with (c) preserved and (d) inversed lattice orientation, respectively. The insets show the crystalline lattice orientation in the two domains separated by the line defect (dashed line). Reprinted figures with permission from [65]. Copyright (2016) by the American Physical Society.

sides of the line defect are symmetrical or antisymmetrical about the line defect [6, 74]. In figure 5.8, if the lattice orientations on both sides of the line defect are identical, $\boldsymbol{a}_i^{L} = \boldsymbol{a}_i^{R}$, the pair integers (n_{Ξ}, m_{Ξ}) satisfy

$$(n_L - m_L) \bmod 3 = (n_R - m_R) \bmod 3 \neq 0, \tag{5.20}$$

for instance, $(n_L, m_L) = (n_R, m_R) = (1, 0)$. In this situation, the valley indices are conserved on both sides of the line defect and the transmission is allowed for hole charge carriers when the spin is additionally conserved, see figure 5.13(c). However, for the inversion domain boundaries, IDB1 and IDB2, as illustrated in the top panel of figure 5.13(d), the lattice vectors on both sides of the line defect are opposite, $\boldsymbol{a}_i^{L} = -\boldsymbol{a}_i^{R}$, and the pair integers satisfy the following relation,

$$0 \neq (n_L - m_L) \bmod 3 \neq (n_R - m_R) \bmod 3 \neq 0. \tag{5.21}$$

For example, $(n_L, m_L) = (1, 0)$ and $(n_R, m_R) = (-1, 0)$. In this situation, the valley indices are swapped on both sides of the line defect, similar to the situation that we discussed in section 5.1.2. This means that the wavefunction of a given valley state is the Hermitian conjugate on the other side of the line defect rather than identical. We set the reference energy, i.e., $\mathcal{E} = 0$, to the VBM in the energy band structure of monolayer TMDC (figure 5.13(a) and (b)). Therefore, considering the spin and momentum conservation, transmission across the line defect is totally blocked for hole charge carriers at low concentrations $(0 > \mathcal{E} > -\Delta\mathcal{E}_{K-\Gamma})$, because in such an energy range, the electronic states from $\boldsymbol{K}_{+}$ and $\boldsymbol{K}_{-}$ valleys are occupied by opposite-spin carriers, see figure 5.13(d). However, the transmission becomes possible when the opposite-spin branches are populated at larger hole charge-carrier concentrations $(\mathcal{E} < -\Delta\mathcal{E}_{K-\Gamma})$. Besides, the spin-degenerate valley at the Γ point will also contribute to the transmission across the line defects when $\mathcal{E} < -\Delta\mathcal{E}_{K-\Gamma}$.

The charge-carrier transmissions $T(k_{\|}, \mathcal{E})$ across the three different kinds of line defects of MoS_2 (figure 5.12) are plotted as a function of momentum $k_{\|}$ and energy $\mathcal{E}$

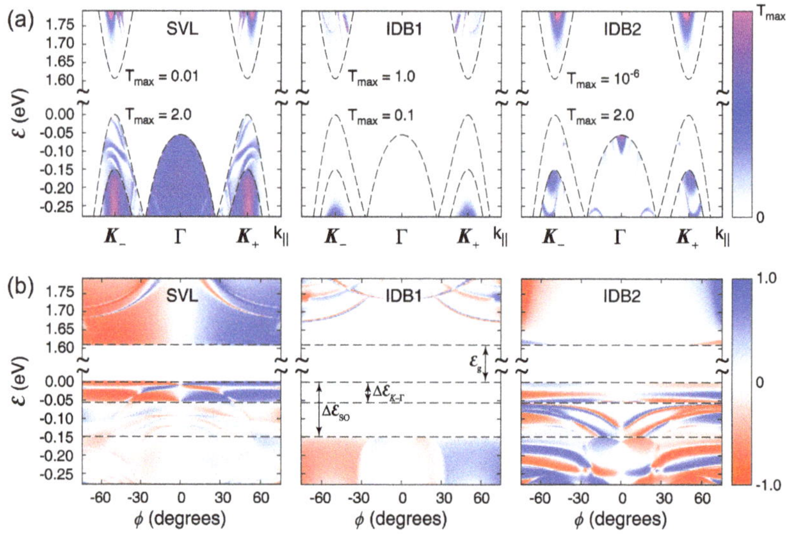

Figure 5.14. (a) Charge-carrier transmissions $T(k_{\|}, \mathcal{E})$ across the three investigated line defects in monolayer MoS_2 as a function of momentum $k_{\|}$ and energy $\mathcal{E}$ relative to the VBM. The dashed lines denote the contours of the bulk bands projected onto the defect direction. The transmissions of electron and hole charge carriers are shown on different scales defined by the upper limit $T_{\max}$. (b) Spin polarization $P_s(\mathcal{E}, \phi)$ as a function of energy $\mathcal{E}$ and incidence angle ϕ calculated for the three studied line defects in monolayer MoS_2. Dashed lines indicate the positions of the band edges. Reprinted figures with permission from [65]. Copyright (2016) by the American Physical Society.

relative to the VBM in figure 5.14(a). The predicted suppression of transmission across the inversion domain boundaries (IDB1 and IDB2) can be clearly seen for the low-energy hole charge carriers ($0 > \mathcal{E} > -\Delta\mathcal{E}_{K-\Gamma}$). In the case of the IDB1 defect, the transmission is strictly zero for $0 > \mathcal{E} > -\Delta\mathcal{E}_{K-\Gamma}$, while for the IDB2 structure a residual transmission not exceeding 10^{-3} was found within the predicted transport gap. The situation in IDB2 structure is enabled by the spin-flip process due to out-of-plane bending at the defect line (right panel of figure 5.12). On the other hand, the maximum transmission of SVL could reach to 2 in such an energy interval due to the conservation of the valley indices on both sides of the line defect. Besides, $T(k_{\|}, \mathcal{E})$ shows strong variations in both $k_{\|}$ and $\mathcal{E}$ with line-like suppression, which is a typical feature of resonant backscattering caused by the localized states around the defects [7, 74]. A similar phenomenon was also reported in graphene with line defects [45].

Another pronounced transmission feature is the electron–hole asymmetry, which exists in all three considered line defects, see figure 5.14(a). For the hole charge carrier, the maximum transmission across the SVL and IDB2 line defects tend to approach 2, the largest number of transmission channels within the investigated energy range. However, the maximum transmission of electron charge carriers across both defects is merely 0.01 for SVL and approaches zero for IDB2. The opposite situation occurs for the IDB1 line defect where the maximum transmission of low-energy electron (hole) charge carriers is 1.0 (0.01). This behaviour can be explained with the electrostatic potential bending at the defect, where the

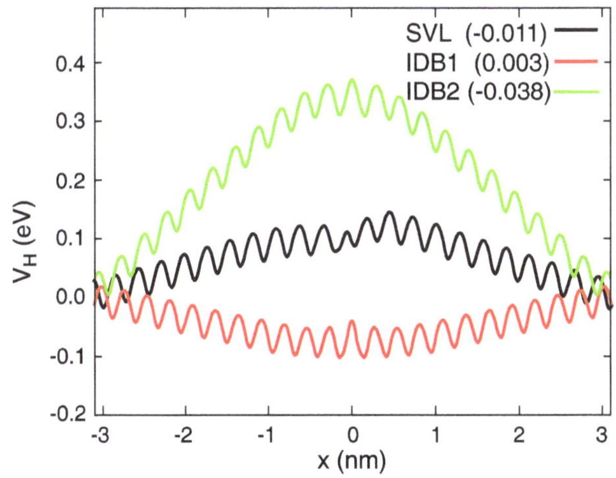

Figure 5.15. Self-consistent Hartree potential V_H averaged over the planes perpendicular to the transport direction for the three line defect models, where the zero potential is set at the bulk monolayer MoS_2 contacts and the line defect is at $x = 0$. Reprinted figures with permission from [65]. Copyright (2016) by the American Physical Society.

electrostatic potential can be described with the self-consistent Hartree potential V_H within the scattering region calculated from the first-principles calculations, as shown in figure 5.15. The zero potential is assumed at the bulk contacts of MoS_2 (far away from the line defect) and the position of the line defect is $x = 0$. The electrostatic potential originates from the charge carriers localized at the line defect. Obviously, the electron transmission across the line defect becomes difficult in the presence of a positive electrostatic potential. For both the SVL and IDB2 line defects, it is found that the positive electrostatic potential gradually strengthens on approaching the line defect ($x = 0$) and it is stronger in the IDB2 than that in the SVL. Therefore, it is not surprising that the electron charge-carrier transmission is considerably small in both IDB2 and SVL line defects while it is less than 10^{-6} for IDB2. On the other hand, the positive electrostatic potential has no influence on the hole charge carriers. The opposite behaviour of charge carriers crossing the IDB1 defect, which was also predicted for the same defect structure in $MoSe_2$ [75], is related to the negative electrostatic potential in the IDB1 line defect.

5.3.2 The VP properties

For the line defect, the transmission of charge carriers are dependent of the valley index κ and incidence angle $\phi \in (-\pi/2, \pi/2)$ [32]. The valley filtering is expected for any periodic line defect which leaves the two valleys separated in $k_\parallel$ space since the following property holds for valley-resolved transmissions for electron with energy $\mathcal{E}$,

$$T_\kappa(\phi) = T_\kappa(k_\parallel) = T_{-\kappa}(-k_\parallel) = T_{-\kappa}(-\phi) \neq T_{-\kappa}(\phi), \tag{5.22}$$

for the incidence angle ϕ bijectively related to momentum $k_{\|}$ in the 1D BZ of the line defect. The corresponding VP of the transmitted charge can be written as

$$P(\mathcal{E}, \phi) = \frac{T_+(\mathcal{E}, \phi) - T_-(\mathcal{E}, \phi)}{T_+(\mathcal{E}, \phi) + T_-(\mathcal{E}, \phi)}. \tag{5.23}$$

At low concentrations, the VP of hole charge carriers P is equivalent to their spin polarization P_s due to the intrinsic spin–valley coupling in monolayer TMDCs [76]. In figure 5.14(b), the spin polarization $P_s(\mathcal{E}, \phi)$ of transmitted charge carriers as a function of the energy $\mathcal{E}$ and incidence angle ϕ are plotted for the three different kinds of line defects of MoS_2. For the SVL line defect, the spin polarization P_s is high in this energy range $(0 > \mathcal{E} > -\Delta\mathcal{E}_{K-\Gamma})$. For instance, at $\mathcal{E} = -10$ meV, spin polarization achieves $P_s = \pm 0.997$ at incidence angles $\phi = \pm 30°$. In fact, the spin polarization is also equivalent to the VP because the hole charge carriers belong only to fully spin polarized valleys $\boldsymbol{K}_\pm$. Importantly, at these charge-carrier energies the transmissions are also high, of the order of 1. At $\mathcal{E} < -\Delta\mathcal{E}_{K-\Gamma}$ the spin polarization dramatically reduces due to a large contribution to conductance of charge carriers in the spin-degenerate Γ valley, which is also characterized by large transmissions. Large values of P_s are also found for the electron charge carriers, although spin-orbit effects in the conduction band are generally weaker and have a complex character. Transmissions are found to be about two orders of magnitude lower.

For the IDB1 line defect, the spin polarization P_s is low across a relevant range of values of $\mathcal{E}$ and ϕ, which is about zero as $0 > \mathcal{E} > -\Delta\mathcal{E}_{SO}$ for arbitrary ϕ. The dominant contribution to the transmission of holes comes from conductance channels involving $\boldsymbol{K}_+$ and $\boldsymbol{K}_-$ valleys (centre panel of figure 5.12). When $\mathcal{E} < -\Delta\mathcal{E}_{SO}$, the spin polarization P_s is reduced rapidly within $\phi = \pm 30°$ because only one of the two indicated channels is realized in this range of incidence angle values [77]. The spin polarization $P_s(\mathcal{E}, \phi)$ of holes transmitted across a IDB2 line defect shows large variations with respect to energy $\mathcal{E}$ and angle ϕ, even in the energy range that corresponds to residual transmission enabled by the spin-flip process. The sign of P_s at a constant incidence angle ϕ changes at $\mathcal{E} \approx -20$ meV. For $\mathcal{E} < -\Delta\mathcal{E}_{K-\Gamma}$ multiple conductance channels compete, resulting in an irregular behaviour of spin polarization.

In the bulk gap of the IDB1 line defect, there exist two spin-degenerate mirror-twin GB (MTGB) bands whose electric states are mostly localized around the line defect [78]. In contrast with the pristine MoS_2, the spin-degenerate bands come from the invariance of IS in the presence of the MTGB and TRS. The degenerate spin states enable the possibility of intervalley scattering. For instance, when the short-range disorders (sulphur vacancies) or the Anderson disorder exist in the region of the MTGB, the transmission enhances prominently at the top of the valence band due to intervalley scattering which occurs at the GB when electrons pass from one grain to the other.

Up to now, we have reviewed the valleytronic studies, which treated the two inequivalent valleys independently, i.e., only intravalley processes are considered. Recently, An *et al* [79] discovered that, at a superlattice barrier in a monolayer graphene, intervalley scattering can selectively block Klein tunnelling in a chosen valley, while retaining perfect transmission in the other. The key to such valley selectivity is the staggered pseudospin gaps created by intervalley backscattering in the superlattice barrier.

References

[1] Biró L P and Lambin P 2013 Grain boundaries in graphene grown by chemical vapor deposition *New J. Phys.* **15** 035024

[2] Yazyev O V and Chen Y P 2014 Polycrystalline graphene and other two-dimensional materials *Nat. Nanotechnol.* **9** 755–67

[3] Cummings A W, Duong D L, Nguyen V L, Tuan D V, Kotakoski J, Barrios Vargas J E, Lee Y H and Roche S 2014 Charge transport in polycrystalline graphene: challenges and opportunities *Adv. Mater.* **26** 5079–94

[4] Huang P Y *et al* 2011 Grains and grain boundaries in single-layer graphene atomic patchwork quilts *Nature* **469** 389–92

[5] Kim K, Lee Z, Regan W, Kisielowski C, Crommie M F and Zettl A 2011 Grain boundary mapping in polycrystalline graphene *ACS Nano* **5** 2142–6

[6] van der Zande A M, Huang P Y, Chenet D A, Berkelbach T C, You Y, Lee G-H, Heinz T F, Reichman D R, Muller D A and Hone J C 2013 Grains and grain boundaries in highly crystalline monolayer molybdenum disulphide *Nat. Mater.* **12** 554–61

[7] Zou X, Liu Y and Yakobson B I 2013 Predicting dislocations and grain boundaries in two-dimensional metal-disulfides from the first principles *Nano Lett.* **13** 253–8

[8] Zhang J *et al* 2014 Scalable growth of high-quality polycrystalline MoS_2 monolayers on SiO_2 with tunable grain sizes *ACS Nano* **8** 6024–30

[9] Wu J, Cao P, Zhang Z, Ning F, Zheng S-S, He J and Zhang Z 2018 Grain-size-controlled mechanical properties of polycrystalline monolayer MoS_2 *Nano Lett.* **18** 1543–52

[10] Najmaei S, Liu Z, Zhou W, Zou X, Shi G, Lei S, Yakobson B I, Idrobo J-C, Ajayan P M and Lou J 2013 Vapour phase growth and grain boundary structure of molybdenum disulphide atomic layers *Nat. Mater.* **12** 754–9

[11] Ma Y, Kolekar S, Coy Diaz H, Aprojanz J, Miccoli I, Tegenkamp C and Batzill M 2017 Metallic twin grain boundaries embedded in $MoSe_2$ monolayers grown by molecular beam epitaxy *ACS Nano* **11** 5130–9

[12] Yazyev O V and Louie S G 2010 Electronic transport in polycrystalline graphene *Nat. Mater.* **9** 806–9

[13] Lin X and Ni J 2011 Half-metallicity in graphene nanoribbons with topological line defects *Phys. Rev.* B **84** 075461

[14] Dai Q Q, Zhu Y F and Jiang Q 2014 Electronic and magnetic properties of armchair graphene nanoribbons with 558 grain boundary *Phys. Chem. Chem. Phys.* **16** 10607–13

[15] Guerra T, Azevedo S and Machado M 2016 Defective graphene and nanoribbons: electronic, magnetic and structural properties *Eur. Phys. J.* B **89** 58

[16] Ayuela A, Jaskólski W, Santos H and Chico L 2014 Electronic properties of graphene grain boundaries *New J. Phys.* **16** 083018

[17] Gao N, Guo Y, Zhou S, Bai Y and Zhao J 2017 Structures and magnetic properties of MoS_2 grain boundaries with antisite defects *J. Phys. Chem.* C **121** 12261–9

[18] Alexandre S S and Nunes R W 2017 Magnetic states of linear defects in graphene monolayers: effects of strain and interaction *Phys. Rev.* B **96** 075445

[19] Luo M, Li B-L and Li D 2019 Effects of divacancy and extended line defects on the thermal transport properties of graphene nanoribbons *Nanomaterials* **9** 1609
[20] Diery W A, Moujaes E A and Nunes R W 2018 Nature of localized phonon modes of tilt grain boundaries in graphene *Carbon* **140** 250–8
[21] Xie Z-X, Zhang Y, Zhang L-F and Fan D-Y 2017 Effect of topological line defects on electron-derived thermal transport in zigzag graphene nanoribbons *Carbon* **113** 292–8
[22] Zhu Z, Yang X, Huang M, He Q, Yang G and Wang Z 2016 Mechanisms governing phonon scattering by topological defects in graphene nanoribbons *Nanotechnology* **27** 055401
[23] Lin C, Chen X and Zou X 2019 Phonon–grain-boundary-interaction-mediated thermal transport in two-dimensional polycrystalline MoS_2 *ACS Appl. Mater. Interfaces* **11** 25547–55
[24] Li X, Zou D, Cui B, Li Y, Wang M, Li D and Liu D 2018 Tuning spin-filtering, rectifying, and negative differential resistance by hydrogenation on topological edge defects of zigzag silicene nanoribbons *Phys. Lett.* A **382** 2475–83
[25] de Oliveira J B, de Oliveira I S S, Padilha J E and Miwa R H 2018 Tunable magnetism and spin-polarized electronic transport in graphene mediated by molecular functionalization of extended defects *Phys. Rev.* B **97** 045107
[26] Phillips M and Mele E J 2017 Charge and spin transport on graphene grain boundaries in a quantizing magnetic field *Phys. Rev.* B **96** 041403
[27] Cummings A W, Cresti A and Roche S 2014 Quantum Hall effect in polycrystalline graphene: the role of grain boundaries *Phys. Rev.* B **90** 161401
[28] Yao H-B, Lü X-L and Zheng Y-S 2013 Quantum Hall boundary state around the line defect in graphene *Phys. Rev.* B **88** 235419
[29] Ribeiro M, Power S R, Roche S, Hueso L E and Casanova F 2017 Scale-invariant large nonlocality in polycrystalline graphene *Nat. Commun.* **8** 2198
[30] Nguyen V H and Charlier J-C 2020 Aharonov–Bohm interferences in polycrystalline graphene *Nanoscale Adv.* **2** 256–63
[31] Wang S, Tan L Z, Wang W, Louie S G and Lin N 2014 Manipulation and characterization of aperiodical graphene structures created in a two-dimensional electron gas *Phys. Rev. Lett.* **113** 196803
[32] Gunlycke D and White C T 2011 Graphene valley filter using a line defect *Phys. Rev. Lett.* **106** 136806
[33] Tian H, Ren C and Wang S 2022 Valleytronics in two-dimensional materials with line defect *Nanotechnology* **33** 212001
[34] Lahiri J, Lin Y, Bozkurt P, Oleynik I I and Batzill M 2010 An extended defect in graphene as a metallic wire *Nat. Nanotechnol.* **5** 326–9
[35] Liu Y, Song J, Li Y, Liu Y and Sun Q-f 2013 Controllable valley polarization using graphene multiple topological line defects *Phys. Rev.* B **87** 195445
[36] Rodrigues J N B, Peres N M R and Lopes dos Santos J M B 2013 Scattering by linear defects in graphene: a tight-binding approach *J. Phys.: Condens. Matter* **25** 075303
[37] Ashcroft N W and Mermin M D 1976 *Solid State Physics* (Boston, MA: Cengage Learning)
[38] Wang S K and Wang J 2015 Valley precession in graphene superlattices *Phys. Rev.* B **92** 075419
[39] Zhou Y-C, Zhang H-L and Deng W-Q 2013 A 3N rule for the electronic properties of doped graphene *Nanotechnology* **24** 225705
[40] Wallace P R 1947 The band theory of graphite *Phys. Rev.* **71** 622–34
[41] Slonczewski J C and Weiss P R 1958 Band structure of graphite *Phys. Rev.* **109** 272–9

[42] Wang S, Ren C, Li Y, Tian H, Lu W and Sun M 2018 Spin and valley filter across line defect in silicene *Appl. Phys. Express* **11** 053004

[43] Jiang L, Lv X and Zheng Y 2011 Valley polarized electronic transport through a line defect in graphene: an analytical approach based on tight-binding model *Phys. Lett.* A **376** 136–41

[44] Jiang L, Zheng Y, Yi C, Li H and Lü T 2009 Analytical study of edge states in a semi-infinite graphene nanoribbon *Phys. Rev.* B **80** 155454

[45] Chen J-H, Autès G, Alem N, Gargiulo F, Gautam A, Linck M, Kisielowski C, Yazyev O V, Louie S G and Zettl A 2014 Controlled growth of a line defect in graphene and implications for gate-tunable valley filtering *Phys. Rev.* B **89** 121407(R)

[46] Lü X-L, Liu Z, Yao H-B, Jiang L-W, Gao W-Z and Zheng Y-S 2012 Valley polarized electronic transmission through a line defect superlattice of graphene *Phys. Rev.* B **86** 045410

[47] Yazyev O V and Louie S G 2010 Topological defects in graphene: dislocations and grain boundaries *Phys. Rev.* B **81** 195420

[48] Yu Q *et al* 2011 Control and characterization of individual grains and grain boundaries in graphene grown by chemical vapour deposition *Nat. Mater.* **10** 443–9

[49] Tsen A W, Brown L, Levendorf M P, Ghahari F, Huang P Y, Havener R W, Ruiz-Vargas C S, Muller D A, Kim P and Park J 2012 Tailoring electrical transport across grain boundaries in polycrystalline graphene *Science* **336** 1143–6

[50] Kochat V, Tiwary C S, Biswas T, Ramalingam G, Hsieh K, Chattopadhyay K, Raghavan S, Jain M and Ghosh A 2016 Magnitude and origin of electrical noise at individual grain boundaries in graphene *Nano Lett.* **16** 562–7

[51] Yao W, Wu B and Liu Y 2020 Growth and grain boundaries in 2D materials *ACS Nano* **14** 9320–46

[52] Nguyen V H, Dechamps S, Dollfus P and Charlier J-C 2016 Valley filtering and electronic optics using polycrystalline graphene *Phys. Rev. Lett.* **117** 247702

[53] Pereira V M, Castro Neto A H and Peres N M R 2009 Tight-binding approach to uniaxial strain in graphene *Phys. Rev.* B **80** 045401

[54] Hung Nguyen V, Hoang T X, Dollfus P and Charlier J-C 2016 Transport properties through graphene grain boundaries: strain effects versus lattice symmetry *Nanoscale* **8** 11658–73

[55] Castro Neto A H, Guinea F, Peres N M R, Novoselov K S and Geim A K 2009 The electronic properties of graphene *Rev. Mod. Phys.* **81** 109–62

[56] Cheianov V V, Fal'ko V and Altshuler B L 2007 The focusing of electron flow and a Veselago lens in graphene p-n junctions *Science* **315** 1252–5

[57] Allain P E and Fuchs J N 2011 Klein tunneling in graphene: optics with massless electrons *Eur. Phys. J.* B **83** 301

[58] Lee G-H, Park G-H and Lee H-J 2015 Observation of negative refraction of Dirac fermions in graphene *Nat. Phys.* **11** 925–9

[59] Chen S *et al* 2016 Electron optics with p–n junctions in ballistic graphene *Science* **353** 1522–5

[60] Tao W-W, Liu B, Dai Q and Wang S-K 2014 Simulation of electronic total-reflection effect in a graphene junction *Commun. Theor. Phys.* **61** 391–6

[61] Rickhaus P, Maurand R, Liu M-H, Weiss M, Richter K and Schönenberger C 2013 Ballistic interferences in suspended graphene *Nat. Commun.* **4** 2342

[62] Rickhaus P, Makk P, Liu M-H, Richter K and Schönenberger C 2015 Gate tuneable beamsplitter in ballistic graphene *Appl. Phys. Lett.* **107** 251901

[63] Rickhaus P, Liu M-H, Makk P, Maurand R, Hess S, Zihlmann S, Weiss M, Richter K and Schönenberger C 2015 Guiding of electrons in a few-mode ballistic graphene channel *Nano Lett.* **15** 5819–25
[64] Wang S, Tian H and Sun M 2023 Valley-polarized and enhanced transmission in graphene with a smooth strain profile *J. Phys.: Condens. Matter* **35** 304002
[65] Pulkin A and Yazyev O V 2016 Spin- and valley-polarized transport across line defects in monolayer MoS_2 *Phys. Rev.* B **93** 041419
[66] Bertoldo F *et al* 2021 Intrinsic defects in MoS_2 grown by pulsed laser deposition: from monolayers to bilayers *ACS Nano* **15** 2858–68
[67] Gali S M, Pershin A, Lherbier A, Charlier J-C and Beljonne D 2020 Electronic and transport properties in defective MoS_2: impact of sulfur vacancies *J. Phys. Chem.* C **124** 15076–84
[68] Zhou M, Wang W, Lu J and Ni Z 2021 How defects influence the photoluminescence of TMDCs *Nano Res.* **14** 29–39
[69] Wang S, Lee G-D, Lee S, Yoon E and Warner J H 2016 Detailed atomic reconstruction of extended line defects in monolayer MoS_2 *ACS Nano* **10** 5419–30
[70] Lin J, Pantelides S T and Zhou W 2015 Vacancy-induced formation and growth of inversion domains in transition-metal dichalcogenide monolayer *ACS Nano* **9** 5189–97
[71] Le D and Rahman T S 2013 Joined edges in MoS_2: metallic and half-metallic wires *J. Phys.: Condens. Matter* **25** 312201
[72] Gibertini M and Marzari N 2015 Emergence of one-dimensional wires of free carriers in transition-metal-dichalcogenide nanostructures *Nano Lett.* **15** 6229–38
[73] Komsa H-P and Krasheninnikov A V 2017 Engineering the electronic properties of two-dimensional transition metal dichalcogenides by introducing mirror twin boundaries *Adv. Electron. Mater.* **3** 1600468
[74] Zhou W, Zou X, Najmaei S, Liu Z, Shi Y, Kong J, Lou J, Ajayan P M, Yakobson B I and Idrobo J-C 2013 Intrinsic structural defects in monolayer molybdenum disulfide *Nano Lett.* **13** 2615–22
[75] Lehtinen O *et al* 2015 Atomic scale microstructure and properties of Se-deficient two-dimensional $MoSe_2$ *ACS Nano* **9** 3274–83
[76] Xiao D, Liu G-B, Feng W, Xu X and Yao W 2012 Coupled spin and valley physics in monolayers of MoS_2 and other group-VI dichalcogenides *Phys. Rev. Lett.* **108** 196802
[77] Habe T and Koshino M 2015 Spin-dependent refraction at the atomic step of transition-metal dichalcogenides *Phys. Rev.* B **91** 201407
[78] Park J, Xue K-H, Mouis M, Triozon F and Cresti A 2019 Electron transport properties of mirror twin grain boundaries in molybdenum disulfide: impact of disorder *Phys. Rev.* B **100** 235403
[79] An X-T and Yao W 2020 Valley-selective Klein tunneling through a superlattice barrier in graphene *Phys. Rev. Appl.* **14** 014039

Sake Wang and Hongyu Tian

Chapter 6

Valley current in multilayer and twisted systems

Similar to silicene and other buckled 2D materials, under the presence of an external perpendicular electric field, a bandgap can be induced in the AB-stacked bilayer graphene. According to the discussions in section 2.8, when there exists a domain wall, across which the Chern number or the effective mass changes sign, the gapless topological kink states appear. Two strategies for generating such topological kink states are mainly adopted and will be discussed in this chapter: voltage kink [1–3] or tilt boundaries [4–7]. We will also show that because periodic alternating regions of AA-, AB- and BA-stacking configurations exist in the moiré crystals, a network of topological kink states can exist at the boundary of the AB- and BA-stacking.

6.1 Effect of the perpendicular electric field on AB-stacked bilayer graphene

In bilayer graphene, the asymmetry between the layers can be generated by the perpendicular external electric field [8–12] or the substrate-induced staggered potential [13, 14], inducing a bandgap in the single-electron spectrum [9–11, 15–23]. The possibility of having two layers at different potentials introduces the concept of a biased bilayer. The ability to open a gap makes bilayer graphene interesting for technological applications.

6.1.1 Biased bilayer (bulk)

Taking into account the applied voltage difference [24–27], we can generalise equation (1.78) in the basis set {A1, B1, A2, B2},

doi:10.1088/978-0-7503-5562-9ch6

$$H_{\mathrm{B:AB}}(\boldsymbol{k},\ V) = \begin{pmatrix} \Delta_z(V)/2 & -\gamma_0 f(\boldsymbol{k}) & 0 & 0 \\ -\gamma_0 f(\boldsymbol{k})^* & \Delta_z(V)/2 & \gamma_1 & 0 \\ 0 & \gamma_1 & -\Delta_z(V)/2 & -\gamma_0 f(\boldsymbol{k}) \\ 0 & 0 & -\gamma_0 f(\boldsymbol{k})^* & -\Delta_z(V)/2 \end{pmatrix}, \tag{6.1}$$

with $\Delta_z(V) = eV$ as the voltage-induced on-site energy difference between the bottom and top layers. Following the treatment in section 1.7, we can obtain two pairs of the four eigenenergies of $H_{\mathrm{B:AB}}(\boldsymbol{k},\ V)$,

$$\begin{aligned} \mathcal{E}_{1,2}(\boldsymbol{k},\ V) &= \pm\sqrt{[\gamma_0|f(\boldsymbol{k})|]^2 + \frac{\gamma_1^2}{2} + \left(\frac{\Delta_z(V)}{2}\right)^2 - \sqrt{[\gamma_0|f(\boldsymbol{k})|]^2\left[\gamma_1^2 + \left(\frac{\Delta_z(V)}{2}\right)^2\right] + \frac{\gamma_1^4}{4}}}, \\ \mathcal{E}_{3,4}(\boldsymbol{k},\ V) &= \pm\sqrt{[\gamma_0|f(\boldsymbol{k})|]^2 + \frac{\gamma_1^2}{2} + \left(\frac{\Delta_z(V)}{2}\right)^2 + \sqrt{[\gamma_0|f(\boldsymbol{k})|]^2\left[\gamma_1^2 + \left(\frac{\Delta_z(V)}{2}\right)^2\right] + \frac{\gamma_1^4}{4}}}. \end{aligned} \tag{6.2}$$

The band structure is shown in figure 6.1. It is noticeable that bands 1 and 2, which touch at the K point in the absence of bias voltage (figure 1.16), shifted apart from the Fermi energy when the bias voltage is applied ($\Delta_z(V) \neq 0$). This means the original gapless semiconductor (Figure 1.16) has developed into an insulator (figure 6.1) whose bandgap is controlled by V. Moreover, close to zero energy, the bands 1 and 2 undergo significant deformation, and their CBM and VBM are no longer located at the K point [11, 23, 32, 33], as in the case of $\Delta_z(V) = 0$. ARPES studies of bilayer graphene films synthesized on SiC substrates confirmed such band structure [20].

To be more intuitive, it is also suggested to expand the energy spectra of 1 and 2. We derive[1]

$$\mathcal{E}_{1,2}(\boldsymbol{q},\ V) \approx \pm\sqrt{\left(\frac{\Delta_z(V)}{2}\right)^2 - \frac{(\hbar v_{\mathrm{F}}\Delta_z(V))^2}{\gamma_1^2}q^2 + \frac{(\hbar v_{\mathrm{F}})^4}{\gamma_1^2}\left[\left(\frac{\Delta_z(V)}{\gamma_1}\right)^2 + 1\right]^2 q^4}. \tag{6.3}$$

One can simply check that the expression under the root sign is always positive when $\Delta_z(V) \neq 0$. Further insight shows the bandgap is close to, but not directly at, the $\boldsymbol{K}_\pm$ point [11, 23, 32]. $\mathcal{E}_{1,2}(\boldsymbol{q},\ V)$ reduces to $\mathcal{E}_{1,2}(\boldsymbol{q})$ in equation (1.81) when $\Delta_z(V) = 0$.

For the effective Hamiltonian close to the Dirac point, we can use the generalized Hamiltonian (1.83), and we have

$$H_{\mathrm{B:AB}}(\boldsymbol{q},\ V) = \begin{pmatrix} \Delta_z(V)/2 & -\frac{(\hbar v_{\mathrm{F}})^2}{\gamma_1}(q_x + \mathrm{i}\kappa q_y)^2 \\ -\frac{(\hbar v_{\mathrm{F}})^2}{\gamma_1}(q_x - \mathrm{i}\kappa q_y)^2 & -\Delta_z(V)/2 \end{pmatrix}. \tag{6.4}$$

[1] Here, we use the Taylor expansion $\sqrt{1+x} \approx 1 + \frac{1}{2}x - \frac{1}{8}x^2$.

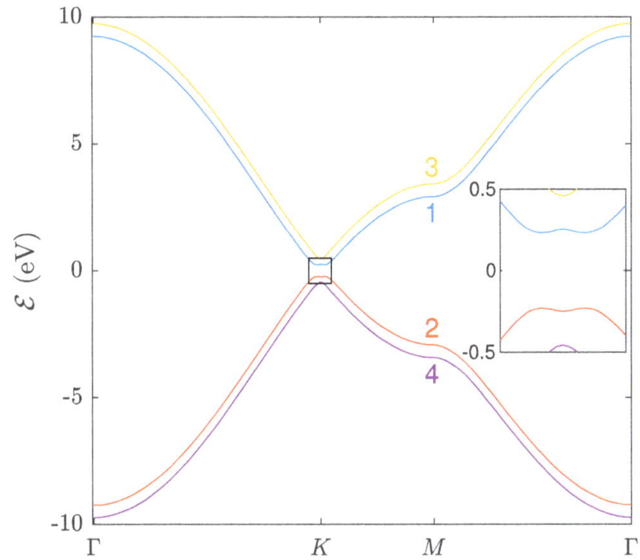

Figure 6.1. Band structure of biased AB bilayer graphene, along the $\boldsymbol{k}$-space trajectory $\Gamma \to K \to M \to \Gamma$. The indices of the bands are annotated by the number nearby. The inset zooms in the area bounded by the black box in the vicinity of the K valley. Plot was made using the parameters $\gamma_0 = 3.16$ eV, $\gamma_1 = 0.381$ eV [28, 29] and $\Delta_z(V) = 0.2\gamma_0$ [30, 31].

The dispersion relation for this Hamiltonian is

$$\mathcal{E}_{\pm}(\boldsymbol{q}, V) = \pm\sqrt{\left(\frac{\Delta_z(V)}{2}\right)^2 + \frac{(\hbar v_{\mathrm{F}} q)^4}{\gamma_1^2}}, \tag{6.5}$$

with the corresponding wavefunction

$$\psi_{\pm}(\boldsymbol{q}, V) = \frac{1}{\sqrt{2|\mathcal{E}_{\pm}(\boldsymbol{q}, V)|\left[|\mathcal{E}_{\pm}(\boldsymbol{q}, V)| \mp \dfrac{\Delta_z(V)}{2}\right]}} \begin{pmatrix} \mp \dfrac{(\hbar v_{\mathrm{F}})^2}{\gamma_1}(q_x + \mathrm{i}\kappa q_y)^2 \\ |\mathcal{E}_{\pm}(\boldsymbol{q}, V)| \mp \dfrac{\Delta_z(V)}{2} \end{pmatrix}. \tag{6.6}$$

The dispersion relation shown in equation (6.5) does not coincide with equation (6.3). However, they both show that under the effect of the perpendicular electric field, the AB-stacked bilayer becomes an insulator with a gap in the order of $|\Delta_z(V)|$. This gap has also been witnessed in DFT studies of free-standing samples [34] or samples on a substrate [35], as well as in experiments [9–11, 28, 36–40]. For example, Kuzmenko *et al* [37] expect that the bandgap can be in the order of 0.1 eV, and spectroscopic measurements have observed a bandgap as large as 0.2 eV [11, 20, 23]. The ability to tune the controlled bandgap appears to be a promising playground with a wide variety of applications.

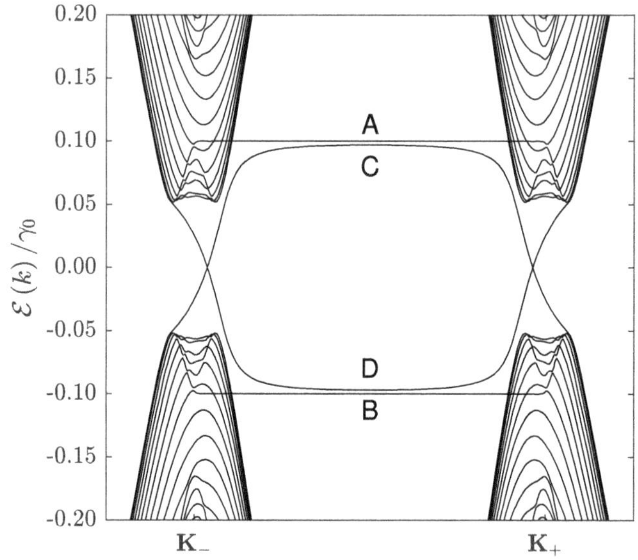

Figure 6.2. Band structure of an AB-stacked bilayer GNR in the presence of bias. Plot was made using the parameters $N = 120$, $\gamma_0 = 3.16$ eV, $\gamma_1 = 0.381$ eV [28, 29] and $\Delta_z(V) = 0.2\gamma_0$ [30, 31].

6.1.2 Biased bilayer nanoribbons with zigzag edges

In monolayer zGNR, zero-energy bands of edge states appear, as we showed in section 1.3. Whether edge states exist in bilayer graphene in the presence of a perpendicular external electric field is the subject of this subsection.

We consider the same ribbon geometry with zigzag edges shown in figure 1.17. The bias generated by the external electric field gives rise to an on-site energy difference between the bottom and top layers. This is parametrised by adding diagonal elements $\frac{\Delta_z(V)}{2}$ and $-\frac{\Delta_z(V)}{2}$ to the blocks $h_{11}(k)$ and $h_{22}(k)$, respectively, in the Hamiltonian in equation (1.86) [41].

Figure 6.2 shows the band structure of a bilayer zGNR under bias. Two partially flat bands, labelled A and B, from $\boldsymbol{K}_-$ to $\boldsymbol{K}_+$, i.e., $\frac{2}{3}\pi \leqslant ka \leqslant \frac{4}{3}\pi$, are clearly visible at $\mathcal{E}(k, V) = \pm\frac{\Delta_z(V)}{2}$. These are the edge states localized at opposite nanoribbon sides [41]. Another important feature is the presence of two dispersive bands, labelled C and D, crossing the bandgap, showing the bilayer zGNR is gapless, even for $\Delta_z(V) \neq 0$. These dispersive bands cross at zero energy near the $\boldsymbol{K}_\pm$ point and approach $\mathcal{E}(k, V) \approx \pm\frac{\Delta_z(V)}{2}$ for $ka \approx \pi$ [41]. They result from the hybridisation of the surface states with delocalised bulk states [31].

6.2 VHE in biased bilayer graphene

As we discussed in section 2.4, the VHE is a generic feature in systems with broken IS [33, 42–44]. Besides monolayer graphene with staggered sublattice potential, here we show another example: the biased bilayer graphene [45, 46].

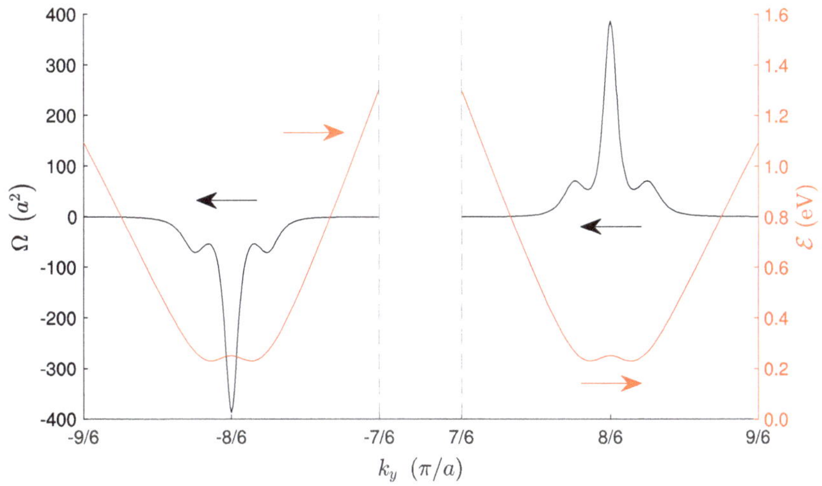

Figure 6.3. Energy band 1 (golden lines) and its corresponding valley-contrasting Berry curvature (black lines). The parameters are chosen to follow figure 6.1.

Firstly, in figure 6.3, we show numerically calculated energy band 1 of biased bilayer graphene and its corresponding Berry curvature. The parameter values are chosen in accordance with figure 6.1. The asymmetric band structure in the vicinity of the $\boldsymbol{K}_{\pm}$ point is because of the TW effect as discussed in section 1.2.2. Further, similar to the monolayer graphene shown in figure 2.3, the Berry curvature peaks at the valleys and has opposite signs in the $\boldsymbol{K}_{+}$ and $\boldsymbol{K}_{-}$ valleys.

Next, following the same procedure given in section 2.4.2, we employ the reduced two-band model, equation (6.4), to analytically calculate the Berry curvature for the basis set {A1, B2} [29, 47] in the vicinity of the $\boldsymbol{K}_{\pm}$ point. One can find that the Berry curvature of the conduction band $\mathcal{E}_{+}(\boldsymbol{q}, V)$ reads,

$$\boldsymbol{\Omega}^{\kappa}_{\mathrm{B:AB}}(\boldsymbol{q}, V) = \kappa\frac{(\hbar v_{\mathrm{F}})^4}{\gamma_1^2}\frac{\Delta_z(V)q^2}{|\mathcal{E}_{\pm}(\boldsymbol{q}, V)|^3}\hat{\boldsymbol{z}}. \tag{6.7}$$

The above expression clearly conveys the valley-contrasting Berry curvature in the biased bilayer graphene system. The corresponding Chern number is [30, 42]

$$\mathcal{C}^{\kappa}_{\mathrm{B:AB}} = \kappa\mathrm{sgn}(\Delta_z(V)). \tag{6.8}$$

The two valleys $\boldsymbol{K}_{+}$ and $\boldsymbol{K}_{-}$ are characterized by opposite Chern numbers 1 and −1, respectively, causing the valley Chern number $\mathcal{C}_{\mathrm{v}}$ to become 2. Further studies [30] show that such a system with a finite Rashba SOC remains a quantum valley Hall insulator with $\mathcal{C}_{\mathrm{v}} = 2$.

We can also calculate the valley Hall conductivity for our system at zero temperature [43, 45, 48]. Noticing $\mathcal{E}_{\pm}(\boldsymbol{q}, V)\mathrm{d}\mathcal{E}_{\pm}(\boldsymbol{q}, V) = 2\frac{(\hbar v_{\mathrm{F}})^4}{\gamma_1^2}q^3\mathrm{d}q$, we obtain

$$\sigma_{\mathrm{V}}(\mathcal{E}_{\mathrm{F}}) = \begin{cases} -\dfrac{2e^2}{h}\dfrac{\Delta_z}{|\mathcal{E}_{\mathrm{F}}|} & \text{if } |\mathcal{E}_{\mathrm{F}}| > \dfrac{\Delta_z}{2} \\ -\dfrac{4e^2}{h} & \text{if } |\mathcal{E}_{\mathrm{F}}| \leqslant \dfrac{\Delta_z}{2}. \end{cases} \tag{6.9}$$

Compared with equation (2.59), the valley Hall conductivity is twice as large as that of the gapped monolayer graphene. This is consistent with:

1. The double of the Chern number in bilayer graphene.
2. The fact that the Berry phase acquired by an electron during one circle around the valley becomes $\pm 2\pi$ in bilayer graphene instead of $\pm\pi$ when the gap closes [49].

Generally, valley Hall conductivity monotonically increases with the number of layers for gapped ABC-stacked multilayer graphene provided the potential differences between nearest layers are constant [43, 50].

For bilayer TMDCs, if one applies an out-of-plane electric field, IS breaks and subsequently the valley physics emerges [51]. We therefore can tune the valley-associated properties by electrical means.

Now we discuss briefly the experimental tuning of the chemical potential [14], which is an important parameter in the detection of VHE in 2D or thin-film systems. To have a comprehensive insight, one needs to be able to tune the Fermi level at each surface of a 2D film. One can achieve this through electrostatic gating by applying a DC voltage across the sample, with a dielectric located above/below. The top and back gates used within the devices for which the measurements are reported were gold metal contact and n-doped silicon, respectively. The generated electric field across hBN and SiO_2 dielectrics produces a finite carrier density at each surface proportional to the voltage applied. By inducing charges of opposite types on each surface, or applying voltages of opposite signs, one can also induce a displacement field. The displacement field in principle should lead to a gap opening at the low energy bands of a rhombohedral graphite system above a critical displacement field [8], as it has been observed in bilayer systems [9–12].

6.3 Domain wall induced with a voltage kink

In 2008, Martin *et al* [1] first proposed the concept of topological confinement in bilayer graphene. The configuration is shown in figure 6.4(a), where two opposite electric fields perpendicular to the bilayer graphene are applied at two sides and further give rise to topological kink states illustrated in figure 6.4(b).

In figure 6.5, we show the intragap dispersions of this system [52]. Let us compare such intragap dispersion with the system where the same biases are applied on both sides of the domain wall as plotted in figure 6.2. Apparently, apart from the originally existing edge state B and dispersive band C, a pair of gapless kink states emerge in the $\boldsymbol{K}_+$ and $\boldsymbol{K}_-$ valleys, respectively, labelled as E, F and G, H. As one can see, the states E and F (G and H) at the $\boldsymbol{K}_-$ ($\boldsymbol{K}_+$) valley only have the

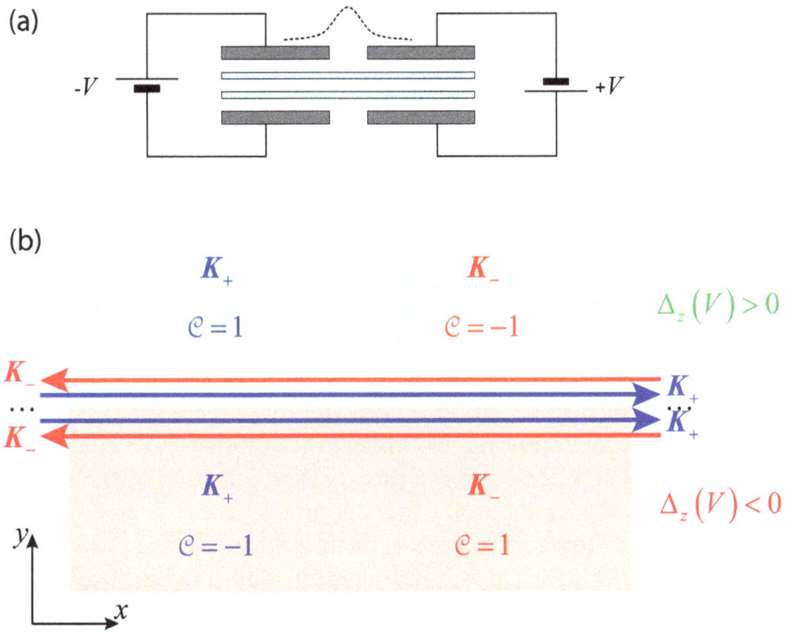

Figure 6.4. (a) Side view of a gated bilayer graphene with a voltage kink. In the central region, where the interlayer voltage changes sign, there exists a domain wall. (b) The schematic top view of (a). The region where the interlayer voltage changes sign supports topological kink states. The corresponding relationship between the kink states and the Chern numbers $\mathcal{C}$ at two valleys is also shown. (a) is a reprinted figure with permission from [1]. Copyright 2008 by the American Physical Society.

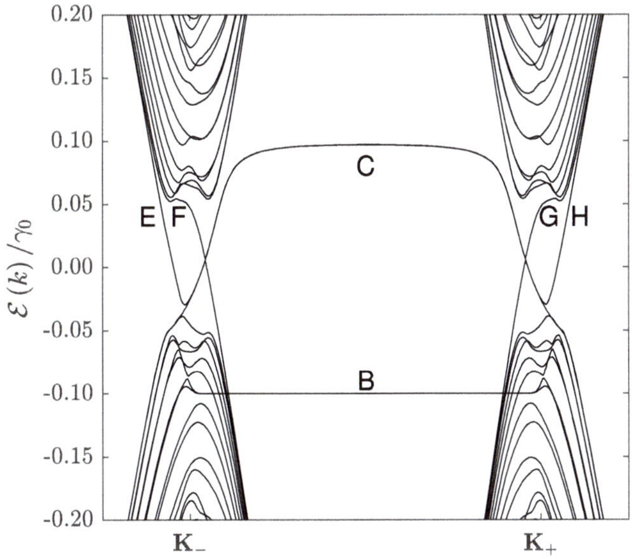

Figure 6.5. Band structure of the bilayer system in figure 6.4, in which the interlayer potentials change sign at the middle of the ribbon. Two left-going states at the K_- valley are labelled as E, F whereas two right-going states are labelled as G, H. Additional flat subband labelled B and curved subband labelled C appear in the bandgap and doubly degenerate because of IS across the ribbon [52, 53]. Subbands B and C can also be found in the biased bilayer graphene in figure 6.2. Parameters are adopted from figure 6.2.

leftward (rightward) group velocity, as can be seen from the negative (positive) slope of the subbands [54]. This means that the valley and momentum degrees become locked.

In order to better understand the origin of the topological kink states, it is instructive to follow the discussion in section 2.8 and study the topological properties of the bulk states. In figure 6.4(b), for the upper side of the domain wall, $\Delta_z(V) > 0$, the Chern numbers $\mathcal{C}$ in the $\boldsymbol{K}_+$ and $\boldsymbol{K}_-$ valleys are 1 and −1 respectively according to equation (6.8). On the other side of the domain wall, the Chern numbers are reversed. Therefore, the differences in the Chern number across the domain wall are $\Delta\mathcal{C} = 2$ and -2 in the $\boldsymbol{K}_+$ and $\boldsymbol{K}_-$ valleys, respectively. Recall the relationship between the topological properties and the number of right (left) moving modes discussed in equation (2.92), the differences in the Chern number across the domain wall will result in the appearance of the topological kink states at the domain wall. The number 2 and the propagating direction of the kink states are both consistent with the change of Chern number $|\Delta\mathcal{C}| = 2$ at each valley [55]. Because in the system shown in figure 6.4, the interlayer voltage changes sign at the domain wall and one can treat that at the domain wall, the bias becomes zero and the bandgap vanishes. Therefore, people also refer to kink states as zero modes, zero-line modes, or zero-energy modes [56].

The kink states can be realized not only with an antisymmetric potential or on domain walls [57], but also on ring-shaped potentials [58] or at a folded bilayer graphene sheet [59]. Similar results are also present in Bernal-stacked multilayer graphene, and the kink states can even occur at boundaries between ribbons with different numbers of layers [50]. The layer localization of kink states [60] and the scattering of kink–antikink states [61] have also been explored. Experimentally, although the bilayer domain walls are robust for topological valley current, the seek for such domain walls is very challenging. According to the original proposal established by Martin *et al* [1], Li *et al* [2] fabricated a dual-split-gate structure in bilayer graphene. By tuning the two pairs of split-gates, they found evidence of the predicted topological kink states. The states possess a mean free path of up to a few hundred nanometres and result in the robust 2D chiral valley currents [2]. This method provides a flexible and tunable method for the realisation of gate-controlled ballistic valley transport [63]. Further, in 2018, Li *et al* [62] demonstrated gate-controlled current transmission in a four-kink router device. As shown in figure 6.6, the '+' and '−' signs represent $\Delta_z(V) > 0$ and <0, respectively. By applying different configurations of the electric field at four discrete regions, one can achieve three configurations of the waveguide, labelled as (a) 'through', (b) 'right turn' and (c) 'left turn'. In all three configurations, the kink states exist only in two of the four channels, and the paths maintain the chirality in each valley. Typically, as shown in figures 6.6(b) and (c), the ability of current to travel around the corner is a direct consequence of the topological nature of the kink states [64]. Similar theoretical [52, 65, 66] and experimental [67–69] findings also exist. Additionally, researches have been able to realize valley filters [70] and interferometers [71] using kink states.

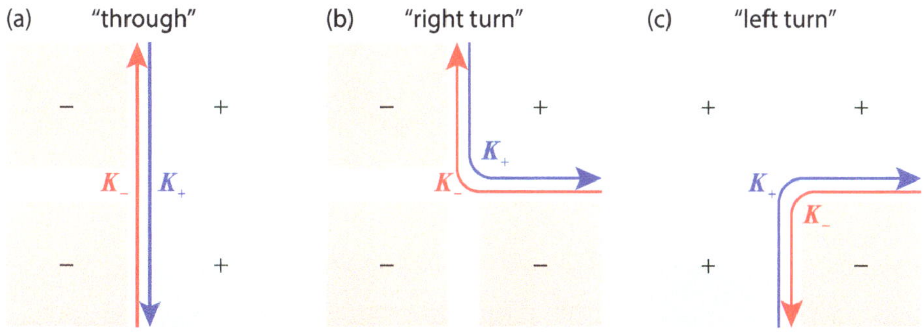

Figure 6.6. Illustration of the (a) 'through', (b) 'right turn' and (c) 'left turn' configurations of the gate-controlled waveguide for the propagation of kink states based on bilayer graphene. $\Delta_z(V) > 0$ and <0 are indicated by '+' and '−' signs, respectively [62].

6.4 Domain wall induced with a tilt-boundary edge states

6.4.1 Bilayer graphene with AB–BA tilt boundary

In this section, different from the model in the previous section with a kink voltage, we will focus on the bilayer graphene with tilt-boundary edge states [4, 5, 72–74]. Specifically, as shown in figure 6.7, a bilayer graphene with an AB–BA tilt boundary at the domain wall is exposed in a perpendicular external electric field, which is unchanged in the entire system. As one can see, this model is essentially the same as the domain wall induced by a kink voltage discussed in the previous section.

In figure 6.8, we show the band structure of the system described in figure 6.7. At the domain wall that separates the AB- and BA-stacking regions, we can find two gapless topological kink states E, F and G, H at $\boldsymbol{K}_-$ and $\boldsymbol{K}_+$, respectively. Therefore, valley current exists and transports along the domain wall. The transport directions are defined by the valley indices.

From equations (1.84) or (6.4), the effective Hamiltonian describing a biased BA-stacked bilayer graphene reads

$$H_{\mathrm{B:BA}}(\boldsymbol{q},\, V) = H_{\mathrm{B:AB}}(\boldsymbol{q},\, V)^* = \begin{pmatrix} \Delta_z(V)/2 & -\dfrac{(\hbar v_{\mathrm{F}})^2}{\gamma_1}(q_x - \mathrm{i}\kappa q_y)^2 \\ -\dfrac{(\hbar v_{\mathrm{F}})^2}{\gamma_1}(q_x + \mathrm{i}\kappa q_y)^2 & -\Delta_z(V)/2 \end{pmatrix}. \quad (6.10)$$

Following the deduction in section 6.2, the Berry curvature of the conduction band for such BA-stacked bilayer graphene can be found. It reads,

$$\boldsymbol{\Omega}^{\kappa}_{\mathrm{B:BA}}(\boldsymbol{q},\, V) = -\kappa\frac{(\hbar v_{\mathrm{F}})^4}{\gamma_1^2}\frac{\Delta_z(V)q^2}{|\mathcal{E}_{\pm}(\boldsymbol{q},\, V)|^3}\hat{\boldsymbol{z}}, \quad (6.11)$$

which is opposite to the AB-stacked bilayer graphene in equation (6.7). Consequently, the corresponding Chern number is

$$C^{\kappa}_{\mathrm{B:BA}} = -\kappa\mathrm{sgn}(\Delta_z(V)). \quad (6.12)$$

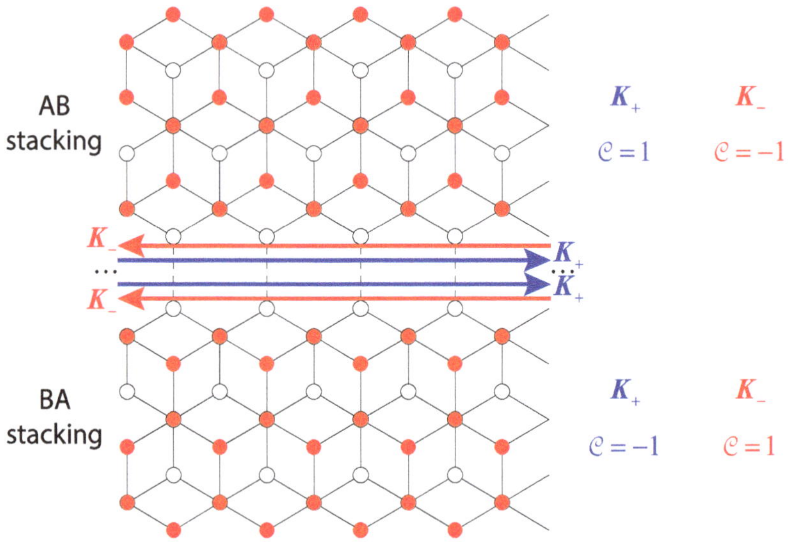

Figure 6.7. Illustration of bilayer graphene with tilt-boundary edge states under the influence of a perpendicular external electric field. At the centre, a domain wall separates AB- and BA-stacking domains and supports topological kink states. The corresponding relationship between the kink states and the Chern number $\mathcal{C}$ at two valleys are also shown.

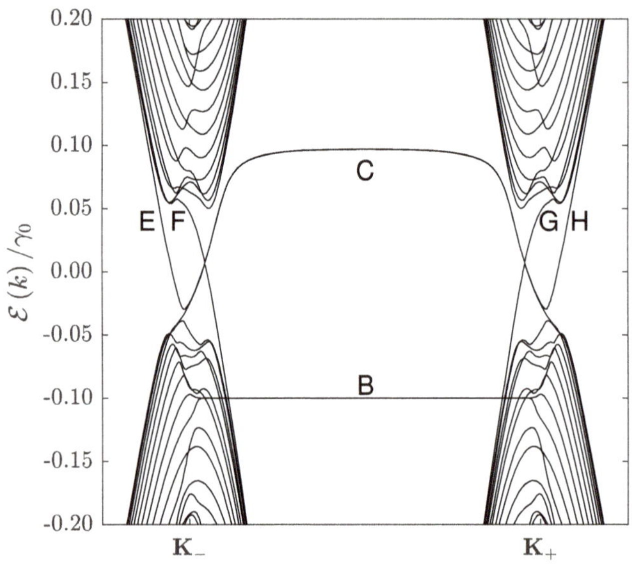

Figure 6.8. Band structure of the bilayer system in figure 6.7, in which the stacking order switches at the middle of the ribbon. The band structure is similar to figure 6.5. Two left-going states at $\boldsymbol{K}_-$ valley are labelled as E, F whereas two right-going states at $\boldsymbol{K}_+$ valley are labelled as G, H. The parameters used in figures 6.2 and 6.5 were adopted.

The two valleys $\boldsymbol{K}_+$ and $\boldsymbol{K}_-$ are characterized by opposite Chern numbers -1 and 1, respectively. We annotate the Chern numbers of $\boldsymbol{K}_+$ and $\boldsymbol{K}_-$ valleys at both sides of the domain wall in figure 6.7. By recalling equation (6.8), one can notice that the Chern numbers for the AB- and BA-stacked bilayer graphene are inverted. The Chern numbers change $\Delta\mathcal{C} = 2$ and -2 for the $\boldsymbol{K}_+$ and $\boldsymbol{K}_-$ valleys, respectively, when passing through the domain wall. Therefore, similar to the bilayer graphene with a kink voltage shown in figure 6.4, it is the $\Delta\mathcal{C}$ that leads to the existence of the topological kink states shown in figure 6.7.

Experimentally, compared with the kink states induced by the kink voltage in the previous section, the different stacking order is easier to realize. In 2015, Ju *et al* [72] first observed topologically protected 1D kink states along the AB–BA domain wall in exfoliated bilayer graphene on SiO_2/Si substrates [69]. The smoothly varying crystal lattice at the atomic-defect-free AB–BA domain wall effectively suppressed the intervalley scattering. Similarly, in buckled honeycomb structures, e.g., silicene, the topological kink states can be formed at the domain wall between two regions with opposite buckling geometries [75]. In addition, besides the setups of voltage kink and the tilt boundary for topological kink states in bilayer graphene mentioned above, a folded bilayer graphene can also host topological kink states, as independently realized by Hou *et al* [76] and Mania *et al* [59]. In these systems, by folding a bilayer graphene sheet (without a line defect), a perpendicular electric field induces reversed bias to the two parts separated by the fold. Similar to the domain wall defining kink voltage, at the curved boundary, a conductance of $4e^2/h$ validated the topological kink states [59].

In summary, the domain walls could be generated in both monolayer and bilayer systems. The generation of the domain wall in monolayer graphene (section 2.8) originates from the staggered on-site sublattice potential [77]. The on-site energy of sublattice A is higher (lower) than that of the B on one (the other) side of the domain wall. For bilayer graphene, the domain wall can be generated in two ways:

1. Add an electric field with a kink voltage (section 6.3) [1].
2. An AB–BA tilt boundary with the homogeneous perpendicular electric field (this section) [4, 5, 72].

6.4.2 Multilayer graphene

In the gapped monolayer, bilayer and multilayer graphene with the domain walls, the IS breaks. For a particular valley, the Chern numbers at both sides of the domain wall are $\mathcal{C} = N_l/2$ and $\mathcal{C} = -N_l/2$, with N_l representing the number of layers. This means that the difference between the Chern numbers across the domain wall is $\Delta\mathcal{C} = \pm N_l$. According to equation (2.92), N_l topological kink states for each valley will correspondingly exist along the domain wall [50, 78]. Furthermore, the direction of the propagation is determined by the valley, which means the topological kink states have a unique characteristic of the valley–momentum locking. When we ignore the intervalley scattering, the backscattering between the different kink states is strictly forbidden. Therefore, electrons always transport ballistically along the edges of the domain wall and each transport channel contributes a quantised conductance of $G_0 = e^2/h$ [79]. As a result, taking the spin DOF into consideration, the conductance of a N_l-layer graphene equals $G = 2N_l G_0$ [53].

6.4.3 Moiré systems

The observation of transport through topologically protected channels suggests the novel possibility that transport physics may be generated. Recent experimental works [80–93] demonstrated that bilayers with small twist angles or small lattice mismatches can atomically reconstruct away from the moiré pattern to form sharp domain walls as a consequence of the periodic alternating regions of AA-, AB- and BA-stacking configurations in the moiré crystal [94–99], as shown in figure 1.19(a). Upon gating, it acts as host to a network of topological kink states, as an extension of figure 6.7. Such networks of topological kink states in the domain walls of TBLG, as schematically illustrated in figure 6.9, are predicted theoretically [100–111] and can be observed experimentally [82–86, 112–114]. If we switch the direction of the gate potential, we would swap the sign of the Chern number according to equations (6.8) and (6.12). Hence, each boundary will have an opposite valley-protected kink state.

Similar to the superlattices with adjacent AB and BA stacking orders supporting topological kink states at the boundaries, topological kink states can also be realized in graphene on top of hBN substrates, hBN wrapped bilayer graphene (unrotated) [115] as a result of lattice mismatch [116]. Moreover, the AB–BA domain and its band structure can be tuned by electrostatic force near the magic angle [114]. The topological network modes have a large coherence length, making it possible to have different interference oscillations. Additionally, the Aharonov–Bohm oscillations under perpendicular magnetic fields are experimentally observed in marginally TBLG

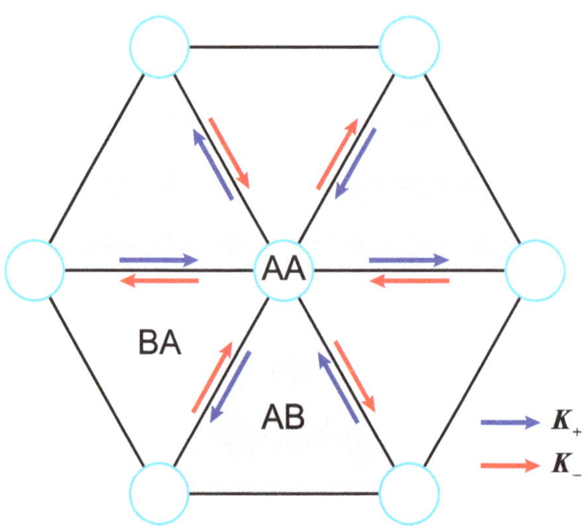

Figure 6.9. The formation of triangular domains with AB and BA stacking is a characteristic feature of the moiré pattern in TBLG. Valley-protected kink states can be found at the AB–BA domain boundaries.

because of the existence of a triangular network of 1D states [13, 82, 86, 104]. It is noted that for the TBLG with a twist angle that is sufficiently close to the commensurate 38.2°, the alternative commensurate configurations, i.e., the gapped exchange symmetry even and the gapless exchange symmetry odd [117], geometrically frustrate the network of topologically protected counter-propagating chiral modes [118].

In monolayer TMDCs, calculations recently revealed that the periodic modulation of moiré patterns can impact the topological order, resulting in intriguing mosaic patterns comprising topological insulating regions and non-topological insulating regions. More interestingly, by applying in-plane strain, the topological insulating/non-topological insulating phases and the kink states at the edges can evolve into a densely arranged 1D array [119], which can transport valley current and may be utilized in valleytronics [120]. Further calculation has predicted that two groups of moiré edge arrays carrying counter-propagating kink states are spatially separate in the out-of-plane direction [121]. However, no experimental evidence has been presented to prove the existence of the predicted dense array of helical channels, which may be because of the challenge of precisely controlling in-plane strains experimentally. With only 1D channels on the boundaries of local stacking zones, valleytronics is still not feasible. Consequently, we need more experimental technical advancements to confirm the usefulness of the array and explore other approaches for its valleytronic applications [120].

References

[1] Martin I, Blanter Y M and Morpurgo A F 2008 Topological confinement in bilayer graphene *Phys. Rev. Lett.* **100** 036804

[2] Li J, Wang K, McFaul K J, Zern Z, Ren Y, Watanabe K, Taniguchi T, Qiao Z and Zhu J 2016 Gate-controlled topological conducting channels in bilayer graphene *Nat. Nanotechnol.* **11** 1060–5

[3] Chen H, Zhou P, Liu J, Qiao J, Oezyilmaz B and Martin J 2020 Gate controlled valley polarizer in bilayer graphene *Nat. Commun.* **11** 1202

[4] Vaezi A, Liang Y, Ngai D H, Yang L and Kim E-A 2013 Topological edge states at a tilt boundary in gated multilayer graphene *Phys. Rev.* X **3** 021018

[5] Zhang F, MacDonald A H and Mele E J 2013 Valley Chern numbers and boundary modes in gapped bilayer graphene *Proc. Natl. Acad. Sci. USA* **110** 10546–51

[6] Koshino M 2013 Electronic transmission through AB-BA domain boundary in bilayer graphene *Phys. Rev.* B **88** 115409

[7] Lin J, Fang W, Zhou W, Lupini A R, Idrobo J C, Kong J, Pennycook S J and Pantelides S T 2013 AC/AB stacking boundaries in bilayer graphene *Nano Lett.* **13** 3262–8

[8] Koshino M 2010 Interlayer screening effect in graphene multilayers with ABA and ABC stacking *Phys. Rev.* B **81** 125304

[9] Castro E V, Novoselov K S, Morozov S V, Peres N M R, Lopes dos Santos J M B, Nilsson J, Guinea F, Geim A K and Castro Neto A H 2007 Biased bilayer graphene: semiconductor with a gap tunable by the electric field effect *Phys. Rev. Lett.* **99** 216802

[10] Oostinga J B, Heersche H B, Liu X, Morpurgo A F and Vandersypen L M K 2008 Gate-induced insulating state in bilayer graphene devices *Nat. Mater.* **7** 151–7

[11] Zhang Y, Tang T-T, Girit C, Hao Z, Martin M C, Zettl A, Crommie M F, Shen Y R and Wang F 2009 Direct observation of a widely tunable bandgap in bilayer graphene *Nature* **459** 820–3

[12] Taychatanapat T and Jarillo-Herrero P 2010 Electronic transport in dual-gated bilayer graphene at large displacement fields *Phys. Rev. Lett.* **105** 166601

[13] Ren Y-N, Zhang Y, Liu Y-W and He L 2020 Twistronics in graphene-based van der Waals structures *Chin. Phys.* B **29** 117303

[14] Ozdemir S 2021 *Electronic Properties of Rhombohedral Graphite* Springer Theses (Cham: Springer Nature)

[15] Rozhkov A V, Sboychakov A O, Rakhmanov A L and Nori F 2016 Electronic properties of graphene-based bilayer systems *Phys. Rep.* **648** 1–104

[16] Guinea F, Castro Neto A H and Peres N M R 2006 Electronic states and Landau levels in graphene stacks *Phys. Rev.* B **73** 245426

[17] Min H, Sahu B, Banerjee S K and MacDonald A H 2007 Ab initio theory of gate induced gaps in graphene bilayers *Phys. Rev.* B **75** 155115

[18] Mikhailov S (ed) 2011 *Physics and Applications of Graphene: Theory* (Rijeka: IntechOpen)

[19] McCann E and Fal'ko V I 2006 Landau-level degeneracy and quantum Hall effect in a graphite bilayer *Phys. Rev. Lett.* **96** 086805

[20] Ohta T, Bostwick A, Seyller T, Horn K and Rotenberg E 2006 Controlling the electronic structure of bilayer graphene *Science* **313** 951–4

[21] Amet F and Finkelstein G 2015 Could use a break *Nat. Phys.* **11** 989–90

[22] Shimazaki Y, Yamamoto M, Borzenets I V, Watanabe K, Taniguchi T and Tarucha S 2015 Generation and detection of pure valley current by electrically induced Berry curvature in bilayer graphene *Nat. Phys.* **11** 1032–6

[23] Mak K F, Lui C H, Shan J and Heinz T F 2009 Observation of an electric-field-induced band gap in bilayer graphene by infrared spectroscopy *Phys. Rev. Lett.* **102** 256405

[24] McCann E 2006 Asymmetry gap in the electronic band structure of bilayer graphene *Phys. Rev.* B **74** 161403(R)

[25] Lu C L, Chang C P, Huang Y C, Chen R B and Lin M L 2006 Influence of an electric field on the optical properties of few-layer graphene with AB stacking *Phys. Rev.* B **73** 144427

[26] Ando T and Koshino M 2009 Field effects on optical phonons in bilayer graphene *J. Phys. Soc. Jpn.* **78** 034709

[27] Ando T and Koshino M 2009 Optical absorption by interlayer density excitations in bilayer graphene *J. Phys. Soc. Jpn.* **78** 104716

[28] Kuzmenko A B, Crassee I, van der Marel D, Blake P and Novoselov K S 2009 Determination of the gate-tunable band gap and tight-binding parameters in bilayer graphene using infrared spectroscopy *Phys. Rev.* B **80** 165406

[29] McCann E and Koshino M 2013 The electronic properties of bilayer graphene *Rep. Prog. Phys.* **76** 056503

[30] Qiao Z, Tse W-K, Jiang H, Yao Y and Niu Q 2011 Two-dimensional topological insulator state and topological phase transition in bilayer graphene *Phys. Rev. Lett.* **107** 256801

[31] Carmelo J M P, Lopes dos Santos J M B, Rocha Vieira V and Sacramento P D (ed) 2007 *Strongly Correlated Systems, Coherence and Entanglement* (Singapore: World Scientific)

[32] Castro Neto A H, Guinea F, Peres N M R, Novoselov K S and Geim A K 2009 The electronic properties of graphene *Rev. Mod. Phys.* **81** 109–62

[33] Xiao D, Yao W and Niu Q 2007 Valley-contrasting physics in graphene: magnetic moment and topological transport *Phys. Rev. Lett.* **99** 236809
[34] Nanda B R K and Satpathy S 2009 Strain and electric field modulation of the electronic structure of bilayer graphene *Phys. Rev.* B **80** 165430
[35] Ramasubramaniam A, Naveh D and Towe E 2011 Tunable band gaps in bilayer graphene–BN heterostructures *Nano Lett.* **11** 1070–5
[36] Henriksen E A and Eisenstein J P 2010 Measurement of the electronic compressibility of bilayer graphene *Phys. Rev.* B **82** 041412
[37] Kuzmenko A B, van Heumen E, van der Marel D, Lerch P, Blake P, Novoselov K S and Geim A K 2009 Infrared spectroscopy of electronic bands in bilayer graphene *Phys. Rev.* B **79** 115441
[38] Jing L, Velasco J Jr, Kratz P, Liu G, Bao W, Bockrath M and Lau C N 2010 Quantum transport and field-induced insulating states in bilayer graphene pnp junctions *Nano Lett.* **10** 4000–4
[39] Jing L, Velasco J Jr, Kratz P, Liu G, Bao W, Bockrath M and Lau C N 2010 Quantum transport and field-induced insulating states in bilayer graphene pnp junctions *Nano Lett.* **10** 4775
[40] Hao Y *et al* 2016 Oxygen-activated growth and bandgap tunability of large single-crystal bilayer graphene *Nat. Nanotechnol.* **11** 426–31
[41] Castro E V, Peres N M R, Lopes dos Santos J M B, Castro Neto A H and Guinea F 2008 Localized states at zigzag edges of bilayer graphene *Phys. Rev. Lett.* **100** 026802
[42] Xiao D, Chang M-C and Niu Q 2010 Berry phase effects on electronic properties *Rev. Mod. Phys.* **82** 1959–2007
[43] Yamamoto M, Shimazaki Y, Borzenets I V and Tarucha S 2015 Valley Hall effect in two-dimensional hexagonal lattices *J. Phys. Soc. Jpn.* **84** 121006
[44] Foa Torres L E F, Roche S and Charlier J-C 2020 *Introduction to Graphene-Based Nanomaterials: From Electronic Structure to Quantum Transport* 2nd edn (Cambridge: Cambridge University Press)
[45] Enoki T and Ando T (ed) 2020 *Physics and Chemistry of Graphene: Graphene to Nanographene* 2nd edn (Singapore: Jenny Stanford Publishing)
[46] Tsymbal E Y and Žutić I (ed) 2019 *Nanoscale Spintronics and Applications* Spintronics Handbook: Spin Transport and Magnetism **vol 3** 2nd edn (Boca Raton, FL: CRC Press)
[47] Mucha-Kruczyński M, McCann E and Fal'ko V I 2010 Electron–hole asymmetry and energy gaps in bilayer graphene *Semicond. Sci. Technol.* **25** 033001
[48] Koshino M 2009 Electronic transport in bilayer graphene *New J. Phys.* **11** 095010
[49] Novoselov K S, McCann E, Morozov S V, Fal'ko V I, Katsnelson M I, Zeitler U, Jiang D, Schedin F and Geim A K 2006 Unconventional quantum Hall effect and Berry's phase of 2π in bilayer graphene *Nat. Phys.* **2** 177–80
[50] Jung J, Zhang F, Qiao Z and MacDonald A H 2011 Valley-Hall kink and edge states in multilayer graphene *Phys. Rev.* B **84** 075418
[51] Wu S *et al* 2013 Electrical tuning of valley magnetic moment through symmetry control in bilayer MoS_2 *Nat. Phys.* **9** 149–53
[52] Qiao Z, Jung J, Niu Q and MacDonald A H 2011 Electronic highways in bilayer graphene *Nano Lett.* **11** 3453–9
[53] Wang Z, Cheng S, Liu X and Jiang H 2021 Topological kink states in graphene *Nanotechnology* **32** 402001

[54] Ferry D K, Goodnick S M and Bird J 2009 *Transport in Nanostructures* 2nd edn (Cambridge: Cambridge University Press)

[55] Jiang B-Y, Ni G-X, Addison Z, Shi J K, Liu X, Zhao S Y F, Kim P, Mele E J, Basov D N and Fogler M M 2017 Plasmon reflections by topological electronic boundaries in bilayer graphene *Nano Lett.* **17** 7080–5

[56] Ren Y, Qiao Z and Niu Q 2016 Topological phases in two-dimensional materials: a review *Rep. Prog. Phys.* **79** 066501

[57] Yin L-J, Jiang H, Qiao J-B and He L 2016 Direct imaging of topological edge states at a bilayer graphene domain wall *Nat. Commun.* **7** 11760

[58] Xavier L J P, Pereira J M Jr, Chaves A, Farias G A and Peeters F M 2010 Topological confinement in graphene bilayer quantum rings *Appl. Phys. Lett.* **96** 212108

[59] Mania E, Cadore A R, Taniguchi T, Watanabe K and Campos L C 2019 Topological valley transport at the curved boundary of a folded bilayer graphene *Commun. Phys.* **2** 6

[60] Jaskólski W, Pelc M, Bryant G W, Chico L and Ayuela A 2018 Controlling the layer localization of gapless states in bilayer graphene with a gate voltage *2D Mater.* **5** 025006

[61] Benchtaber N, Sánchez D and Serra L 2021 Scattering of topological kink-antikink states in bilayer graphene structures *Phys. Rev.* B **104** 155303

[62] Li J, Zhang R-X, Yin Z, Zhang J, Watanabe K, Taniguchi T, Liu C and Zhu J 2018 A valley valve and electron beam splitter *Science* **362** 1149–52

[63] Li J 2017 A transport study of emerging phenomena in bilayer graphene nanostructures *PhD Thesis* The Pennsylvania State University, University Park, PA

[64] Chen X, Zhou Z, Deng B, Wu Z, Xia F, Cao Y, Zhang L, Huang W, Wang N and Wang L 2019 Electrically tunable physical properties of two-dimensional materials *Nano Today* **27** 99–119

[65] Qiao Z, Jung J, Lin C, Ren Y, MacDonald A H and Niu Q 2014 Current partition at topological channel intersections *Phys. Rev. Lett.* **112** 206601

[66] Wang K, Ren Y, Deng X, Yang S A, Jung J and Qiao Z 2017 Gate-tunable current partition in graphene-based topological zero lines *Phys. Rev.* B **95** 245420

[67] Yan M, Lu J, Li F, Deng W, Huang X, Ma J and Liu Z 2018 On-chip valley topological materials for elastic wave manipulation *Nat. Mater.* **17** 993–8

[68] Zhang L *et al* 2019 Valley kink states and topological channel intersections in substrate-integrated photonic circuitry *Laser Photonics Rev.* **13** 1900159

[69] Xue H, Yang Y and Zhang B 2021 Topological valley photonics: physics and device applications *Adv. Photonics Res.* **2** 2100013

[70] da Costa D R, Chaves A, Sena S H R, Farias G A and Peeters F M 2015 Valley filtering using electrostatic potentials in bilayer graphene *Phys. Rev.* B **92** 045417

[71] Cheng S-g, Liu H, Jiang H, Sun Q-F and Xie X C 2018 Manipulation and characterization of the valley-polarized topological kink states in graphene-based interferometers *Phys. Rev. Lett.* **121** 156801

[72] Ju L *et al* 2015 Topological valley transport at bilayer graphene domain walls *Nature* **520** 650–5

[73] Pelc M, Jaskólski W, Ayuela A and Chico L 2015 Topologically confined states at corrugations of gated bilayer graphene *Phys. Rev.* B **92** 085433

[74] Lee C, Kim G, Jung J and Min H 2016 Zero-line modes at stacking faulted domain walls in multilayer graphene *Phys. Rev.* B **94** 125438

[75] Kim Y, Choi K, Ihm J and Jin H 2014 Topological domain walls and quantum valley Hall effects in silicene *Phys. Rev.* B **89** 085429
[76] Hou T, Cheng G, Tse W-K, Zeng C and Qiao Z 2018 Topological zero-line modes in folded bilayer graphene *Phys. Rev.* B **98** 245417
[77] Semenoff G W, Semenoff V and Zhou F 2008 Domain walls in gapped graphene *Phys. Rev. Lett.* **101** 087204
[78] Li X, Qiao Z, Jung J and Niu Q 2012 Unbalanced edge modes and topological phase transition in gated trilayer graphene *Phys. Rev.* B **85** 201404
[79] Ortmann F, Roche S and Valenzuela S O (ed) 2015 *Topological Insulators: Fundamentals and Perspectives* (Singapore: Wiley-VCH)
[80] Sunku S S *et al* 2018 Photonic crystals for nano-light in moiré graphene superlattices *Science* **362** 1153–6
[81] Sunku S S *et al* 2020 Nano-photocurrent mapping of local electronic structure in twisted bilayer graphene *Nano Lett.* **20** 2958–64
[82] Xu S G *et al* 2019 Giant oscillations in a triangular network of one-dimensional states in marginally twisted graphene *Nat. Commun.* **10** 4008
[83] Yoo H *et al* 2019 Atomic and electronic reconstruction at the van der Waals interface in twisted bilayer graphene *Nat. Mater.* **18** 448–53
[84] Huang S, Kim K, Efimkin D K, Lovorn T, Taniguchi T, Watanabe K, MacDonald A H, Tutuc E and LeRoy B J 2018 Topologically protected helical states in minimally twisted bilayer graphene *Phys. Rev. Lett.* **121** 037702
[85] Alden J S, Tsen A W, Huang P Y, Hovden R, Brown L, Park J, Muller D A and McEuen P L 2013 Strain solitons and topological defects in bilayer graphene *Proc. Natl. Acad. Sci. USA* **110** 11256–60
[86] Rickhaus P *et al* 2018 Transport through a network of topological channels in twisted bilayer graphene *Nano Lett.* **18** 6725–30
[87] McGilly L J *et al* 2020 Visualization of moiré superlattices *Nat. Nanotechnol.* **15** 580–4
[88] Sunku S S *et al* 2021 Hyperbolic enhancement of photocurrent patterns in minimally twisted bilayer graphene *Nat. Commun.* **12** 1641
[89] Rosenberger M R, Chuang H-J, Phillips M, Oleshko V P, McCreary K M, Sivaram S V, Hellberg C S and Jonker B T 2020 Twist angle-dependent atomic reconstruction and moiré patterns in transition metal dichalcogenide heterostructures *ACS Nano* **14** 4550–8
[90] Rosenberger M R, Chuang H-J, Phillips M, Oleshko V P, McCreary K M, Sivaram S V, Stephen Hellberg C and Jonker B T 2020 Correction to twist angle-dependent atomic reconstruction and moiré patterns in transition metal dichalcogenide heterostructures *ACS Nano* **14** 14240–2
[91] Weston A *et al* 2020 Atomic reconstruction in twisted bilayers of transition metal dichalcogenides *Nat. Nanotechnol.* **15** 592–7
[92] Li H *et al* 2021 Imaging moiré flat bands in three-dimensional reconstructed WSe_2/WS_2 superlattices *Nat. Mater.* **20** 945–50
[93] Kazmierczak N P, Van Winkle M, Ophus C, Bustillo K C, Carr S, Brown H G, Ciston J, Taniguchi T, Watanabe K and Bediako D K 2021 Strain fields in twisted bilayer graphene *Nat. Mater.* **20** 956–63
[94] Weston A 2022 *Atomic and Electronic Properties of 2D Moiré Interfaces* Springer Theses (Cham: Springer Nature)

[95] Li Y 2019 Quantum transport in gapped graphene *PhD Thesis* University of Cambridge, Cambridge
[96] van Wijk M M, Schuring A, Katsnelson M I and Fasolino A 2015 Relaxation of moiré patterns for slightly misaligned identical lattices: graphene on graphite *2D Mater.* **2** 034010
[97] Nam N N T and Koshino M 2017 Lattice relaxation and energy band modulation in twisted bilayer graphene *Phys. Rev.* B **96** 075311
[98] Gargiulo F and Yazyev O V 2018 Structural and electronic transformation in low-angle twisted bilayer graphene *2D Mater.* **5** 015019
[99] Jain S K, Juričić V and Barkema G T 2017 Structure of twisted and buckled bilayer graphene *2D Mater.* **4** 015018
[100] Anđelković M, Covaci L and Peeters F M 2018 DC conductivity of twisted bilayer graphene: angle-dependent transport properties and effects of disorder *Phys. Rev. Mater.* **2** 034004
[101] Carr S, Massatt D, Torrisi S B, Cazeaux P, Luskin M and Kaxiras E 2018 Relaxation and domain formation in incommensurate two-dimensional heterostructures *Phys. Rev.* B **98** 224102
[102] San-Jose P and Prada E 2013 Helical networks in twisted bilayer graphene under interlayer bias *Phys. Rev.* B **88** 121408
[103] Efimkin D K and MacDonald A H 2018 Helical network model for twisted bilayer graphene *Phys. Rev.* B **98** 035404
[104] De Beule C, Dominguez F and Recher P 2020 Aharonov-Bohm oscillations in minimally twisted bilayer graphene *Phys. Rev. Lett.* **125** 096402
[105] Ramires A and Lado J L 2018 Electrically tunable gauge fields in tiny-angle twisted bilayer graphene *Phys. Rev. Lett.* **121** 146801
[106] Tsim B, Nam N N T and Koshino M 2020 Perfect one-dimensional chiral states in biased twisted bilayer graphene *Phys. Rev.* B **101** 125409
[107] Walet N R and Guinea F 2020 The emergence of one-dimensional channels in marginal-angle twisted bilayer graphene *2D Mater.* **7** 015023
[108] Fleischmann M, Gupta R, Wullschläger F, Theil S, Weckbecker D, Meded V, Sharma S, Meyer B and Shallcross S 2020 Perfect and controllable nesting in minimally twisted bilayer graphene *Nano Lett.* **20** 971–8
[109] Lebedeva I V and Popov A M 2020 Energetics and structure of domain wall networks in minimally twisted bilayer graphene under strain *J. Phys. Chem.* C **124** 2120–30
[110] Hou T, Ren Y, Quan Y, Jung J, Ren W and Qiao Z 2020 Metallic network of topological domain walls *Phys. Rev.* B **101** 201403
[111] Chou Y-Z, Wu F and Das Sarma S 2020 Hofstadter butterfly and Floquet topological insulators in minimally twisted bilayer graphene *Phys. Rev. Res.* **2** 033271
[112] Wolf T M R 2021 Electronic properties of twisted-layer graphene systems *PhD Thesis* ETH Zürich, Zürich
[113] Qiao J-B, Yin L-J and He L 2018 Twisted graphene bilayer around the first magic angle engineered by heterostrain *Phys. Rev.* B **98** 235402
[114] Liu Y-W *et al* 2020 Tunable lattice reconstruction, triangular network of chiral one-dimensional states, and bandwidth of flat bands in magic angle twisted bilayer graphene *Phys. Rev. Lett.* **125** 236102
[115] Endo K *et al* 2019 Topological valley currents in bilayer graphene/hexagonal boron nitride superlattices *Appl. Phys. Lett.* **114** 243105

[116] Kindermann M, Uchoa B and Miller D L 2012 Zero-energy modes and gate-tunable gap in graphene on hexagonal boron nitride *Phys. Rev.* B **86** 115415
[117] Mele E J 2010 Commensuration and interlayer coherence in twisted bilayer graphene *Phys. Rev.* B **81** 161405
[118] Pal H K, Spitz S and Kindermann M 2019 Emergent geometric frustration and flat band in moiré bilayer graphene *Phys. Rev. Lett.* **123** 186402
[119] Tong Q, Yu H, Zhu Q, Wang Y, Xu X and Yao W 2017 Topological mosaics in moiré superlattices of van der Waals heterobilayers *Nat. Phys.* **13** 356–62
[120] Tang K and Qi W 2020 Moiré-pattern-tuned electronic structures of van der Waals heterostructures *Adv. Funct. Mater.* **30** 2002672
[121] Hu C, Michaud-Rioux V, Yao W and Guo H 2018 Moiré valleytronics: realizing dense arrays of topological helical channels *Phys. Rev. Lett.* **121** 186403

Chapter 7

Valley optoelectronics based on monolayer 2D materials

Valley optoelectronics is an emerging field that exploits the valley DOF in monolayer 2D materials, such as TMDCs. This chapter delves into the interaction between valley physics and optoelectronic phenomena in monolayer 2D materials. The chapter begins by exploring the fundamental interaction between electrons and external electromagnetic fields (electron–photon interaction) [1], introducing key concepts such as the Jones vector and its relevance to light polarization. This is followed by a detailed discussion of light absorption in hexagonal lattice systems, with a focus on the optical selection rules governing valley-dependent transitions in IS broken systems and TMDCs.

The chapter also addresses the exciton Hall effect, a novel phenomenon where excitons (bound electron–hole pairs) exhibit valley-dependent deflection under external fields. Additionally, the tuning of VP through the valley Zeeman effect is examined as a means of controlling optical properties in these materials. The chapter further investigates the development of valley-based light-emitting diodes (LEDs) and concludes with an analysis of the experimental demonstration of the VHE in TMDCs, showcasing the practical potential of valley optoelectronics in future technologies.

7.1 Interaction of an electron with an external electromagnetic field

By irradiating a material with electromagnetic waves, with energy greater than the bandgap, the electrons in the material absorb energy from the electromagnetic waves and are excited from the filled valence band to the empty conduction band. Such a process is called the interband absorption process. The excited electron leaves a hole in the valence band, thus the absorption process creates an electron–hole pair. Since both the valence and conduction bands possess a continuous range of energy states, interband transitions are possible over a continuous range of frequencies.

doi:10.1088/978-0-7503-5562-9ch7

The interband absorption process observes the law of energy conservation, which reads

$$\mathcal{E}_{\rm f} = \mathcal{E}_{\rm i} + \hbar\omega, \tag{7.1}$$

where $\mathcal{E}_{\rm i}$ and $\mathcal{E}_{\rm f}$ are the eigenenergies of the initial and final states, respectively, which satisfy $\mathcal{E}_{\rm i} < \mathcal{E}_{\rm f}$. In the case of a semiconductor, the initial (final) state is the state in which the electrons are in the valence (conduction) band. $\hbar\omega > 0$ is the energy of incident light. Obviously, the interband transitions are only possible when the photon energy is larger than the bandgap.

Opposite to absorption, emission is the process in which the excited electron decays into a lower energy level and light with the photon energy $\hbar\omega$ is emitted by the electron. It may occur at any energy and in any direction, provided there are corresponding allowed transitions. Optical emission is the physical basis for important applications, such as lasers and LEDs.

The description of interband optical transition is based on the quantum mechanical treatment of the light–matter interaction to the band states. We will employ a semiclassical approach [1–3], where we treat the electromagnetic field classically and derive it from classical Maxwell's equations [4, 5], while describing the electrons with the quantum mechanical Hamiltonian and Bloch wave functions. This approach generates the same results as the quantum mechanical treatment, which is more rigorous and difficult to understand [3].

Firstly, the Hamiltonian of one electron bound to the periodic potential $U(\boldsymbol{r})$ in the crystal is given by

$$H_0 = \frac{\boldsymbol{p}^2}{2m_{\rm e}} + U(\boldsymbol{r}), \tag{7.2}$$

where $m_{\rm e}$ represents the mass of the electron. When there is a vector potential $\boldsymbol{A}$ defining the electromagnetic field, the momentum $\boldsymbol{p}$ of the electron undergoes a shift to $\boldsymbol{p} - e\boldsymbol{A}$. Thus, the Hamiltonian of one electron when the crystal is irradiated with electromagnetic waves becomes

$$\begin{aligned} H &= \frac{1}{2m_{\rm e}}(\boldsymbol{p} - e\boldsymbol{A})^2 + U(\boldsymbol{r}) \\ &= \frac{1}{2m_{\rm e}}(\boldsymbol{p} - e\boldsymbol{A})\cdot(\boldsymbol{p} - e\boldsymbol{A}) + U(\boldsymbol{r}) \\ &= \frac{1}{2m_{\rm e}}(\boldsymbol{p}^2 - e\boldsymbol{A}\cdot\boldsymbol{p} - e\boldsymbol{p}\cdot\boldsymbol{A} + e^2\boldsymbol{A}^2) + U(\boldsymbol{r}). \end{aligned} \tag{7.3}$$

In the case of normal light intensity, one can ignore the term proportional to $\boldsymbol{A}^2$ [6] since it is smaller than the term proportional to $\boldsymbol{A}$ [7, 8]. If the wavefunction of one electron in the crystal is Ψ and a Coulomb gauge

$$\boldsymbol{\nabla}\cdot\boldsymbol{A} = 0, \tag{7.4}$$

is adopted, by replacing the operator $\boldsymbol{p}$ by $-\mathrm{i}\hbar\nabla$, we have[1]

$$\begin{aligned} H\Psi &= \frac{1}{2m_e}[\boldsymbol{p}^2\Psi + \mathrm{i}\hbar e\boldsymbol{A}\cdot\nabla\Psi + \mathrm{i}\hbar e(\nabla\cdot\boldsymbol{A}\Psi)] + U(\boldsymbol{r})\Psi \\ &= \frac{1}{2m_e}(\boldsymbol{p}^2\Psi + \mathrm{i}\hbar e\boldsymbol{A}\cdot\nabla\Psi + \mathrm{i}\hbar e\Psi\nabla\cdot\boldsymbol{A} + \mathrm{i}\hbar e\boldsymbol{A}\cdot\nabla\Psi) + U(\boldsymbol{r})\Psi \\ &= \frac{1}{2m_e}(\boldsymbol{p}^2\Psi + 2\mathrm{i}\hbar e\boldsymbol{A}\cdot\nabla\Psi) + U(\boldsymbol{r})\Psi \\ &= \left[\frac{\boldsymbol{p}^2}{2m_e} + \mathrm{i}\frac{\hbar e}{m_e}\boldsymbol{A}\cdot\nabla + U(\boldsymbol{r})\right]\Psi. \end{aligned} \tag{7.5}$$

Based on the above, the problem of irradiating electromagnetic waves to electrons in a crystal can be treated as assigning a perturbation Hamiltonian $H_{\mathrm{opt}} = \mathrm{i}\frac{\hbar e}{m_e}\boldsymbol{A}\cdot\nabla$ to the electrons, i.e.,

$$H = H_0 + H_{\mathrm{opt}} = H_0 + \mathrm{i}\frac{\hbar e}{m_e}\boldsymbol{A}\cdot\nabla. \tag{7.6}$$

Next, we express the vector potential $\boldsymbol{A}$ of electromagnetic waves using the electric field $\boldsymbol{E}(\boldsymbol{r}, t)$ given as a solution to Maxwell's equations. For a monochromatic wave,

$$\boldsymbol{E}(\boldsymbol{r}, t) = \mathrm{Re}[\boldsymbol{E}_0\mathrm{e}^{\mathrm{i}(\boldsymbol{k}_0\cdot\boldsymbol{r}-\omega t)}] = \frac{1}{2}\left[\boldsymbol{E}_0\mathrm{e}^{\mathrm{i}(\boldsymbol{k}_0\cdot\boldsymbol{r}-\omega t)} + \boldsymbol{E}_0^*\mathrm{e}^{-\mathrm{i}(\boldsymbol{k}_0\cdot\boldsymbol{r}-\omega t)}\right], \tag{7.7}$$

where $\boldsymbol{k}_0$ and ω are respectively the wavevector and angular frequency of the incident electromagnetic wave. We constrain $\omega > 0$ in our following discussions. In vacuum, where current density vanishes, Ampère's law is given by [9]

$$\nabla\times\boldsymbol{B} = \frac{1}{c^2}\dot{\boldsymbol{E}}, \tag{7.8}$$

where c is the speed of light. By substituting equation (7.7) into the right-hand side of equation (7.8), we have

$$\frac{1}{c^2}\dot{\boldsymbol{E}} = -\mathrm{i}\frac{\omega}{2c^2}\left[\boldsymbol{E}_0\mathrm{e}^{\mathrm{i}(\boldsymbol{k}_0\cdot\boldsymbol{r}-\omega t)} - \boldsymbol{E}_0^*\mathrm{e}^{-\mathrm{i}(\boldsymbol{k}_0\cdot\boldsymbol{r}-\omega t)}\right]. \tag{7.9}$$

Remember the basic formulae of vector analysis and Coulomb gauge in equation (7.4), $\nabla\cdot\boldsymbol{A} = 0$, the left-hand side of Ampère's law is

$$\nabla\times\boldsymbol{B} = \nabla\times(\nabla\times\boldsymbol{A}) = \nabla(\nabla\cdot\boldsymbol{A}) - \nabla^2\boldsymbol{A} = -\nabla^2\boldsymbol{A} = k_0^2\boldsymbol{A}. \tag{7.10}$$

Here we use $\boldsymbol{p} = -\mathrm{i}\hbar\nabla = \hbar\boldsymbol{k}_0$. Finally, from equations (7.8–7.10) and noticing $\omega = c|\boldsymbol{k}_0| = ck_0$, we get [10]

$$\boldsymbol{A} = -\frac{\mathrm{i}}{2\omega}\left[\boldsymbol{E}_0\mathrm{e}^{\mathrm{i}(\boldsymbol{k}_0\cdot\boldsymbol{r}-\omega t)} + \boldsymbol{E}_0^*\mathrm{e}^{-\mathrm{i}(\boldsymbol{k}_0\cdot\boldsymbol{r}-\omega t)}\right]. \tag{7.11}$$

[1] We use the vector identity $\nabla\cdot(f\boldsymbol{A}) = f(\nabla\cdot\boldsymbol{A}) + \boldsymbol{A}\cdot(\nabla f)$.

For circularly polarized light, which we will discuss in this book, the intensity passing through a unit area per unit time doubles compared with linear polarization [11–15]. The intensity of circularly polarized light reads [11]

$$I = \sqrt{\frac{\varepsilon_0}{\mu_0}}\ |\boldsymbol{E}_0|^2 = c\varepsilon_0\ |\boldsymbol{E}_0|^2, \tag{7.12}$$

in units of W m^{-2}. Here we used $c^2 = 1/(\varepsilon_0\mu_0)$, where ε_0 and μ_0 are the permittivity and permeability of free space, respectively. Equation (7.12) can be rewritten as

$$|\boldsymbol{E}_0| = \sqrt{\frac{I}{c\varepsilon_0}}. \tag{7.13}$$

Thus, we can further rewrite equation (7.11) as

$$\begin{aligned} \boldsymbol{A} &= -\frac{\mathrm{i}}{2\omega}|\boldsymbol{E}_0|[\boldsymbol{J}\mathrm{e}^{\mathrm{i}(\boldsymbol{k}_0\cdot\boldsymbol{r}-\omega t)} + \boldsymbol{J}^*\mathrm{e}^{-\mathrm{i}(\boldsymbol{k}_0\cdot\boldsymbol{r}-\omega t)}] \\ &= -\frac{\mathrm{i}}{2\omega}\sqrt{\frac{I}{c\varepsilon_0}}[\boldsymbol{J}\mathrm{e}^{\mathrm{i}(\boldsymbol{k}_0\cdot\boldsymbol{r}-\omega t)} + \boldsymbol{J}^*\mathrm{e}^{-\mathrm{i}(\boldsymbol{k}_0\cdot\boldsymbol{r}-\omega t)}], \end{aligned} \tag{7.14}$$

by defining the Jones vector [16]

$$\boldsymbol{J} = \frac{\boldsymbol{E}_0}{|\boldsymbol{E}_0|}, \tag{7.15}$$

which is a unit vector representing the polarization state of the complex electric field. We will explain in the next section that $\boldsymbol{J}$ is used to describe the polarization state. From the above discussions, we can incorporate the electron–photon interaction and find that

$$H_{\text{opt}} = \frac{\hbar e}{2m_e\omega}\sqrt{\frac{I}{c\varepsilon_0}}[\boldsymbol{J}\mathrm{e}^{\mathrm{i}(\boldsymbol{k}_0\cdot\boldsymbol{r}-\omega t)} + \boldsymbol{J}^*\mathrm{e}^{-\mathrm{i}(\boldsymbol{k}_0\cdot\boldsymbol{r}-\omega t)}]\cdot\boldsymbol{\nabla}. \tag{7.16}$$

To avoid confusion, if we define

$$V = \frac{\hbar e}{2m_e\omega}\sqrt{\frac{I}{c\varepsilon_0}}\mathrm{e}^{\mathrm{i}\boldsymbol{k}_0\cdot\boldsymbol{r}}\boldsymbol{J}\cdot\boldsymbol{\nabla}, \tag{7.17}$$

we can write[2]

$$H_{\text{opt}} = V\mathrm{e}^{-\mathrm{i}\omega t} + V^\dagger\mathrm{e}^{\mathrm{i}\omega t}, \tag{7.18}$$

in the form of time-harmonic perturbation [17, 18]. By using Fermi's golden rule [19–21], which is a corollary of time-dependent perturbation theory, the transition rate or the probability of one electron from the initial state $\Psi_\mathrm{i}(\boldsymbol{k}_\mathrm{i}, \boldsymbol{r})$ (with energy $\mathcal{E}_\mathrm{i}$

[2] Note that since $\langle\Phi|\boldsymbol{\nabla}|\Psi\rangle^\dagger = -\langle\Psi|\boldsymbol{\nabla}|\Phi\rangle$, we formally write $\boldsymbol{\nabla}^\dagger = -\boldsymbol{\nabla}$.

and wavevector $\boldsymbol{k}_\mathrm{i}$) to the final state $\Psi_\mathrm{f}(\boldsymbol{k}_\mathrm{f}, \boldsymbol{r})$ (with energy $\mathcal{E}_\mathrm{f}$ and wavevector $\boldsymbol{k}_\mathrm{f}$) per unit time is [7]

$$R_{\mathrm{i}\to\mathrm{f}} = \frac{2\pi}{\hbar}\,|\langle\Psi_\mathrm{f}(\boldsymbol{k}_\mathrm{f}, \boldsymbol{r})|\,V\,|\Psi_\mathrm{i}(\boldsymbol{k}_\mathrm{i}, \boldsymbol{r})\rangle|^2\delta(\mathcal{E}_\mathrm{f} - \mathcal{E}_\mathrm{i} - \hbar\omega) + \frac{2\pi}{\hbar}\,|\langle\Psi_\mathrm{f}(\boldsymbol{k}_\mathrm{f}, \boldsymbol{r})|\,V^\dagger\,|\Psi_\mathrm{i}(\boldsymbol{k}_\mathrm{i}, \boldsymbol{r})\rangle|^2\delta(\mathcal{E}_\mathrm{f} - \mathcal{E}_\mathrm{i} + \hbar\omega), \tag{7.19}$$

where the delta functions are called the energy conserving delta functions—it requires that the quantum of energy causing the transition (the photon) match the energy difference between the two states [17, 22].

Obviously, in the absorption case, the electron in the valence band absorbs the photon energy and then gets excited into the conduction band. We can disregard the second term because $\mathcal{E}_\mathrm{f} - \mathcal{E}_\mathrm{i} + \hbar\omega > 0$. Comparing equations (7.11) and (7.18) one can see that the term $-\frac{\mathrm{i}}{2\omega}\boldsymbol{E}_0\mathrm{e}^{\mathrm{i}(\boldsymbol{k}_0\cdot\boldsymbol{r}-\omega t)}$ in equation (7.11) is responsible for the process [4]. Further, if we define the transition matrix element $M_{\mathrm{v}\to\mathrm{c}}(\boldsymbol{k}_\mathrm{c}, \boldsymbol{k}_\mathrm{v})$ of the perturbation operator V with the initial state in the valence band $\Psi_\mathrm{v}(\boldsymbol{k}_\mathrm{v}, \boldsymbol{r})$ and final state in the conduction band $\Psi_\mathrm{c}(\boldsymbol{k}_\mathrm{c}, \boldsymbol{r})$ as

$$M_{\mathrm{v}\to\mathrm{c}}(\boldsymbol{k}_\mathrm{c}, \boldsymbol{k}_\mathrm{v}) = \langle\Psi_\mathrm{c}(\boldsymbol{k}_\mathrm{c}, \boldsymbol{r})|\,V\,|\Psi_\mathrm{v}(\boldsymbol{k}_\mathrm{v}, \boldsymbol{r})\rangle, \tag{7.20}$$

we can rewrite the transition rate as

$$R_{\mathrm{v}\to\mathrm{c}} = \frac{2\pi}{\hbar}\,|\langle\Psi_\mathrm{f}(\boldsymbol{k}_\mathrm{f}, \boldsymbol{r})|\,V\,|\Psi_\mathrm{i}(\boldsymbol{k}_\mathrm{i}, \boldsymbol{r})\rangle|^2\delta(\mathcal{E}_\mathrm{f} - \mathcal{E}_\mathrm{i} - \hbar\omega) \equiv \frac{2\pi}{\hbar}\,|M_{\mathrm{v}\to\mathrm{c}}(\boldsymbol{k}_\mathrm{f}, \boldsymbol{k}_\mathrm{i})|^2\delta(\mathcal{E}_\mathrm{c} - \mathcal{E}_\mathrm{v} - \hbar\omega). \tag{7.21}$$

Using dipole vector

$$\boldsymbol{D}_{\mathrm{v}\to\mathrm{c}}(\boldsymbol{k}_\mathrm{c}, \boldsymbol{k}_\mathrm{v}) \equiv \langle\Psi_\mathrm{c}(\boldsymbol{k}_\mathrm{c}, \boldsymbol{r})|\boldsymbol{\nabla}|\Psi_\mathrm{v}(\boldsymbol{k}_\mathrm{v}, \boldsymbol{r})\rangle, \tag{7.22}$$

we have

$$M_{\mathrm{v}\to\mathrm{c}}(\boldsymbol{k}_\mathrm{c}, \boldsymbol{k}_\mathrm{v}) = \langle\Psi_\mathrm{c}(\boldsymbol{k}_\mathrm{c}, \boldsymbol{r})|\,V\,|\Psi_\mathrm{v}(\boldsymbol{k}_\mathrm{v}, \boldsymbol{r})\rangle = \frac{\hbar e}{2m_\mathrm{e}\omega}\sqrt{\frac{I}{c\varepsilon_0}}\,\mathrm{e}^{\mathrm{i}\boldsymbol{k}_0\cdot\boldsymbol{r}}\boldsymbol{J}\cdot\boldsymbol{D}_{\mathrm{v}\to\mathrm{c}}(\boldsymbol{k}_\mathrm{c}, \boldsymbol{k}_\mathrm{v}). \tag{7.23}$$

Since the wavelength of the visible electromagnetic wave is of the order of 500 nm, which is sufficiently larger than the size of the crystal (~0.1 nm), we can take the spatially oscillating part of the electric field out of the integral using the dipole approximation [23]. In this approximation, the electric field perceives the crystal as an electric dipole. In this case, the optical wavevector $\boldsymbol{k}_0$ is much smaller than the size of the BZ, the momentum conservation

$$\hbar\boldsymbol{k}_\mathrm{i} + \hbar\boldsymbol{k}_0 = \hbar\boldsymbol{k}_\mathrm{f}, \tag{7.24}$$

transforms to $\boldsymbol{k}_\mathrm{i} \simeq \boldsymbol{k}_\mathrm{f}$, and we refer to the transition as vertical or direct in reciprocal space [3, 7]. In the case when the wavelength of the electromagnetic field is comparable to the size of the crystal, dipole approximation is not justified.

Conversely, for the emission process, we can ignore the first term in equation (7.19) because $\mathcal{E}_f - \mathcal{E}_i - \hbar\omega < 0$. Compared with the absorption process, the other term in equation (7.11), $-\frac{i}{2\omega}\boldsymbol{E}_0^* e^{-i(\boldsymbol{k}_0 \cdot \boldsymbol{r} - \omega t)}$, is responsible for this process. The probability of relaxation from the high-energy initial state $\Psi_c(\boldsymbol{k}_c, \boldsymbol{r})$ to the low-energy final state $\Psi_v(\boldsymbol{k}_v, \boldsymbol{r})$ is

$$\begin{aligned} R_{c\to v} &= \frac{2\pi}{\hbar} |\langle \Psi_f(\boldsymbol{k}_f, \boldsymbol{r})| V^\dagger |\Psi_i(\boldsymbol{k}_i, \boldsymbol{r})\rangle|^2 \delta(\mathcal{E}_f - \mathcal{E}_i + \hbar\omega) \\ &\equiv \frac{2\pi}{\hbar} |M_{c\to v}(\boldsymbol{k}_f, \boldsymbol{k}_i)|^2 \delta(\mathcal{E}_v - \mathcal{E}_c + \hbar\omega). \end{aligned} \tag{7.25}$$

Again, using dipole approximation, the transition matrix element is

$$\begin{aligned} M_{c\to v}(\boldsymbol{k}_v, \boldsymbol{k}_c) &= \langle \Psi_i(\boldsymbol{k}_i, \boldsymbol{r})| V^\dagger |\Psi_f(\boldsymbol{k}_f, \boldsymbol{r})\rangle \\ &= -\frac{\hbar e}{2m_e\omega}\sqrt{\frac{I}{c\varepsilon_0}} e^{-i\boldsymbol{k}_0 \cdot \boldsymbol{r}} \boldsymbol{J}^* \cdot \boldsymbol{D}_{c\to v}(\boldsymbol{k}_v, \boldsymbol{k}_c) \\ &= \frac{\hbar e}{2m_e\omega}\sqrt{\frac{I}{c\varepsilon_0}} e^{-i\boldsymbol{k}_0 \cdot \boldsymbol{r}} \boldsymbol{J}^* \cdot \boldsymbol{D}_{v\to c}(\boldsymbol{k}_c, \boldsymbol{k}_v)^*. \end{aligned} \tag{7.26}$$

From the above, the results indicate that the transition probability can be determined by finding the dipole vector.

It is noted that the delta functions neglect all mechanisms of spectral broadening due to different scattering processes. A practical way to include the spectral broadening is to replace the delta function with a Lorentzian function [7, 24–26]

$$\delta(\mathcal{E}_c - \mathcal{E}_v - \hbar\omega) \to \mathcal{L}(\mathcal{E}_c - \mathcal{E}_v - \hbar\omega) = \frac{\Gamma_b/\pi}{(\mathcal{E}_c - \mathcal{E}_v - \hbar\omega)^2 + \Gamma_b^2}, \tag{7.27}$$

where Γ_b is a phenomenological factor taking account of the spectral broadening.

7.2 Jones vector

In the early 1940s, Robert Clark Jones developed a concise matrix calculus [27–30] for treating complicated polarization problems, commonly called the Jones matrix calculus. The Jones calculus involves complex quantities contained in 2×1 column matrices (the Jones vector) and 2×2 matrices (the Jones matrices) [31]. The Jones vector is defined by the amplitudes and relative phase shift of the x and y components of the electrical field of fully polarized light [32, 33], but cannot describe unpolarized or partially polarized light [11, 34]. The Jones calculus is powerful for problems related to the coherent superposition of two or more electromagnetic waves because it preserves the phase information [35].

In this section, we will develop and explain the Jones vector [35], which is previously defined in equation (7.15). We define each component of the wavenumber vector as $\boldsymbol{k}_0 = (k_{0x}, k_{0y}, k_{0z})$. When considering an electromagnetic wave travelling in the z-axis, we can write the electric field component of the electromagnetic wave as

$$\boldsymbol{E} = \frac{1}{2}\left[\boldsymbol{E}_0 e^{i(k_{0z}z-\omega t)} + \boldsymbol{E}_0^* e^{-i(k_{0z}z-\omega t)}\right]$$
$$= \frac{1}{2}\begin{pmatrix} E_{0x}e^{i(k_{0z}z-\omega t)} + E_{0x}^* e^{-i(k_{0z}z-\omega t)} \\ E_{0y}e^{i(k_{0z}z-\omega t+\varphi)} + E_{0y}^* e^{-i(k_{0z}z-\omega t+\varphi)} \\ 0 \end{pmatrix}. \tag{7.28}$$

Here, we assume that the x and y components of the electric field are out of phase by φ $(-\pi < \varphi \leqslant \pi)$. The absolute phases of the x and y components are irrelevant, since they can both be multiplied by a complex number of unit modulus, leaving observable quantities of polarization unchanged [35, 36]. By comparing the first and second lines, we have

$$\boldsymbol{E}_0 = \begin{pmatrix} E_{0x} \\ E_{0y}e^{i\varphi} \\ 0 \end{pmatrix} \equiv |\boldsymbol{E}_0|\boldsymbol{J}. \tag{7.29}$$

Therefore, the Jones vector of an electromagnetic wave travelling in the z-axis, in which the x and y components of the electric field are shifted by φ, is

$$\boldsymbol{J} = \frac{\boldsymbol{E}_0}{|\boldsymbol{E}_0|} = \frac{\boldsymbol{E}_0}{\sqrt{\boldsymbol{E}_0^\dagger \boldsymbol{E}_0}} = \frac{1}{\sqrt{E_{0x}^2 + E_{0y}^2}}\begin{pmatrix} E_{0x} \\ E_{0y}e^{i\varphi} \\ 0 \end{pmatrix}. \tag{7.30}$$

The factor $1/\sqrt{E_{0x}^2 + E_{0y}^2}$ is the normalization constant. By defining an angle $\beta = \arctan\frac{E_{0x}}{E_{0y}}$ $(0 \leqslant \beta \leqslant \pi/2)$, we can also represent the normalized Jones vector as [35]

$$\boldsymbol{J} = \begin{pmatrix} \cos\beta \\ \sin\beta e^{i\varphi} \\ 0 \end{pmatrix}. \tag{7.31}$$

Notice that the information on light intensity/amplitudes is skipped by using normalized Jones vectors [37]. The intensity is unrelated to the state of polarization. $\boldsymbol{J}$ is not a vector in the real physical space, instead, it is a vector in an abstract mathematical space [38, 39].

The Jones vector with $\varphi = \pm\pi/2$ represents elliptically or circularly polarized light. Specifically, for the light in which the x and y components of the electric field have equal amplitudes ($E_{0x} = E_{0y}$) and a phase difference of $\varphi = \pi/2$ $(-\pi/2)$, we call it LCP (RCP) light. i.e., for the LCP (RCP), the y-component lags (leads) with respect to the x-component [40]. From the definition, we can find the Jones vector [38, 41]

$$\boldsymbol{J}_\pm = \frac{1}{\sqrt{2}}\begin{pmatrix} 1 \\ e^{i\sigma_\pm\frac{\pi}{2}} \\ 0 \end{pmatrix} = \frac{1}{\sqrt{2}}\begin{pmatrix} 1 \\ i\sigma_\pm \\ 0 \end{pmatrix}, \tag{7.32}$$

where $\sigma_+ = +1$ ($\sigma_- = -1$) denotes LCP (RCP). Here we adopt the convention frequently used in optics. From the point of view of an observer looking at the light head-on (facing the light [42]), left- or right-handedness is determined by pointing one's left or right thumb against the direction of propagation, and then matching the curling of one's fingers to the spatial rotation of the field. For LCP (RCP), the observer will see a counter-clockwise (clockwise) rotation of the electric field vector [40, 43]. In such a convention, the defined handedness of the electromagnetic wave matches the handedness of the screw-type nature of the field in space.

The Jones vectors $\boldsymbol{J}_+$ for LCP and $\boldsymbol{J}_-$ for RCP form a 2D complex orthonormal basis. In other words, they satisfy[3] [32, 35, 44]

$$\boldsymbol{J}_+ \cdot \boldsymbol{J}_- = \boldsymbol{J}_+^\dagger \boldsymbol{J}_- = 0, \tag{7.33}$$

$$|\boldsymbol{J}_+|^2 = |\boldsymbol{J}_-|^2 = 1, \tag{7.34}$$

or compactly [34]

$$\boldsymbol{J}_i \cdot \boldsymbol{J}_j = \delta_{ij}, \tag{7.35}$$

as a single equation.

Further, we can superpose coherent amplitudes in physical reality based on mathematical transformations [42]. The Jones vector of any polarized electromagnetic wave travelling in the z-axis can be expressed as a linear combination of $\boldsymbol{J}_+$ and $\boldsymbol{J}_-$. For example, in definition, the Jones vector $\boldsymbol{J}_\alpha$ of linearly polarized light, whose plane of polarization makes an angle α_p (inclination angle [40]) with the x-axis, is given by [37]

$$\boldsymbol{J}_\alpha = \begin{pmatrix} \cos\alpha_\mathrm{p} \\ \sin\alpha_\mathrm{p} \\ 0 \end{pmatrix}, \tag{7.36}$$

using the definition, but it can also be rewritten as

$$\boldsymbol{J}_\alpha = \frac{1}{\sqrt{2}}(\mathrm{e}^{-\mathrm{i}\alpha_\mathrm{p}}\boldsymbol{J}_+ + \mathrm{e}^{\mathrm{i}\alpha_\mathrm{p}}\boldsymbol{J}_-). \tag{7.37}$$

In particular, 0° linearly (linearly horizontally) polarized light is

$$\boldsymbol{J}_0 = \begin{pmatrix} 1 \\ 0 \\ 0 \end{pmatrix} = \frac{1}{\sqrt{2}}(\boldsymbol{J}_+ + \boldsymbol{J}_-). \tag{7.38}$$

Therefore, 0° linearly polarized light can be obtained by superimposing LCP and RCP of equal amplitudes [34, 42]. If the two opposite circularly polarized lights have unequal amplitudes, the result is an elliptically polarized light [11, 31].

[3] For vectors with complex entries, $\boldsymbol{A} \cdot \boldsymbol{B} = \sum_i A_i B_i^*$.

It should be noted that the transition probability when irradiated with linearly polarized light is not the sum of the transition probability when irradiated with LCP and the transition probability when irradiated with RCP. In general,

$$|\boldsymbol{J}_0 \cdot \boldsymbol{D}|^2 = |\boldsymbol{J}_+ \cdot \boldsymbol{D}|^2 + |\boldsymbol{J}_- \cdot \boldsymbol{D}|^2, \tag{7.39}$$

is false.

Note that, as we saw in the previous section, for relaxation, the initial and final states are interchanged compared with excitation, the dipole vector takes complex conjugation and acquires a minus sign. However, since the Jones vector also takes complex conjugation, when an electron excited by LCP (RCP) relaxes to its original state, it emits light with the same handedness. In the following, for consistency, we will treat only excitation, thus write $\boldsymbol{D}_{\mathrm{v}\to\mathrm{c}}(\boldsymbol{k}_\mathrm{c}, \boldsymbol{k}_\mathrm{v})$ and $\boldsymbol{M}_{\mathrm{v}\to\mathrm{c}}(\boldsymbol{k}_\mathrm{c}, \boldsymbol{k}_\mathrm{v})$ as $\boldsymbol{D}(\boldsymbol{k}_\mathrm{c}, \boldsymbol{k}_\mathrm{v})$ and $\boldsymbol{M}(\boldsymbol{k}_\mathrm{c}, \boldsymbol{k}_\mathrm{v})$, respectively.

7.3 Light absorption in hexagonal lattice system

In this section, we discuss optical absorption in hexagonal lattice systems. Light absorption occurs when visible light is irradiated perpendicularly to an atomic layer and electrons get excited from the valence band to the conduction band. From the discussion in section 7.1, in order to obtain the transition rate, we need to determine the dipole vector and then the transition matrix element.

7.3.1 Dipole vector

The initial state $\Psi_\mathrm{v}(\boldsymbol{k}_\mathrm{v}, \boldsymbol{r})$ is a state in which electrons are in the valence band, and the final state $\Psi_\mathrm{c}(\boldsymbol{k}_\mathrm{c}, \boldsymbol{r})$ is a state in which electrons are in the conduction band. These two states are determined using the TB method explained earlier in section 1.2. We show equation (1.6) again as

$$\begin{aligned}\Psi_\mathrm{v}(\boldsymbol{k}_\mathrm{v}, \boldsymbol{r}) &= c_\mathrm{v}^\mathrm{A}(\boldsymbol{k}_\mathrm{v}) p_z^\mathrm{A}(\boldsymbol{k}_\mathrm{v}, \boldsymbol{r}) + c_\mathrm{v}^\mathrm{B}(\boldsymbol{k}_\mathrm{v}) p_z^\mathrm{B}(\boldsymbol{k}_\mathrm{v}, \boldsymbol{r}),\\ \Psi_\mathrm{c}(\boldsymbol{k}_\mathrm{c}, \boldsymbol{r}) &= c_\mathrm{c}^\mathrm{A}(\boldsymbol{k}_\mathrm{c}) p_z^\mathrm{A}(\boldsymbol{k}_\mathrm{c}, \boldsymbol{r}) + c_\mathrm{c}^\mathrm{B}(\boldsymbol{k}_\mathrm{c}) p_z^\mathrm{B}(\boldsymbol{k}_\mathrm{c}, \boldsymbol{r}).\end{aligned} \tag{7.40}$$

Now we substitute equation (7.40) into the definition of dipole vector in equation (7.22),

$$\begin{aligned}&\boldsymbol{D}(\boldsymbol{k}_\mathrm{c}, \boldsymbol{k}_\mathrm{v})\\ &\equiv \langle \Psi_\mathrm{c}(\boldsymbol{k}_\mathrm{c}, \boldsymbol{r}) | \nabla | \Psi_\mathrm{v}(\boldsymbol{k}_\mathrm{v}, \boldsymbol{r}) \rangle\\ &= \left\langle c_\mathrm{c}^\mathrm{A}(\boldsymbol{k}_\mathrm{c}) p_z^\mathrm{A}(\boldsymbol{k}_\mathrm{c}, \boldsymbol{r}) + c_\mathrm{c}^\mathrm{B}(\boldsymbol{k}_\mathrm{c}) p_z^\mathrm{B}(\boldsymbol{k}_\mathrm{c}, \boldsymbol{r}) \middle| \nabla \middle| c_\mathrm{v}^\mathrm{A}(\boldsymbol{k}_\mathrm{v}) p_z^\mathrm{A}(\boldsymbol{k}_\mathrm{v}, \boldsymbol{r}) + c_\mathrm{v}^\mathrm{B}(\boldsymbol{k}_\mathrm{v}) p_z^\mathrm{B}(\boldsymbol{k}_\mathrm{v}, \boldsymbol{r}) \right\rangle\\ &= c_\mathrm{c}^\mathrm{A}(\boldsymbol{k}_\mathrm{c})^* c_\mathrm{v}^\mathrm{A}(\boldsymbol{k}_\mathrm{v}) \left\langle p_z^\mathrm{A}(\boldsymbol{k}_\mathrm{c}, \boldsymbol{r}) \middle| \nabla \middle| p_z^\mathrm{A}(\boldsymbol{k}_\mathrm{v}, \boldsymbol{r}) \right\rangle + c_\mathrm{c}^\mathrm{B}(\boldsymbol{k}_\mathrm{c})^* c_\mathrm{v}^\mathrm{B}(\boldsymbol{k}_\mathrm{v}) \left\langle p_z^\mathrm{B}(\boldsymbol{k}_\mathrm{c}, \boldsymbol{r}) \middle| \nabla \middle| p_z^\mathrm{B}(\boldsymbol{k}_\mathrm{v}, \boldsymbol{r}) \right\rangle\\ &+ c_\mathrm{c}^\mathrm{A}(\boldsymbol{k}_\mathrm{c})^* c_\mathrm{v}^\mathrm{B}(\boldsymbol{k}_\mathrm{v}) \left\langle p_z^\mathrm{A}(\boldsymbol{k}_\mathrm{c}, \boldsymbol{r}) \middle| \nabla \middle| p_z^\mathrm{B}(\boldsymbol{k}_\mathrm{v}, \boldsymbol{r}) \right\rangle + c_\mathrm{c}^\mathrm{B}(\boldsymbol{k}_\mathrm{c})^* c_\mathrm{v}^\mathrm{A}(\boldsymbol{k}_\mathrm{v}) \left\langle p_z^\mathrm{B}(\boldsymbol{k}_\mathrm{c}, \boldsymbol{r}) \middle| \nabla \middle| p_z^\mathrm{A}(\boldsymbol{k}_\mathrm{v}, \boldsymbol{r}) \right\rangle.\end{aligned} \tag{7.41}$$

In the dipole approximation, since the wavenumber does not change during the transition, we can define $\boldsymbol{k} \equiv \boldsymbol{k}_\mathrm{v} = \boldsymbol{k}_\mathrm{c}$. After some manipulations (see appendix C), in which we restrict ourselves to the interactions between the first NNs only, the dipole vector in equation (7.41) finally becomes a function of $\boldsymbol{k}$. The dipole vector reads

$$\boldsymbol{D}(\boldsymbol{k}) = \frac{\sqrt{3}}{a} m_{\mathrm{o}} \left[c_{\mathrm{c}}^{\mathrm{A}}(\boldsymbol{k})^* c_{\mathrm{v}}^{\mathrm{B}}(\boldsymbol{k}) \sum_{i=1}^{3} \mathrm{e}^{\mathrm{i}\boldsymbol{k}\cdot\boldsymbol{R}_i} \boldsymbol{R}_i - c_{\mathrm{c}}^{\mathrm{B}}(\boldsymbol{k})^* c_{\mathrm{v}}^{\mathrm{A}}(\boldsymbol{k}) \sum_{i=1}^{3} \mathrm{e}^{-\mathrm{i}\boldsymbol{k}\cdot\boldsymbol{R}_i} \boldsymbol{R}_i \right], \tag{7.42}$$

where $\boldsymbol{R}_i$ is defined in equation (1.2), and

$$m_{\mathrm{o}} = \left\langle p_z(\boldsymbol{r} - \boldsymbol{R}_1) \middle| \partial_x \middle| p_z(\boldsymbol{r}) \right\rangle, \tag{7.43}$$

represents the x-component of the atomic dipole vector $\langle p_z(\boldsymbol{r} - \boldsymbol{R}_1)|\boldsymbol{\nabla}|p_z(\boldsymbol{r})\rangle$ between the NNs [23].

7.3.2 System with inversion symmetry (IS)

For pristine graphene where IS is present, the wavefunction coefficients c_ξ^{A} and c_ξ^{B} are given in equation (1.17). Therefore, we can further write the dipole vector as

$$\begin{aligned}
\boldsymbol{D}(\boldsymbol{k}) &= -\frac{\sqrt{3}}{2a} m_{\mathrm{o}} \left[\frac{f(\boldsymbol{k})^*}{w(\boldsymbol{k})} \sum_{i=1}^{3} \mathrm{e}^{\mathrm{i}\boldsymbol{k}\cdot\boldsymbol{R}_i} \boldsymbol{R}_i + \frac{f(\boldsymbol{k})}{w(\boldsymbol{k})} \sum_{i=1}^{3} \mathrm{e}^{-\mathrm{i}\boldsymbol{k}\cdot\boldsymbol{R}_i} \boldsymbol{R}_i \right] \\
&= -\frac{\sqrt{3}}{a w(\boldsymbol{k})} m_{\mathrm{o}} \mathrm{Re} \left[f(\boldsymbol{k}) \sum_{i=1}^{3} \mathrm{e}^{-\mathrm{i}\boldsymbol{k}\cdot\boldsymbol{R}_i} \boldsymbol{R}_i \right] \\
&= -\frac{m_{\mathrm{o}}}{w(\boldsymbol{k})} \begin{pmatrix} \cos\frac{\sqrt{3}k_x a}{2} \cos\frac{k_y a}{2} - \cos k_y a \\ \sqrt{3} \sin\frac{\sqrt{3}k_x a}{2} \sin\frac{k_y a}{2} \\ 0 \end{pmatrix},
\end{aligned} \tag{7.44}$$

which is a real vector. When we calculate the inner product of $\boldsymbol{D}(\boldsymbol{k})$ and the Jones vector $\boldsymbol{J}_\pm$ of circularly polarized light, we get the transition matrix element

$$\begin{aligned}
|M_\pm(\boldsymbol{k})|^2 &\propto |\boldsymbol{J}_\pm \cdot \boldsymbol{D}(\boldsymbol{k})|^2 \\
&= \frac{m_{\mathrm{o}}^2}{2w(\boldsymbol{k})^2} \left[\left(\cos\frac{\sqrt{3}k_x a}{2} \cos\frac{k_y a}{2} - \cos k_y a \right)^2 + 3\sin^2\frac{\sqrt{3}k_x a}{2} \sin^2\frac{k_y a}{2} \right],
\end{aligned} \tag{7.45}$$

which does not depend on the handedness of light ($\sigma_\pm$). Here, we used $\sigma_\pm^2 = 1$. The transition probability remains unchanged regardless of whether the system is irradiated with RCP ($\sigma_- = -1$) or LCP ($\sigma_+ = +1$) light. In addition, equation (7.45) is an even function and results in a valley-independent absorption.

Now we explore in what situation the light absorption for LCP and RCP equals/differs, i.e., $|M_\pm|^2$ is independent of/dependent on $\sigma_\pm$. First, in general, we express $\boldsymbol{D}$ as

$$\boldsymbol{D} = \begin{pmatrix} D_x \\ D_y \\ 0 \end{pmatrix} = \begin{pmatrix} \mathrm{Re}(D_x) + \mathrm{i}\mathrm{Im}(D_x) \\ \mathrm{Re}(D_y) + \mathrm{i}\mathrm{Im}(D_y) \\ 0 \end{pmatrix}. \tag{7.46}$$

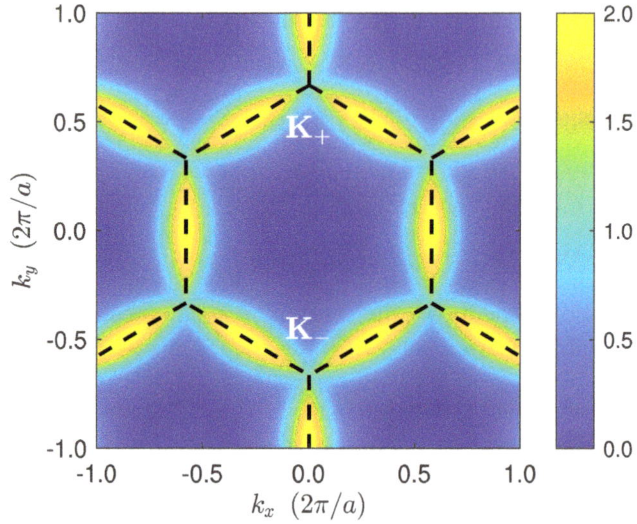

Figure 7.1. Transition matrix element $|M_{\pm}(\boldsymbol{k})|^2$ in unit of m_o^2 for a LCP (RCP) light irradiated onto a hexagonal lattice system with spatial IS. It is maximum at the M points and becomes 0 at the Γ point.

Then,

$$|M_{\pm}|^2 \propto |\boldsymbol{J}_{\pm} \cdot \boldsymbol{D}|^2 = \frac{|D_x|^2 + |D_y|^2}{2} + \sigma_{\pm}[\mathrm{Re}(D_x)\mathrm{Im}(D_y) - \mathrm{Re}(D_y)\mathrm{Im}(D_x)], \qquad (7.47)$$

when we consider $\sigma_{\pm}^2 = 1$. It reveals that light absorption for LCP and RCP light equals, i.e., $|M_{\pm}|^2$ is independent of $\sigma_{\pm}$, when

$$\mathrm{Re}(D_x)\mathrm{Im}(D_y) - \mathrm{Re}(D_y)\mathrm{Im}(D_x) = 0. \qquad (7.48)$$

In this case, the dipole vector $\boldsymbol{D}$ falls into the following three categories:

1. a purely real vector;
2. a purely imaginary vector;
3. the real and imaginary parts of the dipole vector $\boldsymbol{D}$ are parallel or antiparallel.

If the real and imaginary parts of the dipole vector are neither parallel nor antiparallel, equation (7.48) fails, then the light absorption for LCP and RCP light differs. In figure 7.1, we plot the transition matrix element $|M_{\pm}(\boldsymbol{k})|^2$ for LCP (RCP) light as a function of $\boldsymbol{k}$. The $|M_{\pm}(\boldsymbol{k})|^2$ reaches local maxima at the M points ($2m_o^2$). The results show that the plot has six-fold symmetry [45], which also implies that VP does not occur when two atoms in the unit cell are the same. However, breaking the IS of the system [46] results in the emergence of VP, and we will discuss this in the next subsection.

7.3.3 System with broken IS

Next, we will discuss a hexagonal lattice system in which the on-site energies of A and B atoms differ by $\Delta_z > 0$. We will employ the Hamiltonian presented in equation (2.39) to describe this system. As discussed in section 2.4, this system is useful for describing hBN. In contrast to the discussion in the previous section, the energy bandgap Δ_z opens at $\boldsymbol{K}_\pm$ points due to the breaking of the spatial IS, resulting in VP. The energy spectrum reads

$$\mathcal{E}_\pm(\boldsymbol{k}) = \pm\sqrt{\gamma_0^2\,|f(\boldsymbol{k})|^2 + \left(\frac{\Delta_z}{2}\right)^2}, \tag{7.49}$$

with its corresponding eigenstates

$$\begin{aligned} |\psi_+(\boldsymbol{k})\rangle &= \begin{pmatrix} c_c^A(\boldsymbol{k}) \\ c_c^B(\boldsymbol{k}) \end{pmatrix} = \frac{1}{\sqrt{2|\mathcal{E}_\pm(\boldsymbol{k})|\left[|\mathcal{E}_\pm(\boldsymbol{k})| + \frac{\Delta_z}{2}\right]}} \begin{pmatrix} |\mathcal{E}_\pm(\boldsymbol{k})| + \frac{\Delta_z}{2} \\ -\gamma_0 f(\boldsymbol{k})^* \end{pmatrix}, \\ |\psi_-(\boldsymbol{k})\rangle &= \begin{pmatrix} c_v^A(\boldsymbol{k}) \\ c_v^B(\boldsymbol{k}) \end{pmatrix} = \frac{1}{\sqrt{2|\mathcal{E}_\pm(\boldsymbol{k})|\left[|\mathcal{E}_\pm(\boldsymbol{k})| + \frac{\Delta_z}{2}\right]}} \begin{pmatrix} \gamma_0 f(\boldsymbol{k}) \\ |\mathcal{E}_\pm(\boldsymbol{k})| + \frac{\Delta_z}{2} \end{pmatrix}. \end{aligned} \tag{7.50}$$

From equation (7.42), the dipole vector of this system can be further written as

$$\boldsymbol{D}(\boldsymbol{k}) = \frac{\sqrt{3}\,m_o}{2a|\mathcal{E}_\pm(\boldsymbol{k})|\left[|\mathcal{E}_\pm(\boldsymbol{k})| + \frac{\Delta_z}{2}\right]}\left\{\left[|\mathcal{E}_\pm(\boldsymbol{k})| + \frac{\Delta_z}{2}\right]^2 \sum_{i=1}^{3} e^{i\boldsymbol{k}\cdot\boldsymbol{R}_i}\boldsymbol{R}_i + \gamma_0^2 f(\boldsymbol{k})^2 \sum_{i=1}^{3} e^{-i\boldsymbol{k}\cdot\boldsymbol{R}_i}\boldsymbol{R}_i\right\}. \tag{7.51}$$

Unlike equation (7.44), which describes the system with IS, it is impossible to wrap the contents in braces with Re.

At the $\boldsymbol{\Gamma}$ point, from

$$\sum_{i=1}^{3} e^{i\boldsymbol{k}\cdot\boldsymbol{R}_i}\boldsymbol{R}_i = \sum_{i=1}^{3}\boldsymbol{R}_i = 0, \tag{7.52}$$

$\boldsymbol{D}(\boldsymbol{\Gamma}) = 0$. This means that light absorption does not occur at the $\boldsymbol{\Gamma}$ point.

Next, at the $\boldsymbol{M}$ point, using $f(\boldsymbol{M}) = e^{-i\frac{\pi}{3}}$, $|\mathcal{E}_\pm(\boldsymbol{M})| = \sqrt{\gamma_0^2 + \left(\frac{\Delta_z}{2}\right)^2}$, and

$$\sum_{i=1}^{3} e^{i\boldsymbol{M}\cdot\boldsymbol{R}_i}\boldsymbol{R}_i = a\begin{pmatrix} -\frac{1}{\sqrt{3}} + i \\ 0 \\ 0 \end{pmatrix}, \tag{7.53}$$

we obtain

$$\boldsymbol{D}(\boldsymbol{M}) = m_o\begin{pmatrix} -1 + i\sqrt{3} \\ 0 \\ 0 \end{pmatrix}. \tag{7.54}$$

The real part and imaginary part of the dipole vector are antiparallel to each other at the $\boldsymbol{M}$ point. Thus, the optical transition at the $\boldsymbol{M}$ point does not depend on the polarization of circularly polarized light, $\sigma_{\pm}$. This is verified by further calculation on $|M_{\pm}(\boldsymbol{M})|^2 = 2m_o^2$.

Finally, let us focus on the $\boldsymbol{K}_{\pm}$ point. Using $f(\boldsymbol{K}_{\pm}) = 0$, $|\mathcal{E}_{\pm}(\boldsymbol{K}_{\pm})| = \frac{\Delta_z}{2}$, and

$$\sum_{i=1}^{3} \mathrm{e}^{\mathrm{i}\boldsymbol{K}_{\kappa}\cdot\boldsymbol{R}_i}\boldsymbol{R}_i = \frac{\sqrt{3}a}{2}\begin{pmatrix}1\\ \mathrm{i}\kappa\\ 0\end{pmatrix}, \tag{7.55}$$

we get

$$\boldsymbol{D}(\boldsymbol{K}_{\kappa}) = \frac{3}{2}m_o\begin{pmatrix}1\\ \mathrm{i}\kappa\\ 0\end{pmatrix}. \tag{7.56}$$

That is, if $\Delta_z \neq 0$, the real part and the imaginary part of the dipole vector are orthogonal to each other at the exact $\boldsymbol{K}_{\pm}$ point. In other words, the following formula holds,

$$\mathrm{Re}[\boldsymbol{D}(\boldsymbol{K}_{\kappa})]\cdot\mathrm{Im}[\boldsymbol{D}(\boldsymbol{K}_{\kappa})] = 0. \tag{7.57}$$

This is the origin of VP [47]. The square of the absolute value of the transition matrix element is

$$|M_{\pm}(\boldsymbol{K}_{\kappa})|^2 \propto |\boldsymbol{J}_{\pm}\cdot\boldsymbol{D}(\boldsymbol{K}_{\kappa})|^2 = \left|\frac{3}{2\sqrt{2}}m_o(1+\kappa\sigma_{\pm})\right|^2 = \begin{cases}\frac{9}{2}m_o^2 & \text{if } \kappa\sigma_{\pm} = 1\\ 0 & \text{if } \kappa\sigma_{\pm} = -1.\end{cases} \tag{7.58}$$

Therefore, when LCP light ($\sigma_+ = 1$) irradiates, it causes an optical transition to occur at the $\boldsymbol{K}_+$ point ($\kappa = 1$), but fails at the $\boldsymbol{K}_-$ point ($\kappa = -1$). On the other hand, when RCP light ($\sigma_- = -1$) irradiates, it causes an optical transition to occur at the $\boldsymbol{K}_-$ point ($\kappa = -1$), but fails at the $\boldsymbol{K}_+$ point ($\kappa = 1$). VP occurs because there is a preference for the optical transition of valley electrons depending on the polarization state of the circularly polarized light. Therefore, when we fix the energy of LCP (RCP) light to the value of the energy bandgap Δ_z, we always get 100% VP at the $\boldsymbol{K}_+$ ($\boldsymbol{K}_-$) points for any Δ_z.

In addition, note that the energy bandgap Δ_z is not included in the $|M_{\pm}(\boldsymbol{K}_{\kappa})|^2$, since Δ_z is not included in the eigenstates of the conduction and valence bands at the $\boldsymbol{K}_{\kappa}$ point. The eigenstates read

$$\begin{aligned}|\psi_+(\boldsymbol{K}_{\kappa})\rangle &= \begin{pmatrix}c_c^{A}(\boldsymbol{K}_{\kappa})\\ c_c^{B}(\boldsymbol{K}_{\kappa})\end{pmatrix} = \begin{pmatrix}1\\0\end{pmatrix},\\ |\psi_-(\boldsymbol{K}_{\kappa})\rangle &= \begin{pmatrix}c_v^{A}(\boldsymbol{K}_{\kappa})\\ c_v^{B}(\boldsymbol{K}_{\kappa})\end{pmatrix} = \begin{pmatrix}0\\1\end{pmatrix}.\end{aligned} \tag{7.59}$$

Electrons at the CBM (VBM) are localised to A (B) atoms with high (low) on-site energy, regardless of the energy bandgap Δ_z.

Figure 7.2 is a plot of the square of the absolute value of the transition matrix element numerically calculated from the dipole vector shown in equation (7.51), for (a) the LCP and (b) RCP light as a function of $\boldsymbol{k}$ for $\Delta_z = 2$ eV. The optical transition of electrons from $\boldsymbol{K}_+$ and $\boldsymbol{K}_-$ valleys differs when we compare figures (a) and (b). This indicates the emergence of VP, unlike figure 7.1, where the system has IS. Furthermore, we can observe from the figure that the following equation,

$$|M_\pm(\boldsymbol{k})|^2 = |M_\mp(-\boldsymbol{k})|^2, \tag{7.60}$$

holds [45]. Note that the peak of $|M_\pm(\boldsymbol{k})|^2$ is $\frac{9}{2}m_o^2$ at $\boldsymbol{K}_\pm$ point, as given in equation (7.58). In table 7.1, we show a summary of $|M_\pm(\boldsymbol{k})|^2$ at high symmetry points $\boldsymbol{K}_\kappa$, M and Γ.

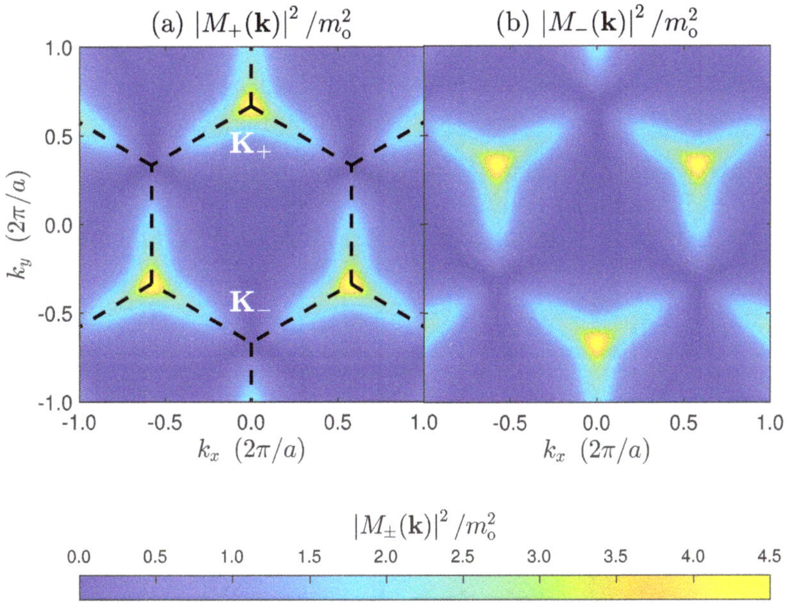

Figure 7.2. The square of the absolute value of the transition matrix element $|M_\pm(\boldsymbol{k})|^2$ when a hexagonal lattice system is irradiated with (a) LCP and (b) RCP light for $\Delta_z = 2$ eV. Electrons in the $\boldsymbol{K}_+$ ($\boldsymbol{K}_-$) valley are easily excited when irradiated with LCP (RCP) light. The peak value is $\frac{9}{2}m_o^2$.

Table 7.1. Square of transition matrix elements $|M_\pm(\boldsymbol{k})|^2$ at high symmetry points.

$\|M_\pm(\boldsymbol{k})\|^2$	$\Delta_z = 0$	$\Delta_z > 0$
$\boldsymbol{K}_\kappa$	0	$\frac{9}{8}m_o^2(1 + \kappa\sigma_\pm)^2$
M	$2m_o^2$	$2m_o^2$
Γ	0	0

7.4 Optical selection rule in TMDCs

7.4.1 Dipole approximation

In the previous section, we have theoretically demonstrated that 2D systems with broken IS absorb/emit contrasting circularly polarised light in different $\boldsymbol{k}$-space region [48]. Since IS has intrinsically broken in monolayer TMDCs due to two different types of atoms (metal and chalcogen), these materials have great promise for valleytronics and valley-dependent physics. In this section, we look at optical interband transition from the spin-split valence-band tops to the conduction band bottoms in monolayer TMDCs. Under the dipole approximation [49], the coupling strength with electric fields of circular polarization $\boldsymbol{J}_{\pm}$ is given by [48–51]

$$\mathcal{P}_{\pm}(\boldsymbol{q}) = \mathcal{P}_x(\boldsymbol{q}) + \mathrm{i}\sigma_{\pm}\mathcal{P}_y(\boldsymbol{q}), \tag{7.61}$$

where

$$\mathcal{P}_{i=x,y}(\boldsymbol{q}) \equiv m_{\mathrm{e}}\left\langle \psi_{\mathrm{c}}(\boldsymbol{q})\left|\frac{1}{\hbar}\partial_{k_i}H\right|\psi_{\mathrm{v}}(\boldsymbol{q})\right\rangle \tag{7.62}$$

is the interband matrix element of the canonical momentum operator [48, 50, 52] and H is given by equation (1.49).

For transitions near $\boldsymbol{K}_{\pm}$ points, we find that the coupling strength for circularly polarised light directly selects the valley [49–51]:

$$|\mathcal{P}_{\pm}(\boldsymbol{q})|^2 = \left[\frac{m_{\mathrm{e}}a\gamma}{\hbar}\left(1 + \kappa\sigma_{\pm}\frac{\Delta_z - \kappa s\lambda_{\mathrm{SO}}}{\sqrt{4a^2\gamma^2q^2 + (\Delta_z - \kappa s\lambda_{\mathrm{SO}})^2}}\right)\right]^2. \tag{7.63}$$

Since $\Delta_z - \kappa s\lambda_{\mathrm{SO}} \gg a\gamma q$, for LCP ($\sigma_+$), $|\mathcal{P}_+(\boldsymbol{q})|^2 = 0$ for $\boldsymbol{K}_-$, while for RCP (σ_-), $|\mathcal{P}_-(\boldsymbol{q})|^2 = 0$ for $\boldsymbol{K}_+$. The interband transitions in the vicinity of the $\boldsymbol{K}_+$ ($\boldsymbol{K}_-$) point couple exclusively to LCP (RCP) light [53, 54].

In addition, the lack of IS together with SOC makes spin and valley dual control in monolayer TMDCs possible. Since TRS requires $\mathcal{E}(\uparrow, \boldsymbol{k}) = \mathcal{E}(\downarrow, -\boldsymbol{k})$, which means opposite valleys have opposite spin at the same energy [50], also known as spin–momentum locking. In figure 7.3, we show the valence and conduction bands in the vicinity of $\boldsymbol{K}_+$ and $\boldsymbol{K}_-$ valleys that are split into the spin-up (solid lines) and spin-down (dotted lines) subbands due to SOC [50, 58–63], together with optical selection rules. For fixed energy, the $\boldsymbol{K}_-$ valley would correspond to optical selection rules of RCP as well as carriers of a fixed spin, while the $\boldsymbol{K}_+$ valley would correspond to opposite conditions [50, 64, 65]. If circularly polarised light of an appropriate energy range is used, electrons and holes can be addressed in these materials that are in a known valley with a known spin [66]. Thus, at these wavelengths there exist spin-dependent optical selection rules [67, 68]. The property of creating carriers with certain spin and valley index via optical injection makes monolayer TMDCs promising candidates for spintronics and valleytronics.

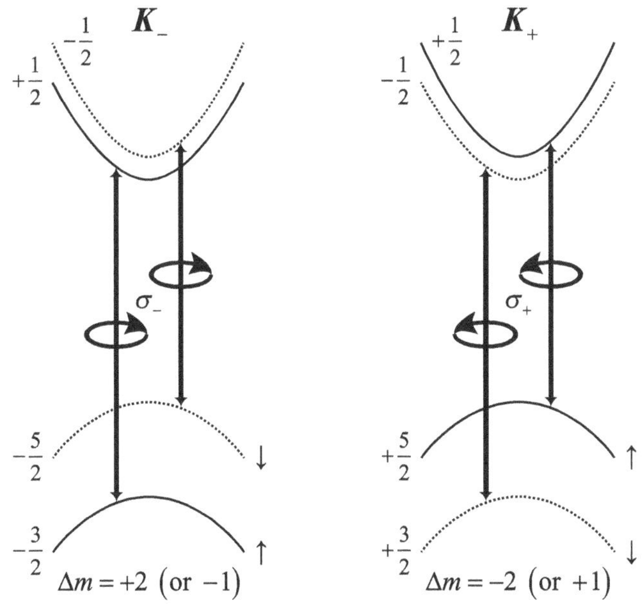

Figure 7.3. Valley and spin optical transition rules in monolayer TMDCs, in which the optical absorption selectively occurs for the LCP (σ_+) and RCP (σ_-) at the $\boldsymbol{K}_+$ and $\boldsymbol{K}_-$ points in the hexagonal BZ, respectively. Spin up (down) bands due to SOC are denoted by solid (dotted) lines. For the transition metal atoms as rotation centre [55, 56], the quantum number m_i for each band is annotated and the valley-dependent optical selection rules are illustrated [57]. At the $\boldsymbol{K}_\kappa$ point, the valence band consists of the $d_{x^2-y^2} + \mathrm{i}\kappa d_{xy}$ orbitals of the transition metal atom whose m value is 2κ. The conduction band consists of d_{z^2} with $m = 0$ at both the $\boldsymbol{K}_+$ and $\boldsymbol{K}_-$ points. Optical dipole transition occurs between the same spin subbands in the $\boldsymbol{K}_\pm$ valley with $\Delta m = \mp 2$ which is equivalent to $\Delta m = \pm 1$ in the 3-fold rotational symmetry with a non-zero $p_1 = -1$ value in equation (7.64). In addition, due to the spin–momentum locking, when a specific energy level is excited in a valley using circularly polarized light, a spin polarization will also be created. Here, the spin-splitting in the conduction band is assumed to be negligible.

7.4.2 Conservation law of the pseudoangular momentum

In monolayer TMDCs that observes N-fold rotational symmetry around the z-axis perpendicular to the material in the xOy plane, the valley-dependent optical selection rules can also be inferred from the selection rule for the circularly polarised vortex light [57] or the conservation law of the pseudoangular momentum [48, 69–74], which investigates the quantum number m associated to the electronic states near the $\boldsymbol{K}_\pm$ points [55, 56, 67, 75, 76]

$$m_\mathrm{c} - m_\mathrm{v} = (\sigma_\pm + \ell) + Np_1 \quad (p_1 \in \mathbb{Z}), \tag{7.64}$$

or [77]

$$\mathrm{mod}(m_\mathrm{c} - m_\mathrm{v}, N) = \sigma_\pm + \ell, \tag{7.65}$$

where ℓ is the orbital angular momentum of light, m_v (m_c) is the eigenvalue of the N-fold rotational symmetry for the valence (conduction) band that depends on the $\boldsymbol{K}_+$ or $\boldsymbol{K}_-$ valley. It is noted that the value of m_v (m_c) gives an opposite sign between $\boldsymbol{K}_+$ and $\boldsymbol{K}_-$ valleys because of TRS.

As shown in equation (1.47), in monolayer TMDCs, at the $\boldsymbol{K}_\kappa$ point, the conduction (valence) band consists of d_{z^2} ($d_{x^2-y^2} + i\kappa d_{xy}$) orbitals from transition metal atoms [50, 58, 59, 78, 79] with

$$m = \begin{cases} 0 & \text{for conduction band} \\ 2\kappa & \text{for valence band,} \end{cases} \tag{7.66}$$

specified for rotation centre: transition metal atoms [55, 56][4]. It is noted that the phase factor, $e^{i\boldsymbol{k}\cdot\boldsymbol{r}}$ [57], of the Bloch wavefunction at the $\boldsymbol{k} = \boldsymbol{K}_\pm$ point does not change the shape by operating C_3 ($2\pi/3$) rotation to $\boldsymbol{r}$, that is, $e^{i\boldsymbol{k}\cdot(C_3\boldsymbol{r})} = e^{i\boldsymbol{k}\cdot\boldsymbol{r}}$. Only the atomic orbitals at each $\boldsymbol{r}$ are rotated $2\pi/3$ around $C_3\boldsymbol{r}$ by the C_3 rotation [70].

The half-integer values of quantum number m_i for each subband in figure 7.3 are given by summing m_v (m_c) and $\frac{1}{2}s$ with spin index s due to SOC [70]. Consequently, at the $\boldsymbol{K}_-$ valley, the valence (conduction) bands are assigned $m = -5/2$ and $-3/2$ ($m = -1/2$ and $+1/2$) for the spin down and up, respectively. At the $\boldsymbol{K}_+$ valley, by contrast, the quantum numbers have opposite signs because of TRS. The dipole optical transition occurs from the valence to the conduction band with keeping the spin direction unchanged, which gives $m_c - m_v = \Delta m = \mp 2$ for the $\boldsymbol{K}_\pm$ points. If we adopt $p_1 = \mp 1$ in equation (7.64), $\Delta m = \mp 2$ is equivalent to $\Delta m = \pm 1$ in the 3-fold rotational symmetry around the z-axis [48, 72, 81]. This means that the angular momentum of the photon is transferred to the crystal, and the optical transition occurs with $\Delta m = \pm 1$ that is equivalent to $\Delta m = \mp 2$ from the $d_{x^2-y^2} + i\kappa d_{xy}$ to the d_{z^2}. Therefore, equation (7.64) can be rewritten as [57]

$$\Delta m \equiv m_c - m_v = \begin{cases} +1 & \text{for } \boldsymbol{K}_+ \\ -1 & \text{for } \boldsymbol{K}_-. \end{cases} \tag{7.67}$$

The results are consistent with those shown in reference [82], which did not show the derivation of the results. Eventually, when we consider a non-vortex light, which is absent from the orbital angular momentum ($\ell = 0$), we further have

$$\sigma_\pm \rightarrow \begin{cases} \sigma_+ = +1 & \text{for } \boldsymbol{K}_+ \\ \sigma_- = -1 & \text{for } \boldsymbol{K}_-, \end{cases} \tag{7.68}$$

which is consistent with the valley-selection rule. This means that at the $\boldsymbol{K}_-$ point, interband transitions from valence bands to conduction bands can occur through absorption of RCP (σ_-) that carries an angular momentum $\Delta m = -1$, while at the $\boldsymbol{K}_+$ point, interband transition can occur through absorption of LCP (σ_+) with $\Delta m = +1$. The optical selection rule means that each valley can be addressed by optical signals independently, allowing for the opto-valleytronic application.

[4] Note that though both m_c and m_v depend on the choice of rotation centre, mod($m_c - m_v$, N) does not [77]. Transition metal atoms is used as the rotation centre in the symmetry analysis in references [59, 79], while the hollow centre of the hexagon formed by transition metal and chalcogen is used in references [75, 80].

7.4.3 Excitons

While the above discussion focuses valley-dependent physics phenomena originating from the VP of free carriers, due to geometric perpendicular confinement [83] and reduced dielectric screening [84–86] in monolayer materials, electrons and holes created in optical excitation have strong Coulomb interaction with higher binding energies (200–900 meV [87, 88–91]) compared with traditional III–V semiconductors (<10 meV) [92]. This results in new bound states [67, 93–95] like hydrogen atoms and ions, named excitons [96, 97]. Consequently, unlike in the bulk case, the optical response of the monolayer TMDC is dominated by excitons, which enables direct optical observations even at room temperature [98, 99]. Excitons are of the Wannier type [91]: the wavefunction of the exciton extends over multiple unit cells [100, 101], electron and hole of valley exciton are confined in the same $\boldsymbol{K}_{\pm}$ in the momentum space, corresponding to circularly polarized electroluminescence (EL), endowing them with a binary valley pseudospin [102–104] as well as valley-dependent optical selection, which broadens the ideas in the design of optical and polarization switches. These excitonic states are the key to understanding most optical properties of semiconductors and therefore play an important role in a variety of photonics and optoelectronic applications.

Five types of excitons are diagrammatically illustrated in figure 7.4. We can distinguish different types of excitons using PL spectra due to their distinct binding energies. A neutral exciton (X^0) is a bound state of a negatively charged electron in the conduction band and a positively charged hole in the valence band bound by Coulomb force, analogous to a hydrogen atom [98, 105]. X^0 can be divided into bright [106] and dark excitons resulting from spin momentum conservation. Bright excitons consist of an electron and a hole with opposite spins, and they can easily recombine and result in the emission of photons, while dark excitons consist of an electron and a hole with the same spins, and they cannot easily recombine due to the absence of spin momentum conservation [91]. An X^0 can easily capture an additional electron (hole) to form a three-body bound state called a trion X^- (X^+) [94, 95, 107, 108], analogous to H^- (H_2^+) [93]. Trions can also be observed, albeit at a lower temperature due to their smaller binding energy ($\sim$20 meV [95]). When the density of excitons is large, the large wavefunction overlap between X^0 makes it possible to form biexcitons (XX) [109–113].

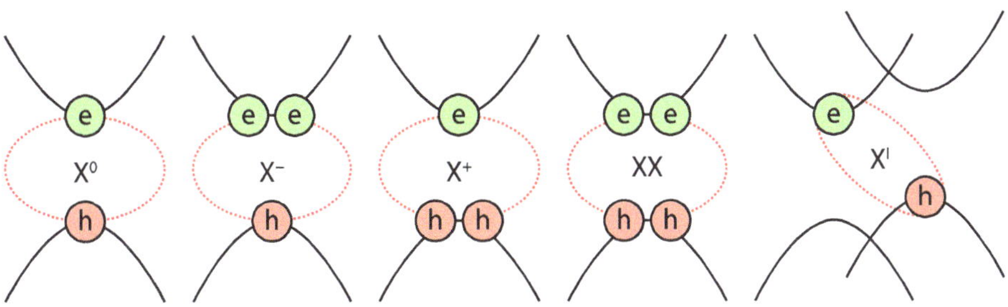

Figure 7.4. Types of exciton in atomically thin TMDCs and heterostructures: neutral exciton (X^0), trion (X^-, X^+), biexciton (XX), and interlayer exciton (X^I).

The rapid development of transfer techniques has opened up a vast parameter space of artificial vdW heterostructures, where different 2D materials are deterministically stacked upon each other [114]. An early theoretical work [115] showed that several combinations of 2D semiconductor bilayer heterostructures have resulted in type-II (staggered) band alignment [116, 117]. In type-II heterostructures, the CBM is in one layer, and the VBM is in the other layer, as shown in the rightmost panel of figure 7.4. One example of a type-II heterostructure is $MoSe_2$–WSe_2, in which the CBM and VBM are in the $MoSe_2$ and WSe_2 layers, respectively [118]. The injected electrons and holes undergo rapid charge transfer [119] and are separated to the $MoSe_2$ and WSe_2 layers, respectively. If the attractive Coulomb interaction between the electrons and holes in opposite layers is sufficiently strong, an interlayer exciton (IX), X^I, can form [77, 118, 120–137]. The energy of the IX is determined by the band offsets between layers and the binding energy of the IX. Also, due to the spatial separation, the electron–hole wavefunction overlap in the out-of-plane direction is reduced, leading to a dramatically decreased electron–hole exchange interaction and exciton recombination rate, facilitating the potential for significantly longer lifetimes [118, 129, 138, 139] (up to 5 orders of magnitude, $\sim$1.7 μs [138]) of the IXs in 2D heterostructures [118, 140, 141], compared to 1 ps for the intralayer exciton [141–143]. Type-II heterostructures are particularly important in electronics and optoelectronics because these have applications with photoelectronic devices. The long lifetime of IXs provides an opportunity to realize spatial transport of IXs before radiative decay. IXs in heterostructures will be discussed intensively in the next chapter.

7.4.4 Valley excitons: valley-polarized photoluminescence (PL)

The excitons live in the Dirac valleys and contain valley information. Optical excitations with energies close to the band edge will generate electrons in the conduction band and holes in the valence band in the same valley ($\boldsymbol{K}_+$ or $\boldsymbol{K}_-$). These electrons and holes will then relax to the band edges, forming excitons [103]. Intervalley scattering is suppressed under low temperature in pure samples because the two unequivalent valleys are located far away from each other in momentum space [144–146]. The bright excitons recombine efficiently, which suggests that excitons consist of electrons and holes confined at either the $\boldsymbol{K}_+$ or $\boldsymbol{K}_-$ valley, and are correspondingly named valley excitons [103, 106]. The intervalley excitons, i.e., electron and hole are separated in unequivalent valleys, are dark since they require assistance from phonons to emit [147].

Based on the analysis of IS breaking and introduction of valley excitons, one can conclude that we can address and read out the valley information optically using circularly polarized light. We define the degree of circular polarization [148]

$$P = \frac{I_+ - I_-}{I_+ + I_-}, \tag{7.69}$$

where I_+ and I_- is the intensity of the LCP (σ_+) and RCP (σ_-) PL, respectively. P is determined by the absorption process and the subsequent depolarization process.

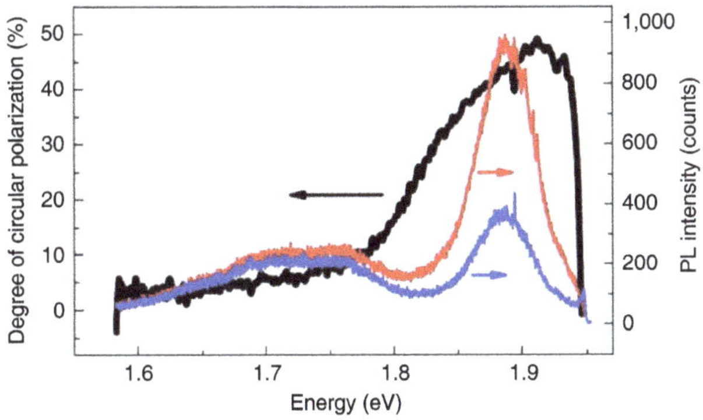

Figure 7.5. Circularly polarized PL of monolayer MoS_2 at 83 K, along with the degree of circular polarization of the PL spectra, as a function of photon energy. The red and blue curves correspond to the intensities of LCP and RCP lights, respectively, in the luminescence spectrum (right-hand scale). The black curve is the net degree of polarization (left-hand scale). Reproduced from [75]. CC BY 3.0.

For perfectly circularly polarized PL, $P = 1$ (LCP) or $P = -1$ (RCP). Experimentally, due to the valley-contrasting optical selection rule, P further reflects the degree of VP, characterizing the population difference of the excited carriers between the $\boldsymbol{K}_+$ and $\boldsymbol{K}_-$ valleys (figure 7.5).

As we discussed, the valley-dependent optical selection rules suggest PL spectrum observes optical circular dichroism [149, 150], which means that the helicity of the luminescence should exactly follow that of the excitation light. In other words, the RCP (LCP) excitation generates RCP (LCP) luminescence. The ability to control carrier spin and carrier confinement within a specific valley with circularly polarized light in TMDCs was first experimentally demonstrated in 2012 [66, 75, 151–154] and has been crucial to the research progress of TMDCs ever since. An example [75] of such experiments is illustrated in figure 7.5, in which the PL intensities with two circular polarizations are shown (right-hand scale) and the corresponding degree of polarization (left-hand scale) as a function of photon energy, under a LCP irradiation at wavelength 633 nm from a HeNe laser. The degree of polarization reaches 50%. Meanwhile, Zeng *et al* [151] studied the circularly polarized luminescence spectra of a pristine monolayer MoS_2 (peak around 1.9 eV) with LCP and RCP excitation (HeNe laser, 1.96 eV) at a near-resonant condition at 10 K. The luminescence spectra demonstrated a symmetric polarization for excitation with opposite helicities, i.e., $P = 32\%$ (-32%) under LCP (RCP) excitation for the most monolayer MoS_2 samples. The relatively low values of P compared with theoretical prediction for the validity of full circular polarization might result from the poor sample quality and the effect of substrates [149], which act as atomic scattering centres. The atomic scattering centres dramatically increase the intervalley scattering, which induces valley depolarisation.

In other studies, Mak *et al* [66] prepared monolayer MoS_2 on an hBN substrate and reported nearly 100% VP for photon energies in the range 1.90–1.95 eV on resonance excitation with A exciton at 14 K. Wu *et al* [148] found that physical vapour-grown high optical quality monolayer MoS_2 yields almost 100% VP at 30 K, and remarkably, even about 30% residual polarization at room temperature. This suggested that not only the circular dichroism is unity, but also the valley excited excitons experience almost no depolarization process before their emission [93]. Later, Jones *et al* [155] and Dai *et al* [49] demonstrated the circularly polarized optical selection rules in monolayer WSe_2 and WS_2, respectively. Park *et al* [156] encapsulated monolayer MoS_2 with two hBN flakes to promote the polarisation up to 70%. These results showed that due to SOC-induced spin–valley locking, together with large momentum separation of $\boldsymbol{K}_-$ and $\boldsymbol{K}_+$ valleys, the emitted PL preserves a large PL handedness P, which suggests well-maintained VP [64, 68]. It would be possible to address the valley in a precise and deterministic way through optical stimuli.

Generally, because of intervalley scattering due to impurities and vacancies, the P of natural mining TMDC products is much smaller than 1. Increasing the temperature also reduces the P through the increasing intervalley phonon scattering [151]. Furthermore, by increasing the layer number of TMDCs, the IS is preserved and the valley splitting vanishes [151]. Nevertheless, VP could reappear if the symmetry is broken [148, 157], for example, by applying sufficient strain, or an out-of-plane electric field which could come from external electric gating or interface charging. The AB-stacked bilayer [158, 159] and interactions with the substrates including charge transfers or photo-doping could also induce a finite circular dichroism to some extent [149, 160]. However, in the case of $MoSe_2$, no such observation has been made so far, which still challenges our understanding of this particular system [93].

7.5 Exciton Hall effect

The findings of valley exciton and VHE are further reinforced by the spatially resolved measurements of valley-polarized exciton diffusion in monolayer MoS_2 [161], which concludes that spin and valley currents can be generated perpendicular to the applied temperature gradient in the 2D plane of the monolayer TMDCs.

Figure 7.6(a) shows the schematics of the exciton Hall effect. When a circularly polarized light is focused on one end of the long and narrow device, the density of excitons in this endpoint increases rapidly and excitons flow to the other endpoint because of the statistical forces, i.e., gradients of the temperature and/or exciton density in the longitudinal direction. Meanwhile, a longitudinal electric field develops due to the Seebeck effect [162]. Besides this parallel effect, in the presence of the temperature gradient, the holes (or electrons) experience a Lorentz-like force and thus move in the direction perpendicular to the diffusion current, which is the anomalous Nernst effect [163] induced by the intrinsic non-vanishing Berry curvature [164]. The measured current originates thus from a single valley with a single spin component. Therefore, by lowering the Fermi level, a single spin current carrying single valley information can be generated by a temperature gradient (∇T), which hints at possible applications in spin caloritronics and valleytronics.

In the experiment, as shown in figure 7.6(b), a linearly polarized pump at 1.96 eV was centred at one endpoint of a mechanically exfoliated elongated monolayer

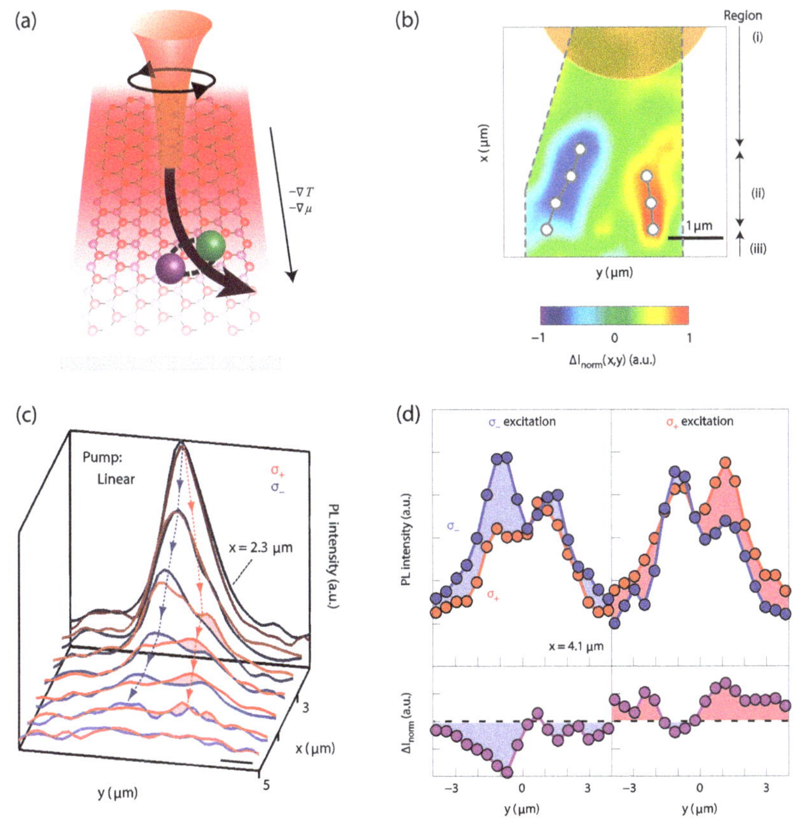

Figure 7.6. (a) Schematic depiction of the exciton Hall effect under the illumination of the laser that creates a longitudinal chemical potential ($\nabla\mu$) and temperature gradient (∇T), which drives the transverse diffusion of the excitons. (b) Colour plot of the normalised intensity for the difference of LCP (σ_+) and RCP (σ_-) excitation, $\Delta I_{\text{norm}}(x, y)$ defined in equation (7.70), in real space for the exciton Hall effect. White dots correspond to the peak positions of $\Delta I_{\text{norm}}(x, y)$. The grey dashed lines mark the edges of the monolayer and the orange circle indicates the laser spot. The sign reversal at the centre of the monolayer represents the occurrence of the valley-contrasting exciton Hall effect. (c) Polarization-resolved PL intensity cross-sections taken along y, for every step of $x \approx 0.46$ μm, demonstrating transverse diffusion of valley-polarised excitons. The scale bar along the y-axis is 1 μm. (d) Upper panel: Cross-sectional profiles of PL intensities under circularly polarised excitation, signifying selective spatial transport of valley-polarised excitons. Lower panel: $\Delta I_{\text{norm}}(x, y)$ versus y under σ_+ or σ_- excitation. Reproduced from [161], with permission from Springer Nature.

MoS_2 at 30 K [165] to generate high density of $\boldsymbol{K}_+$ and $\boldsymbol{K}_-$ valley excitons, resulting in chiefly unidirectional thermal and chemical potential gradients. As the excitons diffuse over a distance of several microns in the monolayer [166], the excitons from $\boldsymbol{K}_+$ and $\boldsymbol{K}_-$ valleys with opposite Berry curvature deflect and accumulate to opposite sides of the channel without breaking the bound states. Therefore, excitons in opposite sides recombine separately and emit opposite circularly polarised light simultaneously. As a matter of fact, separated light emitting is difficult to detect because the lifetime of excitons is short. Figure 7.6(b) shows the real-space polarization-resolved PL mapping of the exciton trajectory by helicity-selective laser probing under the excitation of linearly polarized light [167], which reveals the opposite transverse diffusion of the excitons in opposite valleys. As the diffusion distance increases, the luminescence gets weaker. To present the spatial separation of excitons in $\boldsymbol{K}_+$ and $\boldsymbol{K}_-$ valley, the normalized intensity difference are defined as

$$\Delta I_{\text{norm}}(x, y) = \frac{I_+(x, y) - I_-(x, y)}{\sum_y [I_+(x, y) + I_-(x, y)]}, \tag{7.70}$$

where I_+ and I_- are the LCP (σ_+) and RCP (σ_-) components of PL intensity, respectively. Clearly, as shown in figures 7.6(b) and (c), the PL peaks of LCP and RCP components are obvious and were separated on a spatial scale of about 2 μm. Subsequently, in figure 7.6(d), the selective special transport of $\boldsymbol{K}_+$ and $\boldsymbol{K}_-$ valley-polarized excitons can be realized under the LCP and RCP excitation, respectively [165].

The advantage of this method is the observation of the exciton Hall effect instead of individual transport of electrons or holes induced via VHE. The extracted valley diffusion length ℓ_v was of $\approx$ 2.0 μm, consistent with the spin–valley free path of electrons and holes in monolayer WS_2 [168]. Also, the ℓ_v is comparable to the exciton diffusion length ℓ_x, suggesting the robust preservation of VP during transport. Moreover, the Hall angle is defined by the ratio of the spin Hall current to the total dissipative current [169]. If we define the exciton Hall angle $\alpha_E = L_{xy}/L_{xx}$, where L_{xy} is half of the distance between the separated peaks (σ_+ and σ_-) and $L_{xx} = x$ is the distance from the excitation centre ($x = 0$). The extracted exciton Hall angle (α_E) was up to 0.20 ± 0.09, two orders of magnitude larger than the conventional single-particle dominated valley Hall angle (described in section 2.4.3 and will be discussed in section 7.7.2) of monolayer MoS_2 [161, 170], which may arise from the internal topological nature of the excitonic states [104].

Such transverse splitting indicates the occurrence of the exciton Hall effect in monolayer MoS_2 and is promising for novel valleytronic application in 2D semiconductors. This theoretical and experimental proof of VP and the anomalous Hall effect driven by the internal topological nature of monolayer TMDCs reflect the potential ability to control spin and valleys as information carriers. Therefore, monolayer TMDCs are the best candidates for using the valley DOF to explore new classes of electronic and optoelectronic device applications, yielding the concept of valleytronics [171].

7.6 Tuning VP: valley Zeeman effect

In the conventional Zeeman effect [172, 173], a spin-dependent energy shift of the electronic energy levels takes place in the presence of an external magnetic field, thereby increasing the degeneracy. Similarly, the magnetic moments associated with $\boldsymbol{K}_+$ and $\boldsymbol{K}_-$ valleys that we discussed in sections 2.5 and 7.4.2 make it possible to result in energy splitting and lift the valley degeneracy by applying an external out-of-plane magnetic field [174], producing valley Zeeman effect [175, 176]. The valley Zeeman splitting measures the difference between the Zeeman shifts of the conduction and valence band edges in $\boldsymbol{K}_+$ and $\boldsymbol{K}_-$ valleys, thereby providing an additional degree of external control over the valley DOF and developing valleytronic applications [160].

Figure 7.7(a) displays energy level diagrams for the three contributions to the magnetic moment of electrons in a single-particle picture for the W-based TMDC compounds [175, 177–179]. For the Mo-based TMDC compounds, the spin magnetic moment for the conduction bands reverses its direction [180] because the W-based and Mo-based compounds have opposite spin splitting order for the conduction bands [59–62, 79, 87, 90, 181–183]. The conduction and valence bands under a zero (positive) magnetic field are denoted by light thin (dark thick) lines. At zero magnetic field ($B = 0$), the TRS works and the two valleys are degenerate. Because of the spin splitting of ~0.4 eV in the valence band, the $\boldsymbol{K}_+$ ($\boldsymbol{K}_-$) valley has only the spin-up (down) state. For the conduction band, the splitting has opposite signs in the two valleys with smaller magnitude, compared with the valence band [59, 79, 80]. After the application of an out-of-plane magnetic field $\boldsymbol{B}$ ($B_z \neq 0$) which

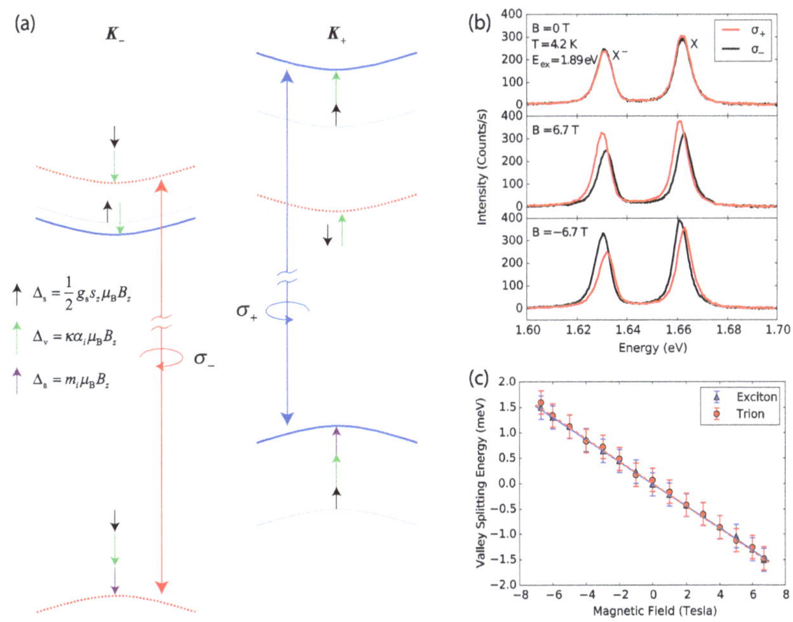

Figure 7.7. (a) Schematic band diagram of WSe_2 under zero (light and thin lines) and finite out-of-plane magnetic field $\boldsymbol{B}$ (dark and thick lines). The spin-split conduction bands and only the highest valence band are shown. Blue solid and red dotted lines represent spin up and down, respectively. Black, green, and purple arrows denote the Zeeman shift in energy from spin (Δ_s), valley orbital (Δ_v), and intra-atomic d orbital magnetic moment (Δ_a) contributions, respectively. The σ_- (σ_+) polarized transition energy increases (decreases) in an applied magnetic field with $B_z > 0$, lifting the degeneracy of the two valleys. Reproduced from [175], with permission from Springer Nature. (b) Polarization-resolved PL spectra of $MoSe_2$ under out-of-plane magnetic fields of 0, 6.7 and −6.7 T at 4.2 K, showing a small but observable energy shift of the exciton emission. The two peaks correspond to the neutral exciton X and trion X^-. (c) The linear energy shift of valley Zeeman splitting as a function of the magnetic field for the neutral exciton and trion. Figures (b) and (c) are reprinted with permission from [177]. Copyright (2015) by the American Physical Society.

breaks TRS, the total Zeeman splitting of each band is determined by the superposition of the following three independent magnetic moment contributions [175, 177, 179]:

1. Spin magnetic moment: A free electron, spin 1/2 particle, with its associated degenerate spin up and down states, placed in a magnetic field $\boldsymbol{B}$ will couple to the magnetic field due to its intrinsic magnetic moment [1, 184]. Spin may be described as an intrinsic relativistic quantum dipole moment with an associated magnetic moment

$$\boldsymbol{\mu}_{\mathrm{s}} = \frac{g_{\mathrm{s}}\mu_{\mathrm{B}}}{\hbar}\boldsymbol{S}, \tag{7.71}$$

where μ_{B} is Bohr magneton, $\boldsymbol{S} = \frac{\hbar}{2}\boldsymbol{s}$ with Pauli matrix $\boldsymbol{s}$ shown in equation (1.26), and g_{s} is the electron gyromagnetic factor or g-factor. For an electron in a vacuum, the electron g-factor is approximately $g_{\mathrm{s}} = 2$ [1]. This magnetic moment may couple to an external magnetic field, causing a shift in the energy of the electrons spin states relative by

$$\Delta_{\mathrm{s}} = \boldsymbol{\mu}_{\mathrm{s}} \cdot \boldsymbol{B}. \tag{7.72}$$

Here the energy shift Δ_{s} is known as the Zeeman shift. Thus, for an external out-of-plane magnetic field B, the associated Zeeman shift (black arrows in figure 7.7) is [185]

$$\Delta_{\mathrm{s}} = \frac{1}{2}g_{\mathrm{s}}s_z\mu_{\mathrm{B}}B. \tag{7.73}$$

For the spin magnetic moment, the resultant contribution on Zeeman energy splitting is cancelled since optical transition conserves spin.

2. Valley/interatomic/intercellular/Bloch electron orbital magnetic moment [90]: As we discussed in section 2.5, in inversion-asymmetric 2D materials described by the massive Dirac model, electrons acquire valley magnetic moments, which reads

$$\mu_{\mathrm{v}} = \kappa\alpha_i\mu_{\mathrm{B}} \tag{7.74}$$

with $\alpha_i = \frac{m_{\mathrm{e}}}{m_i^*}$ (i = c, v). Here, m_{c}^* (m_{v}^*) stands for the effective mass of the conduction (valence) band. Further, the valley magnetic moment component-contributed Zeeman shift is

$$\Delta_{\mathrm{v}} = \kappa\alpha_i\mu_{\mathrm{B}}B. \tag{7.75}$$

Therefore, in the $\boldsymbol{K}_\kappa$ valley, the interband transition acquires a net shift of $\kappa\mu_{\mathrm{B}}B\left(\frac{m_{\mathrm{e}}}{m_{\mathrm{c}}^*} - \frac{m_{\mathrm{e}}}{m_{\mathrm{v}}^*}\right)$, which stems from the variation in effective mass of conduction and valence bands [64, 68]. The time-reversal pairs of states in the $\boldsymbol{K}_+$ and $\boldsymbol{K}_-$ valleys move toward opposite directions. Under the $\boldsymbol{k} \cdot \boldsymbol{p}$

approximation model, in the vicinity of $\boldsymbol{K}_{\pm}$ points, the effective masses of conduction and valence band edges are normally equal [50, 59][5] due to electron–hole symmetry [186]. As a result, the total contribution of the valley magnetic moment on Zeeman splitting can also be zero.

3. (Intra-)atomic d orbital magnetic moment: In monolayer TMDCs, the conduction (valence) band in the $\boldsymbol{K}_{\kappa}$ valley mainly consists of d_{z^2} ($d_{x^2-y^2} + \mathrm{i}\kappa d_{xy}$) orbitals from metal atoms with the magnetic quantum number $m_c = 0$ ($m_v = 2\kappa$) [50, 68, 75, 76, 78, 175, 186, 187] and atomic orbital magnetic moment [188]

$$\mu_a = \begin{cases} 0 & \text{for conduction band} \\ 2\kappa\mu_B & \text{for valence band.} \end{cases} \tag{7.76}$$

This contributes to a Zeeman shift (purple arrows in figure 7.7) of

$$\Delta_a = \begin{cases} 0 & \text{for conduction band} \\ 2\kappa\mu_B B & \text{for valence band,} \end{cases} \tag{7.77}$$

at the $\boldsymbol{K}_{\kappa}$ valley. This further results in a net shift of $-2\kappa\mu_B B$ for interband transitions in the $\boldsymbol{K}_{\kappa}$ valley for σ_{κ}. Therefore, the intra-atomic d orbital contribution can dominantly give rise to the Zeeman splitting energy of $-4\mu_B B$.

To summarise, while spin and valley orbital magnetic moments do not affect the optical transition energies, the intra-atomic d orbital moment has a significant contribution to the Zeeman shift, together with the optical selection rules discussed earlier, the Zeeman shift turns out to be [87]

$$\Delta\mathcal{E}_Z = \mathcal{E}(\sigma_+) - \mathcal{E}(\sigma_-) = -4\mu_B B. \tag{7.78}$$

This can be rewritten as $\Delta\mathcal{E}_Z = g\mu_B B$ and the resultant g-factor is 4 (corresponding to a valley splitting of about 0.23 meV T^{-1} [189]), where $\mathcal{E}(\sigma_{\kappa})$ is the excitonic transition energies for excitons corresponding to σ_{κ} polarized PL, in the $\boldsymbol{K}_{\kappa}$ valley. Thus, in the magnetic field modulated polarized PL spectra, the $\boldsymbol{K}_+$ valley exciton should be redshifted (blueshifted) compared with the $\boldsymbol{K}_-$ valley exciton for $B > 0$ ($B < 0$).

Now we discuss the VP of excitons. The magnetic-field-induced valley Zeeman effect pushes specific valleys in the energy spectrum up and down, thus expanding the differences in exciton populations between two valleys. Moreover, the band offset reduces the intervalley scattering and results in an increase in VP. It has been reported that magnetic-field-modulated VP of excitons or IXs linearly increases at a rate from 1%/T to 7.7%/T in nonmagnetic monolayer and bilayer TMDCs [130–132, 190–193].

The valley-dependent Zeeman splitting in the optical transition in the $\boldsymbol{K}_{\pm}$ valley can be probed by polarization-resolved PL experiments in an external magnetic field

[5] Corrections beyond the two-band model result in a finite difference for the electron and hole magnetic moment.

[175, 191, 194, 195]. Generally, without a magnetic field, the TRS satisfies and determines that the polarization-resolved valley-exciton PL excited by the RCP light in the $\boldsymbol{K}_-$ valley is identical to that of the LCP light in the $\boldsymbol{K}_+$ valley, thus no splitting occurs. If $\boldsymbol{B} \neq 0$, valley Zeeman splitting is confirmed with an observable split of the LCP and RCP PL, as a result of the broken TRS.

To resolve the splitting between two valley excitons, the sample was both excited and detected with the same polarized light. The splitting can be determined by addressing one valley at a time and comparing the PL peak positions, which correspond to the exciton resonances for two circular polarizations. Figure 7.7(b) shows PL spectra of $MoSe_2$ for σ_+ and σ_- detection at low temperature (4.2 K), the magnetic-field-dependent splitting of X^- and X^0 peaks were meanwhile observed [179]. In the case of a zero magnetic field, the PL peaks associated with both valleys, i.e. $\boldsymbol{K}_+$ (red, σ_+) and $\boldsymbol{K}_-$ (black, σ_-) are identical (top panel). However, under a high magnetic field (6.7 T, middle panel), both components' spectra are split, and σ_+ has a slightly lower energy than σ_-. On the other hand, under a low magnetic field (−6.7 T, bottom panel), the σ_+ spectral component is shifted to a higher energy than that of the σ_- component [175, 196]. Figure 7.7(c) shows that under finite magnetic field, valley Zeeman splitting emerges and it scales linearly with applied magnetic field. The slope of valley Zeeman splitting versus B_z (tuning strength [197]), or g-factor, is extracted to be $-(3.8 \pm 0.2)\mu_B$ [177, 178], which matches the theoretical results surprisingly well. Further studies [176, 198] reveal that the linear Zeeman shift extends up to the magnetic fields as high as $B \sim 65$ T, highlighting the robust nature of the valley Zeeman effect.

Such a Zeeman effect has been theoretically discussed in monolayer TMDCs [199, 200], typically MoS_2 [201] and also observed in diverse 2D materials, such as (bilayer [190, 202]) MoS_2 [176, 203–206], $MoSe_2$ [178, 180, 203, 207], (bilayer [193]) $MoTe_2$ [191], WS_2 [176, 192, 205, 207–210], WSe_2 [175, 179, 180, 195, 198, 207, 211–216], $ZrTe_5$ [217] and (TMDC [218, 219]) heterostructures [220]. Based on the analysis of Zeeman splitting, researchers are also able to attain the exciton parameters by the nonlinear terms [176] in the Zeeman effect, such as exciton binding energy and radius [176, 196, 211, 212, 221]. The magnetic-field-dependent valley splitting of the PL is on the order of 0.07–0.36 meV T^{-1}, corresponding to g-factors with magnitudes of 1–6 [160, 169, 176–180, 198]. The deviation of g-factors from 4 [175, 179] in equation (7.78) might be attributed to the valley magnetic moment, in which corrections beyond the two-band model result in electron–hole mass asymmetry [59, 79, 222] with a finite difference for the electron and hole magnetic moment, and it is this difference, as well as the intra-atomic d orbital contribution or other higher-order effects [223, 224], that is measured by the splitting of the σ_+ and σ_- PL peaks [175, 179, 186].

The measurement of valley Zeeman splitting did motivate the magnetic control of tuning valley-associated states. In 2016, Wang *et al* [195] went another step further to demonstrate that the exciton valley coherence in monolayer WSe_2 could be controlled by an external magnetic field. As discussed in equation (7.37), linearly polarised light may be described as a linear combination of LCP and RCP light. By using a PL setup, excitons excited with linearly polarised light are created in a coherent superposition of spin and valley indices. Then the authors demonstrated the possibility to rotate the coherent superposition of valley states by as much as 30° using perpendicular fields up to 9 T. For applications requiring quantum state manipulation (e.g., qubits), this represents the essential capability to rotate the exciton spin around the equatorial plane of the Bloch sphere.

It is noted that the current research is based on magnetic fields of tens of tesla with substantial energy cost to manipulate the VP, which hinders the practical application of devices based on valley DOF, especially at high temperature. It also contradicts to the energy-efficient computing principle of valleytronics. Further research is needed to enhance the magnetic field response of 2D TMDCs to achieve low magnetic field-tunable VP. Several approaches have been proposed as follows.

7.6.1 Defect states

Wang *et al* [225] exhibited valley pseudospin of defect states in CVD-grown monolayer MoS_2. The intensive valley Zeeman splitting is attributed to the presence of five *d* orbitals in the Mo atom and the significant effective mass of electrons in defect states. This result is also applicable to other TMDCs, such as $MoSe_2$, WS_2, and WSe_2.

7.6.2 Electron doping

Electron doping can engineer Coulomb interactions [169, 226] to drive the electron–hole symmetry breaking [227, 228] and first-order magnetic phase transition [229, 230], giving rise to strongly enhanced valley Zeeman splitting with a net intercellular contribution from phase winding of the wavepacket of Bloch electrons [189, 231, 232, 233]. The g-factor in monolayer WSe_2 by reducing the hole density approaches 12, two times higher than the predicted value by single-particle model [226, 227]. Also, by examining the doping dependence, Wang *et al* [229] have verified the presence of zero-energy LLs [234–237] and determined that each LL is both valley- and spin-polarised. The experiment is done by observing inter-LL transitions by performing polarisation-resolved magneto-optical spectroscopy on dual-gated field-effect devices of monolayer WSe_2.

7.6.3 Optical Stark effect

The conventional Stark effect is the alteration to the wavelength of light emitted by atoms due to the presence of an electric field [1]. For the optical Stark effect, a shift in energy of the exciton resonance can be induced by a circularly polarized pulsed laser with below bandgap radiation [87, 238, 239]. In this case, the electric dipole interaction between the exciton and the optical field serves to shift the energy of the exciton resonance. This optical Stark-like energy shift becomes valley-selective in monolayer TMDCs, consistent with the established valley-dependent selection rules, thus the handedness of the polarised pump laser can be used to independently shift the resonance energy of a single valley. The induced effective Zeeman splitting can be up to $\approx$20 meV, corresponding to effective magnetic fields of tens of tesla [240, 241]. The effective magnetic field created by the Stark effect can also be employed to rotate a coherent superposition of valley states [102]. It is noted that a similar technical problem appears in the optical Stark effect, that is, a huge amplitude of oscillating electric field is required for the VP [242].

7.6.4 Composition ratios in monolayer TMDC alloys

For the case of monolayer TMDC alloys, the effective masses of electrons and holes are significantly tunable with the composition ratios [243]. The different effective masses between valence and conduction bands can result in a *g*-factor not exactly equal to 4. Therefore, corrections need to be carried out by taking different effective masses into account [64, 244].

In addition, Li *et al* [245] experimentally found an enhanced valley Zeeman splitting in CVD-grown Fe-doped monolayer MoS_2, which means further lifting of valley degeneracy. The introduction of Fe atoms results in the Heisenberg exchange interaction produced by the *d* orbital hybridization between Fe and Mo atoms, which leads to the enhancement of the valley Zeeman splitting and further enhances the valley degeneracy. The temperature dependence of the valley Zeeman splitting has also been studied [245], in order to gain further insights. The experimental data reveal a clear and monotonic increase in the splitting with decreasing temperature. Meanwhile, Zhou *et al* [246] successfully synthesized monolayer MoS_2 doped with magnetic Co atoms by CVD, in which Co atoms replace Mo sites. Thanks to the internal magnetic moments induced by magnetic dopants [247, 248], the magnitude of the valley splitting can be controlled by changing the concentration of the dopant. The results of polarization-resolved PL spectroscopy showed that in Co-doped monolayer MoS_2 with Co concentrations of 0.8%, 1.7% and 2.5%, the valley splitting reached 3.9, 5.2 and 6.15 meV at 7 T, respectively.

The valley splitting for various excitons are also pronouncedly different and can be greatly enhanced by changing the dielectric environment [249].

The magnetic control of valley pseudospin in the valley Zeeman effect lifts the degeneracy of LCP and RCP emission in the energy scale and allows the further tunability of exciton formation, emission, VP and coherence in 2D TMDCs [174], together with valley-contrasting optical selection rules, opening a new field for magneto-optical applications [197].

7.7 Harness valley degree of freedom in optoelectronic devices

The main challenge in developing valleytronic devices is to achieve permanent VP. In systems with broken IS, optical excitonic transitions are associated with the valley DOF [64]. Therefore, it is possible to interconvert VP with the optical polarization of emitted and absorbed photons in 2D systems. This means that VP can be reflected by photons. Such valleytronic devices can be realized by developing a valley-polarized optoelectronic emitter and absorber (a valley optical interconnection).

Based on VP, novel valleytronic devices have been designed and demonstrated. In this part, we discuss the latest developments in the optoelectronic devices of valley LEDs and valleytronic transistors.

7.7.1 Valley light-emitting diodes (LEDs)

PN junctions are the essential element for the fabrication of traditional LEDs, in which electrons and holes are injected into n- and p-doping regions under an electric field, respectively, and then radiative recombination happens at the intrinsic (i) region in the middle [130, 251, 252].

7.7.1.1 Original idea: difference of TW effect between conduction and valence bands

Zhang *et al* [250] reported an electrically switchable circularly polarized light source based on valley excitons in monolayer WSe_2. They realized subtle tuning of EL using ambipolar monolayer WSe_2 in a spatially defined ion-gel gating [160] as the field-effect transistor (FET). Monolayer WSe_2 exhibits ambipolar conductive properties, which are electronically conductive in the case of a positive gate voltage and hole-conducting in the case of a negative gate voltage. Based on this, the WSe_2 device is applied to two gate voltages so that the material is divided into three regions, as shown in figure 7.8(a). The left (magenta) region is electronically conductive under the action of gate–source voltage V_{GS}, and is called the n region. The right (green) region is hole-conducting under the action of gate–drain voltage V_{GD}, and is called the p region. The middle (white) region is intrinsically conductive, thus forming a p–i–n junction [253]. The polarity of the p–i–n junction can also be changed by adjusting the source, drain and gate voltages. Electrons and holes in the p–i–n junction are driven by source–drain voltage V_{DS} and combine to form EL. Figure 7.8 (b) shows the consequent polarization-resolved PL spectra, which increases with increasing gate voltage [250], measured at 100 K. Interestingly, in the case of forwarding current, the EL exhibits obvious circular polarization and RCP emission dominates. The degree of circular polarization P reaches as high as 45%. After exchanging the source–drain bias and the direction of the current, i.e., the device is operated in n–i–p mode, it is found that the sign of P is reversed. The degree of polarization of the EL is comparable with that of the PL in monolayer TMDCs [151, 155]. This is a significant demonstration of electrical control of circularly polarized luminescence.

The result was explained by the relative orientation of the bias field and the channel material (TMDC), as well as electron–hole overlap controlled by an in-plane electric field [250, 254, 255]. Remember the band structure of TMDC at $\boldsymbol{K}_{\pm}$ points discussed in section 1.5.3, the TW effect of the valance band is more pronounced than that of the conduction band. According to the Drude model [26, 256, 257], under an in-plane electric field $\boldsymbol{E}$ at the i region, the hole (electron) distribution shifted parallel (antiparallel) to the field in the momentum space by [250, 258]

$$\Delta \boldsymbol{k} = \pm \frac{e}{\hbar} \tau \boldsymbol{E}, \tag{7.79}$$

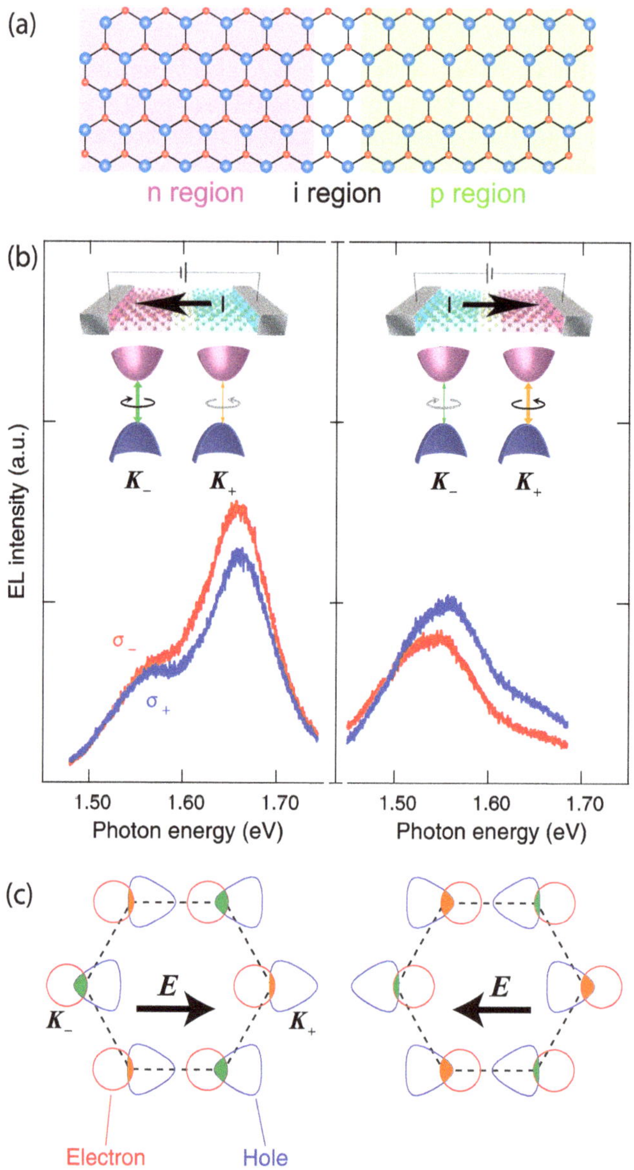

Figure 7.8. (a) A TMDC p–i–n junction consists of 'p' (green), 'i' (white), and 'n' (magenta) regions from right to left. In the top view, a metal (blue) and two stacking chalcogen atoms (red) are shown [26]. (b) Top: Two opposite configurations of source–drain bias and the direction of the current. Middle: valley-dependent optical selection rule with dominating (secondary) helicity drawn in black (grey) arrow. Bottom: Corresponding circularly polarized EL spectra. (c) Schematic illustrations of electron (red lines) and hole distributions (blue lines) shifted by the electric fields originating from the built-in potential. Orange and green areas represent the electron–hole overlap for $\boldsymbol{K}_{+}$ and $\boldsymbol{K}_{-}$ valleys, respectively. The TW effect has been exaggerated to make the difference of overlapping area apparent. Figures (b) and (c) are from [250], reprinted with permission from the AAAS.

where τ is the relaxation time of hole (electron). This situation is similar to that when the water surface in a PET bottle is tilted under acceleration [26]. As a result, as shown in figure 7.8(c), the electron–hole overlap differs between $\boldsymbol{K}_+$ and $\boldsymbol{K}_-$ valleys (different overlapping areas) because of the different shapes of Fermi circles between conduction and valence bands induced by TW. Since excitonic PL reflects the carrier population of the individual valleys, consequently, the momentum-conserving electron–hole recombination becomes unbalanced between $\boldsymbol{K}_+$ and $\boldsymbol{K}_-$, resulting in circularly polarized EL [130, 165]. Reversing the field direction switches the dominant polarization in the emitted light.

The authors [250] also demonstrated the above phenomena in multilayer WSe_2 p–i–n junctions. They formed the junction using electrical double-layer (EDL) gating with multilayer WSe_2. EDL gating provides a very large gate field due to the accumulation of carriers inside the 2D channel. There are two major difficulties with multilayer TMDCs: 1) An indirect bandgap [259], which is undesirable for optoelectronic device applications; 2) Restored IS decreases the circular polarization. EDL gating overcomes these difficulties, as the large gate field result from electrolyte–solid interface [254] produces higher carrier densities in the direct gap $\boldsymbol{K}_\pm$ valleys and breaks IS [148, 260]. Hence, the multilayer EDL transistor operates in a similar way to the monolayer TMDC transistor [260]. Additionally, the authors of the original study [250] also reported similar behaviour in MoS_2 and $MoSe_2$ [160, 250, 261].

Hereafter, Yang *et al* [262] constructed a p^+-Si/i-WS_2/n-ITO heterojunction with monolayer WS_2. They observed circularly polarized EL decreases as the forward bias increases, with a maximum degree of circular polarization of 81%. A decreased tendency of VP is mainly attributed to the rising intervalley scattering with the increasing doping.

7.7.1.2 *Improvement: difference of strain effect between the* **k***-space shift of CBM and VBM*

Valley LEDs typically work at low temperatures due to the exponential decay of temperature-dependent polarization induced by phonon-mediated intervalley scattering. Pu *et al* [263] realized a room-temperature chiral LED in monolayer WS_2 and WSe_2 on plastic substrates by introducing uniaxial strain to the device with a bending stage. To evaluate the circularly polarized light emissions in the LED, the authors introduced CVD-grown monolayer WS_2 and WSe_2 LEDs to uniform and transparent spin-coating ion-gel films [264, 265]. Before the EL measurements of the devices, in figure 7.9(a), PL peak-energy mapping has been initially performed. Inside the monolayer flakes, low-energy shifts (redshifts) of the PL are partially observed in the red dashed rectangular area, whereas no energy shifts are observed in the green dashed rectangular area. Figure 7.9(b) presents the corresponding PL spectra recorded at the two areas, the redshifts occurred without any modifications of the spectral shapes. The origins of these redshifts can be considered as local strains induced inside the crystals because of the difference between TMDCs and substrates, including lattice mismatch and thermal expansion coefficients [263, 265, 266]. Figure 7.9(c) shows an EL image of the same device. Comparing figures (b) and

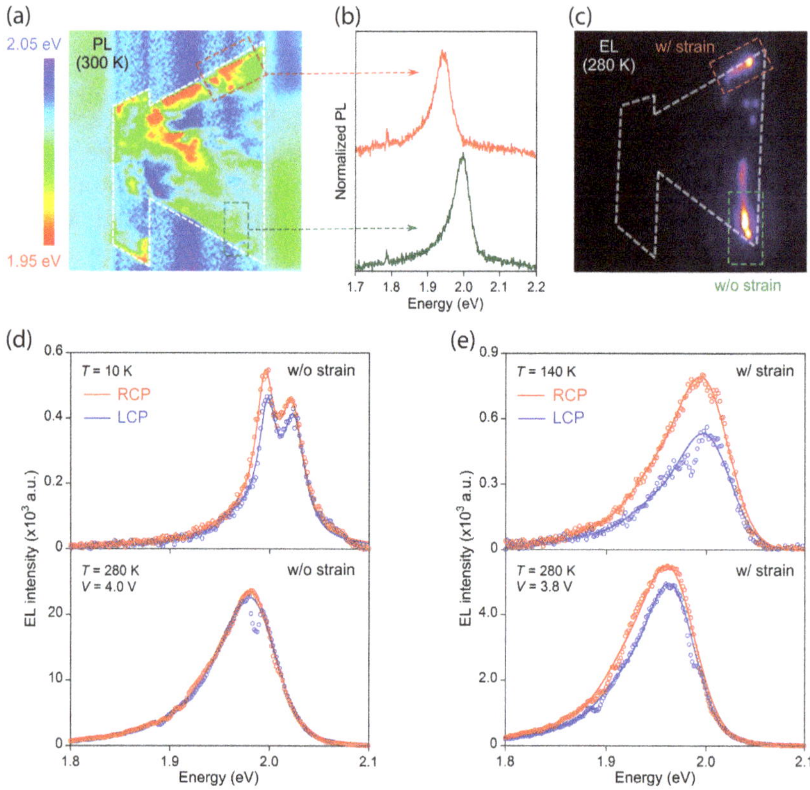

Figure 7.9. (a) PL peak-energy mapping for WS_2 LEDs measured at room temperature (300 K). The colourbar indicates the photon energy ranging from 1.95 to 2.05 eV. (b) PL spectra measured in the red and green areas, respectively, in (a). A redshift due to strain was obtained without modifications of the spectral shapes. (c) EL image of WS_2 LEDs measured at 280 K. Comparing the PL and EL, the EL was generated in both strained and unstrained regions. (d) Polarization-resolved EL spectra for unstrained regions at 10 K (top) and 280 K (bottom). Circular polarization was only obtained at 10 K, in which the RCP EL is larger than that of the LCP. (e) Polarization-resolved EL spectra for strained regions at 140 K (top) and 280 K (bottom). In contrast to unstrained regions, high EL polarization was acquired for the strained EL [263] John Wiley & Sons. [© 2021 Wiley-VCH GmbH].

(c), the EL originated from both strained and unstrained regions. Therefore, strain effects on circularly polarized EL can be directly evaluated via spatially resolved polarized spectroscopy. Figure 7.9(d) shows the polarization-resolved EL spectra measured in unstrained regions at 10 and 280 K. Almost 10% degrees of circular polarization P was observed at 10 K, while $P \approx 0$ at 280 K. These results are consistent with previous reports, in which circular polarization was observed due to TW effects [250]. On the other hand, as shown in figure 7.9(e), a larger P was obtained at a higher temperature of 140 K in the strained regions, compared with 0 K in the unstrained region. Significantly, circular polarization at nearly room temperature (280 K) in the strained region has been observed. The authors also summarized the temperature dependence of the circular polarization for the strained and unstrained regions recorded from WS_2 and WSe_2 LEDs. P vanished above 50 K in the unstrained regions, while evident in the strained regions, and P survived until 280 K. Based on the spatial-polarized EL spectra, one can conclude that the strain effect could play a crucial role in generating room-temperature circularly polarized EL. Later, the authors investigated the circularly polarized EL from the strained and unstrained regions by reversing the source–drain bias. Circularly polarized EL was still only observed in strained areas at room temperature. Besides, the intensity difference between LCP and RCP flips, indicating the electrical control of room-temperature circularly polarized EL in LED based on strained TMDCs.

The result is originated from the effect of the coexistence of strain-mediated valley drifts and electric-field-driven Fermi pocket deviations in the momentum space [263, 267, 268]. We will follow the original theoretical calculation introduced in reference [26]. According to equation (3.1), if we ignore shear strain, the strain tensor can be given by [269, 270]

$$\varepsilon = \begin{pmatrix} \varepsilon_{xx} & 0 \\ 0 & \varepsilon_{yy} \end{pmatrix}. \tag{7.80}$$

Johari *et al* [271] and Kumar *et al* [272] pointed out from DFT calculations that the band structure of monolayer MoS_2—for the tensile strain up to 10% in the zigzag (x) direction and same magnitude of compressive strain in the armchair (y) direction, the CBM and VBM are still located around the $\boldsymbol{K}_\pm$ points. Therefore, in our discussion, we set a fixed strain ratio $0 \leqslant \varepsilon_{xx} = -\varepsilon_{yy} = \varepsilon_b \leqslant 0.1$.

Since the EL process occurs around the band edges in the $\boldsymbol{K}_\pm$ valleys, we employ an effective model [273] which reproduces the energy dispersion of the conduction and valence bands near the $\boldsymbol{K}_\pm$ points for strained TMDC. According to equation (7) in reference [273], on the bases of the conduction and valence low-energy bands, the strain-dependent, spinless effective 2 × 2 model Hamiltonian around the $\boldsymbol{K}_\pm$ points can be written up to the second order in strain and momentum, as

$$\begin{aligned} H_s^\kappa(\boldsymbol{q}, \varepsilon_0) = {} & \frac{\Delta_z}{2}\sigma_z + \mathsf{S} + \gamma_s a_0[(\kappa q_x + A_1)\sigma_x + q_y\sigma_y] \\ & + \frac{\hbar^2}{4m_e}\left\{\alpha\left[(q_x + \kappa A_2)^2 + q_y^2\right] + \beta\left[(q_x + \kappa A_3)^2 + q_y^2\right]\sigma_z\right\}. \end{aligned} \tag{7.81}$$

Here, $\sigma_x, \sigma_y, \sigma_z$ are the Pauli matrices in the 2 × 2 'band' space and $\mathsf{S} = \mathrm{diag}(S_+, S_-)$; other variables will be explained later. For better interpretation of the above Hamiltonian, we can rewrite it as a general Hamiltonian described by the identity matrix σ_0 and the Pauli matrix $\boldsymbol{\sigma}$ [274], as follows

$$\begin{aligned} H_s^\kappa(\boldsymbol{q}, \varepsilon_b) &= h_0(\boldsymbol{q}, \varepsilon_b)\sigma_0 + h_1(\boldsymbol{q}, \varepsilon_b)\sigma_x + h_2(\boldsymbol{q}, \varepsilon_b)\sigma_y + h_3(\boldsymbol{q}, \varepsilon_b)\sigma_z \\ &= \begin{pmatrix} h_0(\boldsymbol{q}, \varepsilon_b) + h_3(\boldsymbol{q}, \varepsilon_b) & h_1(\boldsymbol{q}, \varepsilon_b) - \mathrm{i}h_2(\boldsymbol{q}, \varepsilon_b) \\ h_1(\boldsymbol{q}, \varepsilon_b) + \mathrm{i}h_2(\boldsymbol{q}, \varepsilon_b) & h_0(\boldsymbol{q}, \varepsilon_b) - h_3(\boldsymbol{q}, \varepsilon_b) \end{pmatrix}, \end{aligned} \tag{7.82}$$

where $h_i(\boldsymbol{q}, \varepsilon_b)$ ($i = 0$, 1, 2 and 3) are defined by [26]

$$\begin{aligned}
h_0(\boldsymbol{q}, \varepsilon_b) &= \frac{\mathrm{tr(S)}}{2} + \frac{\hbar^2\alpha}{4m_e}\left[(q_x + \kappa A_2)^2 + q_y^2\right],\\
h_1(\boldsymbol{q}, \varepsilon_b) &= \gamma_s a_0(\kappa q_x + A_1),\\
h_2(\boldsymbol{q}, \varepsilon_b) &= \gamma_s a_0 q_y,\\
h_3(\boldsymbol{q}, \varepsilon_b) &= \frac{1}{2}\left\{\Delta_z + S_+ - S_- + \frac{\hbar^2\beta}{2m_e}\left[(q_x + \kappa A_3)^2 + q_y^2\right]\right\}.
\end{aligned} \tag{7.83}$$

Here Δ_z represents the bandgap of the unstrained TMDC. α and β are dimensionless parameters to reproduce the strain effect in TMDC. $S_\pm \equiv 4\varsigma_\pm\varepsilon_b^2$ are strain-dependent scalar potentials, in which $\varsigma_\pm$ are coefficients of the strain for the lowest conduction band (+) and for the highest valence band (−). In the presence of the strain, q_x in the κ valley is shifted by the pseudo-vector potential defined by $A_i = \frac{2g_i}{a_0}\varepsilon_b$ ($i = 1$, 2, 3), in which g_i is the dimensionless parameter and a_0 is the metal–chalcogenide bond length projected on the xy-plane [273]. The A_1 term has been frequently discussed in graphene [275–284] in section 3.1. The A_2 and A_3 terms are newly introduced in the strained TMDC [285]. It is important to note that in the zigzag nanoribbon as shown in figure 7.8(a), the shift in k_y does not appear since the pseudo-vector potential for k_y, A_y, does not exist in equation (7.81). It is because the A_y is proportional to the difference of the bond lengths for the two symmetric NN bonds to the y axis (see equation (3.9) [276]), and that the difference does not occur for the present strain $\varepsilon_x = -\varepsilon_y \equiv \varepsilon_b$. Thus the energy band does not shift in the k_y direction but only in the k_x direction when the strain is applied.

For the typical monolayer TMDC, MoS_2, the dimensionless parameters, g_i ($i = 1$, 2, 3), α, and β are fitted to $g_1 = 0.002$, $g_2 = -56.551$, $g_3 = 1.635$, $\alpha = -0.01$ and $\beta = -1.54$; while the other energy parameters are $\gamma_s = 2.34$, $\varsigma_+ = 15.99$, $\varsigma_- = 15.92$, $\Delta_z = 1.82$ in eV. According to the DFT result [271], even for $\varepsilon_b = 0.1$, the lattice constants of strained monolayer TMDCs change within 5% from the unstrained ones. Thus, for all ε_b, we safely set $a_0 = \frac{a}{\sqrt{3}}$, where the lattice constant for unstrained monolayer MoS_2 is given by $a = 3.18$Å [271].

By solving the effective Hamiltonian in equation (7.81), we get the energy dispersion as a function of $\boldsymbol{q}$ and ε_b in the $\boldsymbol{K}_\kappa$ valley as follows:

$$\mathcal{E}_\xi^\kappa(\boldsymbol{q}, \varepsilon_b) = h_0(\boldsymbol{q}, \varepsilon_b) + \xi\sqrt{h_1(\boldsymbol{q}, \varepsilon_b)^2 + h_2(\boldsymbol{q}, \varepsilon_b)^2 + h_3(\boldsymbol{q}, \varepsilon_b)^2}. \tag{7.84}$$

It is important to point out that since the $h_0(\boldsymbol{q}, \varepsilon_b)$ is a function of $\boldsymbol{q}$, the energy dispersion of the conduction and valence bands are not symmetric in energy. On the other hand, if the h_0 is a constant, the two energy bands would be symmetric. Further, the extrema of $h_0(\boldsymbol{q}, \varepsilon_b)$ and the term inside the square root are not at the same q_x. This gives the different locations of the CBM and the VBM on the k_x-axis [273].

In figure 7.10(a), we plot the energy dispersion of MoS_2 for the conduction (up) and valence (down) bands on the q_x axis near the $\boldsymbol{K}_-$ (left) and $\boldsymbol{K}_+$ (right) points in

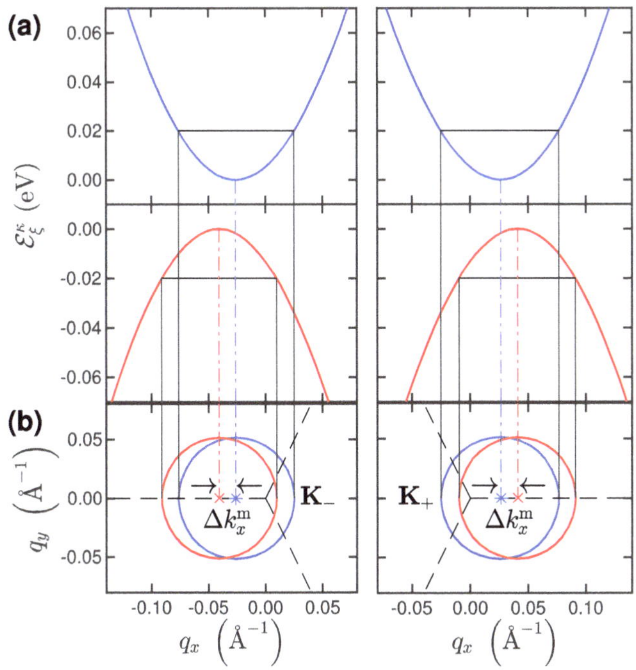

Figure 7.10. (a) Energy dispersion of MoS_2 for the conduction band (up) and the valence band (down) on the k_x axis near the $\boldsymbol{K}_-$ (left) and $\boldsymbol{K}_+$ (right) point in the presence of the strain ($\varepsilon_x = -\varepsilon_y = \varepsilon_b = 0.05$). The strained TMDC has an indirect gap. (b) Equienergy contours of the electron (blue loop) and the hole (red loop) distributions in the $\boldsymbol{k}$-space in the $\boldsymbol{K}_-$ (left) and $\boldsymbol{K}_+$ (right) valleys. Here, we assume the pseudo Fermi energies for the electrons and the holes are taken to be $\mathcal{E}_F = 0.02$ eV. Reproduced from [26]. CC BY 4.0.

the presence of the strain ($\varepsilon_x = -\varepsilon_y = \varepsilon_b = 0.05$). As shown in figure 7.10(a), the strained TMDC shows an indirect gap both for the $\boldsymbol{K}_+$ and $\boldsymbol{K}_-$ valleys. In figure 7.10(b), we plot equienergy contours for the pseudo Fermi energy ($\mathcal{E}_F = 0.02$ eV) for the electron (blue loop) measured from the bottom of the conduction band and the hole (red loop) from the top of the valence band. Here, the pseudo Fermi energy is defined by the maximum energy of the electrons or the holes in which the electrons or the holes are, respectively, injected in the conduction or the valence band in the i region. Since the inside of the red (blue) loop corresponds to the states occupied by the electrons (the holes), the EL occurs in the overlapping regions of the two loops. It is noted that the relative portion of the overlapping area to the loop area decreases with increasing $\mathcal{E}_F$ for a given value of the shift.

The relative shift of the two loops are symmetric for the $\boldsymbol{K}_+$ and $\boldsymbol{K}_-$ valleys since the pesudo-vector potential does not break the TRS. The relative shift at each valley occurs via the strain effect. In fact, the shift does not occur when we select $\varepsilon_b = 0$ in the Hamiltonian (7.81). When we discuss the strain effect, since the relative shift made by the strain is more effective than that by the TW effect even for a small $\varepsilon_b = 0.01$, thus we neglect the TW effect in the present Hamiltonian. Since we cannot find analytical expression of the shift, we plot in figure 7.11 the difference of the

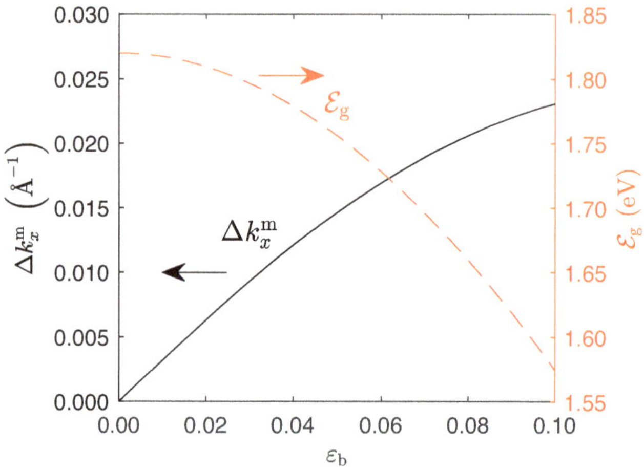

Figure 7.11. The difference of the k_x and energy gap, Δk_x^{m} and $\mathcal{E}_{\mathrm{g}}$, respectively, between the CBM and the VBM at the $\boldsymbol{K}_-$ valley are plotted as a function of ε_{b}. Reproduced from [26]. CC BY 4.0.

k_x coordinates of CBM ($k_{x,\mathrm{c}}^{\mathrm{m}}$) and VBM ($k_{x,\mathrm{v}}^{\mathrm{m}}$) in the $\boldsymbol{K}_-$ valley, $\Delta k_x^{\mathrm{m}} = k_{x,\mathrm{c}}^{\mathrm{m}} - k_{x,\mathrm{v}}^{\mathrm{m}}$, as a function of ε_{b}. As we can see, Δk_x^{m} increases with increasing ε_{b} due to the term of A_2. Thus, the electron–hole asymmetry increases with increasing strain. We also plot the value of the indirect gap $\mathcal{E}_{\mathrm{g}}$ as a function of ε_{b} in figure 7.11. The bandgap decreases from $\Delta_z = 1.82$ eV (unstrained) to $\sim$1.57 eV ($\varepsilon_{\mathrm{b}} = 0.1$) and shows the same trend as studied by the DFT result [271, 272], but the calculated value does not agree well with the DFT result for large strain ($\varepsilon_{\mathrm{b}} \geqslant 0.06$). This might be originated from the limitations of the low-energy Hamiltonian.

When we apply the in-plane electric field $\boldsymbol{E}$ in the i region, the occupied $\boldsymbol{k}$ region of the electrons (or holes) shifts from the equi-enegy loop according to the Drude model in equation (7.79). This means that the charge occupation of electrons (or holes) is shifted in the fixed energy dispersion in the presence of $\boldsymbol{E}$. In figure 7.12(a), we illustrate the tilted occupied $\boldsymbol{k}$ states of electrons and holes in the presence of $\boldsymbol{E}$ along the k_x direction. In the plot, we adopt $\tau E = 5 \times 10^{-8}$ N $\cdot$ s/C and the other parameters are taken to be the same as those in figure 7.10. Since the loops of electron and hole are shifted in opposite directions to each other for a given $\boldsymbol{E}$ and the loops of electron (or hole) are shifted in the same direction for the $\boldsymbol{K}_+$ and $\boldsymbol{K}_-$ valleys, the overlapping regions in the $\boldsymbol{K}_+$ and $\boldsymbol{K}_-$ valleys become inequivalent, as shown in figure 7.12(b), which is the origin of circularly polarized EL emission in the strained TMDC.

This work proposed a mechanism for electrically controlling the circularly polarized EL in strained TMDCs and optimized the degree of circular polarisation to realize electrically switchable room-temperature valley LEDs. The results provide a new pathway for using LEDs in optoelectronic device applications and investigating the fundamental quantum properties of materials.

Such TMDC monolayer p–n junctions can also be used for light detection and energy harvesting [286]. Additionally, excitonic valley coherence has also been demonstrated [155]. This type of dynamic control of the valley index opens up the

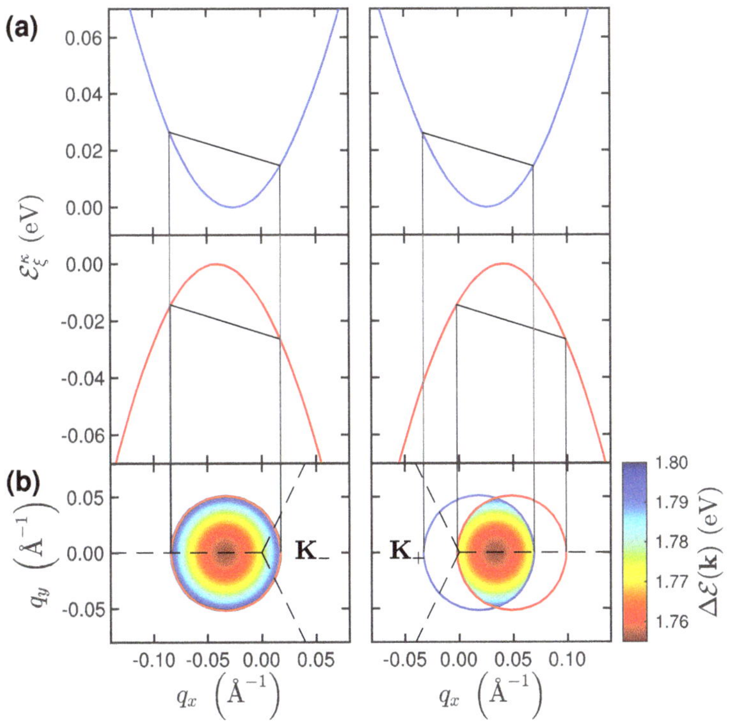

Figure 7.12. (a) In the presence of $\boldsymbol{E}$, the occupied states of electrons (top) and holes (bottom) are shifted in the opposite direction to each other, compared with figure 7.10(a). (b) The shifted loops of electrons (blue line) and holes (red line) are plotted in the $\boldsymbol{k}$-space. The overlapping regions are in the colour that represents the energy difference between the conduction and valence bands $\Delta\mathcal{E}(\boldsymbol{k})$ with the corresponding colour scale. We set $\tau E = 5 \times 10^{-8}$ N · s/C. Other parameters are the same as those in figure 7.10. Reproduced from [26]. CC BY 4.0.

exciting possibility of valleytronics—optoelectronic devices and systems based on the manipulation of the electrons' valley index [287].

7.7.2 Valleytronic transistor—experimental demonstration of VHE

In the previous sections, we reviewed the conversion of the VP into circularly polarized light in TMDC layers [75, 250]. In this section, we will introduce the conversion of the circularly polarized light into VP, which is made manifest in VHE.

As we have discussed in section 2.6, for monolayer TMDC, an applied in-plane electrical field will bring forth a traditional longitudinal charge current as well as a transverse valley current where electrons in the $\boldsymbol{K}_+$ and $\boldsymbol{K}_-$ valleys flow in opposite transverse directions. This phenomenon is known as the VHE. Similar to the intrinsic anomalous Hall effect in ferromagnets and the intrinsic spin Hall effect in heavy metal systems, the VHE originates from non-zero Berry curvature in momentum space and leads to VP accumulating on the two transverse edges of the sample (figure 2.8(b)). Several major experimental breakthroughs [170, 288–290] have been achieved to demonstrate the VHE. The materials are monolayer MoS_2 [170], single layer graphene

[288] and bilayer graphene with the IS broken property [289, 290]. The VHE was first demonstrated experimentally by Mak *et al* using optical injection and electrical readout in a monolayer MoS_2 FET with Hall bar geometry at 77 K (figures 7.13(a) and (b)) [170, 291]. As shown in figure 7.13(a), the direct energy gaps are located at the two valleys, and the valley-polarized carriers can be optically pumped by circularly polarized photons.

In figure 7.13(b), a Hall bar device based on monolayer MoS_2 is prepared by mechanical exfoliation [292], a commercially available circularly polarized pump laser (near the exciton resonance at 1.9 eV [293]) injected valley-polarized electrons and holes by utilizing the optical selection rules. Specifically, they selectively applied right-to-left- (R-L)/right-to-left-handed (L-R) excitation to preferentially inject electrons and holes in the $\boldsymbol{K}_{\pm}$ valley, thus creating an imbalance of carrier population between two different valleys [104]. A source–drain voltage V_x drove a current (I_x)

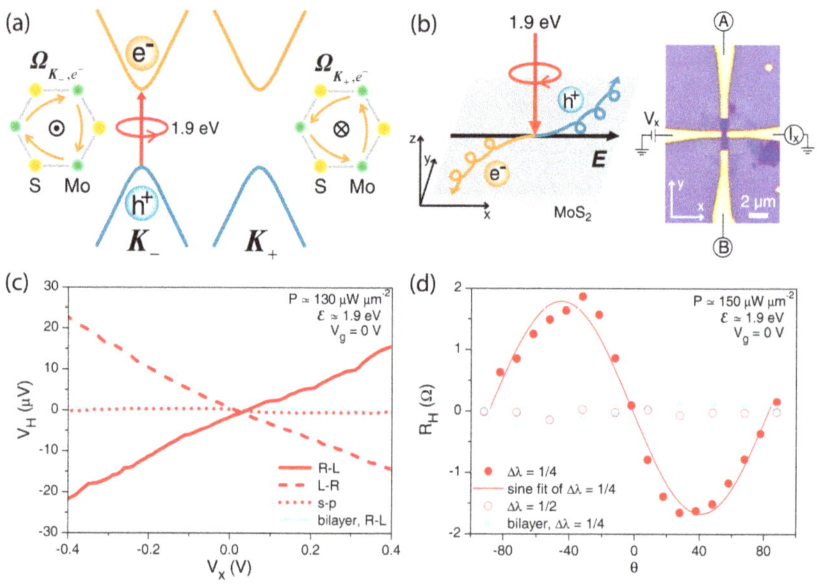

Figure 7.13. Monolayer MoS_2 Hall bar device for demonstrating VHE. (a) Schematics of the valley-dependent optical selection rules and the electrons at the $\boldsymbol{K}_+$ and $\boldsymbol{K}_-$ valleys that possess opposite Berry curvatures $\boldsymbol{\Omega}_{\boldsymbol{K}_{\pm},e^-}$. The orange arrows represent the clockwise and anticlockwise hopping motions of the $\boldsymbol{K}_-$ and $\boldsymbol{K}_+$ electrons, respectively. (b) Left: Schematics of a photo-induced VHE driven by a net VP and in-plane electric field, right: an optical image of the Hall bar device to detect VHE. (c) Hall voltage (V_H) as a function of the source–drain voltage bias (V_x) that is applied perpendicularly. The red dotted line shows the result for monolayer samples excited by a linearly polarized laser (s-p), and the blue line shows that for a bilayer sample excited by a circularly polarized laser with R-L modulation. In both cases, zero Hall voltage was observed. The solid and dashed red lines show linear dependence of V_H on V_x for monolayer samples excited by a circularly polarized laser with opposite handedness (R-L and L-R modulations). (d) The anomalous Hall resistance of the Hall bar device as a function of θ, which is the angle between the polarisation direction of the incident linearly polarized light and the fast axis of the quarter-waveplate ($\Delta\lambda = 1/4$, solid red circles) and half-waveplate ($\Delta\lambda = 1/2$, open red circles). Results from the bilayer Hall bar device under quarter-wave modulation are shown with solid blue circles. From [170]. Reprinted with permission from the AAAS.

and yielded an electric field, which resulted in a photoexcited hole and electron travelling oppositely in traditional longitudinal directions as well as transverse directions due to VHE [figure 2.5(b)]. The resulting valley Hall current in the transverse direction arises a preferential flow of holes (electrons) toward the contact A (B), which yielded a transverse Hall voltage (V_H) between contacts A and B, enabling electrical detection. The magnitude of the VHE is evaluated by the Hall conductivity. In addition to the intrinsic Hall conductivity in equation (2.36) induced by Berry curvature, carriers on MoS_2 are also subjected to the extrinsic side-jump effect [294–296], which is twice as big as the intrinsic effect and has the opposite sign [170, 297]. The predicted Hall conductivity is [170]

$$\sigma_H \approx -\frac{\hbar^2 \pi \Delta n_v}{m^* \mathcal{E}_g} \frac{e^2}{h}, \tag{7.85}$$

where m^* is the effective mass of carriers on monolayer MoS_2 [298], $\mathcal{E}_g \approx 1.9$ eV is the bandgap of monolayer MoS_2 [98], and $\Delta n_v \equiv n_+ - n_-$ is the density difference between the photoexcited carriers in the $\boldsymbol{K}_+$ and $\boldsymbol{K}_-$ valleys [297].

During the measurement, V_H is simultaneously recorded as a function of V_x of the MoS_2 channel. As shown in figure 7.13(c), a finite V_H in the order of 100 μV [299] that scales linearly with V_x is observed under R-L modulations, while the sign of V_H flips under L-R modulation. In addition, no V_H was observed in the MoS_2 monolayer under linear (s–p) polarized light modulation. This is the signature of the helicity-dependent characteristic of valleys and a photo-induced VHE driven by the net VP [130].

It is also possible to continuously tune the polarization of the irradiated light by the angle θ of the incident linearly polarized light with respect to the fast axis of a quarter-wave ($\Delta\lambda = 1/4$) modulator, which produces circularly polarized light. We find that the Hall resistance ($R_H = V_H/I_x$) sinusoidally changes with θ (figure 7.13(d)), the circular polarization of the resultant light is maximised when $\theta = 45°$. This confirms the linear relation between the anomalous Hall voltage, degree of VP [295] and the excitation ellipticity [50, 300]. On the other hand, no signal is probed under half-wave ($\Delta\lambda = 1/2$) modulation, which preserves linearly polarized behaviour of incident light.

These demonstrations are qualitatively consistent with equation (7.85), including the relative dominance of the extrinsic effect over the intrinsic Berry phase effect confirmed by the sign of the V_H [295]. Such demonstrations involve a transverse component of the electric current, although a transverse nondissipative pure valley current without the electric current cannot be demonstrated in this method. To investigate the pure valley current, one needs to employ a material with high crystal quality, which has not yet been achieved for TMDC layers. The sensible answer to this is to employ graphene [288], as narrated in section 2.4.4.

For a bilayer MoS_2, the VHE is inapplicable due to IS [301], but revokes with an perpendicular electric field added to break the IS in the bilayer and makes the VHE effect controllable with a top gate, which is also observed experimentally [148, 302–304]. We will discuss this in detail in section 8.1. These results demonstrated that the Hall signal was originated mainly from VHE [305].

Further, the dichroic spin–valley-coupled photogalvanic current has been demonstrated in devices based on CVD-grown MoS_2 [306] and WSe_2 monolayers [307, 308], thus extending the applications of the valleytronics of TMDC monolayers and opening up new avenues of approach to photoelectrical conversion in optoelectronics [150].

References

[1] Gasiorowicz S 2003 *Quantum Physics* 3rd edn (New York: Wiley)

[2] Schiff L I 1968 *Quantum Mechanics* International Series in Pure and Applied Physics 3rd edn (New York: McGraw-Hill)

[3] Yu P Y and Cardona M 2010 *Fundamentals of Semiconductors: Physics and Materials Properties* Graduate Texts in Physics 4th edn (Heidelberg: Springer)

[4] Lüth H 2015 *Quantum Physics in the Nanoworld: Schrödinger's Cat and the Dwarfs* Graduate Texts in Physics 2nd edn (Heidelberg: Springer)

[5] Sharma A C 2021 *A Textbook on Modern Quantum Mechanics* (Boca Raton, FL: CRC Press)

[6] Li M-F 1995 *Modern Semiconductor Quantum Physics* International Series on Advances in Solid State Electronics and Technology (Singapore: World Scientific)

[7] Toropov A A and Shubina T V 2015 *Plasmonic Effects in Metal–Semiconductor Nanostructures* Series on Semiconductor Science and Technology (Oxford: Oxford University Press)

[8] Basu P K, Mukhopadhyay B and Basu R 2022 *Semiconductor Nanophotonics* Series on Semiconductor Science and Technology (Oxford: Oxford University Press)

[9] Griffiths D J 2017 *Introduction to Electrodynamics* 4th edn (Cambridge: Cambridge University Press)

[10] Grundmann M 2021 *The Physics of Semiconductors: An Introduction Including Nanophysics and Applications* Graduate Texts in Physics 4th edn (Cham: Springer)

[11] Bottacchi S 2014 *Theory and Design of Terabit Optical Fiber Transmission Systems* (Cambridge: Cambridge University Press)

[12] Pollack G L and Stump D R 2002 *Electromagnetism* (San Francisco, CA: Addison-Wesley)

[13] Santamato E, Daino B, Romagnoli M, Settembre M and Shen Y R 1986 Collective rotation of molecules driven by the angular momentum of light in a nematic film *Phys. Rev. Lett.* **57** 2423–6

[14] Zel'dovich B Ya and Tabiryan N V 1982 Orientational effect of a light wave on a cholesteric mesophase *Zh. Eksp. Teor. Fiz.* **82** 167–76

[15] Yamanouchi K and Katsumi M (ed) 2012 *Multiphoton Processes and Attosecond Physics: Proc. 12th Int. Conf. on Multiphoton Processes (ICOMP12) and the 3rd Int. Conf. on Attosecond Physics (ATTO3)* Springer Proceedings in Physics **vol 125** (Heidelberg: Springer Science)

[16] Born M and Wolf E 2013 *Principles of Optics: Electromagnetic Theory of Propagation, Interference and Diffraction of Light* 7th edn (Cambridge: Cambridge University Press)

[17] McIntyre D H 2022 *Quantum Mechanics: A Paradigms Approach* (Cambridge: Cambridge University Press)

[18] Ballentine L E 2014 *Quantum Mechanics: A Modern Development* 2nd edn (Singapore: World Scientific)

[19] Fermi E 1932 Quantum theory of radiation *Rev. Mod. Phys.* **4** 87–132

[20] Coldren L A, Corzine S W and Mašanović M L 2012 *Diode Lasers and Photonic Integrated Circuits* Wiley Series in Microwave and Optical Engineering 2nd edn (Hoboken, NJ: Wiley)

[21] Hung N T, Nugraha A R T and Saito R 2022 *Quantum ESPRESSO Course for Solid-State Physics* 1st edn (New York: Jenny Stanford Publishing)

[22] Bechstedt F 2003 *Principles of Surface Physics* Advanced Texts in Physics (Heidelberg: Springer)

[23] Grüneis A, Saito R, Samsonidze G G, Kimura T, Pimenta M A, Jorio A, Souza Filho A G, Dresselhaus G and Dresselhaus M S 2003 Inhomogeneous optical absorption around the *K* point in graphite and carbon nanotubes *Phys. Rev.* B **67** 165402

[24] Ridley B K 2013 *Quantum Processes in Semiconductors* 5th edn (Oxford: Oxford University Press)

[25] Levi A F J 2023 *Applied Quantum Mechanics* 3rd edn (Cambridge: Cambridge University Press)

[26] Wang S, Ukhtary M S and Saito R 2020 Strain effect on circularly polarized electroluminescence in transition metal dichalcogenides *Phys. Rev. Res.* **2** 033340

[27] Jones R C 1941 A new calculus for the treatment of optical systems: I. Description and discussion of the calculus *J. Opt. Soc. Am.* **31** 488–93

[28] Hurwitz H and Jones R C 1941 A new calculus for the treatment of optical systems: II. Proof of three general equivalence theorems *J. Opt. Soc. Am.* **31** 493–9

[29] Jones R C 1941 A new calculus for the treatment of optical systems: III. The Sohncke theory of optical activity *J. Opt. Soc. Am.* **31** 500–3

[30] Jones R C 1942 A new calculus for the treatment of optical systems: IV. *J. Opt. Soc. Am.* **32** 486–93

[31] Goldstein D H 2017 *Polarized Light* 3rd edn (Boca Raton, FL: CRC Press)

[32] Yao X S and Chen X J 2022 *Polarization Measurement and Control in Optical Fiber Communication and Sensor Systems* (Hoboken, NJ: Wiley)

[33] Menzel R 2007 *Photonics: Linear and Nonlinear Interactions of Laser Light and Matter* Advanced Texts in Physics 2nd edn (Heidelberg: Springer)

[34] Collett E 2005 *Field Guide to Polarization* SPIE Field Guides **vol FG05** (Bellingham, WA: SPIE Press)

[35] Sharma K K 2006 *Optics: Principles and Applications* (San Diego, CA: Elsevier)

[36] Tompkins H and Irene E A (ed) 2005 *Handbook of Ellipsometry* (Norwich: William Andrew Publishing)

[37] Kuch W, Schäfer R, Fischer P and Hillebrecht F U 2015 *Magnetic Microscopy of Layered Structures* Springer Series in Surface Sciences **vol 57** (Heidelberg: Springer)

[38] Yeh P and Gu C 2009 *Optics of Liquid Crystal Displays* Wiley Series in Pure and Applied Optics 2nd edn (New York: Wiley)

[39] Azzam R M A and Bashara N M 1988 *Ellipsometry and Polarized Light* (New York: North-Holland)

[40] Pedrotti F L, Pedrotti L M and Pedrotti L S 2017 *Introduction to Optics* 3rd edn (Cambridge: Cambridge University Press)

[41] Lakshminarayanan V, Ghalila H, Ammar A and Varadharajan L S 2018 *Understanding Optics with Python* CRC Series on Multidisciplinary and Applied Optics (Boca Raton, FL: CRC Press)

[42] Reider G A 2016 *Photonics: An Introduction* (Cham: Springer)

[43] Khoo I-C 2022 *Liquid Crystals* Wiley Series in Pure and Applied Optics 3rd edn (New York: Wiley)

[44] Hecht E 2016 *Optics* 5th edn (London: Pearson)

[45] Ghalamkari K, Tatsumi Y and Saito R 2018 Energy band gap dependence of valley polarization of the hexagonal lattice *J. Phys. Soc. Jpn.* **87** 024710

[46] Ezawa M 2012 Spin-valley optical selection rule and strong circular dichroism in silicene *Phys. Rev.* B **86** 161407(R)

[47] Tatsumi Y, Ghalamkari K and Saito R 2016 Laser energy dependence of valley polarization in transition-metal dichalcogenides *Phys. Rev.* B **94** 235408

[48] Yao W, Xiao D and Niu Q 2008 Valley-dependent optoelectronics from inversion symmetry breaking *Phys. Rev.* B **77** 235406

[49] Wang Z M (ed) 2014 *MoS_2: Materials, Physics, and Devices* Lecture Notes in Nanoscale Science and Technology **vol 21** (Cham: Springer)

[50] Xiao D, Liu G-B, Feng W, Xu X and Yao W 2012 Coupled spin and valley physics in monolayers of MoS_2 and other group-VI dichalcogenides *Phys. Rev. Lett.* **108** 196802

[51] Diebold A and Hofmann T 2021 *Optical and Electrical Properties of Nanoscale Materials* Springer Series in Materials Science **vol 318** (Cham: Springer)

[52] Souza I and Vanderbilt D 2008 Dichroic f-sum rule and the orbital magnetization of crystals *Phys. Rev.* B **77** 054438

[53] Berghäuser G and Malic E 2014 Analytical approach to excitonic properties of MoS_2 *Phys. Rev.* B **89** 125309

[54] Bouet L 2015 Valley dynamics and excitonic properties in monolayer transition metal dichalcogenides *PhD Thesis* Université de Toulouse, Toulouse

[55] Sie E J 2017 *Coherent Light-Matter Interactions in Monolayer Transition-Metal Dichalcogenides* Springer Theses (Cham: Springer)

[56] Liu G-B, Xiao D, Yao Y, Xu X and Yao W 2015 Electronic structures and theoretical modelling of two-dimensional group-VIB transition metal dichalcogenides *Chem. Soc. Rev.* **44** 2643–63

[57] Saito R, Ukhtary M S, Wang S and Hung N T 2021 Selection rule for Raman spectra of two-dimensional materials using circularly-polarized vortex light *Phys. Chem. Chem. Phys.* **23** 17271–8

[58] Zhu Z Y, Cheng Y C and Schwingenschlögl U 2011 Giant spin-orbit-induced spin splitting in two-dimensional transition-metal dichalcogenide semiconductors *Phys. Rev.* B **84** 153402

[59] Liu G-B, Shan W-Y, Yao Y, Yao W and Xiao D 2013 Three-band tight-binding model for monolayers of group-VIB transition metal dichalcogenides *Phys. Rev.* B **88** 085433

[60] Kośmider K, González J W and Fernández-Rossier J 2013 Large spin splitting in the conduction band of transition metal dichalcogenide monolayers *Phys. Rev.* B **88** 245436

[61] Kormányos A, Zólyomi V, Drummond N D and Burkard G 2014 Spin-orbit coupling, quantum dots, and qubits in monolayer transition metal dichalcogenides *Phys. Rev.* X **4** 011034

[62] Kormányos A, Zólyomi V, Drummond N D and Burkard G 2014 Erratum: Spin-orbit coupling, quantum dots, and qubits in monolayer transition metal dichalcogenides [Phys. Rev. X 4, 011 034 (2014)] *Phys. Rev.* X **4** 039901

[63] Roldán R, López-Sancho M P, Guinea F, Cappelluti E, Silva-Guillén J A and Ordejón P 2014 Momentum dependence of spin–orbit interaction effects in single-layer and multi-layer transition metal dichalcogenides *2D Mater.* **1** 034003

[64] Xu X, Yao W, Xiao D and Heinz T F 2014 Spin and pseudospins in layered transition metal dichalcogenides *Nat. Phys.* **10** 343–50

[65] Yu H, Liu G-B and Yao W 2018 Brightened spin-triplet interlayer excitons and optical selection rules in van der Waals heterobilayers *2D Mater.* **5** 035021
[66] Mak K F, He K, Shan J and Heinz T F 2012 Control of valley polarization in monolayer MoS_2 by optical helicity *Nat. Nanotechnol.* **7** 494–8
[67] Mak K F and Shan J 2016 Photonics and optoelectronics of 2D semiconductor transition metal dichalcogenides *Nat. Photon.* **10** 216–26
[68] Mak K F, Xiao D and Shan J 2018 Light–valley interactions in 2D semiconductors *Nat. Photon.* **12** 451–60
[69] Tatsumi Y, Kaneko T and Saito R 2018 Conservation law of angular momentum in helicity-dependent Raman and Rayleigh scattering *Phys. Rev.* B **97** 195444
[70] Tatsumi Y 2017 Polarization dependence of optical absorption and Raman spectra for atomic layer materials *PhD Thesis* Tohoku University, Sendai
[71] Simon H J and Bloembergen N 1968 Second-harmonic light generation in crystals with natural optical activity *Phys. Rev.* **171** 1104–14
[72] Visser J, Eliel E R and Nienhuis G 2002 Polarization entanglement in a crystal with threefold symmetry *Phys. Rev.* A **66** 033814
[73] Bloembergen N 1980 Conservation laws in nonlinear optics *J. Opt. Soc. Am.* **70** 1429–36
[74] Zhang X, Shan W-Y and Xiao D 2018 Optical selection rule of excitons in gapped chiral fermion systems *Phys. Rev. Lett.* **120** 077401
[75] Cao T *et al* 2012 Valley-selective circular dichroism of monolayer molybdenum disulphide *Nat. Commun.* **3** 887
[76] Xie H 2019 Probing excitonic mechanics in suspended and strained transition metal dichalcogenides monolayers *PhD Thesis* The Pennsylvania State University, University Park, PA
[77] Rivera P, Yu H, Seyler K L, Wilson N P, Yao W and Xu X 2018 Interlayer valley excitons in heterobilayers of transition metal dichalcogenides *Nat. Nanotechnol.* **13** 1004–15
[78] Cappelluti E, Roldán R, Silva-Guillén J A, Ordejón P and Guinea F 2013 Tight-binding model and direct-gap/indirect-gap transition in single-layer and multilayer MoS_2 *Phys. Rev.* B **88** 075409
[79] Liu G-B, Shan W-Y, Yao Y, Yao W and Xiao D 2014 Erratum: Three-band tight-binding model for monolayers of group-VIB transition metal dichalcogenides [Phys. Rev. B 88, 085433 (2013)] *Phys. Rev.* B **89** 039901
[80] Kormányos A, Zólyomi V, Drummond N D, Rakyta P, Burkard G and Fal'ko V I 2013 Monolayer MoS_2: trigonal warping, the Γ valley, and spin-orbit coupling effects *Phys. Rev.* B **88** 045416
[81] Higuchi T, Kanda N, Tamaru H and Kuwata-Gonokami M 2011 Selection rules for light-induced magnetization of a crystal with threefold symmetry: the case of antiferromagnetic NiO *Phys. Rev. Lett.* **106** 047401
[82] Ishii S, Yokoshi N and Ishihara H 2019 Optical selection rule of monolayer transition metal dichalcogenide by an optical vortex *J. Phys.: Conf. Ser.* **1220** 012056
[83] Yang X L, Guo S H, Chan F T, Wong K W and Ching W Y 1991 Analytic solution of a two-dimensional hydrogen atom. I. Nonrelativistic theory *Phys. Rev.* A **43** 1186–96
[84] Hüser F, Olsen T and Thygesen K S 2013 How dielectric screening in two-dimensional crystals affects the convergence of excited-state calculations: monolayer MoS_2 *Phys. Rev.* B **88** 245309

[85] Cudazzo P, Tokatly I V and Rubio A 2011 Dielectric screening in two-dimensional insulators: implications for excitonic and impurity states in graphane *Phys. Rev.* B **84** 085406

[86] Latini S, Olsen T and Thygesen K S 2015 Excitons in van der Waals heterostructures: the important role of dielectric screening *Phys. Rev.* B **92** 245123

[87] Wang G, Chernikov A, Glazov M M, Heinz T F, Marie X, Amand T and Urbaszek B 2018 Colloquium: excitons in atomically thin transition metal dichalcogenides *Rev. Mod. Phys.* **90** 021001

[88] Chernikov A, Berkelbach T C, Hill H M, Rigosi A, Li Y, Aslan B, Reichman D R, Hybertsen M S and Heinz T F 2014 Exciton binding energy and nonhydrogenic Rydberg series in monolayer WS_2 *Phys. Rev. Lett.* **113** 076802

[89] Zhu B, Chen X and Cui X 2015 Exciton binding energy of monolayer WS_2 *Sci. Rep.* **5** 9218

[90] Rasmita A and Gao W-B 2021 Opto-valleytronics in the 2D van der Waals heterostructure *Nano Res.* **14** 1901–11

[91] Ye Z, Cao T, O'Brien K, Zhu H, Yin X, Wang Y, Louie S G and Zhang X 2014 Probing excitonic dark states in single-layer tungsten disulphide *Nature* **513** 214–8

[92] Kübler J K 1969 The exciton binding energy of III–V semiconductor compounds *Phys. Stat. Sol. (b)* **35** 189–95

[93] Wu S 2016 Device physics of two-dimensional crystalline materials *PhD Thesis* University of Washington, Seattle, WA

[94] Ross J S *et al* 2013 Electrical control of neutral and charged excitons in a monolayer semiconductor *Nat. Commun.* **4** 1474

[95] Mak K F, He K, Lee C, Lee G H, Hone J, Heinz T F and Shan J 2013 Tightly bound trions in monolayer MoS_2 *Nat. Mater.* **12** 207–11

[96] Pei J, Yang J, Yildirim T, Zhang H and Lu Y 2019 Many-body complexes in 2D semiconductors *Adv. Mater.* **31** 1706945

[97] Thygesen K S 2017 Calculating excitons, plasmons, and quasiparticles in 2D materials and van der Waals heterostructures *2D Mater.* **4** 022004

[98] Mak K F, Lee C, Hone J, Shan J and Heinz T F 2010 Atomically thin MoS_2: a new direct-gap semiconductor *Phys. Rev. Lett.* **105** 136805

[99] Hu Z, Liu X, Hernández-Martínez P L, Zhang S, Gu P, Du W, Xu W, Demir H V, Liu H and Xiong Q 2022 Interfacial charge and energy transfer in van der Waals heterojunctions *InfoMat* **4** e12290

[100] Qiu D Y, da Jornada F H and Louie S G 2013 Optical spectrum of MoS_2: many-body effects and diversity of exciton states *Phys. Rev. Lett.* **111** 216805

[101] Qiu D Y, da Jornada F H and Louie S G 2015 Erratum: Optical spectrum of MoS_2: many-body effects and diversity of exciton states [Phys. Rev. Lett. 111, 216805 (2013)] *Phys. Rev. Lett.* **115** 119901

[102] Ye Z, Sun D and Heinz T F 2017 Optical manipulation of valley pseudospin *Nat. Phys.* **13** 26–9

[103] Yu H, Cui X, Xu X and Yao W 2015 Valley excitons in two-dimensional semiconductors *Natl. Sci. Rev.* **2** 57–70

[104] Yi Y, Chen Z, Yu X-F, Zhou Z-K and Li J 2019 Recent advances in quantum effects of 2D materials *Adv. Quantum Technol.* **2** 1800111

[105] Xu L, Zhao L, Wang Y, Zou M, Zhang Q and Cao A 2019 Analysis of photoluminescence behavior of high-quality single-layer MoS_2 *Nano Res.* **12** 1619–24

[106] Yu H, Liu G-B, Gong P, Xu X and Yao W 2014 Dirac cones and Dirac saddle points of bright excitons in monolayer transition metal dichalcogenides *Nat. Commun.* **5** 3876

[107] Drüppel M, Deilmann T, Krüger P and Rohlfing M 2017 Diversity of trion states and substrate effects in the optical properties of an MoS_2 monolayer *Nat. Commun.* **8** 2117

[108] Plechinger G, Nagler P, Arora A, Schmidt R, Chernikov A, del Águila A G, Christianen P C M, Bratschitsch R, Schüller C and Korn T 2016 Trion fine structure and coupled spin–valley dynamics in monolayer tungsten disulfide *Nat. Commun.* **7** 12715

[109] You Y, Zhang X-X, Berkelbach T C, Hybertsen M S, Reichman D R and Heinz T F 2015 Observation of biexcitons in monolayer WSe_2 *Nat. Phys.* **11** 477–81

[110] He Z, Xu W, Zhou Y, Wang X, Sheng Y, Rong Y, Guo S, Zhang J, Smith J M and Warner J H 2016 Biexciton formation in bilayer tungsten disulfide *ACS Nano* **10** 2176–83

[111] Hao K *et al* 2017 Neutral and charged inter-valley biexcitons in monolayer $MoSe_2$ *Nat. Commun.* **8** 15552

[112] Li Z *et al* 2018 Revealing the biexciton and trion-exciton complexes in BN encapsulated WSe_2 *Nat. Commun.* **9** 3719

[113] Sie E J, Frenzel A J, Lee Y-H, Kong J and Gedik N 2015 Intervalley biexcitons and many-body effects in monolayer MoS_2 *Phys. Rev.* B **92** 125417

[114] Geim A K and Grigorieva I V 2013 Van der Waals heterostructures *Nature* **499** 419–25

[115] Kang J, Tongay S, Zhou J, Li J and Wu J 2013 Band offsets and heterostructures of two-dimensional semiconductors *Appl. Phys. Lett.* **102** 012111

[116] Wang S, Ren C, Tian H, Yu J and Sun M 2018 MoS_2/ZnO van der Waals heterostructure as a high-efficiency water splitting photocatalyst: a first-principles study *Phys. Chem. Chem. Phys.* **20** 13394–9

[117] Wang S, Tian H, Ren C, Yu J and Sun M 2018 Electronic and optical properties of heterostructures based on transition metal dichalcogenides and graphene-like zinc oxide *Sci. Rep.* **8** 12009

[118] Rivera P *et al* 2015 Observation of long-lived interlayer excitons in monolayer $MoSe_2$–WSe_2 heterostructures *Nat. Commun.* **6** 6242

[119] Hong X, Kim J, Shi S-F, Zhang Y, Jin C, Sun Y, Tongay S, Wu J, Zhang Y and Wang F 2014 Ultrafast charge transfer in atomically thin MoS_2/WS_2 heterostructures *Nat. Nanotechnol.* **9** 682–6

[120] Fogler M M, Butov L V and Novoselov K S 2014 High-temperature superfluidity with indirect excitons in van der Waals heterostructures *Nat. Commun.* **5** 4555

[121] Gong Y *et al* 2014 Vertical and in-plane heterostructures from WS_2/MoS_2 monolayers *Nat. Mater.* **13** 1135–42

[122] Huang S, Ling X, Liang L, Kong J, Terrones H, Meunier V and Dresselhaus M S 2014 Probing the interlayer coupling of twisted bilayer MoS_2 using photoluminescence spectroscopy *Nano Lett.* **14** 5500–8

[123] Lee C-H *et al* 2014 Atomically thin p–n junctions with van der Waals heterointerfaces *Nat. Nanotechnol.* **9** 676–81

[124] Kunstmann J *et al* 2018 Momentum-space indirect interlayer excitons in transition-metal dichalcogenide van der Waals heterostructures *Nat. Phys.* **14** 801–5

[125] Ceballos F, Bellus M Z, Chiu H-Y and Zhao H 2014 Ultrafast charge separation and indirect exciton formation in a MoS_2–$MoSe_2$ van der Waals heterostructure *ACS Nano* **8** 12717–24

[126] Zheng W *et al* 2019 Direct vapor growth of 2D vertical heterostructures with tunable band alignments and interfacial charge transfer behaviors *Adv. Sci.* **6** 1802204

[127] Li L *et al* 2020 Wavelength-tunable interlayer exciton emission at the near-infrared region in van der Waals semiconductor heterostructures *Nano Lett.* **20** 3361–8
[128] Fang H *et al* 2014 Strong interlayer coupling in van der Waals heterostructures built from single-layer chalcogenides *Proc. Natl. Acad. Sci. USA* **111** 6198–202
[129] Rivera P, Seyler K L, Yu H, Schaibley J R, Yan J, Mandrus D G, Yao W and Xu X 2016 Valley-polarized exciton dynamics in a 2D semiconductor heterostructure *Science* **351** 688–91
[130] Ma H, Zhu Y, Liu Y, Bai R, Zhang X, Ren Y and Jiang C 2023 Valley polarization in transition metal dichalcogenide layered semiconductors: Generation, relaxation, manipulation and transport *Chin. Phys.* B **32** 107201
[131] Cadiz F *et al* 2017 Excitonic linewidth approaching the homogeneous limit in MoS_2-based van der Waals heterostructures *Phys. Rev.* X **7** 021026
[132] Nagler P *et al* 2017 Giant magnetic splitting inducing near-unity valley polarization in van der Waals heterostructures *Nat. Commun.* **8** 1551
[133] Chen Y J, Koteles E S, Elman B S and Armiento C A 1987 Effect of electric fields on excitons in a coupled double-quantum-well structure *Phys. Rev.* B **36** 4562–5
[134] Nagler P *et al* 2017 Interlayer exciton dynamics in a dichalcogenide monolayer heterostructure *2D Mater.* **4** 025112
[135] Gao S, Yang L and Spataru C D 2017 Interlayer coupling and gate-tunable excitons in transition metal dichalcogenide heterostructures *Nano Lett.* **17** 7809–13
[136] Hanbicki A T, Chuang H-J, Rosenberger M R, Hellberg C S, Sivaram S V, McCreary K M, Mazin I I and Jonker B T 2018 Double indirect interlayer exciton in a $MoSe_2/WSe_2$ van der Waals heterostructure *ACS Nano* **12** 4719–26
[137] Kamban H C and Pedersen T G 2020 Interlayer excitons in van der Waals heterostructures: binding energy, Stark shift, and field-induced dissociation *Sci. Rep.* **10** 5537
[138] Jiang C, Xu W, Rasmita A, Huang Z, Li K, Xiong Q and Gao W-B 2018 Microsecond dark-exciton valley polarization memory in two-dimensional heterostructures *Nat. Commun.* **9** 753
[139] Palummo M, Bernardi M and Grossman J C 2015 Exciton radiative lifetimes in two-dimensional transition metal dichalcogenides *Nano Lett.* **15** 2794–800
[140] Miller B, Steinhoff A, Pano B, Klein J, Jahnke F, Holleitner A and Wurstbauer U 2017 Long-lived direct and indirect interlayer excitons in van der Waals heterostructures *Nano Lett.* **17** 5229–37
[141] Baranowski M *et al* 2017 Probing the interlayer exciton physics in a $MoS_2/MoSe_2/MoS_2$ van der Waals heterostructure *Nano Lett.* **17** 6360–5
[142] Moody G *et al* 2015 Intrinsic homogeneous linewidth and broadening mechanisms of excitons in monolayer transition metal dichalcogenides *Nat. Commun.* **6** 8315
[143] Moody G, Schaibley J and Xu X 2016 Exciton dynamics in monolayer transition metal dichalcogenides [invited] *J. Opt. Soc. Am.* B **33** C39–49
[144] Morozov S V, Novoselov K S, Katsnelson M I, Schedin F, Ponomarenko L A, Jiang D and Geim A K 2006 Strong suppression of weak localization in graphene *Phys. Rev. Lett.* **97** 016801
[145] Morpurgo A F and Guinea F 2006 Intervalley scattering, long-range disorder, and effective time-reversal symmetry breaking in graphene *Phys. Rev. Lett.* **97** 196804
[146] Gorbachev R V, Tikhonenko F V, Mayorov A S, Horsell D W and Savchenko A K 2007 Weak localization in bilayer graphene *Phys. Rev. Lett.* **98** 176805

[147] Li Z *et al* 2019 Momentum-dark intervalley exciton in monolayer tungsten diselenide brightened via chiral phonon *ACS Nano* **13** 14107–13
[148] Wu S *et al* 2013 Electrical tuning of valley magnetic moment through symmetry control in bilayer MoS_2 *Nat. Phys.* **9** 149–53
[149] Zeng H and Cui X 2015 An optical spectroscopic study on two-dimensional group-VI transition metal dichalcogenides *Chem. Soc. Rev.* **44** 2629–42
[150] Lu J, Liu H, Tok E S and Sow C-H 2016 Interactions between lasers and two-dimensional transition metal dichalcogenides *Chem. Soc. Rev.* **45** 2494–515
[151] Zeng H, Dai J, Yao W, Xiao D and Cui X 2012 Valley polarization in MoS_2 monolayers by optical pumping *Nat. Nanotechnol.* **7** 490–3
[152] Kioseoglou G, Hanbicki A T, Currie M, Friedman A L, Gunlycke D and Jonker B T 2012 Valley polarization and intervalley scattering in monolayer MoS_2 *Appl. Phys. Lett.* **101**
[153] Sallen G *et al* 2012 Robust optical emission polarization in MoS_2 monolayers through selective valley excitation *Phys. Rev.* B **86** 081301(R)
[154] Sallen G *et al* 2014 Erratum: Robust optical emission polarization in MoS_2 monolayers through selective valley excitation [Phys. Rev. B 86, 081301(R) (2012)] *Phys. Rev.* B **89** 079903
[155] Jones A M *et al* 2013 Optical generation of excitonic valley coherence in monolayer WSe_2 *Nat. Nanotechnol.* **8** 634–8
[156] Park S, Arscott S, Taniguchi T, Watanabe K, Sirotti F and Cadiz F 2022 Efficient valley polarization of charged excitons and resident carriers in molybdenum disulfide monolayers by optical pumping *Commun. Phys.* **5** 73
[157] Yan A, Ong C S, Qiu D Y, Ophus C, Ciston J, Merino C, Louie S G and Zettl A 2017 Dynamics of symmetry-breaking stacking boundaries in bilayer MoS_2 *J. Phys. Chem.* C **121** 22559–66
[158] Wang Y, Cong C, Shang J, Eginligil M, Jin Y, Li G, Chen Y, Peimyoo N and Yu T 2019 Unveiling exceptionally robust valley contrast in AA- and AB-stacked bilayer WS_2 *Nanoscale Horiz.* **4** 396–403
[159] Jones A M, Yu H, Ross J S, Klement P, Ghimire N J, Yan J, Mandrus D G, Yao W and Xu X 2014 Spin–layer locking effects in optical orientation of exciton spin in bilayer WSe_2 *Nat. Phys.* **10** 130–4
[160] Schaibley J R, Yu H, Clark G, Rivera P, Ross J S, Seyler K L, Yao W and Xu X 2016 Valleytronics in 2D materials *Nat. Rev. Mater.* **1** 16055
[161] Onga M, Zhang Y, Ideue T and Iwasa Y 2017 Exciton Hall effect in monolayer MoS_2 *Nat. Mater.* **16** 1193–7
[162] Konabe S and Yamamoto T 2014 Valley photothermoelectric effects in transition-metal dichalcogenides *Phys. Rev.* B **90** 075430
[163] Ziman J M 2001 *Electrons and Phonons: The Theory of Transport Phenomena in Solids* Oxford Classic Series (Oxford: Oxford University Press)
[164] Yu X-Q, Zhu Z-G, Su G and Jauho A-P 2015 Thermally driven pure spin and valley currents via the anomalous nernst effect in monolayer group-VI dichalcogenides *Phys. Rev. Lett.* **115** 246601
[165] Liu Y, Gao Y, Zhang S, He J, Yu J and Liu Z 2019 Valleytronics in transition metal dichalcogenides materials *Nano Res.* **12** 2695–711
[166] Chang C *et al* 2021 Recent progress on two-dimensional materials *Acta Phys.-Chim. Sin.* **37** 2108017
[167] LaMountain T, Lenferink E J, Chen Y-J, Stanev T K and Stern N P 2018 Environmental engineering of transition metal dichalcogenide optoelectronics *Front. Phys.* **13** 138114

[168] Xie L and Cui X 2016 Manipulating spin-polarized photocurrents in 2D transition metal dichalcogenides *Proc. Natl. Acad. Sci. USA* **113** 3746–50

[169] Rong R, Liu Y, Nie X, Zhang W, Zhang Z, Liu Y and Guo W 2023 The interaction of 2D materials with circularly polarized light *Adv. Sci.* **10** 2206191

[170] Mak K F, McGill K L, Park J and McEuen P L 2014 The valley Hall effect in MoS_2 transistors *Science* **344** 1489

[171] Pu J and Takenobu T 2018 Monolayer transition metal dichalcogenides as light sources *Adv. Mater.* **30** 1707627

[172] Zeeman P 1896 Over de invloed eener magnetisatie op den aard van het door een stof uitgezonden licht *Versl. Kon. Akad. Wetensch. Amsterdam* **5** 181–4, 242–8

[173] Preston T 1898 XXXVI. Radiation phenomena in the magnetic field *Lond. Edinb. Dubl. Phil. Mag.* **45** 325–39

[174] Su L, Fan X, Yin T, Wang H, Li Y, Liu F, Li J, Zhang H and Xie H 2020 Inorganic 2D luminescent materials: structure, luminescence modulation, and applications *Adv. Opt. Mater.* **8** 1900978

[175] Aivazian G, Gong Z, Jones A M, Chu R-L, Yan J, Mandrus D G, Zhang C, Cobden D, Yao W and Xu X 2015 Magnetic control of valley pseudospin in monolayer WSe_2 *Nat. Phys.* **11** 148–52

[176] Stier A V, McCreary K M, Jonker B T, Kono J and Crooker S A 2016 Exciton diamagnetic shifts and valley Zeeman effects in monolayer WS_2 and MoS_2 to 65 Tesla *Nat. Commun.* **7** 10643

[177] MacNeill D, Heikes C, Mak K F, Anderson Z, Kormányos A, Zólyomi V, Park J and Ralph D C 2015 Breaking of valley degeneracy by magnetic field in monolayer $MoSe_2$ *Phys. Rev. Lett.* **114** 037401

[178] Li Y *et al* 2014 Valley splitting and polarization by the Zeeman effect in monolayer $MoSe_2$ *Phys. Rev. Lett.* **113** 266804

[179] Srivastava A, Sidler M, Allain A V, Lembke D S, Kis A and Imamoğlu A 2015 Valley Zeeman effect in elementary optical excitations of monolayer WSe_2 *Nat. Phys.* **11** 141–7

[180] Wang G, Bouet L, Glazov M M, Amand T, Ivchenko E L, Palleau E, Marie X and Urbaszek B 2015 Magneto-optics in transition metal diselenide monolayers *2D Mater.* **2** 034002

[181] Wang Z, Zhao L, Mak K F and Shan J 2017 Probing the spin-polarized electronic band structure in monolayer transition metal dichalcogenides by optical spectroscopy *Nano Lett.* **17** 740–6

[182] Kormányos A, Burkard G, Gmitra M, Fabian J, Zólyomi V, Drummond N D and Fal'ko V 2015 $\mathbf{k}\cdot\mathbf{p}$ Theory for two-dimensional transition metal dichalcogenide semiconductors *2D Mater.* **2** 022001

[183] Feng W, Yao Y, Zhu W, Zhou J, Yao W and Xiao D 2012 Intrinsic spin Hall effect in monolayers of group-VI dichalcogenides: a first-principles study *Phys. Rev.* B **86** 165108

[184] Gywat O, Krenner H J and Berezovsky J 2009 *Spins in Optically Active Quantum Dots: Concepts and Methods* (Weinheim: Wiley-VCH)

[185] Leisgang N M 2022 Electrical control of excitons in a gated two-dimensional semiconductor *PhD Thesis* University of Basel, Basel

[186] Avouris P, Heinz T F and Low T (ed) 2017 *2D Materials: Properties and Devices* (Cambridge: Cambridge University Press)

[187] Pfennig B W 2021 *Principles of Inorganic Chemistry* 2nd edn (Hoboken, NJ: Wiley)

[188] Wang Z 2019 Spin- and valley-dependent excitons in atomically thin transition metal dichalcogenides *PhD Thesis* The Pennsylvania State University, University Park, PA

[189] Du L, Hasan T, Castellanos-Gomez A, Liu G-B, Yao Y, Lau C N and Sun Z 2021 Engineering symmetry breaking in 2D layered materials *Nat. Rev. Phys.* **3** 193–206

[190] Zhao Y *et al* 2022 Interlayer exciton complexes in bilayer MoS_2 *Phys. Rev.* B **105** L041411

[191] Arora A, Schmidt R, Schneider R, Molas M R, Breslavetz I, Potemski M and Bratschitsch R 2016 Valley Zeeman splitting and valley polarization of neutral and charged excitons in monolayer $MoTe_2$ at high magnetic fields *Nano Lett.* **16** 3624–9

[192] Nagler P *et al* 2018 Zeeman splitting and inverted polarization of biexciton emission in monolayer WS_2 *Phys. Rev. Lett.* **121** 057402

[193] Jiang C, Liu F, Cuadra J, Huang Z, Li K, Rasmita A, Srivastava A, Liu Z and Gao W-B 2017 Zeeman splitting via spin-valley-layer coupling in bilayer $MoTe_2$ *Nat. Commun.* **8** 802

[194] Soni A and Pal S K 2022 Valley degree of freedom in two-dimensional van der Waals materials *J. Phys. D: Appl. Phys.* **55** 303003

[195] Wang G, Marie X, Liu B L, Amand T, Robert C, Cadiz F, Renucci P and Urbaszek B 2016 Control of exciton valley coherence in transition metal dichalcogenide monolayers *Phys. Rev. Lett.* **117** 187401

[196] Lan T, Ding B and Liu B 2020 Magneto-optic effect of two-dimensional materials and related applications *Nano Select* **1** 298–310

[197] Zheng W, Jiang Y, Hu X, Li H, Zeng Z, Wang X and Pan A 2018 Light emission properties of 2D transition metal dichalcogenides: fundamentals and applications *Adv. Opt. Mater.* **6** 1800420

[198] Mitioglu A A, Plochocka P, Granados del Aguila Á, Christianen P C M, Deligeorgis G, Anghel S, Kulyuk L and Maude D K 2015 Optical investigation of monolayer and bulk tungsten diselenide (WSe_2) in high magnetic fields *Nano Lett.* **15** 4387–92

[199] Rybkovskiy D V, Gerber I C and Durnev M V 2017 Atomically inspired $\boldsymbol{k}\cdot\boldsymbol{p}$ approach and valley Zeeman effect in transition metal dichalcogenide monolayers *Phys. Rev.* B **95** 155406

[200] Liu W, Luo C, Tang X, Peng X and Zhong J 2019 Valleytronic properties of monolayer WSe_2 in external magnetic field *AIP Adv.* **9** 045222

[201] Rostami H and Asgari R 2015 Valley Zeeman effect and spin-valley polarized conductance in monolayer MoS_2 in a perpendicular magnetic field *Phys. Rev.* B **91** 075433

[202] Lorchat E, Selig M, Katsch F, Yumigeta K, Tongay S, Knorr A, Schneider C and Höfling S 2021 Excitons in bilayer MoS_2 displaying a colossal electric field splitting and tunable magnetic response *Phys. Rev. Lett.* **126** 037401

[203] Mitioglu A A, Galkowski K, Surrente A, Klopotowski L, Dumcenco D, Kis A, Maude D K and Plochocka P 2016 Magnetoexcitons in large area CVD-grown monolayer MoS_2 and $MoSe_2$ on sapphire *Phys. Rev.* B **93** 165412

[204] Klein J *et al* 2021 Controlling exciton many-body states by the electric-field effect in monolayer MoS_2 *Phys. Rev. Res.* **3** L022009

[205] Cong C *et al* 2018 Intrinsic excitonic emission and valley Zeeman splitting in epitaxial MS_2 (M = Mo and W) monolayers on hexagonal boron nitride *Nano Res.* **11** 6227–36

[206] Wu Y J, Shen C, Tan Q H, Shi J, Liu X F, Wu Z H, Zhang J, Tan P H and Zheng H Z 2018 Valley Zeeman splitting of monolayer MoS_2 probed by low-field magnetic circular dichroism spectroscopy at room temperature *Appl. Phys. Lett.* **112** 153105

[207] Koperski M, Molas M R, Arora A, Nogajewski K, Bartos M, Wyzula J, Vaclavkova D, Kossacki P and Potemski M 2019 Orbital, spin and valley contributions to Zeeman splitting of excitonic resonances in $MoSe_2$, WSe_2 and WS_2 Monolayers *2D Mater.* **6** 015001
[208] Schmidt R *et al* 2016 Magnetic-field-induced rotation of polarized light emission from monolayer WS_2 *Phys. Rev. Lett.* **117** 077402
[209] Qu F, Bragança H, Vasconcelos R, Liu F, Xie S-J and Zeng H 2019 Controlling valley splitting and polarization of dark- and bi-excitons in monolayer WS_2 by a tilted magnetic field *2D Mater.* **6** 045014
[210] Plechinger G *et al* 2016 Excitonic valley effects in monolayer WS_2 under high magnetic fields *Nano Lett.* **16** 7899–904
[211] Li Z *et al* 2019 Emerging photoluminescence from the dark-exciton phonon replica in monolayer WSe_2 *Nat. Commun.* **10** 2469
[212] Li Z *et al* 2019 Author correction: emerging photoluminescence from the dark-exciton phonon replica in monolayer WSe_2 *Nat. Commun.* **10** 4649
[213] Lyons T P, Dufferwiel S, Brooks M, Withers F, Taniguchi T, Watanabe K, Novoselov K S, Burkard G and Tartakovskii A I 2019 The valley Zeeman effect in inter- and intra-valley trions in monolayer WSe_2 *Nat. Commun.* **10** 2330
[214] Zou C *et al* 2018 Probing magnetic-proximity-effect enlarged valley splitting in monolayer WSe_2 by photoluminescence *Nano Res.* **11** 6252–9
[215] Smoleński T, Goryca M, Koperski M, Faugeras C, Kazimierczuk T, Bogucki A, Nogajewski K, Kossacki P and Potemski M 2016 Tuning valley polarization in a WSe_2 monolayer with a tiny magnetic field *Phys. Rev.* X **6** 021024
[216] Zou C, Zhang H, Chen Y, Feng S, Wu L, Zhang J, Yu T, Shang J and Cong C 2020 Spatial variations of valley splitting in monolayer transition metal dichalcogenide *InfoMat* **2** 585–92
[217] Liu Y *et al* 2016 Zeeman splitting and dynamical mass generation in Dirac semimetal $ZrTe_5$ *Nat. Commun.* **7** 12516
[218] Surrente A *et al* 2018 Intervalley scattering of interlayer excitons in a $MoS_2/MoSe_2/MoS_2$ heterostructure in high magnetic field *Nano Lett.* **18** 3994–4000
[219] Wang T *et al* 2020 Giant valley-Zeeman splitting from spin-singlet and spin-triplet interlayer excitons in $WSe_2/MoSe_2$ heterostructure *Nano Lett.* **20** 694–700
[220] Zihlmann S, Cummings A W, Garcia J H, Kedves M, Watanabe K, Taniguchi T, Schönenberger C and Makk P 2018 Large spin relaxation anisotropy and valley-Zeeman spin-orbit coupling in WSe_2/graphene/h-BN heterostructures *Phys. Rev.* B **97** 075434
[221] Qu F, Dias A C, Fu J, Villegas-Lelovsky L and Azevedo D L 2017 Tunable spin and valley dependent magneto-optical absorption in molybdenum disulfide quantum dots *Sci. Rep.* **7** 41044
[222] Rostami H, Moghaddam A G and Asgari R 2013 Effective lattice Hamiltonian for monolayer MoS_2: tailoring electronic structure with perpendicular electric and magnetic fields *Phys. Rev.* B **88** 085440
[223] Deilmann T, Krüger P and Rohlfing M 2020 Ab initio studies of exciton g factors: monolayer transition metal dichalcogenides in magnetic fields *Phys. Rev. Lett.* **124** 226402
[224] Woźniak T, Faria P E Jr, Seifert G, Chaves A and Kunstmann J 2020 Exciton g factors of van der Waals heterostructures from first-principles calculations *Phys. Rev.* B **101** 235408
[225] Wang Y *et al* 2020 Spin-valley locking effect in defect states of monolayer MoS_2 *Nano Lett.* **20** 2129–36

[226] Wang Z, Mak K F and Shan J 2018 Strongly interaction-enhanced valley magnetic response in monolayer WSe_2 *Phys. Rev. Lett.* **120** 066402
[227] Gustafsson M V, Yankowitz M, Forsythe C, Rhodes D, Watanabe K, Taniguchi T, Hone J, Zhu X and Dean C R 2018 Ambipolar Landau levels and strong band-selective carrier interactions in monolayer WSe_2 *Nat. Mater.* **17** 411–5
[228] Movva H C P, Fallahazad B, Kim K, Larentis S, Taniguchi T, Watanabe K, Banerjee S K and Tutuc E 2017 Density-dependent quantum Hall states and Zeeman splitting in monolayer and bilayer WSe_2 *Phys. Rev. Lett.* **118** 247701
[229] Wang Z, Shan J and Mak K F 2017 Valley- and spin-polarized Landau levels in monolayer WSe_2 *Nat. Nanotechnol.* **12** 144–9
[230] Roch J G, Miserev D, Froehlicher G, Leisgang N, Sponfeldner L, Watanabe K, Taniguchi T, Klinovaja J, Loss D and Warburton R J 2020 First-order magnetic phase transition of mobile electrons in monolayer MoS_2 *Phys. Rev. Lett.* **124** 187602
[231] Back P, Sidler M, Cotlet O, Srivastava A, Takemura N, Kroner M and Imamoğlu A 2017 Giant paramagnetism-induced valley polarization of electrons in charge-tunable monolayer $MoSe_2$ *Phys. Rev. Lett.* **118** 237404
[232] Zhang J *et al* 2019 Enhancing and controlling valley magnetic response in MoS_2/WS_2 heterostructures by all-optical route *Nat. Commun.* **10** 4226
[233] Chu J *et al* 2021 2D polarized materials: ferromagnetic, ferrovalley, ferroelectric materials, and related heterostructures *Adv. Mater.* **33** 2004469
[234] Cui X *et al* 2015 Multi-terminal transport measurements of MoS_2 using a van der Waals heterostructure device platform *Nat. Nanotechnol.* **10** 534–40
[235] Fallahazad B, Movva H C P, Kim K, Larentis S, Taniguchi T, Watanabe K, Banerjee S K and Tutuc E 2016 Shubnikov–de Haas oscillations of high-mobility holes in monolayer and bilayer WSe_2: Landau level degeneracy, effective mass, and negative compressibility *Phys. Rev. Lett.* **116** 086601
[236] Cai T, Yang S A, Li X, Zhang F, Shi J, Yao W and Niu Q 2013 Magnetic control of the valley degree of freedom of massive Dirac fermions with application to transition metal dichalcogenides *Phys. Rev.* B **88** 115140
[237] Li X, Zhang F and Niu Q 2013 Unconventional quantum Hall effect and tunable spin Hall effect in Dirac materials: application to an isolated MoS_2 trilayer *Phys. Rev. Lett.* **110** 066803
[238] Joffre M, Hulin D, Migus A and Combescot M 1989 Laser-induced exciton splitting *Phys. Rev. Lett.* **62** 74–7
[239] Press D, Ladd T D, Zhang B and Yamamoto Y 2008 Complete quantum control of a single quantum dot spin using ultrafast optical pulses *Nature* **456** 218–21
[240] Kim J, Hong X, Jin C, Shi S-F, Chang C-Y-S, Chiu M-H, Li L-J and Wang F 2014 Ultrafast generation of pseudo-magnetic field for valley excitons in WSe_2 monolayers *Science* **346** 1205–8
[241] Sie E J, McIver J W, Lee Y-H, Fu L, Kong J and Gedik N 2015 Valley-selective optical Stark effect in monolayer WS_2 *Nat. Mater.* **14** 290–4
[242] Liu W and Xu Y (ed) 2020 *Spintronic 2D Materials: Fundamentals and Applications* Materials Today (Amsterdam: Elsevier)
[243] Xi J, Zhao T, Wang D and Shuai Z 2014 Tunable electronic properties of two-dimensional transition metal dichalcogenide alloys: a first-principles prediction *J. Phys. Chem. Lett.* **5** 285–91

[244] Wu L *et al* 2021 Observation of strong valley magnetic response in monolayer transition metal dichalcogenide alloys of $Mo_{0.5}W_{0.5}Se_2$ and $Mo_{0.5}W_{0.5}Se_2/WS_2$ heterostructures *ACS Nano* **15** 8397–406

[245] Li Q *et al* 2020 Enhanced valley Zeeman splitting in Fe-doped monolayer MoS_2 *ACS Nano* **14** 4636–45

[246] Zhou J *et al* 2020 Synthesis of Co-Doped MoS_2 monolayers with enhanced valley splitting *Adv. Mater.* **32** 1906536

[247] Wang S and Yu J 2018 Magnetic behaviors of 3d transition metal-doped silicane: a first-principle study *J. Supercond. Nov. Magn.* **31** 2789–95

[248] He W, Zhang S, Luo Y and Wang S 2024 Exploring monolayer GaN doped with transition metals: insights from first-principles studies *J. Supercond. Nov. Magn.* **37** 157–63

[249] Stier A V, Wilson N P, Clark G, Xu X and Crooker S A 2016 Probing the influence of dielectric environment on excitons in monolayer WSe_2: insight from high magnetic fields *Nano Lett.* **16** 7054–60

[250] Zhang Y J, Oka T, Suzuki R, Ye J T and Iwasa Y 2014 Electrically switchable chiral light-emitting transistor *Science* **344** 725–8

[251] Baugher B W H, Churchill H O H, Yang Y and Jarillo-Herrero P 2014 Optoelectronic devices based on electrically tunable p–n diodes in a monolayer dichalcogenide *Nat. Nanotechnol.* **9** 262–7

[252] Ross J S *et al* 2014 Electrically tunable excitonic light-emitting diodes based on monolayer WSe_2 p–n junctions *Nat. Nanotechnol.* **9** 268–72

[253] Minden H 1965 P–N junction electroluminescence and diode lasers *IEEE Trans. Parts Mater. Packag.* **1** 40–7

[254] Arul N S and Nithya V D (ed) 2019 *Two Dimensional Transition Metal Dichalcogenides: Synthesis, Properties, and Applications* (Singapore: Springer)

[255] Liu F, Zhou J, Zhu C and Liu Z 2017 Electric field effect in two-dimensional transition metal dichalcogenides *Adv. Funct. Mater.* **27** 1602404

[256] Drude P 1900 Zur elektronentheorie der metalle *Ann. Phys.* **306** 566–613

[257] Tanner D B 2019 *Optical Effects in Solids* (Cambridge: Cambridge University Press)

[258] Ramasubramaniam A 2012 Large excitonic effects in monolayers of molybdenum and tungsten dichalcogenides *Phys. Rev.* B **86** 115409

[259] Splendiani A, Sun L, Zhang Y, Li T, Kim J, Chim C-Y, Galli G and Wang F 2010 Emerging photoluminescence in monolayer MoS_2 *Nano Lett.* **10** 1271–5

[260] Yuan H *et al* 2013 Zeeman-type spin splitting controlled by an electric field *Nat. Phys.* **9** 563–9

[261] Onga M, Zhang Y, Suzuki R and Iwasa Y 2016 High circular polarization in electro-luminescence from $MoSe_2$ *Appl. Phys. Lett.* **108** 073107

[262] Yang W *et al* 2016 Electrically tunable valley-light emitting diode (vLED) based on CVD-grown monolayer WS_2 *Nano Lett.* **16** 1560–7

[263] Pu J, Zhang W, Matsuoka H, Kobayashi Y, Takaguchi Y, Miyata Y, Matsuda K, Miyauchi Y and Takenobu T 2021 Room-temperature chiral light-emitting diode based on strained monolayer semiconductors *Adv. Mater.* **33** 2100601

[264] Pu J, Fujimoto T, Ohasi Y, Kimura S, Chen C-H, Li L-J, Sakanoue T and Takenobu T 2017 A versatile and simple approach to generate light emission in semiconductors mediated by electric double layers *Adv. Mater.* **29** 1606918

[265] Kobayashi Y, Sasaki S, Mori S, Hibino H, Liu Z, Watanabe K, Taniguchi T, Suenaga K, Maniwa Y and Miyata Y 2015 Growth and optical properties of high-quality monolayer WS_2 on graphite *ACS Nano* **9** 4056–63
[266] Ahn G H, Amani M, Rasool H, Lien D-H, Mastandrea J P, Ager J W III, Dubey M, Chrzan D C, Minor A M and Javey A 2017 Strain-engineered growth of two-dimensional materials *Nat. Commun.* **8** 608
[267] Lee J, Wang Z, Xie H, Mak K F and Shan J 2017 Valley magnetoelectricity in single-layer MoS_2 *Nat. Mater.* **16** 887–91
[268] Pu J and Takenobu T 2021 Recent advances in light-emitting electrochemical cells with low-dimensional quantum materials *J. Imaging Soc. Jpn.* **60** 656–72
[269] Nye J F 1985 *Physical Properties of Crystals: Their Representation by Tensors and Matrices* (Oxford: Clarendon Press)
[270] Sadd M H 2020 *Elasticity: Theory, Applications, and Numerics* 4th edn (Oxford: Academic Press)
[271] Johari P and Shenoy V B 2012 Tuning the electronic properties of semiconducting transition metal dichalcogenides by applying mechanical strains *ACS Nano* **6** 5449–56
[272] Kumar A and Ahluwalia P K 2013 Mechanical strain dependent electronic and dielectric properties of two-dimensional honeycomb structures of MoX_2 (X=S, Se, Te) *Physica B: Condens. Matter* **419** 66–75
[273] Rostami H, Roldán R, Cappelluti E, Asgari R and Guinea F 2015 Theory of strain in single-layer transition metal dichalcogenides *Phys. Rev.* B **92** 195402
[274] Ghalamkari K, Tatsumi Y and Saito R 2018 Perfect circular dichroism in the Haldane model *J. Phys. Soc. Jpn.* **87** 063708
[275] Pereira V M and Castro Neto A H 2009 Strain engineering of graphene's electronic structure *Phys. Rev. Lett.* **103** 046801
[276] Sasaki K-i and Saito R 2008 Pseudospin and deformation-induced gauge field in graphene *Prog. Theor. Phys. Suppl.* **176** 253–78
[277] Sasaki K-i, Saito R, Dresselhaus M S and Wakabayashi K 2010 Soliton trap in strained graphene nanoribbons *New J. Phys.* **12** 103015
[278] Kitt A L, Pereira V M, Swan A K and Goldberg B B 2012 Lattice-corrected strain-induced vector potentials in graphene *Phys. Rev.* B **85** 115432
[279] Kitt A L, Pereira V M, Swan A K and Goldberg B B 2013 Erratum: Lattice-corrected strain-induced vector potentials in graphene [Phys. Rev. B 85, 115432 (2012)] *Phys. Rev.* B **87** 159909(E)
[280] Fujita T, Jalil M B A and Tan S G 2010 Valley filter in strain engineered graphene *Appl. Phys. Lett.* **97** 043508
[281] Niu Z 2012 Spin and valley dependent electronic transport in strain engineered graphene *J. Appl. Phys.* **111** 103712
[282] Wang S-K and Wang J 2015 Spin and valley filter in strain engineered silicene *Chin. Phys.* B **24** 037202
[283] Defo R K, Fang S, Shirodkar S N, Tritsaris G A, Dimoulas A and Kaxiras E 2016 Strain dependence of band gaps and exciton energies in pure and mixed transition-metal dichalcogenides *Phys. Rev.* B **94** 155310
[284] Wang S, Tian H and Sun M 2023 Valley-polarized and enhanced transmission in graphene with a smooth strain profile *J. Phys.: Condens. Matter* **35** 304002

[285] Pearce A J, Mariani E and Burkard G 2016 Tight-binding approach to strain and curvature in monolayer transition-metal dichalcogenides *Phys. Rev.* B **94** 155416

[286] Pospischil A, Furchi M M and Mueller T 2014 Solar-energy conversion and light emission in an atomic monolayer p–n diode *Nat. Nanotechnol.* **9** 257–61

[287] Xia F, Wang H, Xiao D, Dubey M and Ramasubramaniam A 2014 Two-dimensional material nanophotonics *Nat. Photon.* **8** 899–907

[288] Gorbachev R V *et al* 2014 Detecting topological currents in graphene superlattices *Science* **346** 448

[289] Sui M *et al* 2015 Gate-tunable topological valley transport in bilayer graphene *Nat. Phys.* **11** 1027–31

[290] Shimazaki Y, Yamamoto M, Borzenets I V, Watanabe K, Taniguchi T and Tarucha S 2015 Generation and detection of pure valley current by electrically induced Berry curvature in bilayer graphene *Nat. Phys.* **11** 1032–6

[291] Bussolotti F, Maddumapatabandi T D and Goh K E J 2023 Band structure and spin texture of 2D materials for valleytronics: insights from spin and angle-resolved photoemission spectroscopy *Mater. Quantum. Technol.* **3** 032001

[292] Zhao S 2015 Preparation and investigation of intrinsic electronic and optical properties of individual structure-identified nanostructures: double-wall carbon nanotubes and two-dimensional atomic layers *PhD Thesis* Nagoya University, Nagoya

[293] Bao Q and Hoh H Y (ed) 2019 *2D Materials for Photonic and Optoelectronic Applications* Woodhead Publishing Series in Electronic and Optical Materials (Duxford: Woodhead Publishing)

[294] Nagaosa N, Sinova J, Onoda S, MacDonald A H and Ong N P 2010 Anomalous Hall effect *Rev. Mod. Phys.* **82** 1539–92

[295] Yamamoto M, Shimazaki Y, Borzenets I V and Tarucha S 2015 Valley Hall effect in two-dimensional hexagonal lattices *J. Phys. Soc. Jpn.* **84** 121006

[296] Berger L 1970 Side-jump mechanism for the Hall effect of ferromagnets *Phys. Rev.* B **2** 4559–66

[297] Xiao D, Yao W and Niu Q 2007 Valley-contrasting physics in graphene: magnetic moment and topological transport *Phys. Rev. Lett.* **99** 236809

[298] Cheiwchanchamnangij T and Lambrecht W R L 2012 Quasiparticle band structure calculation of monolayer, bilayer, and bulk MoS_2 *Phys. Rev.* B **85** 205302

[299] Tang J, Zheng Y, Jiang K, You Q, Yin Z, Xie Z, Li H, Han C, Zhang X and Shi Y 2023 Interlayer exciton dynamics of transition metal dichalcogenide heterostructures under electric fields *Nano Res.* **17** 4555–72

[300] Tsymbal E Y and Žutić I (ed) 2019 *Nanoscale Spintronics and Applications* Spintronics Handbook: Spin Transport and Magnetism **vol 3** 2nd edn (Boca Raton, FL: CRC Press)

[301] Goh K E J, Wong C P Y and Wang T (ed) 2023 *Valleytronics in 2D Materials* (Singapore: World Scientific)

[302] Lee J, Mak K F and Shan J 2016 Electrical control of the valley Hall effect in bilayer MoS_2 transistors *Nat. Nanotechnol.* **11** 421–5

[303] Wang Q, Lai J and Sun D 2016 Review of photo response in semiconductor transition metal dichalcogenides based photosensitive devices *Opt. Mater. Express* **6** 2313–27

[304] Conder K 2018 Optimisation of transition metal dichalcogenide devices for measurement of spin-Hall voltages generated by optical spin orientation *PhD Thesis* University of Exeter, Exeter

[305] Kang S, Lee D, Kim J, Capasso A, Kang H S, Park J-W, Lee C-H and Lee G-H 2020 2D semiconducting materials for electronic and optoelectronic applications: potential and challenge *2D Mater.* **7** 022003

[306] Eginligil M, Cao B, Wang Z, Shen X, Cong C, Shang J, Soci C and Yu T 2015 Dichroic spin–valley photocurrent in monolayer molybdenum disulphide *Nat. Commun.* **6** 7636
[307] Yuan H *et al* 2014 Generation and electric control of spin-valley-coupled circular photogalvanic current in WSe_2 *Nat. Nanotechnol.* **9** 851–7
[308] Pospischil A and Mueller T 2016 Optoelectronic devices based on atomically thin transition metal dichalcogenides *Appl. Sci.* **6** 78

Chapter 8

Valley optoelectronics based on vdW materials

Parts of this chapter have been reprinted from [33], copyright (2020), with permission from Elsevier.

Valley optoelectronics, leveraging the valley DOF in 2D materials, offers novel opportunities for controlling electronic and optical processes. This chapter delves into the detection and manipulation of valley-dependent phenomena in vdW materials, with a focus on bilayer TMDCs and heterostructures. Key topics include the experimental detection of the valley Hall effect in bilayer TMDCs, which highlights the interplay between valley physics and external fields. The chapter also examines band alignment and the formation of interlayer excitons in heterostructures, where electron–hole pairs are spatially separated between layers. We explore valley-polarized interlayer excitons and the role of moiré patterns in creating moiré excitons with unique optical properties. At last, we will introduce some proposals for the enhancement of the valley Zeeman effect based on layered materials. These topics are discussed in the context of their impact on device functionality and the future of valleytronic applications. Challenges and opportunities for implementing these phenomena in practical devices are also addressed, offering insights into their potential for next-generation optoelectronic systems.

8.1 Detection of VHE in bilayer TMDCs

Because the crystal symmetry and PL polarization can be controlled by the external electric field, the VHE, first experimentally observed in monolayer MoS_2 [1], could also be tuned in the bilayer MoS_2 transistor. In 2016, Lee *et al* [2] demonstrated the electrical generation of valley current (which is non-emissive) in a bilayer MoS_2 transistor, and optical detection of the VHE using magneto-optical Kerr rotation (KR) microscopy [3, 4]. KR spectroscopy is based on the magneto-optical Kerr effect where linearly polarised light experiences a small rotation (θ_K) in the polarisation plane when interacting with either a magnetic material or a material under an external magnetic field, because LCP and RCP light have different refractive indices

doi:10.1088/978-0-7503-5562-9ch8

in such materials [5]. The amount of rotation, θ_K, is proportional to the net magnetisation of the material in the out-of-plane direction [6]. As discussed in section 2.5, in 2D materials with a TRS but a broken lattice IS, the sign-inequivalent Berry curvature in inequivalent valleys acts as an anti-parallel effective out-of-plane magnetic moments without the requirement of any external magnetic field [6]. This further leads to experimental detection of the opposite KR for reflected light from $\boldsymbol{K}_+$ and $\boldsymbol{K}_-$ valley states [7].

Monolayer MoS_2 has broken IS, which is a requirement to observe valley-contrasting physical effects. Bilayer 2H-MoS_2, on the other hand, is inversion-symmetric (in contrast, 3R-MoS_2 is non-centrosymmetric for any layer thicknesses [8]), so the VHE should be forbidden. However, it was shown earlier [9] that a sufficiently large out-of-plane electric field E (provided by a vertical gate voltage V_g) can be applied to introduce a potential difference between the two layers and break the IS of bilayer MoS_2 (figure 8.1(a)) [10], similar to the case of bilayer graphene in section 6.2. Therefore, by varying the gate voltage, which breaks the IS, the VHE magnitude and the resulting transverse valley current can be controlled.

As shown in figure 8.1(b), due to the VHE, a longitudinal electrical field induces a valley current density j_v in the transverse direction, which leads to an accumulation of VP along the sample edges. Here, the magnetic moment due to VP causes a disparity in the refractive index for LCP and RCP light, which, in turn, produces a finite readable KR signal (that is, polarization rotation) when focusing linearly polarized light on the sample at normal incidence near the fundamental exciton resonance [8, 11].

By scanning the laser beam over the sample area to spatially resolve the KR signal, specifically, at $V_g = 20$ V, as shown in the bottom left panel of figure 8.1(c), one can find that a clear KR signal appeared only near the two edges within the valley relaxation length and maximised at the level of $\sim$100 µrad. More importantly, the KR signal has opposite signs for the two edges of the sample, further confirming that the accumulation of $\boldsymbol{K}_+$ and $\boldsymbol{K}_-$ VPs manifest on opposite edges, and the signature of VHE is observed. Since the magnitude of IS breaking is regulated by the gate bias, the measured VP signal is strongly dependent on the gate voltage. The bottom and top panels of figure 8.1(c) show and summarise the experimental results, respectively [6]. By reversing the direction of the vertical electric field [2] from $E > 0$ to $E < 0$, the direction of the valley current can be switched, suggesting the electrical tunability of Berry curvature and VHE in bilayer MoS_2 [12]. Meanwhile, the current vanishes at a voltage approaching but not exactly zero (4 V [2]; bottom right panel of figure 8.1(c)), suggesting substrate effects on symmetry [7]. The spatial and gate voltage dependence of the KR signal indicates a clear VHE in bilayer MoS_2 with a field-induced broken IS [13].

In addition, as shown in figure 8.1(d), the KR angle and longitudinal current density j were found to follow the same dependence on bias voltage V_x, indicating that the observed KR is driven by the longitudinal current. This confirms that the valley current was generated through the VHE and KR imaging, allowing direct visualisation of the $\boldsymbol{K}_+$ and $\boldsymbol{K}_-$ valley currents at opposite edges of the bilayer MoS_2 channel. Further control experiments show the θ_K sign flips with the reversal of the

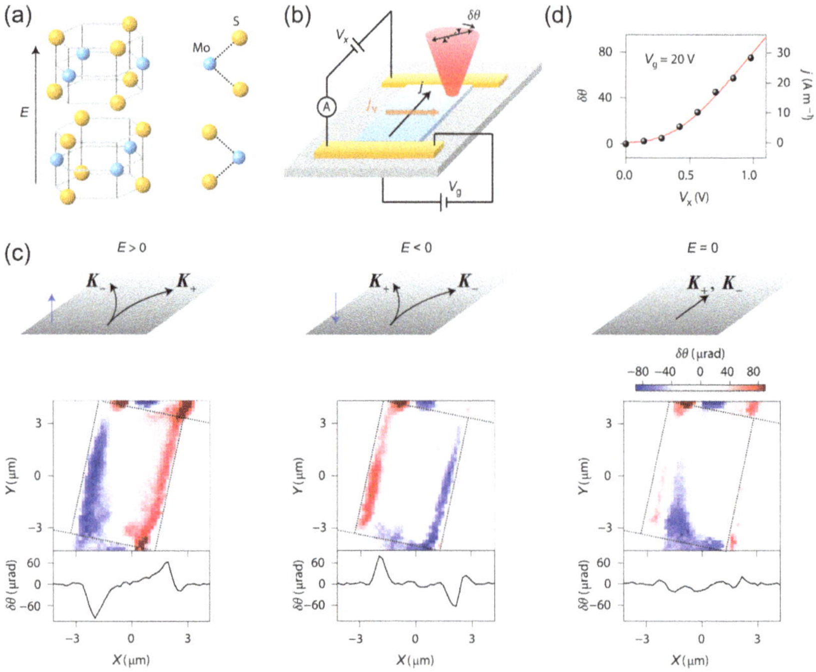

Figure 8.1. Experimental demonstration of VHE in bilayer MoS_2 device. (a) Bilayer MoS_2 crystal structure. Vertical electric field E breaks the IS of the bilayer. (b) Schematics of magneto-optical KR microscopy of MoS_2 FET with a bias voltage (V_x) applied on the source–drain electrodes and gate voltage (V_g) applied through the Si/SiO_2 substrate. A longitudinal electron current j (black arrow) drives a transverse valley current j_V (orange arrow) by the VHE, giving rise to a KR angle ($\delta\theta$, defined as the angle at which the linearly polarised probing light is rotated after being reflected by the sample area) on sample edges. (c) Gate dependence of the VHE in bilayer MoS_2. Top panel: Illustrations of the VHE in bilayer MoS_2 devices under positive (left), negative (middle) and near zero (right) [4] out-of-plane electric fields. Blue arrows indicate the direction of vertical electric fields. Bottom panel: Spatial 2D mapping and selected horizontal linecut of KR signal in microradians of linearly polarised light near the exciton resonance for different E values corresponding to the top panel. All measurements were performed at the same bias (2.5 V peak-to-peak). Black dashed lines show the boundary of the channel determined from the reflection image. The results show VP accumulation on sample edges. (d) Bias voltage (V_x) dependence of $\delta\theta$ on one edge of the channel (black symbols, left y axis) and the longitudinal current density j (red line, right y axis) under $V_g = 20$ V. Reproduced from [2], with permission from Springer Nature.

longitudinal current direction [3]. These observations are consistent with the symmetry-dependent Berry curvature and valley Hall conductivity in bilayer MoS_2 [6].

The VHE observation in monolayer, bilayer, and even multilayer [14] TMDCs paves the way for circularly-polarised-light-tunable Hall devices for future valleytronics [15]. The experimental reports of the VHE are important first steps towards valleytronic devices because they demonstrate that electrical manipulation of VP (as opposed to optical manipulation) is possible [11].

It is also worth mentioning that Low *et al* [16] theoretically studied the valley current in symmetry-broken black phosphorus thin films. In symmetry-broken black phosphorus, LCP or RCP light couples to states that have a certain handedness of Berry curvature. However, in black phosphorus, due to its different band structure, the properties of topological valley current were predicted to be distinctively different compared to 2D hexagonal lattice systems like graphene and TMDCs. Observing finite transverse Hall voltage directly under a longitudinal drive current may be possible, even without circularly polarized light excitation. As a result, this will generate finite topological currents with components along both the longitudinal and transverse directions. This behaviour was not possible in graphene or TMDCs. Therefore, the new effects of coupling between black phosphorus and circularly polarized light can also be interesting [17].

8.2 2D layered materials

8.2.1 Fabrication of 2D layered materials

Thanks to the improved transfer techniques, one of the most exciting opportunities in 2D layered materials is the ability to stack disparate layers together to realize vdW heterostructures [18–27], which play an important role in controlling VP. The most common approach to create vdW heterostructures involves isolating individual monolayers either through exfoliation from 'bulk flakes' using scotch tape [28] or through direct monolayer growth, achieved by CVD [29]. Using both methods, it is quite common to obtain a single crystal with areas of 10×10 μm or larger. To identify single layers, researchers often rely on optical contrast on 285-nm SiO_2/Si substrates, which provide a favourable contrast for many 2D materials [30, 31]. The most common method in use is based on a dry polymer stamping technique using a polydimethylsiloxane (PDMS) stamp on a glass slide to pick up the layers. Once all the layers are picked up, these are transferred onto a target substrate, resulting in the desired device [32, 33].

In section 2.4.4, we mentioned one of the first 2D heterostructures to be studied, graphene on hBN [34]. Subsequent research [35] found that topological currents in graphene–hBN heterostructures offered initial evidence of the VHE in 2D systems. This capability to stack disparate 2D layers together has provided many new opportunities, ranging from creating ultrathin p–n junctions [36, 37] to inducing superconducting states in bilayer heterostructures [38]. An important feature of these 2D heterostructures is that when layers are stacked together, the relative crystallographic orientation between layers (twist angle) can be controlled, allowing for the creation of a periodically changing electronic bandgap [39] and further manipulation of these artificial materials. This is crucial for assembled (homo) bilayers of the same material because a small twist angle results in a long-range periodic moiré structure [33].

8.2.2 Types and applications of 2D heterostructures

Band alignment is one of the most fundamental characteristics of 2D heterostructures. The energy band structure of 2D heterostructures is responsible for the charge transfer mechanism, illumination properties, and charge-trapping phenomena. Generally, according to the band alignment, heterostructures formed by two

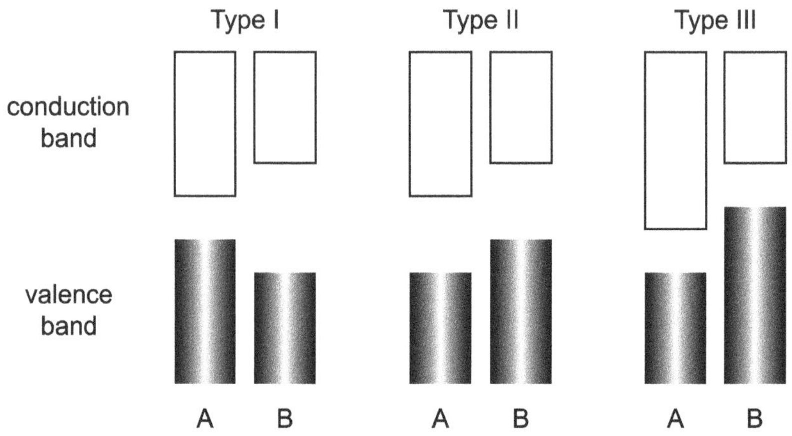

Figure 8.2. Schematic band diagram of type-I, II, and III heterostructures [40, 41]. The ranges of conduction (valence) bands are shown as (unfilled) filled rectangles.

different 2D materials can be divided into three types: type I (symmetric/straddling), type II (staggered), or type III (broken) [40–43], as shown in figure 8.2. The band alignments of a semicouducting vdW heterostructure are found to be vital for different varieties of behaviours and applications. We now discuss each type specifically.

1. In a type-I heterostructure, the bandgap of one semiconductor is entirely encompassed within the bandgap of the other one. The CBM and the VBM of two composite layers (A and B) obey the following rule:

$$VBM_B < VBM_A < CBM_A < CBM_B. \tag{8.1}$$

   Since the VBM and CBM of a type-I heterostructure are located in one constituent layer, efficient recombination of the photogenerated electrons and holes can occur when it is irradiated by light. Therefore, Type-I heterostructures have seen extensive use in optical devices, such as LEDs [21, 40, 41, 44] and lasers [45].

2. In a type-II heterostructure, the bandgaps from the two constituent layers overlaps, but the lower CBM and higher VBM belong to different layers, formulated as

$$VBM_A < VBM_B < CBM_A < CBM_B. \tag{8.2}$$

   The photogenerated electron–hole pairs can be spatially separated at the interface, with electrons transferred to one layer and the holes to the other with extremely strong carrier confinement [28, 29, 46–49]. Therefore, type-II heterostructures facilitate the prolongation of the interlayer excitation lifetime and are the desired structures for the electrons/holes separation, and consequently, have been demonstrated to be a fundamental component of photoelectronic related devices, such as photovoltaic devices and photodetectors [50].

3. In a type-III heterostructure, the bandgaps do not overlap at all. Therefore, the CBM and the VBM of two composite layers obeys

$$VBM_A < CBM_A < VBM_B < CBM_B. \quad (8.3)$$

From the diagram, one can realize that the valence band of one constituent layer overlaps with the conduction band of the other layer [51]. This type of heterostructure allows the band-to-band tunnelling effect of carriers and enables the operation of the tunnel field effect transistors [52, 53]. The speed of electron and hole transportation in type-III heterostructures is noticeably faster than in type-II heterostructures. This causes a large number of electrons/holes to separate at the interfaces, which promotes semi-metallic characteristics in type-III heterostructures. This phenomenon generates a strong built-in electric field at the type-III heterostructures making this type of heterostructure the ideal candidate for photodetectors and a new generation of thermal photovoltaic cells [54].

8.3 Interlayer excitions (IXs) in type-II van der Walls (vdW) heterostructures

In valleytronics based on 2D systems, type-II heterostructures can be used to encode long-lived VPs, and thus are crucial in electronics and optoelectronics because they have applications to devices such as diodes and LEDs. Early theoretical studies using first-principles calculations [55–60] demonstrated that staggered type-II band alignment can be achieved in various combinations of 2D semiconductor bilayer heterostructures, like group VI TMDCs MoX_2–WX_2 ($X=$ S, Se, Te) [36, 61–64]. Furthermore, several experiments have been carried out to measure the band alignment in various heterobilayers [65, 66]. Sub-micrometer ARPES (μ-ARPES) was used to determine the valence band offset of 300 meV in $MoSe_2$–WSe_2, with the valence band edge located at $\boldsymbol{K}_\pm$ in the WSe_2 layer [62]. Using microbeam x-ray photoelectron spectroscopy (μ-XPS), in conjunction with STM/STS measurements, a type-II alignment was also found for the MoS_2–WSe_2 heterobilayer with finite twist angle, with valence (conduction) band offsets of 0.83 (0.76) eV at $\boldsymbol{K}_\pm$ [63]. Furthermore, the rotationally aligned MoS_2–WSe_2 heterobilayer, grown by CVD, also has type-II band alignment with the band edges located at $\boldsymbol{K}_\pm$ [19]. Meanwhile, similar measurements on MoS_2–WS_2 show type-II band alignment with the conduction band edge at $\boldsymbol{K}_\pm$ in MoS_2, but the valence band edge at the nearly degenerate $\boldsymbol{K}_\pm$ or Γ in WS_2 [67].

The direct consequence of type-II band alignment is to form an IX due to charge transfer [67]. The measurement of IXs can be experimentally challenging because they emit at lower energy compared with the constituent monolayer and could lie in the infrared region, which usually is out of the measurement range for conventional transient absorption spectroscopy and time-resolve PL [68]. Furthermore, due to momentum mismatch, IXs can be weakly emissive or even dark, requiring detection at low temperature [39]. Fang *et al* [46] have demonstrated a strong interlayer coupling between 2D interfaces, causing a spatially direct absorption but indirect emission, suggesting the formation of IXs after the transfer of electrons (holes) from WSe_2 (MoS_2) to MoS_2 (WSe_2) in the type-II heterostructure.

Using PL and PL excitation spectroscopies [28, 70–73], Rivera *et al* [28] first extensively reported the observations of IXs, which is featured as a reduced energy peak distinct from constituent monolayers in the PL spectrum associated with a much longer lifetime [74, 75]. Figures 8.3(a and b) illustrate the manufactured type-II $MoSe_2$–WSe_2 heterostructure, in which mechanically exfoliated monolayer WSe_2 is merely stacked on top of monolayer $MoSe_2$.

Figure 8.3(c) shows the energy band diagram of the heterostructure. Due to the type-II band alignment, the lowest conduction band is in the $MoSe_2$ layer, and the highest valence band is in the WSe_2 layer. Upon injection of electrons and holes via optical excitation, electrons undergo charge transfer to the $MoSe_2$ layer, and holes undergo charge transfer to the WSe_2 layer. Charge transfer dynamics in type-II band structure is probed by ultrafast pump–probe spectroscopy [49, 76–78] and found that charge transfer occurs within nearly 50 fs after optical excitation [47, 79]. Therefore, the electron (hole) predominantly resides in the $MoSe_2$ (WSe_2) layer. If the attractive

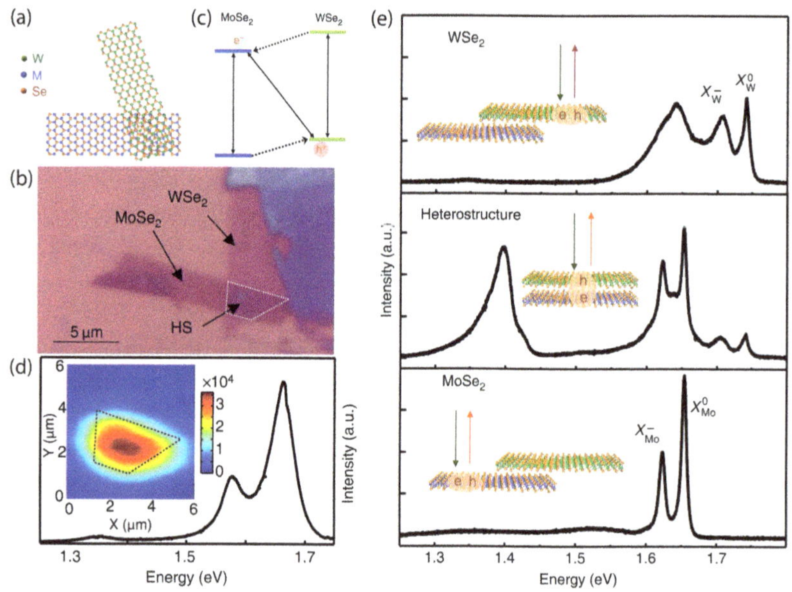

Figure 8.3. The properties of IXs in $MoSe_2$–WSe_2 vertical heterostructure. (a) Depiction, (b) optical microscopy image, and (c) schematic diagram of type-II band alignment of a typical $MoSe_2$–WSe_2 vertical heterostructure. (d) Room-temperature PL spectrum of the heterostructure under 20 μW laser excitation at 2.33 eV. Inset shows spatial map of integrated PL intensity from the low-energy peak (1.273–1.400 eV), which can only be observed in the heterostructure area. In (b) and (d), heterostructure region is outlined by the white and black dashed line, respectively. (e) Spatially resolved PL spectra of individual constituent monolayers (top and bottom panels) and the heterostructure (middle panel) at 20 K under 20 μW excitation at 1.88 eV (plotted on the same scale), featuring an IX peak next to peaks of individual constituents. Notations X_W^0 (X_W^-) and X_{Mo}^0 (X_{Mo}^-) stand for neutral exciton (trion/charged exciton [69]) peak emissions due to WSe_2 and $MoSe_2$, respectively. The PL emission from IX emerges at ~1.4 eV in the middle panel. Reproduced from [28], with permission from Springer Nature.

Coulomb interaction between the electrons and holes in opposite layers is sufficiently strong, an IX can form. The energy of the IX is determined by the band offsets between layers and the binding energy of the IX [62, 80–83]. The binding energy of IXs is calculated and given by 150 meV for the case of $MoSe_2$–WSe_2 heterostructure on SiO_2 substrate [84]. The IX binding energy of 90 meV is also observed in the MoS_2–$MoSe_2$ heterostructure, as shown by the temperature-dependent PL data [83].

As shown in figure 8.3(d), the heterostructure was studied by spatially dependent PL measurements, which are taken from the isolated WSe_2, $MoSe_2$ and overlapping heterostructure regions, at room temperature. The measurement reveals three main peaks in the spectrum. While the emission peaks at 1.65 eV and 1.57 eV are correlated to the exciton emission for $MoSe_2$ and WSe_2, respectively, the other one at 1.35 eV is assigned as the emission of IX because of type-II band structure. In the inset of figure 8.3(d), the PL mapping around IX further confirmed this [79]. Then the structures were studied by low-temperature PL experiments at 20 K under 20 μW excitation at 1.88 eV as shown in figure 8.3(e). PL from the isolated $MoSe_2$ and WSe_2 regions show monolayer (intralayer) excitons well known from earlier studies. i.e., for WSe_2 and $MoSe_2$, the exciton peaks all lie above 1.6 eV. However, PL from the heterostructure region shows an additional pronounced redshifted low-energy peak around 1.4 eV [85], which does not belong to intralayer emission species but is consistent with the prediction of a lower-energy IX [86]. The PL from the IX increased when the excitation energy was resonant with the emission of the two consituent layers, which is also consistent with charge transfer at the type-II interface [28]. Furthermore, since the large binding energies (~100 meV) of IXs [87–89] with spatially separated electron and hole, an out-of-plane dipole moment exhibits, the exciton properties, such as emission energy, luminescence intensity, diffusion length and lifetime, can be controlled by applying an out-of-plane electric field (controlled by external gating bias) via the DC Stark effect [28, 65, 72, 90, 91]. The demonstrated IX energy shift is ~70 meV [92], aligns with performed first-principles calculations [65, 91]. The spatially indirect nature of the IX also results in a significantly longer lifetime of 1–1000 ns for the IX [73, 93] (1.8 ns in this study [28]), compared to 1 ps for the intralayer exciton [94, 95]. This long lifetime provides an opportunity to realize spatial transport of IXs before these radiatively decay.

The IX was later observed by different groups in the other heterostructures [96], i.e., MoS_2–$MoSe_2$ [48, 83], MoS_2–WS_2 [49, 97–99], MoS_2–WSe_2 [19, 46, 68, 100], $MoSe_2$–WS_2 [101, 102], $MoSe_2$–WSe_2 [28, 62, 70, 73, 93, 97, 99, 103–108], and WS_2–WSe_2 [76, 109, 110]. The brightness of IXs heavily relies on the orientation of the two layers [93]. This aligns with the finding that misorientation of the layers makes the interlayer species indirect in both space and momentum [111].

8.4 Valley-polarized exciton

As we discussed in section 7.4.4, the exciton of monolayer TMDC carries information encoded in valleys. For heterostructures, non-degenerate continuous-wave pump–probe spectroscopy discovered that the charge transfer of electrons and holes in $MoSe_2$–WSe_2 is both directional and conserves spin–valley polarization, because of the momentum conservation [112] and spin-flipping requirement for carriers to transfer to the opposite valley. Specifically, the spin–valley polarization of excited electrons transfers from WSe_2 to $MoSe_2$, while the hole spin–valley polarization transfers in the opposite direction (figure 8.3 [28]) [103]. Similar phenomena were also observed in the MoS_2–WSe_2 heterobilayer [113]. These experiments showed that the spin of charge carriers remains stable despite interlayer charge transfer [65, 103], the IXs are still located at the $\boldsymbol{K}_+$ and $\boldsymbol{K}_-$ valleys of the two constituent layers in the heterostructure, inheriting the valley DOF [103, 111, 114]. Further, theories predicted that IXs in the light cones have valley-dependent optical selection rules with opposite handedness for opposite valleys [93, 111, 114, 115]. This means that the electrons and holes that make up the IX can still be used to encode VP and can be read out either by monitoring polarization of the IX PL [93, 106, 116–119] or by resonantly probing the intralayer exciton resonances using polarized light [103, 120].

In general, the selection rule is dependent on the twist angle between the two monolayers and has a periodic spatial dependence that results from the moiré pattern between the layers [121–123]. However, nearly aligned monolayers with slight twist and lattice mismatch in real space—zero (AA-stacking) and 60°-twist angle (AB-stacking), allow the $\boldsymbol{K}_+$ and $\boldsymbol{K}_-$ valleys of one layer to communicate, respectively, with the $\boldsymbol{K}_+$ valley of the other. Rivera *et al* [93] first studied the VP dynamics in mechanically stacked heterostructure WSe_2–$MoSe_2$ by circularly polarized optical pumping at room temperature. They demonstrate that the $MoSe_2$–WSe_2 heterostructures retain a PL polarization selection rule similar to that of single monolayers. With close lattice constants and a small interlayer twist angle between the two TMDC layers, the valleys in their BZ align almost perfectly [124], as shown in figure 8.4(a). The whole PL process is schematically illustrated in figure 8.4(b):

1. σ_+ excitation produces valley-polarized photocarriers in the $+\boldsymbol{K}_W$ valley of WSe_2 and the $+\boldsymbol{K}_M$ valley of $MoSe_2$.
2. Ultrafast charge transfer process: The carriers diffuse to the edge of the heterostructure in real space.
3. IXs are formed when charges are transferred between layers in sub-picosecond time scales.
4. The huge variance in momentum significantly restrains the interlayer transition between $+\boldsymbol{K}_W$ and $-\boldsymbol{K}_M$ or $-\boldsymbol{K}_W$ and $+\boldsymbol{K}_M$ valleys. On the contrary, the momentum difference between $+\boldsymbol{K}_W$ and $+\boldsymbol{K}_M$ valleys is small, and the spin conservation between these valleys makes such interlayer relaxation become the main pathway to emit photons with σ_+ handedness.

At zero applied gate voltage, as shown in Fig. 8.4(c), PL measurements show ~30% valley-polarized IX emission with a lifetime of 10 ns [125], which is dominantly co-polarized (retained the same polarization [39]) with circularly polarized excitation resonant with the WSe_2 intralayer exciton [93, 116, 117]. The IX emits at an energy lower than $MoSe_2$ and WSe_2 exciton energy, resulting from the reduced bandgap shown in figure 8.4(b).

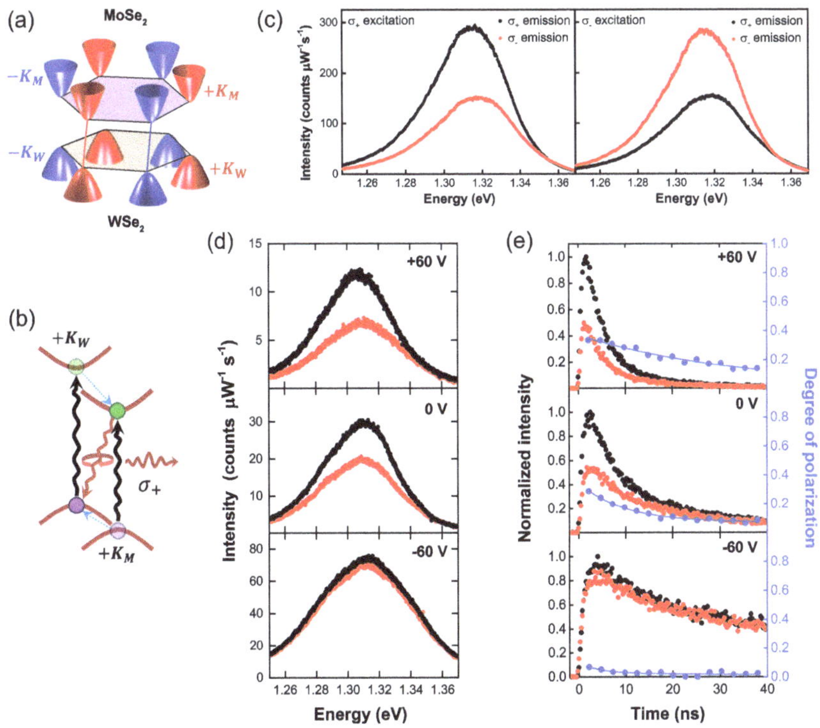

Figure 8.4. (a) Illustration showing the alignment of the hexagonal BZ of $MoSe_2$ and WSe_2 in a $MoSe_2$–WSe_2 heterostructure with small twist angle. $\boldsymbol{K}_+$ ($+\boldsymbol{K}_M$ and $+\boldsymbol{K}_W$) bands are marked red, while $\boldsymbol{K}_-$ ($-\boldsymbol{K}_M$ and $-\boldsymbol{K}_W$) bands are marked blue. (b) Schematic drawing of the IX excitation/emission in the $\boldsymbol{K}_+$ valley of the heterostructure. Intralayer excitons are excited by σ_+ polarized optical pump (black wavy lines) and form IXs in the $\boldsymbol{K}_+$ valley through fast interlayer charge transfer (blue dotted arrows). The light selection rule in the $\boldsymbol{K}_+$ valley produces co-polarized PL. (c) For certain twist angles, the IX can exhibit valley-conserved PL, similar to that of monolayers. (d) Circular polarization-resolved PL spectra of the IX that can be modulated by gate voltages. All the data were obtained under σ_+ pulsed circularly polarized laser excitation, with the co-polarized (σ_+) and cross-polarized (σ_-) PL spectra shown in black and red, respectively. (e) Time-resolved IX circular PL (left axis) for co-polarized and cross-polarized emission data at selected gate voltages, demonstrating control of IX population and VP dynamics. The blue points (right axis) show the decay of VP and solid lines are the single exponential fits with extracted valley lifetimes of 39 ± 2, 10 ± 1, and 5 ± 2 ns at gate voltages of +60, 0, and −60 V, respectively. From [93]. Reprinted with permission from the AAAS.

8.4.1 Manipulation of VP

Compared with intralayer excitons, in heterostructure, the electron and hole are separated in both real and momentum spaces. Under illumination, the photo-generated electrons and holes are confined in MoS_2 and WSe_2, respectively, the electron and hole wavefunctions have much smaller overlap in IXs than in intralayer excitons. They will therefore have a weaker exchange interaction (which serves as a source of valley depolarization in monolayer TMDCs [103]) and exhibit a slower recombination process (both radiative and nonradiative) [23]. On the other hand, due to the lattice mismatch and the twisted angle between two constituent layers caused by the stacking process, the $\boldsymbol{K}_{\pm}$ valleys of the two constituent layers are separated in momentum space [28]. These advantages allow for enhanced VP times [93, 103, 113, 126], which can be prolonged by several orders of magnitude under an external electric field (up to microsecond) [74, 93, 117], far surpassing the VP lifetimes in the order of picoseconds for intralayer excitons in most neat monolayer TMDCs [127–132]. Figures 8.4(d) and (e) show the strong gate-dependent VP and its lifetime of the IXs, respectively. The heterostructure is exposed under a 50 ps laser pulse and σ_+ excitation. The VP lifetime is determined by fitting a single exponential decay [51]. By applying a gate voltage of 60 V at 30 K, the VP has a peak value of approximately 40%. The VP of IXs decays with lifetimes of about 39 ± 2, 10, and 5 ns for gate voltages of +60, 0, and −60 V, respectively, which can be seen in figure 8.4(e) [12]. As we can see, the VP and its lifetime was extremely suppressed at the gate voltage of −60 V. We note that the valley lifetimes of IX in the WS_2–WSe_2 heterobilayer can reach up to 20 μs [120] and the single particle VP lifetime in the heterobilayer can even be much longer than that of IXs, with the hole VP time in WSe_2–MoS_2 exceeding 40 μs at low temperature [113]. Later, up to 80% VP was reported for IXs in a $MoSe_2$–WSe_2 heterostructure (without externally applied electric or magnetic fields) [119].

Future experimental and theoretical studies [106, 113, 120, 125, 133, 134] are required to fully understand the microscopic mechanism for the observed gate-dependent PL dynamics of the IX [125]. In addition, it is interesting to find that the increase in PL lifetime accompanies the decrease in the VP lifetime as the gate voltage goes negative [75]. The loss of VP was ascribed to valley depolarization of excitons on the femtosecond to picosecond timescale preceding photoinduced charge transfer [69]. This provides a readout channel to measure VP of the IX.

8.4.2 Diffusion of valley-polarized IXs

Since the IX is a net-charge neutral quasiparticle, its spatial transport allows for the realization of pure valley current (without a charge current). Recall that due to the spin–valley locking effect [135], a VP in a monolayer TMDC results in an accompanying spin polarization, so a valley current can also accurately be called a spin–valley current. The long interlayer lifetimes [117] have motivated scientists to explore the transport properties and spatial mapping of lateral drift and diffusion of valley-polarized IX over micron-length scales [69, 93] in a $MoSe_2$–WSe_2 heterostructure [93] and make vdW heterostructures an exciting platform for valley optoelectronic applications [11, 111].

Rivera *et al* [93] used the $MoSe_2$–WSe_2 IXs described previously [93] and first demonstrated the valley-polarized IX spatial transport. In the experiments, the spatial profile of the IX's PL was imaged under circularly polarized excitation. The co- and cross-polarized PL spatial maps were measured as a function of excitation laser intensity (1–60 μW). As shown in figure 8.5(a), the spatial pattern of the VP maxima for the IXs evolved into a ring with a diameter that increased with the excitation intensity. This demonstrated the different drift velocity of IXs in opposite valleys [75] and it becomes the signature of valley-polarized exciton transport, as

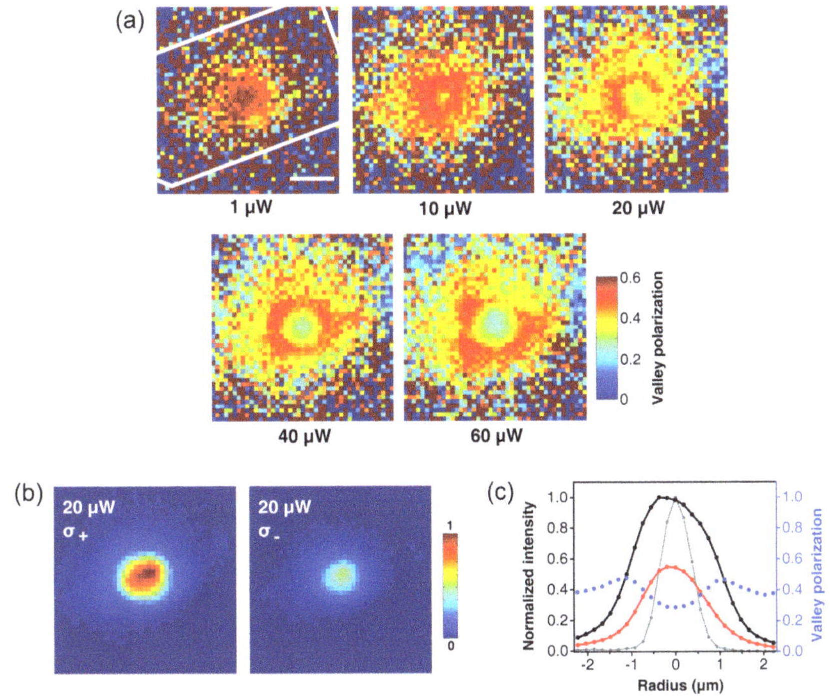

Figure 8.5. (a) The valley expansion of IXs, visualised by spatial mapping of valley-polarized IXs lateral diffusion under 1–60 μW σ_+ excitation, showing the formation of a ring with larger diameter with increased excitation power. The white frame outlines the region of the heterostructure [65]. The image includes a scale bar that indicates a length of 2 micrometres. (b) Spatial map of co-polarised (σ_+) and cross-polarised (σ_-) IXs diffusion and PL. (c) The spatial profile of PL co-polarized (cross-polarized) with the σ_+ excitation is shown in solid black (red) lines. The laser intensity profile is in grey, and the degree of circular polarization is in blue. From [93]. Reprinted with permission from the AAAS.

identified in figure 8.5(b). Such an observation can possibly be explained by valley-dependent many-body interactions [12] and a drift diffusion model for valley-polarized IXs [93] considering density. The observed ring pattern arises because the majority of polarized IXs (in one single valley) are subject to two different repulsive forces:

1. The valley-independent out-of-plane dipole–dipole interaction between IXs;
2. The valley-dependent electron–electron and hole–hole exchange interaction in individual layers [33].

The latter one (exchange effect) is existed in the same valley but is suppressed in opposite valleys. Therefore, the majority of valley excitons experience a stronger repulsive force and thus undergo more rapid expansion than the minority excitons, leading to valley-asymmetric transport of the IXs with the resulting ring pattern in the spatial distribution [65, 114]. Further, since the IX density is higher with increasing excitation intensity, an increasing expansion of VP maxima due to increased repulsion can be observed away from the excitation spot. These phenomena imply that the valley-dependent exchange interaction strength dominates over the valley-independent dipole–dipole repulsion, and the IX gas can expand well beyond the excitation laser profile without losing valley information, as shown in figure 8.5(c) [65]. Later, experimental research has demonstrated the valley diffusion length could exceed 20 μm in the WS_2–WSe_2 heterobilayer, which is drastically promising for the exciton devices [120].

The spin- and valley-contrasting emission of IXs has been demonstrated in various 2D bilayer systems. Apart from the comparable exciton lifetime and valley depolarization time to their monolayer counterparts, the reported VP of IXs in TMDCs-based heterostructure is somehow unsatisfying. The limiting factors may lie in sample inhomogeneity, indirect bandgap transitions [106, 118], and slow charge separation [93]. More research efforts are demanded to reveal the underlying mechanisms and push the magnitude of VP close to the theoretical value [15]. Nevertheless, this finding allows optical addressing of both the spin and valley configuration of IXs, making TMDC heterostructures promising in device technologies based on valleytronics [136]. Moreover, IXs are not confined to the bilayers. In multilayer heterostructures constructed by stacking monolayer and few-layer TMDC, emission from IXs can be obtained and engineered [137].

8.5 Moiré IXs

In general, as discussed in section 1.9, stacking two monolayers into a vdW heterostructure forms a varying interlayer atomic alignment, namely, a long-period moiré pattern, owing to the misalignment θ between the crystalline axes (excluding angle $\theta \leqslant 1°$ [138, 139]) and/or lattice mismatch $\delta_a = \frac{|a - a'|}{a} \approx 0.1\%$, where a and a' are the lattice constants of the two constituent monolayers [122, 123, 140–142]. The interlayer atomic registry varies periodically on a length scale much larger than the lattice constant a, and the introduced periodically modulated landscape of interlayer potential [19, 122, 143] can be as large as 150 meV [19], which generates localised states and provides exotic emission features [72, 93, 122, 143–148]. The moiré wavelength is given as [149–154]

$$\lambda = \frac{(1 + \delta_a)a}{\sqrt{2(1 + \delta_a)(1 - \cos\theta) + \delta_a^2}} \approx \frac{a}{\sqrt{\theta^2 + \delta_a^2}}, \tag{8.4}$$

which depends sensitively on the twist angle θ and the lattice mismatch δ_a. The moiré wavelength ranges from several to several hundred nanometres [114, 155]. For example, in carefully aligned $MoSe_2$–WSe_2 heterobilayers with $\delta_a < 0.2\%$, the moiré wavelength can exceed 100 nm.

In a long-period moiré pattern, the interlayer stacking smoothly varies across the supercell. The different locales in the moiré superlattice can therefore be characterized by the interlayer translation $\boldsymbol{R}_0(\boldsymbol{r})$, which changes with a smooth and periodic rhythm with position $\boldsymbol{r}$. The vdW interaction depends on $\boldsymbol{R}_0(\boldsymbol{r})$, resulting in a finely-tuned lateral modulation of the interlayer distance across a single moiré supercell. This effect was indicated by various first-principles calculations [121, 123, 156, 157] and has been confirmed by STM/STS measurements [19]. Meanwhile, the position-dependent local bandgap of each monolayer and the heterostructure is also shown in first-principles calculations [123, 153, 158–160] and observed in the STM/STS measurements [19].

In addition, as we showed in section 6.4.3, the moiré superlattice strongly affects the transport properties of the twisted graphene bilayers. In the case of heterobilayer TMDCs, an additional level of complexity is added, as moiré–exciton interactions [19, 115, 123, 161]. IXs in the spatially periodic inhomogeneous potential introduced by moiré superlattice can be described by wavepackets moving adiabatically in the slowly varying periodic superlattice potential [123]. The real-space spatial extension of IXs is small compared to the moiré period. Both the energy and optical properties of an IX wavepacket are therefore determined by the local atomic registry within a length scale $\ell < \lambda$ but $\ell \gg a$, which is locally almost indistinguishable from a lattice-matched heterobilayer with certain relative interlayer translation [65]. Specifically, moiré superlattices with 0° and 60° (or 180°) interlayer twist angles correspond to the 3R (rhombohedral) and 2H (hexagonal) bulk polytypes, thus these stackings are named R-type and H-type stackings, respectively [61, 123]. The high-symmetry sites of the moiré superlattice preserve the threefold rotation symmetry of their component monolayers. In R-type stacking, as schematically illustrated in figure 8.6(a), there are three high-symmetry sites which are commonly denoted as R_h^h, R_h^X, and R_h^M local atomic registries [145, 146]. Here, R_h^r refers to R-type stacking (with 0° twist angle) and the hole-layer h sites vertically align to the electron-layer r sites. X, M and h stand for chalcogen, transition metal and hollow centre of the hexagon (formed by M and X), respectively [154]. These three high-symmetry points are schematically illustrated in the upper panel of figure 8.6(b) and they are local energy extrema, which play a role in trapping IXs in the moiré supercell [123, 153]. These high-symmetry sites result in the spatial modulation of interlayer distance and affect the spin–valley, electronic [162–166] and excitonic properties [106, 107, 115, 119, 122, 123, 142, 143, 145, 147, 153, 155, 159, 160, 166, 167].

The difference in the atomic registry at each high-symmetry site determines a unique valley optical selection rule of excitons for each atomic registry [115, 123, 153]. Early experimental methods and sample quality cannot provide sufficient conditions for the observation and further investigation of moiré modulated exciton properties. Whereas, several theoretical predictions [115, 123, 153, 159, 160, 162] on the physics of excitons in moiré superlattice have been reported.

In order to infer the valley optical selection rule of the IXs in heterobilayer, one needs to understand the symmetry dictated selection rules for interband optical transition [65]. The quantum number m_i depends on the band index as well as the choice of rotation centre, as summarised in table 8.1. As we shown in equation (7.65) [169], if the orbital angular momentum ℓ is not considered, the C_3 symmetry requires that a circularly polarised photon with handedness $\sigma_\pm$ can excite an electron from the valence band to the conduction band only when[1]

[1] Readers can check the well-known valley optical selection rule in monolayer TMDCs (section 7.4) [135] by choosing a certain rotation centre. Though both m_v and m_c depend on the choice of rotation centre, equation (8.5) does not.

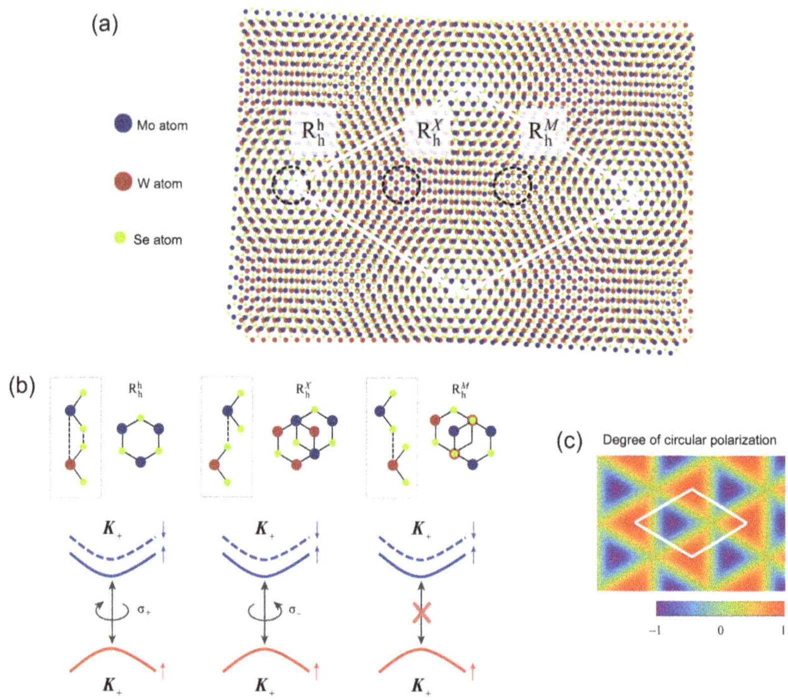

Figure 8.6. (a) A top–down view of long-period moiré superlattice (white rhombus) formed by a $MoSe_2$–WSe_2 vertical heterostructure, showing three highlighted regions in the moiré superlattice with threefold rotational symmetry, where the atomic registries resemble three different R-type commensurate bilayers. (b) Upper panel: The side view (left, boxed) and top view (right) of the three different R-type commensurate bilayers named R_h^r, which means R-type stacking with r site of the MoX_2 layer aligning with the hexagon centre of WX_2 layer [12]. Lower panel: Optical selection rules of different atomic registries in $\boldsymbol{K}_+$ valley. (c) Calculated optical selection rule for the spin-up $\boldsymbol{K}_+$ valley IX at different wavepacket centres in the moiré pattern. Reproduced from [143], with permission from Springer Nature.

Table 8.1. A list of the quantum numbers m_i for the conduction (i= c) and valence bands (i= v) at $\boldsymbol{K}_+$ valley with different rotation centres (h, X or M) of monolayer TMDCs [65, 168]. That of the $\boldsymbol{K}_-$ valley can be obtained by a time reversal, i.e., $m_i(\boldsymbol{K}_-) = -m_i(\boldsymbol{K}_+)$.

Rotation centre	…	m_v	m_c	…
h	…	0	+1	…
X	…	+1	−1	…
M	…	−1	0	…

$$\mathrm{mod}(m_c - m_v, 3) = \sigma_\pm. \tag{8.5}$$

Then the interband optical transition from valence to conduction bands of different layers can be concluded. For in-plane light polarisation, an IX wavepacket from the $\boldsymbol{K}_+$ valley in R_h^h and R_h^X heterobilayer couples with LCP and RCP light, respectively, as a consequence of mod(+1 − 0, 3) = +1 and mod(−1 − 0, 3) = −1. On the other hand, circularly polarised light coupling is forbidden at R_h^M site (optically dark) [111] because mod(0 − 0, 3) = 0. The above conclusions are demonstrated in the lower panel of figure 8.6(b). For a general stacking location that lacks the C_3 symmetry, the IX wavepackets couple to photons with elliptical polarisation [123]. The landscape of the degree of circular polarization in the heterobilayer has been plotted in figure 8.6(c), which shows that the IX wavepackets located at R_h^h and R_h^X sites couple to light with opposite circular polarization.

In experiment [106, 107, 143, 170], these multiple IX states and the corresponding optical selection rules at different moiré sites have been demonstrated in a $MoSe_2$–WSe_2 heterobilayer with a twist angle of 1° and a WSe_2–WS_2 moiré superlattice [143, 167]. Moreover, Seyler *et al* [122] have astonishingly found that quantum emitters in a $MoSe_2$–WSe_2 heterobilayer also present high VP. The degree of VP for a heterobilayer with a twist angle of 57° is over 70%, while the selection rule is reversed for a heterobilayer with a twist angle of 2°. The presence of strong circular polarization in different handedness implies that the IX is confined in moiré potentials. In addition, the moiré pattern depends on the twist angle between the two layers; therefore, the optical response of the heterostructures can be controlled by tuning the relative orientation of the layers [153, 171, 172].

In brief, the valley optical selection rule remains valid for the interlayer exciton, but it varies depending on where and at what energy the emission takes place. Additionally, in an inhomogeneous sample, the combination of the inhomogeneous broadening and the valley optical selection rule may result in a low degree of circular polarization in the PL emission. To avoid such broadening, an encapsulation of the TMDC sample using hBN is usually used [173, 174].

8.6 Enhancement of valley Zeeman effect in vdW heterostructures

In section 7.6, we covered the valley Zeeman effect as a magnetic method for tuning VP, and we presented several approaches to enhance the VP. i.e., defect states, electron doping, optical stark effect, etc. In this section, we will explore the valley Zeeman effect in vdW heterostructures.

8.6.1 Magnetic proximity effect from magnetic layered materials or substrates

The TRS of nonmagnetic TMDC semiconductors can be broken by proximity to magnetic layered materials or substrates [175, 176]. For graphene, early experiments with magnetic insulating substrates like yttrium iron garnet or EuS demonstrated the proximity-induced anomalous Hall effect [177] and additional effective magnetic fields reaching ~14 T [178]. Meanwhile, the interfacial magnetic exchange field in TMDC semiconductors offers an effective method for valley control [12]. Theoretical estimations predict that such fields can be as high as a few thousand teslas [179], quickly surpassing the strengths of state-of-the-art laboratory magnets. By transferring the monolayer WS_2 onto the ferromagnetic $LaMnO_3$ substrate, a considerably improved VP of up to 80% in monolayer WS_2 under nonresonant excitation at 4.2 K can be observed [180]. The temperature dependence of the VP indicates the exciton–magnon coupling between WS_2 and $LaMnO_3$. The authors proposed that the nonradiative transfer of photoexcited energy to the $LaMnO_3$ or the formation of IXs decrease intervalley scattering and thus boost VP.

The VP of the TMDC monolayer could be further modulated via the magnetic proximity effect by applying gate voltage or adjusting the optical excitation power [12]. Theoretical calculations revealed that the interfacial interaction between $MoTe_2$ and ferromagnetic EuO substrate could simultaneously induce a valley splitting of 300 meV, shedding light on valleytronics without external polarization at room temperature [179, 181, 182]. By using a ferromagnet EuS substrate instead of a SiO_2/Si substrate, the valley splitting in monolayer WSe_2 is largely enhanced to 2.5 meV T^{-1} under an out-of-plane magnetic field of 1 T [183, 184]. However, one of the limiting factors of the exchange coupling is the poor interface quality in heterostructures of 3D substrates and 2D vdW materials [183]. Recently re-discovered vdW magnets like CrI_3 and $Cr_2Ge_2Te_6$ offer ultimately smooth surfaces and ideal coupling for 2D-on-2D vdW heterostructures [6]. For a ferromagnetic vertical heterostructure composed of a monolayer WSe_2 and ultrathin ferromagnetic CrI_3 semiconductor layers [185], when the sample temperature is cooled down gradually below the Curie temperature of CrI_3 (61 K [186]), the valley splitting becomes visible based on the polarized PL spectra. At 5 K, the valley splitting of WSe_2 resulted from CrI_3 layers is estimated to be 3.5 meV, equivalent to a magnetic exchange field of $\sim$13 T [187]. Further measurements reveal that a valley splitting rate of over 150 meV T^{-1} can be reached [188]. In addition, fast switching of valley splitting can be realised via the change of magnetisation of CrI_3, which can be manipulated by external electric field (induced from gate voltage) [189] or minor variations in laser excitation power [187], providing new opportunities to optical manipulation of the valley. The latter is attributed to the optical control of the magnetic exchange field of CrI_3 over a range of 20 T. It should be noted that CrI_3 is a very unstable 2D material requiring a high-purity oxygen-free glove box for vdW assembly. Improved VP (80% at 4.2 K and 53% at 160 K) is also found in the monolayer $MoSe_2$ proximitized by a 2D ferromagnetic semiconductor $Cr_2Ge_2Te_6$ via a back-gate voltage [190].

These findings shed fresh light on VP manipulation in nonmagnetic TMDC by means of ferromagnetic semiconductor proximitizing. Furthermore, an ideal vdW coupling allows studying of newly discovered ultra-thin magnets through the optical response of TMDC materials and a better understanding of the properties of all-2D magnetic heterostructures [191, 192]. We envision that future valleytronic platforms could rely heavily on the Zeeman effect as well as interfacial interactions with ferromagnetic substrates to tune the valley properties in pursuit of non-volatility, exceptional energy-efficient computing [11] and robust cycling durability. Future work in this 2D semiconductor valleytronics could be enhanced by the discovery and development of air-stable 2D FMIs with higher Curie temperatures and larger bandgaps [33].

8.6.2 IXs

In 2D semiconductors, heterostructures play an important role in controlling VP. As discussed in section 7.4, heterostructures offer a possibility to overcome the limitations of monolayer systems, which have extremely short lifetimes of excitons at the order of picoseconds [92, 94, 193] and the fast polarisation dephasing mechanisms [194], facilitating its application in valleytronics requiring long-range spatial transport. It is important to remember that the conduction and valence band valleys that give rise to the IX are still located at the $\boldsymbol{K}_-$ and $\boldsymbol{K}_+$ valleys of the two different layers. Moreover, in analogy to intralayer excitons, VP of IXs can be injected via optical excitation [93]. This means that the electrons and holes that make up the IX can still be used to encode VP and can be read out either by monitoring polarization of the IX PL [93] or by resonantly probing the intralayer exciton resonances using polarized light [103, 120]. Another advantage of IXs over intralayer excitons for valleytronics is that since the electron and hole are spatially separated, the electron–hole exchange interaction (which causes depolarization in monolayer TMDCs) is suppressed, which allows for enhanced VP times. In general, the selection rule is dependent on the twist angle between the two monolayers and has a periodic spatial dependence that results from the moiré pattern between the layers [55, 122, 123], as discussed in section 8.5. However, for a nearly aligned monolayer (near zero-twist angle), typically AA and AB stackings, negligible momentum mismatch allows radiative recombination of charge carriers at the $\boldsymbol{K}_\pm$ points.

As we show in section 8.4, $MoSe_2$–WSe_2 heterostructures show a pronounced light emission from IXs below the energies of the individual monolayer transitions [195], as well as a PL polarization selection rule similar to that of single monolayers. Here, polarization helicity of the interlayer PL is shown to be dominantly co-polarized with the excitation laser's circular helicity, with long-lived VP (up to 40 ns) [93]. By exposing such zero-twisted AB-stacked vdW heterostructure to intense magnetic fields (30 T), Nagler *et al* observed the Zeeman splittings with a high degree of VP over $\approx$ 80% and the lifetime of IXs approaching 70 ns, corresponding to a much higher effective g-factor of around 15 [195] for the IX [196], which can be understood as arising from the sum of the valence and conduction valley magnetic moments [197]. In a MoS_2–$MoSe_2$–MoS_2 heterostructure, $\sim$100% VP is demonstrated if the magnetic field is strong enough (>20 T) [198]. These findings provide a readout channel to measure VP of the IX, which is ideally suited for valleytronic applications.

Also, nearly two orders of magnitude enhanced valley magnetic response has been recently observed in WSe_2–WS_2 moiré superlattices by taking advantage of the antiferromagnetic interactions between localized excitons and holes around half-filling of the first hole moiré superlattice band [165], opening up exciting possibilities for valleytronic applications [199].

In 2019, Park [200] proposed a model for a valley filter and valley valve device. The device is based on the valley-dependent Zeeman interaction between the applied magnetic field and the orbital magnetic moment that can exist in a system with broken IS.

References

[1] Mak K F, McGill K L, Park J and McEuen P L 2014 The valley Hall effect in MoS_2 transistors *Science* **344** 1489

[2] Lee J, Mak K F and Shan J 2016 Electrical control of the valley Hall effect in bilayer MoS_2 transistors *Nat. Nanotechnol.* **11** 421–5

[3] Lee J, Wang Z, Xie H, Mak K F and Shan J 2017 Valley magnetoelectricity in single-layer MoS_2 *Nat. Mater.* **16** 887–91

[4] Lee J 2020 Valley dependent electronics in 2D materials *Vac. Mag.* **6** 13–7

[5] Sugano S and Kojima N (ed) 2000 *Magneto-Optics* Springer Series in Solid-State Sciences **vol 128** (Heidelberg: Springer)

[6] Goh K E J, Wong C P Y and Wang T (ed) 2023 *Valleytronics in 2D Materials* (Singapore: World Scientific)

[7] Bussolotti F, Kawai H, Ooi Z E, Chellappan V, Thian D, Pang A L C and Goh K E J 2018 Roadmap on finding chiral valleys: screening 2D materials for valleytronics *Nano Futures* **2** 032001

[8] Mak K F, Xiao D and Shan J 2018 Light–valley interactions in 2D semiconductors *Nat. Photon.* **12** 451–60

[9] Wu S *et al* 2013 Electrical tuning of valley magnetic moment through symmetry control in bilayer MoS_2 *Nat. Phys.* **9** 149–53

[10] Tuan Hung N, Nguyen T, Van Thanh V, Wang S, Saito R and Li M 2024 Symmetry breaking in 2D materials for optimizing second-harmonic generation *J. Phys. D: Appl. Phys.* **57** 333002

[11] Schaibley J R, Yu H, Clark G, Rivera P, Ross J S, Seyler K L, Yao W and Xu X 2016 Valleytronics in 2D materials *Nat. Rev. Mater.* **1** 16055

[12] Ma H, Zhu Y, Liu Y, Bai R, Zhang X, Ren Y and Jiang C 2023 Valley polarization in transition metal dichalcogenide layered semiconductors: generation, relaxation, manipulation and transport *Chin. Phys.* B **32** 107201

[13] Liu F, Zhou J, Zhu C and Liu Z 2017 Electric field effect in two-dimensional transition metal dichalcogenides *Adv. Funct. Mater.* **27** 1602404

[14] Guan H, Tang N, Huang H, Zhang X, Su M, Liu X, Liao L, Ge W and Shen B 2019 Inversion symmetry breaking induced valley Hall effect in multilayer WSe_2 *ACS Nano* **13** 9325–31

[15] Rong R, Liu Y, Nie X, Zhang W, Zhang Z, Liu Y and Guo W 2023 The interaction of 2D materials with circularly polarized light *Adv. Sci.* **10** 2206191

[16] Low T, Jiang Y and Guinea F 2015 Topological currents in black phosphorus with broken inversion symmetry *Phys. Rev.* B **92** 235447

[17] Deng B, Frisenda R, Li C, Chen X, Castellanos-Gomez A and Xia F 2018 Progress on black phosphorus photonics *Adv. Opt. Mater.* **6** 1800365

[18] Geim A K and Grigorieva I V 2013 Van der Waals heterostructures *Nature* **499** 419–25

[19] Zhang C, Chuu C-P, Ren X, Li M-Y, Li L-J, Jin C, Chou M-Y and Shih C-K 2017 Interlayer couplings, moiré patterns, and 2D electronic superlattices in MoS_2/WSe_2 hetero-bilayers *Sci. Adv.* **3** e1601459

[20] Xu W *et al* 2018 Determining the optimized interlayer separation distance in vertical stacked 2D WS_2:hBN:MoS_2 heterostructures for exciton energy transfer *Small* **14** 1703727

[21] Withers F *et al* 2015 Light-emitting diodes by band-structure engineering in van der Waals heterostructures *Nat. Mater.* **14** 301–6

[22] Novoselov K S, Mishchenko A, Carvalho A and Castro Neto A H 2016 2D materials and van der Waals heterostructures *Science* **353** aac9439

[23] Jin C, Ma E Y, Karni O, Regan E C, Wang F and Heinz T F 2018 Ultrafast dynamics in van der Waals heterostructures *Nat. Nanotechnol.* **13** 994–1003

[24] Hsu W-T *et al* 2018 Negative circular polarization emissions from $WSe_2/MoSe_2$ commensurate heterobilayers *Nat. Commun.* **9** 1356
[25] Chen H *et al* 2016 Ultrafast formation of interlayer hot excitons in atomically thin MoS_2/WS_2 heterostructures *Nat. Commun.* **7** 12512
[26] Cha S *et al* 2018 Generation, transport and detection of valley-locked spin photocurrent in WSe_2–graphene–Bi_2Se_3 heterostructures *Nat. Nanotechnol.* **13** 910–4
[27] Wang S, Chou J-P, Ren C, Tian H, Yu J, Sun C, Xu Y and Sun M 2019 Tunable Schottky barrier in graphene/graphene-like germanium carbide van der Waals heterostructure *Sci. Rep.* **9** 5208
[28] Rivera P *et al* 2015 Observation of long-lived interlayer excitons in monolayer $MoSe_2$–WSe_2 heterostructures *Nat. Commun.* **6** 6242
[29] Gong Y *et al* 2014 Vertical and in-plane heterostructures from WS_2/MoS_2 monolayers *Nat. Mater.* **13** 1135–42
[30] Blake P, Hill E W, Castro Neto A H, Novoselov K S, Jiang D, Yang R, Booth T J and Geim A K 2007 Making graphene visible *Appl. Phys. Lett.* **91** 063124
[31] Rubio-Bollinger G, Guerrero R, De Lara D P, Quereda J, Vaquero-Garzon L, Agraït N, Bratschitsch R and Castellanos-Gomez A 2015 Enhanced visibility of MoS_2, $MoSe_2$, WSe_2 and black-phosphorus: making optical identification of 2D semiconductors easier *Electronics* **4** 847–56
[32] Zomer P J, Guimarães M H D, Brant J C, Tombros N and van Wees B J 2014 Fast pick up technique for high quality heterostructures of bilayer graphene and hexagonal boron nitride *Appl. Phys. Lett.* **105** 013101
[33] Bao Q and Hoh H Y (ed) 2019 *2D Materials for Photonic and Optoelectronic Applications* Woodhead Publishing Series in Electronic and Optical Materials (Duxford: Woodhead Publishing)
[34] Dean C R *et al* 2010 Boron nitride substrates for high-quality graphene electronics *Nat. Nanotechnol.* **5** 722–6
[35] Gorbachev R V *et al* 2014 Detecting topological currents in graphene superlattices *Science* **346** 448
[36] Cheng R, Li D, Zhou H, Wang C, Yin A, Jiang S, Liu Y, Chen Y, Huang Y and Duan X 2014 Electroluminescence and photocurrent generation from atomically sharp WSe_2/MoS_2 heterojunction p–n diodes *Nano Lett.* **14** 5590–7
[37] Ross J S *et al* 2017 Interlayer exciton optoelectronics in a 2D heterostructure p–n junction *Nano Lett.* **17** 638–43
[38] Cao Y, Fatemi V, Fang S, Watanabe K, Taniguchi T, Kaxiras E and Jarillo-Herrero P 2018 Unconventional superconductivity in magic-angle graphene superlattices *Nature* **556** 43–50
[39] Shrestha S and Cotlet M 2023 Interfacial charge transfer in atomically thin 2D transition-metal dichalcogenide heterostructures *ACS Appl. Opt. Mater.* **1** 1192–207
[40] Özçelik V O, Azadani J G, Yang C, Koester S J and Low T 2016 Band alignment of two-dimensional semiconductors for designing heterostructures with momentum space matching *Phys. Rev.* B **94** 035125
[41] Wang S, Tian H, Ren C, Yu J and Sun M 2018 Electronic and optical properties of heterostructures based on transition metal dichalcogenides and graphene-like zinc oxide *Sci. Rep.* **8** 12009

[42] Jit S and Das S (ed) 2020 *2D Nanoscale Heterostructured Materials: Synthesis, Properties, and Applications* Micro and Nano Technologies (Amsterdam: Elsevier)

[43] Cai Z, Liu B, Zou X and Cheng H-M 2018 Chemical vapor deposition growth and applications of two-dimensional materials and their heterostructures *Chem. Rev.* **118** 6091–133

[44] Xu W, Liu W, Schmidt J F, Zhao W, Lu X, Raab T, Diederichs C, Gao W, Seletskiy D V and Xiong Q 2017 Correlated fluorescence blinking in two-dimensional semiconductor heterostructures *Nature* **541** 62–7

[45] Zang Z, Cai W and Zhou Y 2023 *Metal Oxide Semiconductors: Synthesis, Properties, and Devices* (Weinheim: Wiley-VCH)

[46] Fang H *et al* 2014 Strong interlayer coupling in van der Waals heterostructures built from single-layer chalcogenides *Proc. Natl. Acad. Sci. USA* **111** 6198–202

[47] Hong X, Kim J, Shi S-F, Zhang Y, Jin C, Sun Y, Tongay S, Wu J, Zhang Y and Wang F 2014 Ultrafast charge transfer in atomically thin MoS_2/WS_2 heterostructures *Nat. Nanotechnol.* **9** 682–6

[48] Ceballos F, Bellus M Z, Chiu H-Y and Zhao H 2014 Ultrafast charge separation and indirect exciton formation in a MoS_2–$MoSe_2$ van der Waals heterostructure *ACS Nano* **8** 12717–24

[49] Heo H *et al* 2015 Interlayer orientation-dependent light absorption and emission in monolayer semiconductor stacks *Nat. Commun.* **6** 7372

[50] Zhang T, Xu X, Huang B, Dai Y, Kou L and Ma Y 2023 Layer-polarized anomalous Hall effects in valleytronic van der Waals bilayers *Mater. Horiz.* **10** 483–90

[51] Liu Y, Zhang S, He J, Wang Z M and Liu Z 2019 Recent progress in the fabrication, properties, and devices of heterostructures based on 2D materials *Nano-Micro Lett.* **11** 13

[52] Shim J *et al* 2016 Phosphorene/rhenium disulfide heterojunction-based negative differential resistance device for multi-valued logic *Nat. Commun.* **7** 13413

[53] Ferdous N, Islam Md S and Park J 2024 A resilient type-III broken gap Ga_2O_3/SiC van der Waals heterogeneous bilayer with band-to-band tunneling effect and tunable electronic property *Sci. Rep.* **14** 12748

[54] Zhong H, Xu Z, Feng C, Wan X, Li J, Wang H and Tang G 2023 Broken-gap type-III band alignment in monolayer halide perovskite/antiperovskite oxide van der Waals heterojunctions *Nanoscale* **15** 11560–8

[55] Kang J, Tongay S, Zhou J, Li J and Wu J 2013 Band offsets and heterostructures of two-dimensional semiconductors *Appl. Phys. Lett.* **102** 012111

[56] Kang S, Lee D, Kim J, Capasso A, Kang H S, Park J-W, Lee C-H and Lee G-H 2020 2D semiconducting materials for electronic and optoelectronic applications: potential and challenge *2D Mater.* **7** 022003

[57] Kośmider K and Fernández-Rossier J 2013 Electronic properties of the MoS_2-WS_2 heterojunction *Phys. Rev.* B **87** 075451

[58] Terrones H, López-Urías F and Terrones M 2013 Novel hetero-layered materials with tunable direct band gaps by sandwiching different metal disulfides and diselenides *Sci. Rep.* **3** 1549

[59] Komsa H-P and Krasheninnikov A V 2013 Electronic structures and optical properties of realistic transition metal dichalcogenide heterostructures from first principles *Phys. Rev.* B **88** 085318

[60] Chaves A, Azadani J G, Özçelik V O, Grassi R and Low T 2018 Electrical control of excitons in van der Waals heterostructures with type-II band alignment *Phys. Rev.* B **98** 121302

[61] Tang J, Zheng Y, Jiang K, You Q, Yin Z, Xie Z, Li H, Han C, Zhang X and Shi Y 2023 Interlayer exciton dynamics of transition metal dichalcogenide heterostructures under electric fields *Nano Res.* **17** 4555–72

[62] Wilson N R *et al* 2017 Determination of band offsets, hybridization, and exciton binding in 2D semiconductor heterostructures *Sci. Adv.* **3** e1601832

[63] Chiu M-H, Zhang C, Shiu H-W, Chuu C-P, Chen C-H, Chang C-Y S, Chou M-Y, Shih C-K and Li L-J 2015 Determination of band alignment in the single-layer MoS_2/WSe_2 heterojunction *Nat. Commun.* **6** 7666

[64] Liu Y, Weiss N O, Duan X, Cheng H-C, Huang Y and Duan X 2016 Van der Waals heterostructures and devices *Nat. Rev. Mater.* **1** 16042

[65] Rivera P, Yu H, Seyler K L, Wilson N P, Yao W and Xu X 2018 Interlayer valley excitons in heterobilayers of transition metal dichalcogenides *Nat. Nanotechnol.* **13** 1004–15

[66] Zheng S, Sun L, Yin T, Dubrovkin A M, Liu F, Liu Z, Shen Z X and Fan H J 2015 Monolayers of $W_xMo_{1-x}S_2$ alloy heterostructure with in-plane composition variations *Appl. Phys. Lett.* **106** 063113

[67] Hill H M, Rigosi A F, Rim K T, Flynn G W and Heinz T F 2016 Band alignment in MoS_2/WS_2 transition metal dichalcogenide heterostructures probed by scanning tunneling microscopy and spectroscopy *Nano Lett.* **16** 4831–7

[68] Karni O *et al* 2019 Infrared interlayer exciton emission in MoS_2/WSe_2 heterostructures *Phys. Rev. Lett.* **123** 247402

[69] Sulas-Kern D B, Miller E M and Blackburn J L 2020 Photoinduced charge transfer in transition metal dichalcogenide heterojunctions – towards next generation energy technologies *Energy Environ. Sci.* **13** 2684–740

[70] Nagler P *et al* 2017 Interlayer exciton dynamics in a dichalcogenide monolayer heterostructure *2D Mater.* **4** 025112

[71] Fogler M M, Butov L V and Novoselov K S 2014 High-temperature superfluidity with indirect excitons in van der Waals heterostructures *Nat. Commun.* **5** 4555

[72] Jauregui L A *et al* 2019 Electrical control of interlayer exciton dynamics in atomically thin heterostructures *Science* **366** 870–5

[73] Miller B, Steinhoff A, Pano B, Klein J, Jahnke F, Holleitner A and Wurstbauer U 2017 Long-lived direct and indirect interlayer excitons in van der Waals heterostructures *Nano Lett.* **17** 5229–37

[74] Hu Z, Liu X, Hernández-Martínez P L, Zhang S, Gu P, Du W, Xu W, Demir H V, Liu H and Xiong Q 2022 Interfacial charge and energy transfer in van der Waals heterojunctions *InfoMat* **4** e12290

[75] Liu Y, Gao Y, Zhang S, He J, Yu J and Liu Z 2019 Valleytronics in transition metal dichalcogenides materials *Nano Res.* **12** 2695–711

[76] Wang K *et al* 2016 Interlayer coupling in twisted WSe_2/WS_2 bilayer heterostructures revealed by optical spectroscopy *ACS Nano* **10** 6612–22

[77] Ji Z *et al* 2017 Robust stacking-independent ultrafast charge transfer in MoS_2/WS_2 bilayers *ACS Nano* **11** 12020–6

[78] Zhu H, Wang J, Gong Z, Kim Y D, Hone J and Zhu X-Y 2017 Interfacial charge transfer circumventing momentum mismatch at two-dimensional van der Waals heterojunctions *Nano Lett.* **17** 3591–8
[79] Huang L, Krasnok A, Alú A, Yu Y, Neshev D and Miroshnichenko A E 2022 Enhanced light–matter interaction in two-dimensional transition metal dichalcogenides *Rep. Prog. Phys.* **85** 046401
[80] Latini S, Winther K T, Olsen T and Thygesen K S 2017 Interlayer excitons and band alignment in MoS_2/hBN/WSe_2 van der Waals heterostructures *Nano Lett.* **17** 938–45
[81] Zhu X, Monahan N R, Gong Z, Zhu H, Williams K W and Nelson C A 2015 Charge transfer excitons at van der Waals interfaces *J. Am. Chem. Soc.* **137** 8313–20
[82] Zhu X, Monahan N R, Gong Z, Zhu H, Williams K W and Nelson C A 2015 Correction to "Charge transfer excitons at van der Waals interfaces" *J. Am. Chem. Soc.* **137** 14230
[83] Mouri S, Zhang W, Kozawa D, Miyauchi Y, Eda G and Matsuda K 2017 Thermal dissociation of inter-layer excitons in MoS_2/$MoSe_2$ hetero-bilayers *Nanoscale* **9** 6674–9
[84] Ovesen S, Brem S, Linderälv C, Kuisma M, Korn T, Erhart P, Selig M and Malic E 2019 Interlayer exciton dynamics in van der Waals heterostructures *Commun. Phys.* **2** 23
[85] Dresselhaus M, Dresselhaus G, Cronin S B and Souza Filho A G 2018 *Solid State Properties: From Bulk to Nano* Graduate Texts in Physics (Heidelberg: Springer)
[86] Nakato T, Kawamata J and Takagi S (ed) 2017 *Inorganic Nanosheets and Nanosheet-Based Materials: Fundamentals and Applications of Two-Dimensional Systems* Nanostructure Science and Technology (Tokyo: Springer)
[87] Kamban H C and Pedersen T G 2020 Interlayer excitons in van der Waals heterostructures: binding energy, Stark shift, and field-induced dissociation *Sci. Rep.* **10** 5537
[88] Li Y *et al* 2014 Valley splitting and polarization by the Zeeman effect in monolayer $MoSe_2$ *Phys. Rev. Lett.* **113** 266804
[89] Van der Donck M and Peeters F M 2018 Interlayer excitons in transition metal dichalcogenide heterostructures *Phys. Rev.* B **98** 115104
[90] Unuchek D, Ciarrocchi A, Avsar A, Watanabe K, Taniguchi T and Kis A 2018 Room-temperature electrical control of exciton flux in a van der Waals heterostructure *Nature* **560** 340–4
[91] Gao S, Yang L and Spataru C D 2017 Interlayer coupling and gate-tunable excitons in transition metal dichalcogenide heterostructures *Nano Lett.* **17** 7809–13
[92] Robert C *et al* 2016 Exciton radiative lifetime in transition metal dichalcogenide monolayers *Phys. Rev.* B **93** 205423
[93] Rivera P, Seyler K L, Yu H, Schaibley J R, Yan J, Mandrus D G, Yao W and Xu X 2016 Valley-polarized exciton dynamics in a 2D semiconductor heterostructure *Science* **351** 688–91
[94] Moody G *et al* 2015 Intrinsic homogeneous linewidth and broadening mechanisms of excitons in monolayer transition metal dichalcogenides *Nat. Commun.* **6** 8315
[95] Moody G, Schaibley J and Xu X 2016 Exciton dynamics in monolayer transition metal dichalcogenides [invited] *J. Opt. Soc. Am.* B **33** C39–49
[96] Calman E V, Fogler M M, Butov L V, Hu S, Mishchenko A and Geim A K 2018 Indirect excitons in van der Waals heterostructures at room temperature *Nat. Commun.* **9** 1895
[97] Rigosi A F, Hill H M, Li Y, Chernikov A and Heinz T F 2015 Probing interlayer interactions in transition metal dichalcogenide heterostructures by optical spectroscopy: MoS_2/WS_2 and $MoSe_2$/WSe_2 *Nano Lett.* **15** 5033–8

[98] Tongay S *et al* 2014 Tuning interlayer coupling in large-area heterostructures with CVD-grown MoS_2 and WS_2 monolayers *Nano Lett.* **14** 3185–90
[99] Torun E, Miranda H P C, Molina-Sánchez A and Wirtz L 2018 Interlayer and intralayer excitons in MoS_2/WS_2 and $MoSe_2/WSe_2$ heterobilayers *Phys. Rev.* B **97** 245427
[100] Chiu M-H, Li M-Y, Zhang W, Hsu W-T, Chang W-H, Terrones M, Terrones H and Li L-J 2014 Spectroscopic signatures for interlayer coupling in MoS_2–WSe_2 van der Waals stacking *ACS Nano* **8** 9649–56
[101] Bellus M Z, Ceballos F, Chiu H-Y and Zhao H 2015 Tightly bound trions in transition metal dichalcogenide heterostructures *ACS Nano* **9** 6459–64
[102] Ceballos F, Bellus M Z, Chiu H-Y and Zhao H 2015 Probing charge transfer excitons in a $MoSe_2$–WS_2 van der Waals heterostructure *Nanoscale* **7** 17523–8
[103] Schaibley J R, Rivera P, Yu H, Seyler K L, Yan J, Mandrus D G, Taniguchi T, Watanabe K, Yao W and Xu X 2016 Directional interlayer spin-valley transfer in two-dimensional heterostructures *Nat. Commun.* **7** 13747
[104] Nayak P K *et al* 2017 Probing evolution of twist-angle-dependent interlayer excitons in $MoSe_2/WSe_2$ van der Waals heterostructures *ACS Nano* **11** 4041–50
[105] Nayak P K *et al* 2017 Correction to probing evolution of twist-angle-dependent interlayer excitons in $MoSe_2/WSe_2$ van der Waals heterostructures *ACS Nano* **11** 9566
[106] Ciarrocchi A, Unuchek D, Avsar A, Watanabe K, Taniguchi T and Kis A 2019 Polarization switching and electrical control of interlayer excitons in two-dimensional van der Waals heterostructures *Nat. Photon.* **13** 131–6
[107] Wang T *et al* 2020 Giant valley-Zeeman splitting from spin-singlet and spin-triplet interlayer excitons in $WSe_2/MoSe_2$ heterostructure *Nano Lett.* **20** 694–700
[108] Calman E V, Fowler-Gerace L H, Choksy D J, Butov L V, Nikonov D E, Young I A, Hu S, Mishchenko A and Geim A K 2020 Indirect excitons and trions in $MoSe_2/WSe_2$ van der Waals heterostructures *Nano Lett.* **20** 1869–75
[109] Wu K *et al* 2021 Identification of twist-angle-dependent excitons in WS_2/WSe_2 heterobilayers *Natl. Sci. Rev.* **9** nwab135
[110] Zhu M, Zhang Z, Zhang T, Liu D, Zhang H, Zhang Z, Li Z, Cheng Y and Huang W 2022 Exchange between interlayer and intralayer exciton in WSe_2/WS_2 heterostructure by interlayer coupling engineering *Nano Lett.* **22** 4528–34
[111] Yu H, Wang Y, Tong Q, Xu X and Yao W 2015 Anomalous light cones and valley optical selection rules of interlayer excitons in twisted heterobilayers *Phys. Rev. Lett.* **115** 187002
[112] Dey P, Yang L, Robert C, Wang G, Urbaszek B, Marie X and Crooker S A 2017 Gate-controlled spin-valley locking of resident carriers in WSe_2 monolayers *Phys. Rev. Lett.* **119** 137401
[113] Kim J *et al* 2017 Observation of ultralong valley lifetime in WSe_2/MoS_2 heterostructures *Sci. Adv.* **3** e1700518
[114] Jiang Y, Chen S, Zheng W, Zheng B and Pan A 2021 Interlayer exciton formation, relaxation, and transport in TMD van der Waals heterostructures *Light Sci. Appl.* **10** 72
[115] Yu H, Liu G-B and Yao W 2018 Brightened spin-triplet interlayer excitons and optical selection rules in van der Waals heterobilayers *2D Mater.* **5** 035021
[116] Baranowski M *et al* 2017 Probing the interlayer exciton physics in a $MoS_2/MoSe_2/MoS_2$ van der Waals heterostructure *Nano Lett.* **17** 6360–5
[117] Jiang C, Xu W, Rasmita A, Huang Z, Li K, Xiong Q and Gao W-B 2018 Microsecond dark-exciton valley polarization memory in two-dimensional heterostructures *Nat. Commun.* **9** 753

[118] Hanbicki A T, Chuang H-J, Rosenberger M R, Hellberg C S, Sivaram S V, McCreary K M, Mazin I I and Jonker B T 2018 Double indirect interlayer exciton in a $MoSe_2/WSe_2$ van der Waals heterostructure *ACS Nano* **12** 4719–26

[119] Zhang L, Gogna R, Burg G W, Horng J, Paik E, Chou Y-H, Kim K, Tutuc E and Deng H 2019 Highly valley-polarized singlet and triplet interlayer excitons in van der Waals heterostructure *Phys. Rev.* B **100** 041402

[120] Jin C *et al* 2018 Imaging of pure spin-valley diffusion current in WS_2-WSe_2 heterostructures *Science* **360** 893–6

[121] Kang J, Li J, Li S-S, Xia J-B and Wang L-W 2013 Electronic structural moiré pattern effects on $MoS_2/MoSe_2$ 2D heterostructures *Nano Lett.* **13** 5485–90

[122] Seyler K L, Rivera P, Yu H, Wilson N P, Ray E L, Mandrus D G, Yan J, Yao W and Xu X 2019 Signatures of moiré-trapped valley excitons in $MoSe_2/WSe_2$ heterobilayers *Nature* **567** 66–70

[123] Yu H, Liu G-B, Tang J, Xu X and Yao W 2017 Moiré excitons: from programmable quantum emitter arrays to spin-orbit–coupled artificial lattices *Sci. Adv.* **3** e1701696

[124] Wang Y, Nie Z and Wang F 2020 Modulation of photocarrier relaxation dynamics in two-dimensional semiconductors *Light Sci. Appl.* **9** 192

[125] Chen W, Zheng C, Pei J and Zhan H 2023 External field regulation strategies for exciton dynamics in 2D TMDs *Opt. Mater. Express* **13** 1007–30

[126] Surrente A, Dumcenco D, Yang Z, Kuc A, Jing Y, Heine T, Kung Y-C, Maude D K, Kis A and Plochocka P 2017 Defect healing and charge transfer-mediated valley polarization in $MoS_2/MoSe_2/MoS_2$ trilayer van der Waals heterostructures *Nano Lett.* **17** 4130–6

[127] Glazov M M, Ivchenko E L, Wang G, Amand T, Marie X, Urbaszek B and Liu B L 2015 Spin and valley dynamics of excitons in transition metal dichalcogenide monolayers *Phys. Stat. Sol. (b)* **252** 2349–62

[128] Hao K *et al* 2016 Direct measurement of exciton valley coherence in monolayer WSe_2 *Nat. Phys.* **12** 677–82

[129] Jakubczyk T, Delmonte V, Koperski M, Nogajewski K, Faugeras C, Langbein W, Potemski M and Kasprzak J 2016 Radiatively limited dephasing and exciton dynamics in $MoSe_2$ monolayers revealed with four-wave mixing microscopy *Nano Lett.* **16** 5333–9

[130] Wang Q, Ge S, Li X, Qiu J, Ji Y, Feng J and Sun D 2013 Valley carrier dynamics in monolayer molybdenum disulfide from helicity-resolved ultrafast pump–probe spectroscopy *ACS Nano* **7** 11087–93

[131] Schmidt R, Berghäuser G, Schneider R, Selig M, Tonndorf P, Malić E, Knorr A, Michaelis de Vasconcellos S and Bratschitsch R 2016 Ultrafast Coulomb-induced intervalley coupling in atomically thin WS_2 *Nano Lett.* **16** 2945–50

[132] Mai C, Barrette A, Yu Y, Semenov Y G, Kim K W, Cao L and Gundogdu K 2014 Many-body effects in valleytronics: direct measurement of valley lifetimes in single-layer MoS_2 *Nano Lett.* **14** 202–6

[133] Scuri G *et al* 2020 Electrically tunable valley dynamics in twisted WSe_2/WSe_2 bilayers *Phys. Rev. Lett.* **124** 217403

[134] Li X *et al* 2022 Nonvolatile electrical valley manipulation in WS_2 by ferroelectric gating *ACS Nano* **16** 20598–606

[135] Xiao D, Liu G-B, Feng W, Xu X and Yao W 2012 Coupled spin and valley physics in monolayers of MoS_2 and other group-VI dichalcogenides *Phys. Rev. Lett.* **108** 196802

[136] Li H, Ling J, Lin J, Lu X and Xu W 2023 Interface engineering in two-dimensional heterostructures towards novel emitters *J. Semicond.* **44** 011001

[137] Tan Q, Rasmita A, Li S, Liu S, Huang Z, Xiong Q, Yang S A, Novoselov K S and Gao W-B 2021 Layer-engineered interlayer excitons *Sci. Adv.* **7** eabh0863

[138] Rosenberger M R, Chuang H-J, Phillips M, Oleshko V P, McCreary K M, Sivaram S V, Hellberg C S and Jonker B T 2020 Twist angle-dependent atomic reconstruction and moiré patterns in transition metal dichalcogenide heterostructures *ACS Nano* **14** 4550–8

[139] Rosenberger M R, Chuang H-J, Phillips M, Oleshko V P, McCreary K M, Sivaram S V, Hellberg C S and Jonker B T 2020 Correction to twist angle-dependent atomic reconstruction and moiré patterns in transition metal dichalcogenide heterostructures *ACS Nano* **14** 14240–2

[140] Andrei E Y, Efetov D K, Jarillo-Herrero P, MacDonald A H, Mak K F, Senthil T, Tutuc E, Yazdani A and Young A F 2021 The marvels of moiré materials *Nat. Rev. Mater.* **6** 201–6

[141] Brem S, Linderälv C, Erhart P and Malic E 2020 Tunable phases of moiré excitons in van der Waals heterostructures *Nano Lett.* **20** 8534–40

[142] Liu Y, Zeng C, Yu J, Zhong J, Li B, Zhang Z, Liu Z, Wang Z M, Pan A and Duan X 2021 Moiré superlattices and related moiré excitons in twisted van der Waals heterostructures *Chem. Soc. Rev.* **50** 6401–22

[143] Tran K *et al* 2019 Evidence for moiré excitons in van der Waals heterostructures *Nature* **567** 71–5

[144] Liu E *et al* 2021 Signatures of moiré trions in $WSe_2/MoSe_2$ heterobilayers *Nature* **594** 46–50

[145] Jin C *et al* 2019 Observation of moiré excitons in WSe_2/WS_2 heterostructure superlattices *Nature* **567** 76–80

[146] Jin C *et al* 2019 Author correction: observation of moiré excitons in WSe_2/WS_2 heterostructure superlattices *Nature* **569** E7

[147] Alexeev E M *et al* 2019 Resonantly hybridized excitons in moiré superlattices in van der Waals heterostructures *Nature* **567** 81–6

[148] Kunstmann J *et al* 2018 Momentum-space indirect interlayer excitons in transition-metal dichalcogenide van der Waals heterostructures *Nat. Phys.* **14** 801–5

[149] Yankowitz M, Xue J, Cormode D, Sanchez-Yamagishi J D, Watanabe K, Taniguchi T, Jarillo-Herrero P, Jacquod P and LeRoy B J 2012 Emergence of superlattice Dirac points in graphene on hexagonal boron nitride *Nat. Phys.* **8** 382–6

[150] Holwill M 2019 *Nanomechanics in van der Waals Heterostructures* Springer Theses (Cham: Springer)

[151] Weston A 2022 *Atomic and Electronic Properties of 2D Moiré Interfaces* Springer Theses (Cham: Springer Nature)

[152] Tan P-H (ed) 2018 *Raman Spectroscopy of Two-Dimensional Materials* Springer Series in Materials Science **vol 276** 1st edn (Singapore: Springer)

[153] Wu F, Lovorn T and MacDonald A H 2018 Theory of optical absorption by interlayer excitons in transition metal dichalcogenide heterobilayers *Phys. Rev.* B **97** 035306

[154] Li S, Wei K, Liu Q, Tang Y and Jiang T 2024 Twistronics and moiré excitonic physics in van der Waals heterostructures *Front. Phys.* **19** 42501

[155] Huang D, Choi J, Shih C-K and Li X 2022 Excitons in semiconductor moiré superlattices *Nat. Nanotechnol.* **17** 227–38

[156] Liu K, Zhang L, Cao T, Jin C, Qiu D, Zhou Q, Zettl A, Yang P, Louie S G and Wang F 2014 Evolution of interlayer coupling in twisted molybdenum disulfide bilayers *Nat. Commun.* **5** 4966

[157] van der Zande A M *et al* 2014 Tailoring the electronic structure in bilayer molybdenum disulfide via interlayer twist *Nano Lett.* **14** 3869–75
[158] Wu M, Qian X and Li J 2014 Tunable exciton funnel using moiré superlattice in twisted van der Waals bilayer *Nano Lett.* **14** 5350–7
[159] Wu F, Lovorn T and MacDonald A H 2017 Topological exciton bands in moiré heterojunctions *Phys. Rev. Lett.* **118** 147401
[160] Wang Y, Wang Z, Yao W, Liu G-B and Yu H 2017 Interlayer coupling in commensurate and incommensurate bilayer structures of transition-metal dichalcogenides *Phys. Rev.* **B 95** 115429
[161] Ciarrocchi A, Tagarelli F, Avsar A and Kis A 2022 Excitonic devices with van der Waals heterostructures: valleytronics meets twistronics *Nat. Rev. Mater.* **7** 449–64
[162] Wu F, Lovorn T, Tutuc E and MacDonald A H 2018 Hubbard model physics in transition metal dichalcogenide moiré bands *Phys. Rev. Lett.* **121** 026402
[163] Wu F, Lovorn T, Tutuc E, Martin I and MacDonald A H 2019 Topological insulators in twisted transition metal dichalcogenide homobilayers *Phys. Rev. Lett.* **122** 086402
[164] Regan E C *et al* 2020 Mott and generalized Wigner crystal states in WSe_2/WS_2 moiré superlattices *Nature* **579** 359–63
[165] Tang Y *et al* 2020 Simulation of Hubbard model physics in WSe_2/WS_2 moiré superlattices *Nature* **579** 353–8
[166] Shimazaki Y, Schwartz I, Watanabe K, Taniguchi T, Kroner M and Imamoğlu A 2020 Strongly correlated electrons and hybrid excitons in a moiré heterostructure *Nature* **580** 472–7
[167] Jin C *et al* 2019 Identification of spin, valley and moiré quasi-angular momentum of interlayer excitons *Nat. Phys.* **15** 1140–4
[168] Liu G-B, Xiao D, Yao Y, Xu X and Yao W 2015 Electronic structures and theoretical modelling of two-dimensional group-VIB transition metal dichalcogenides *Chem. Soc. Rev.* **44** 2643–63
[169] Saito R, Ukhtary M S, Wang S and Hung N T 2021 Selection rule for Raman spectra of two-dimensional materials using circularly-polarized vortex light *Phys. Chem. Chem. Phys.* **23** 17271–8
[170] Unuchek D, Ciarrocchi A, Avsar A, Sun Z, Watanabe K, Taniguchi T and Kis A 2019 Valley-polarized exciton currents in a van der Waals heterostructure *Nat. Nanotechnol.* **14** 1104–9
[171] Patel H, Huang L, Kim C-J, Park J and Graham M W 2019 Stacking angle-tunable photoluminescence from interlayer exciton states in twisted bilayer graphene *Nat. Commun.* **10** 1445
[172] Zuo L *et al* 2017 Polymer-modified halide perovskite films for efficient and stable planar heterojunction solar cells *Sci. Adv.* **3** e1700106
[173] Wang G, Chernikov A, Glazov M M, Heinz T F, Marie X, Amand T and Urbaszek B 2018 Colloquium: excitons in atomically thin transition metal dichalcogenides *Rev. Mod. Phys.* **90** 021001
[174] Rasmita A and Gao W-B 2021 Opto-valleytronics in the 2D van der Waals heterostructure *Nano Res.* **14** 1901–11
[175] Chu J *et al* 2021 2D polarized materials: ferromagnetic, ferrovalley, ferroelectric materials, and related heterostructures *Adv. Mater.* **33** 2004469

[176] Castro E C, Brandão D S, Bragança H, Martins A S, Riche F, Dias A C, Zhao J H, Fonseca A L A and Qu F 2023 Mechanisms of interlayer exciton emission and giant valley polarization in van der Waals heterostructures *Phys. Rev.* B **107** 035439
[177] Wang Z, Tang C, Sachs R, Barlas Y and Shi J 2015 Proximity-induced ferromagnetism in graphene revealed by the anomalous Hall effect *Phys. Rev. Lett.* **114** 016603
[178] Wei P *et al* 2016 Strong interfacial exchange field in the graphene/EuS heterostructure *Nat. Mater.* **15** 711–6
[179] Qi J, Li X, Niu Q and Feng J 2015 Giant and tunable valley degeneracy splitting in $MoTe_2$ *Phys. Rev.* B **92** 121403
[180] Dang J, Yang M, Xie X, Yang Z, Dai D, Zuo Z, Wang C, Jin K and Xu X 2022 Enhanced valley polarization in $WS_2/LaMnO_3$ heterostructure *Small* **18** 2106029
[181] Zhang Q, Yang S A, Mi W, Cheng Y and Schwingenschlögl U 2016 Large spin-valley polarization in monolayer $MoTe_2$ on top of EuO(111) *Adv. Mater.* **28** 959–66
[182] Zhang Q, Yang S A, Mi W, Cheng Y and Schwingenschlögl U 2016 Large spin-valley polarization in monolayer $MoTe_2$ on top of EuO(111) *Adv. Mater.* **28** 7043–7
[183] Zhao C *et al* 2017 Enhanced valley splitting in monolayer WSe_2 due to magnetic exchange field *Nat. Nanotechnol.* **12** 757–62
[184] Su L, Fan X, Yin T, Wang H, Li Y, Liu F, Li J, Zhang H and Xie H 2020 Inorganic 2D luminescent materials: structure, luminescence modulation, and applications *Adv. Opt. Mater.* **8** 1900978
[185] Zhong D *et al* 2017 Van der Waals engineering of ferromagnetic semiconductor heterostructures for spin and valleytronics *Sci. Adv.* **3** e1603113
[186] Huang B *et al* 2017 Layer-dependent ferromagnetism in a van der Waals crystal down to the monolayer limit *Nature* **546** 270–3
[187] Seyler K L *et al* 2018 Valley manipulation by optically tuning the magnetic proximity effect in WSe_2/CrI_3 heterostructures *Nano Lett.* **18** 3823–8
[188] Zheng W, Jiang Y, Hu X, Li H, Zeng Z, Wang X and Pan A 2018 Light emission properties of 2D transition metal dichalcogenides: fundamentals and applications *Adv. Opt. Mater.* **6** 1800420
[189] Li L, Jiang S, Wang Z, Watanabe K, Taniguchi T, Shan J and Mak K F 2020 Electrical switching of valley polarization in monolayer semiconductors *Phys. Rev. Mater.* **4** 104005
[190] Zhang T, Zhao S, Wang A, Xiong Z, Liu Y, Xi M, Li S, Lei H, Han Z V and Wang F 2022 Electrically and magnetically tunable valley polarization in monolayer $MoSe_2$ proximitized by a 2D ferromagnetic semiconductor *Adv. Funct. Mater.* **32** 2204779
[191] Ciorciaro L, Kroner M, Watanabe K, Taniguchi T and Imamoglu A 2020 Observation of magnetic proximity effect using resonant optical spectroscopy of an electrically tunable $MoSe_2/CrBr_3$ heterostructure *Phys. Rev. Lett.* **124** 197401
[192] Ciorciaro L, Kroner M, Watanabe K, Taniguchi T and Imamoglu A 2023 Erratum: Observation of magnetic proximity effect using resonant optical spectroscopy of an electrically tunable $MoSe_2/CrBr_3$ heterostructure [Phys. Rev. Lett. 124, 197401 (2020)] *Phys. Rev. Lett.* **130** 019901
[193] Poellmann C, Steinleitner P, Leierseder U, Nagler P, Plechinger G, Porer M, Bratschitsch R, Schüller C, Korn T and Huber R 2015 Resonant internal quantum transitions and femtosecond radiative decay of excitons in monolayer WSe_2 *Nat. Mater.* **14** 889–93

[194] Glazov M M, Amand T, Marie X, Lagarde D, Bouet L and Urbaszek B 2014 Exciton fine structure and spin decoherence in monolayers of transition metal dichalcogenides *Phys. Rev.* B **89** 201302

[195] Nagler P *et al* 2017 Giant magnetic splitting inducing near-unity valley polarization in van der Waals heterostructures *Nat. Commun.* **8** 1551

[196] Zhao S, Li X, Dong B, Wang H, Wang H, Zhang Y, Han Z and Zhang H 2021 Valley manipulation in monolayer transition metal dichalcogenides and their hybrid systems: status and challenges *Rep. Prog. Phys.* **84** 026401

[197] LaMountain T, Lenferink E J, Chen Y-J, Stanev T K and Stern N P 2018 Environmental engineering of transition metal dichalcogenide optoelectronics *Front. Phys.* **13** 138114

[198] Surrente A *et al* 2018 Intervalley scattering of interlayer excitons in a $MoS_2/MoSe_2/MoS_2$ heterostructure in high magnetic field *Nano Lett.* **18** 3994–4000

[199] Du L, Hasan T, Castellanos-Gomez A, Liu G-B, Yao Y, Lau C N and Sun Z 2021 Engineering symmetry breaking in 2D layered materials *Nat. Rev. Phys.* **3** 193–206

[200] Park C 2019 Magnetoelectrically controlled valley filter and valley valve in bilayer graphene *Phys. Rev. Appl.* **11** 044033

IOP Publishing

Two-dimensional Valleytronic Materials
From principles to device applications
Sake Wang and Hongyu Tian

Appendix A

The matrix elements of the graphene TB Hamiltonian

From equation (1.11), for $H_{\mathrm{AA}}(\boldsymbol{k})$, we have

$$\begin{aligned} H_{\mathrm{AA}}(\boldsymbol{k}) &= \frac{1}{N}\sum_{\boldsymbol{R}_{\mathrm{A}}}\sum_{\boldsymbol{R}'_{\mathrm{A}}}\left\langle \mathrm{e}^{\mathrm{i}\boldsymbol{k}\cdot\boldsymbol{R}'_{\mathrm{A}}}p_z(\boldsymbol{r}-\boldsymbol{R}'_{\mathrm{A}})|H|\mathrm{e}^{\mathrm{i}\boldsymbol{k}\cdot\boldsymbol{R}_{\mathrm{A}}}p_z(\boldsymbol{r}-\boldsymbol{R}_{\mathrm{A}})\right\rangle \\ &= \frac{1}{N}\sum_{\boldsymbol{R}_{\mathrm{A}}}\sum_{\boldsymbol{R}'_{\mathrm{A}}}\mathrm{e}^{\mathrm{i}\boldsymbol{k}\cdot(\boldsymbol{R}_{\mathrm{A}}-\boldsymbol{R}'_{\mathrm{A}})}\left\langle p_z\left(\boldsymbol{r}'+\boldsymbol{R}_{\mathrm{A}}-\boldsymbol{R}'_{\mathrm{A}}\right)|H|p_z(\boldsymbol{r}')\right\rangle \\ &= \frac{1}{N}\sum_{\boldsymbol{R}_{\mathrm{A}}}\sum_{\boldsymbol{R}'_{\mathrm{A}}}\mathrm{e}^{\mathrm{i}\boldsymbol{k}\cdot(\boldsymbol{R}_{\mathrm{A}}-\boldsymbol{R}'_{\mathrm{A}})}\left\langle p_z(\boldsymbol{r}+\boldsymbol{R}_{\mathrm{A}}-\boldsymbol{R}'_{\mathrm{A}})|H|p_z(\boldsymbol{r})\right\rangle \\ &= \frac{1}{N}\sum_{\boldsymbol{R}_{\mathrm{A}}}\sum_{\Delta\boldsymbol{R}_{\mathrm{A}}}\mathrm{e}^{\mathrm{i}\boldsymbol{k}\cdot\Delta\boldsymbol{R}_{\mathrm{A}}}\left\langle p_z(\boldsymbol{r}+\Delta\boldsymbol{R}_{\mathrm{A}})|H|p_z(\boldsymbol{r})\right\rangle. \end{aligned} \tag{A.1}$$

We perform $\boldsymbol{r}-\boldsymbol{R}_{\mathrm{A}}=\boldsymbol{r}'$ in the second line. Also, we define $\Delta\boldsymbol{R}_{\mathrm{A}}=\boldsymbol{R}_{\mathrm{A}}-\boldsymbol{R}'_{\mathrm{A}}$ in the last line. If we only consider $\Delta R_{\mathrm{A}}=0$, thus

$$H_{\mathrm{AA}}(\boldsymbol{k}) \simeq \frac{1}{N}\sum_{\boldsymbol{R}_{\mathrm{A}}}\sum_{\Delta\boldsymbol{R}_{\mathrm{A}}}\left\langle p_z(\boldsymbol{r})|H|p_z(\boldsymbol{r})\right\rangle = \mathcal{E}_{\mathrm{C}}. \tag{A.2}$$

This is the on-site potential of the A carbon atom. The B sublattice has the same structure as the A sublattice, and the two sublattices are chemically identical, so we have

$$H_{\mathrm{BB}}(\boldsymbol{k}) \simeq \mathcal{E}_{\mathrm{C}}. \tag{A.3}$$

Similarly, we find $H_{\mathrm{AB}}(\boldsymbol{k})$,

doi:10.1088/978-0-7503-5562-9ch9

$$\begin{aligned} H_{\mathrm{AB}}(\boldsymbol{k}) &= \frac{1}{N}\sum_{\boldsymbol{R}_\mathrm{A}}\sum_{\boldsymbol{R}_\mathrm{B}}\left\langle \mathrm{e}^{\mathrm{i}\boldsymbol{k}\cdot\boldsymbol{R}_\mathrm{A}} p_z(\boldsymbol{r}-\boldsymbol{R}_\mathrm{A})\right| H \left|\mathrm{e}^{\mathrm{i}\boldsymbol{k}\cdot\boldsymbol{R}_\mathrm{B}} p_z(\boldsymbol{r}-\boldsymbol{R}_\mathrm{B})\right\rangle \\ &= \frac{1}{N}\sum_{\boldsymbol{R}_\mathrm{A}}\sum_{\boldsymbol{R}_\mathrm{B}} \mathrm{e}^{\mathrm{i}\boldsymbol{k}\cdot(\boldsymbol{R}_\mathrm{B}-\boldsymbol{R}_\mathrm{A})}\left\langle p_z(\boldsymbol{r}'+\boldsymbol{R}_\mathrm{B}-\boldsymbol{R}_\mathrm{A})\right| H \left| p_z(\boldsymbol{r}')\right\rangle \\ &= \frac{1}{N}\sum_{\boldsymbol{R}_\mathrm{A}}\sum_{\boldsymbol{R}_\mathrm{B}} \mathrm{e}^{\mathrm{i}\boldsymbol{k}\cdot(\boldsymbol{R}_\mathrm{B}-\boldsymbol{R}_\mathrm{A})}\left\langle p_z(\boldsymbol{r}+\boldsymbol{R}_\mathrm{B}-\boldsymbol{R}_\mathrm{A})\right| H \left| p_z(\boldsymbol{r})\right\rangle \\ &= \frac{1}{N}\sum_{\boldsymbol{R}_\mathrm{A}}\sum_{\Delta\boldsymbol{R}_\mathrm{AB}} \mathrm{e}^{\mathrm{i}\boldsymbol{k}\cdot\Delta\boldsymbol{R}_\mathrm{AB}}\left\langle p_z(\boldsymbol{r}+\Delta\boldsymbol{R}_\mathrm{AB})\right| H \left| p_z(\boldsymbol{r})\right\rangle. \end{aligned} \tag{A.4}$$

Again, we perform $\boldsymbol{r} - \boldsymbol{R}_\mathrm{B} = \boldsymbol{r}'$ in the second line. Also, we define $\Delta\boldsymbol{R}_\mathrm{AB} = \boldsymbol{R}_\mathrm{B} - \boldsymbol{R}_\mathrm{A}$ in the last line. We restrict ourselves to the interactions to the first NNs only by considering the three vectors $\Delta\boldsymbol{R}_\mathrm{AB} = \boldsymbol{R}_1, \boldsymbol{R}_2, \boldsymbol{R}_3$ between the NNs. Hence,

$$H_{\mathrm{AB}}(\boldsymbol{k}) = -\gamma_0\left(\mathrm{e}^{\mathrm{i}\boldsymbol{k}\cdot\boldsymbol{R}_1} + \mathrm{e}^{\mathrm{i}\boldsymbol{k}\cdot\boldsymbol{R}_2} + \mathrm{e}^{\mathrm{i}\boldsymbol{k}\cdot\boldsymbol{R}_3}\right) = -\gamma_0 f(\boldsymbol{k}), \tag{A.5}$$

where

$$\gamma_0 = -\left\langle p_z(\boldsymbol{r}+\boldsymbol{R}_i)\right| H \left| p_z(\boldsymbol{r})\right\rangle \quad (i = 1, 2, 3) \tag{A.6}$$

is the NN hopping energy. Since the Hamiltonian is Hermitian, $H_{\mathrm{AB}}(\boldsymbol{k}) = H_{\mathrm{BA}}(\boldsymbol{k})^*$ holds.

IOP Publishing

Two-dimensional Valleytronic Materials
From principles to device applications
Sake Wang and Hongyu Tian

Appendix B

Hermite polynomials

Hermite polynomials are named after French mathematician Charles Hermite (1822–1901). They are a classical orthogonal polynomial sequence that arises in physics, as the eigenstates of the quantum harmonic oscillator. The recurrence relation is

$$\mathcal{H}_{n+1}(x) = 2x\mathcal{H}_n(x) - 2n\mathcal{H}_{n-1}(x), \tag{B.1}$$

and the first few Hermite polynomials are

$$\mathcal{H}_0(x) = 1, \tag{B.2}$$

$$\mathcal{H}_1(x) = 2x, \tag{B.3}$$

$$\mathcal{H}_2(x) = 4x^2 - 2, \tag{B.4}$$

$$\mathcal{H}_3(x) = 8x^3 - 12x, \tag{B.5}$$

$$\mathcal{H}_4(x) = 16x^4 - 48x^2 + 12, \tag{B.6}$$

$$\mathcal{H}_5(x) = 32x^5 - 160x^3 + 120x, \tag{B.7}$$

$$\mathcal{H}_6(x) = 64x^6 - 480x^4 + 720x^2 - 120. \tag{B.8}$$

doi:10.1088/978-0-7503-5562-9ch10

Appendix C

The dipole vector of hexagonal systems

According to equation (7.41), the dipole vector of the monolayer graphene reads

$$\begin{aligned}\boldsymbol{D}(\boldsymbol{k}_\mathrm{c}, \boldsymbol{k}_\mathrm{V})\\ &= c_\mathrm{c}^\mathrm{A}(\boldsymbol{k}_\mathrm{c})^* c_\mathrm{V}^\mathrm{A}(\boldsymbol{k}_\mathrm{V})\left\langle p_z^\mathrm{A}(\boldsymbol{k}_\mathrm{c}, \boldsymbol{r})\middle|\boldsymbol{\nabla}\middle|p_z^\mathrm{A}(\boldsymbol{k}_\mathrm{V}, \boldsymbol{r})\right\rangle + c_\mathrm{c}^\mathrm{B}(\boldsymbol{k}_\mathrm{c})^* c_\mathrm{V}^\mathrm{B}(\boldsymbol{k}_\mathrm{V})\left\langle p_z^\mathrm{B}(\boldsymbol{k}_\mathrm{c}, \boldsymbol{r})\middle|\boldsymbol{\nabla}\middle|p_z^\mathrm{B}(\boldsymbol{k}_\mathrm{V}, \boldsymbol{r})\right\rangle \\ &+ c_\mathrm{c}^\mathrm{A}(\boldsymbol{k}_\mathrm{c})^* c_\mathrm{V}^\mathrm{B}(\boldsymbol{k}_\mathrm{V})\left\langle p_z^\mathrm{A}(\boldsymbol{k}_\mathrm{c}, \boldsymbol{r})\middle|\boldsymbol{\nabla}\middle|p_z^\mathrm{B}(\boldsymbol{k}_\mathrm{V}, \boldsymbol{r})\right\rangle + c_\mathrm{c}^\mathrm{B}(\boldsymbol{k}_\mathrm{c})^* c_\mathrm{V}^\mathrm{A}(\boldsymbol{k}_\mathrm{V})\left\langle p_z^\mathrm{B}(\boldsymbol{k}_\mathrm{c}, \boldsymbol{r})\middle|\boldsymbol{\nabla}\middle|p_z^\mathrm{A}(\boldsymbol{k}_\mathrm{V}, \boldsymbol{r})\right\rangle.\end{aligned} \tag{C.1}$$

First, we show that the first and second terms in the above equation become 0. By substituting the Bloch orbit definition in equations (1.7) and (1.8), we have

$$\begin{aligned}\left\langle p_z^\mathrm{A}(\boldsymbol{k}_\mathrm{c}, \boldsymbol{r})\middle|\boldsymbol{\nabla}\middle|p_z^\mathrm{A}(\boldsymbol{k}_\mathrm{V}, \boldsymbol{r})\right\rangle &= \frac{1}{N}\sum_{\boldsymbol{R}_\mathrm{A}}\sum_{\boldsymbol{R}'_\mathrm{A}} \mathrm{e}^{-\mathrm{i}\boldsymbol{k}_\mathrm{c}\cdot\boldsymbol{R}_\mathrm{A}}\mathrm{e}^{\mathrm{i}\boldsymbol{k}_\mathrm{V}\cdot\boldsymbol{R}'_\mathrm{A}}\left\langle p_z(\boldsymbol{r}-\boldsymbol{R}_\mathrm{A})\middle|\boldsymbol{\nabla}\middle|p_z(\boldsymbol{r}-\boldsymbol{R}'_\mathrm{A})\right\rangle \\ &= \frac{1}{N}\sum_{\boldsymbol{R}_\mathrm{A}}\sum_{\boldsymbol{R}'_\mathrm{A}} \mathrm{e}^{-\mathrm{i}\boldsymbol{k}_\mathrm{c}\cdot\boldsymbol{R}_\mathrm{A}}\mathrm{e}^{\mathrm{i}\boldsymbol{k}_\mathrm{V}\cdot\boldsymbol{R}'_\mathrm{A}}\left\langle p_z(\boldsymbol{r}'+\boldsymbol{R}'_\mathrm{A}-\boldsymbol{R}_\mathrm{A})\middle|\boldsymbol{\nabla}\middle|p_z(\boldsymbol{r}')\right\rangle \\ &= \frac{1}{N}\sum_{\boldsymbol{R}_\mathrm{A}}\sum_{\boldsymbol{R}'_\mathrm{A}} \mathrm{e}^{-\mathrm{i}\boldsymbol{k}_\mathrm{c}\cdot\boldsymbol{R}_\mathrm{A}}\mathrm{e}^{\mathrm{i}\boldsymbol{k}_\mathrm{V}\cdot\boldsymbol{R}'_\mathrm{A}}\left\langle p_z(\boldsymbol{r}'+\boldsymbol{R}'_\mathrm{A}-\boldsymbol{R}_\mathrm{A})\middle|\boldsymbol{\nabla}\middle|p_z(\boldsymbol{r})\right\rangle \\ &= \frac{1}{N}\sum_{\Delta\boldsymbol{R}_\mathrm{A}}\sum_{\boldsymbol{R}'_\mathrm{A}} \mathrm{e}^{\mathrm{i}\boldsymbol{k}_\mathrm{c}\cdot\Delta\boldsymbol{R}_\mathrm{A}}\mathrm{e}^{\mathrm{i}(\boldsymbol{k}_\mathrm{V}-\boldsymbol{k}_\mathrm{c})\cdot\boldsymbol{R}'_\mathrm{A}}\left\langle p_z(\boldsymbol{r}+\Delta\boldsymbol{R}_\mathrm{A})\middle|\boldsymbol{\nabla}\middle|p_z(\boldsymbol{r})\right\rangle \\ &= \sum_{\Delta\boldsymbol{R}_\mathrm{A}} \mathrm{e}^{\mathrm{i}\boldsymbol{k}_\mathrm{c}\cdot\Delta\boldsymbol{R}_\mathrm{A}}\left\langle p_z(\boldsymbol{r}+\Delta\boldsymbol{R}_\mathrm{A})\middle|\boldsymbol{\nabla}\middle|p_z(\boldsymbol{r})\right\rangle\delta_{\boldsymbol{k}_\mathrm{V},\boldsymbol{k}_\mathrm{c}} \\ &\simeq \left\langle p_z(\boldsymbol{r})\middle|\boldsymbol{\nabla}\middle|p_z(\boldsymbol{r})\right\rangle\delta_{\boldsymbol{k}_\mathrm{V},\boldsymbol{k}_\mathrm{c}}.\end{aligned} \tag{C.2}$$

We perform $\boldsymbol{r}-\boldsymbol{R}'_\mathrm{A}=\boldsymbol{r}'$ in the second line. Also, we define $\Delta\boldsymbol{R}_\mathrm{A}=\boldsymbol{R}'_\mathrm{A}-\boldsymbol{R}_\mathrm{A}$ in the fourth line. In approximation, we consider the integral to the NN and $\Delta\boldsymbol{R}_\mathrm{A}=0$. The integral $\langle p_z(\boldsymbol{r})|\boldsymbol{\nabla}|p_z(\boldsymbol{r})\rangle$ vanishes because the $\boldsymbol{\nabla}$ gives odd symmetry in the x, y and z. Therefore,

$$\left\langle p_z^\mathrm{A}(\boldsymbol{k}_\mathrm{c}, \boldsymbol{r})\middle|\boldsymbol{\nabla}\middle|p_z^\mathrm{A}(\boldsymbol{k}_\mathrm{V}, \boldsymbol{r})\right\rangle = 0. \tag{C.3}$$

doi:10.1088/978-0-7503-5562-9ch11

In the same way, we obtain

$$\left\langle p_z^{\mathrm{B}}(\boldsymbol{k}_{\mathrm{c}}, \boldsymbol{r}) \middle| \nabla \middle| p_z^{\mathrm{B}}(\boldsymbol{k}_{\mathrm{v}}, \boldsymbol{r}) \right\rangle = 0. \tag{C.4}$$

Next, we calculate the remaining two terms,

$$\begin{aligned}
\left\langle p_z^{\mathrm{A}}(\boldsymbol{k}_{\mathrm{c}}, \boldsymbol{r}) \middle| \nabla \middle| p_z^{\mathrm{B}}(\boldsymbol{k}_{\mathrm{v}}, \boldsymbol{r}) \right\rangle &= \frac{1}{N} \sum_{\boldsymbol{R}_{\mathrm{A}}} \sum_{\boldsymbol{R}_{\mathrm{B}}} \mathrm{e}^{-\mathrm{i}\boldsymbol{k}_{\mathrm{c}} \cdot \boldsymbol{R}_{\mathrm{A}}} \mathrm{e}^{\mathrm{i}\boldsymbol{k}_{\mathrm{v}} \cdot \boldsymbol{R}_{\mathrm{B}}} \left\langle p_z(\boldsymbol{r} - \boldsymbol{R}_{\mathrm{A}}) \middle| \nabla \middle| p_z(\boldsymbol{r} - \boldsymbol{R}_{\mathrm{B}}) \right\rangle \\
&= \frac{1}{N} \sum_{\boldsymbol{R}_{\mathrm{A}}} \sum_{\boldsymbol{R}_{\mathrm{B}}} \mathrm{e}^{-\mathrm{i}\boldsymbol{k}_{\mathrm{c}} \cdot \boldsymbol{R}_{\mathrm{A}}} \mathrm{e}^{\mathrm{i}\boldsymbol{k}_{\mathrm{v}} \cdot \boldsymbol{R}_{\mathrm{B}}} \left\langle p_z(\boldsymbol{r}' + \boldsymbol{R}_{\mathrm{B}} - \boldsymbol{R}_{\mathrm{A}}) \middle| \nabla \middle| p_z(\boldsymbol{r}') \right\rangle \\
&= \frac{1}{N} \sum_{\boldsymbol{R}_{\mathrm{A}}} \sum_{\boldsymbol{R}_{\mathrm{B}}} \mathrm{e}^{-\mathrm{i}\boldsymbol{k}_{\mathrm{c}} \cdot \boldsymbol{R}_{\mathrm{A}}} \mathrm{e}^{\mathrm{i}\boldsymbol{k}_{\mathrm{v}} \cdot \boldsymbol{R}_{\mathrm{B}}} \left\langle p_z(\boldsymbol{r} + \boldsymbol{R}_{\mathrm{B}} - \boldsymbol{R}_{\mathrm{A}}) \middle| \nabla \middle| p_z(\boldsymbol{r}) \right\rangle \\
&= \frac{1}{N} \sum_{\boldsymbol{R}_{\mathrm{A}}} \sum_{\Delta\boldsymbol{R}_{\mathrm{AB}}} \mathrm{e}^{\mathrm{i}(\boldsymbol{k}_{\mathrm{v}} - \boldsymbol{k}_{\mathrm{c}}) \cdot \boldsymbol{R}_{\mathrm{A}}} \mathrm{e}^{-\mathrm{i}\boldsymbol{k}_{\mathrm{v}} \cdot \Delta\boldsymbol{R}_{\mathrm{AB}}} \left\langle p_z(\boldsymbol{r} - \Delta\boldsymbol{R}_{\mathrm{AB}}) \middle| \nabla \middle| p_z(\boldsymbol{r}) \right\rangle \\
&= \sum_{\Delta\boldsymbol{R}_{\mathrm{AB}}} \mathrm{e}^{-\mathrm{i}\boldsymbol{k}_{\mathrm{v}} \cdot \Delta\boldsymbol{R}_{\mathrm{AB}}} \left\langle p_z(\boldsymbol{r} - \Delta\boldsymbol{R}_{\mathrm{AB}}) \middle| \nabla \middle| p_z(\boldsymbol{r}) \right\rangle \delta_{\boldsymbol{k}_{\mathrm{v}}, \boldsymbol{k}_{\mathrm{c}}} \\
&= \sum_{\Delta\boldsymbol{R}_{\mathrm{BA}}} \mathrm{e}^{\mathrm{i}\boldsymbol{k}_{\mathrm{v}} \cdot \Delta\boldsymbol{R}_{\mathrm{BA}}} \left\langle p_z(\boldsymbol{r} + \Delta\boldsymbol{R}_{\mathrm{BA}}) \middle| \nabla \middle| p_z(\boldsymbol{r}) \right\rangle \delta_{\boldsymbol{k}_{\mathrm{v}}, \boldsymbol{k}_{\mathrm{c}}}.
\end{aligned} \tag{C.5}$$

Once again, we perform $\boldsymbol{r}' = \boldsymbol{r} - \boldsymbol{R}_{\mathrm{B}}$ in the second line. Also, we define $\Delta\boldsymbol{R}_{\mathrm{AB}} = \boldsymbol{R}_{\mathrm{A}} - \boldsymbol{R}_{\mathrm{B}}$ and $\Delta\boldsymbol{R}_{\mathrm{BA}} = -\Delta\boldsymbol{R}_{\mathrm{AB}}$ in the fourth and last line, respectively. Notice that $\Delta\boldsymbol{R}_{\mathrm{BA}}$ is $\boldsymbol{R}_i$ in Fig. 1.1(a) if we only consider the NNs. Using the following identity,

$$\begin{aligned}
\left\langle p_z(\boldsymbol{r} - \Delta\boldsymbol{R}_{\mathrm{BA}}) \middle| \nabla \middle| p_z(\boldsymbol{r}) \right\rangle^* &= - \left\langle p_z(\boldsymbol{r}) \middle| \nabla \middle| p_z(\boldsymbol{r} - \Delta\boldsymbol{R}_{\mathrm{BA}}) \right\rangle \\
&= - \left\langle p_z(\boldsymbol{r}' + \Delta\boldsymbol{R}_{\mathrm{BA}}) \middle| \nabla \middle| p_z(\boldsymbol{r}') \right\rangle \\
&= - \left\langle p_z(\boldsymbol{r} + \Delta\boldsymbol{R}_{\mathrm{BA}}) \middle| \nabla \middle| p_z(\boldsymbol{r}) \right\rangle,
\end{aligned} \tag{C.6}$$

where we define $\boldsymbol{r}' = \boldsymbol{r} - \Delta\boldsymbol{R}_{\mathrm{BA}}$, we can rewrite equation (C.5), as

$$\left\langle p_z^{\mathrm{A}}(\boldsymbol{k}_{\mathrm{c}}, \boldsymbol{r}) \middle| \nabla \middle| p_z^{\mathrm{B}}(\boldsymbol{k}_{\mathrm{v}}, \boldsymbol{r}) \right\rangle = - \sum_{\Delta\boldsymbol{R}_{\mathrm{BA}}} \mathrm{e}^{\mathrm{i}\boldsymbol{k}_{\mathrm{v}} \cdot \Delta\boldsymbol{R}_{\mathrm{BA}}} \left\langle p_z(\boldsymbol{r} - \Delta\boldsymbol{R}_{\mathrm{BA}}) \middle| \nabla \middle| p_z(\boldsymbol{r}) \right\rangle^* \delta_{\boldsymbol{k}_{\mathrm{v}}, \boldsymbol{k}_{\mathrm{c}}}. \tag{C.7}$$

Similarly,

$$\left\langle p_z^{\mathrm{B}}(\boldsymbol{k}_{\mathrm{c}}, \boldsymbol{r}) \middle| \nabla \middle| p_z^{\mathrm{A}}(\boldsymbol{k}_{\mathrm{v}}, \boldsymbol{r}) \right\rangle = \sum_{\Delta\boldsymbol{R}_{\mathrm{BA}}} \mathrm{e}^{-\mathrm{i}\boldsymbol{k}_{\mathrm{v}} \cdot \Delta\boldsymbol{R}_{\mathrm{BA}}} \left\langle p_z(\boldsymbol{r} - \Delta\boldsymbol{R}_{\mathrm{BA}}) \middle| \nabla \middle| p_z(\boldsymbol{r}) \right\rangle \delta_{\boldsymbol{k}_{\mathrm{v}}, \boldsymbol{k}_{\mathrm{c}}}. \tag{C.8}$$

As can be seen from the emergence of the Kronecker delta $\delta_{\boldsymbol{k}_{\mathrm{v}}, \boldsymbol{k}_{\mathrm{c}}}$ in equations (C.7) and (C.8), the wavenumber does not change during the transition in the dipole approximation. In other words, only transition $\boldsymbol{k}_{\mathrm{v}} = \boldsymbol{k}_{\mathrm{c}}$ (vertical transition) is allowed. Here, if we define $\boldsymbol{k} \equiv \boldsymbol{k}_{\mathrm{v}} = \boldsymbol{k}_{\mathrm{c}}$, the dipole vector $\boldsymbol{D}(\boldsymbol{k}_{\mathrm{c}}, \boldsymbol{k}_{\mathrm{v}})$ is a function of $\boldsymbol{k}$ and reads

$$\begin{aligned}
\boldsymbol{D}(\boldsymbol{k}) = &- c_{\mathrm{c}}^{\mathrm{A}}(\boldsymbol{k})^* c_{\mathrm{v}}^{\mathrm{B}}(\boldsymbol{k}) \sum_{\Delta\boldsymbol{R}_{\mathrm{BA}}} \mathrm{e}^{\mathrm{i}\boldsymbol{k} \cdot \Delta\boldsymbol{R}_{\mathrm{BA}}} \left\langle p_z(\boldsymbol{r} - \Delta\boldsymbol{R}_{\mathrm{BA}}) \middle| \nabla \middle| p_z(\boldsymbol{r}) \right\rangle^* \\
&+ c_{\mathrm{c}}^{\mathrm{B}}(\boldsymbol{k})^* c_{\mathrm{v}}^{\mathrm{A}}(\boldsymbol{k}) \sum_{\Delta\boldsymbol{R}_{\mathrm{BA}}} \mathrm{e}^{-\mathrm{i}\boldsymbol{k} \cdot \Delta\boldsymbol{R}_{\mathrm{BA}}} \left\langle p_z(\boldsymbol{r} - \Delta\boldsymbol{R}_{\mathrm{BA}}) \middle| \nabla \middle| p_z(\boldsymbol{r}) \right\rangle.
\end{aligned} \tag{C.9}$$

Here, since the hexagonal lattice system shown in figure 1.1(a) has three-fold rotational symmetry around one atom, by using an appropriate constant d, the atomic dipole vector can be written as

$$\left\langle p_z(\boldsymbol{r}-\boldsymbol{R}_i)\middle|\nabla\middle|p_z(\boldsymbol{r})\right\rangle = -d\boldsymbol{R}_i \quad (i=1,2,3). \tag{C.10}$$

If we define

$$m_o = \left\langle p_z(\boldsymbol{r}-\boldsymbol{R}_1)\middle|\partial_x\middle|p_z(\boldsymbol{r})\right\rangle, \tag{C.11}$$

as the x-component of the atomic dipole vector $\langle p_z(\boldsymbol{r}-\boldsymbol{R}_1)|\nabla|p_z(\boldsymbol{r})\rangle$ between the NNs, we have

$$\sum_{\Delta\boldsymbol{R}_{\mathrm{BA}}} \mathrm{e}^{-\mathrm{i}\boldsymbol{k}\cdot\Delta\boldsymbol{R}_{\mathrm{BA}}}\left\langle p_z(\boldsymbol{r}-\Delta\boldsymbol{R}_{\mathrm{BA}})\middle|\nabla\middle|p_z(\boldsymbol{r})\right\rangle = -\frac{\sqrt{3}}{a}m_o\sum_{i=1}^{3}\mathrm{e}^{-\mathrm{i}\boldsymbol{k}\cdot\boldsymbol{R}_i}\boldsymbol{R}_i. \tag{C.12}$$

Therefore, the dipole vector in equation (C.9) becomes

$$\boldsymbol{D}(\boldsymbol{k}) = \frac{\sqrt{3}}{a}m_o\left[c_{\mathrm{c}}^{\mathrm{A}}(\boldsymbol{k})^* c_{\mathrm{v}}^{\mathrm{B}}(\boldsymbol{k})\sum_{i=1}^{3}\mathrm{e}^{\mathrm{i}\boldsymbol{k}\cdot\boldsymbol{R}_i}\boldsymbol{R}_i - c_{\mathrm{c}}^{\mathrm{B}}(\boldsymbol{k})^* c_{\mathrm{v}}^{\mathrm{A}}(\boldsymbol{k})\sum_{i=1}^{3}\mathrm{e}^{-\mathrm{i}\boldsymbol{k}\cdot\boldsymbol{R}_i}\boldsymbol{R}_i\right]. \tag{C.13}$$

www.ingramcontent.com/pod-product-compliance
Ingram Content Group UK Ltd.
Pitfield, Milton Keynes, MK11 3LW, UK
UKHW051901100926
8556IPUK00005B/720